International Conference on Engineering and Technology for Sustainable Development 2015

IOP Conference Series: Materials Science and Engineering
Volume 105

Yogyakarta, Indonesia
11 - 12 November 2015

ISBN: 978-1-5108-1956-6
ISSN: 1757-8981

Table of contents

Volume 105

International Conference on Engineering and Technology for Sustainable Development (ICET4SD) 201511–12 November 2015, Yogyakarta, Indonesia

Accepted papers received: 2 December 2015
Published online: 28 January 2016

Paper

IOP Conf. Series: Materials Science and Engineering **105** (2016) 011001 doi:10.1088/1757-899X/105/1/011001

International Conference on Engineering and Technology for Sustainable Development 2015

Collaboration:

Industrial Engineering Department-Informatics Department-Chemical Engineering Department

Electrical Engineering Department-Mechanical Engineering Department

Rajamanggala University of Technology Tawan-OK, Thailand

Hanyang University of Korea

IOP Publishing ISI

Keynote Speakers:

Prof. Reyer Zwiggelaar

Aberystwyth University, Wales UK

Prof. Zahari Taha

Universiti Malaysia Pahang, Malaysia

Assoc. Prof. Dr. Suzana Yusup

Universiti Teknologi Petronas, Malaysia

The Phoenix Hotel Yogyakarta

11-12 November 2015

Preface for Publication of:

"International Conference on Engineering & Technology for Sustainable Development"

IOP Conf. Series: Materials Science and Engineering **105** (2016) 011001 doi:10.1088/1757-899X/105/1/011001

Welcome Note

On behalf of the Organizing committee of International Conference on Engineering and Technology for Sustainable Development 2015 (ICET4SD 2015), we would like to express our gratitude and welcome all participants to beautiful Yogyakarta, Indonesia. This conference had been held for the first time by the Faculty of Industrial Technology Universitas Islam Indonesia (FTI UII) with collaboration from our strategic partners and reputable universities, such as RMUTT-OK Thailand and Hanyang University of Korea.

The aim of this international conference is as an international academic forum for exchanging ideas, knowledge and expertise on various fields, for example industrial engineering and management systems, informatics, chemical engineering, mechanical engineering, electrical engineering and other engineering disciplines. This international conference with the theme of "Sustainable Development" is a continuation of a national seminar that had been held for 14 consecutive years.

With the vision of UII to actualize UII as '*rahmatan lil'alamin*', committed to perfection (excellence), '*risalah islamiyah*', in the field of teaching, research, community development and Islamic mission, at a developed countries' university level, therefore, FTI UII believes in contributing through the international conference.

Industrial development is improving fast due to the development of science, knowledge and technology. It is inevitable that from this industrial development there will be development of other economical sectors. In addition, controlling and developing science and technology would support sustainable development of industry.

In this international conference, there are various full papers submitted from various regions/nations (Japan, Malaysia and Indonesia). However, only some papers are qualified for being presented in ICET4SD 2015 and published in the Open Access IOP Conference Series: Materials Science and Engineering (MSE) which is indexed by SCOPUS/ISI web of science.

Special gratitude for all the committee, reviewers, keynote speakers, and the Rector of Universitas Islam Indonesia as well as the Dean of Faculty of Industrial Technology for all the hard work, support and encouragement. Hopefully this international conference will be sustainably hosted and develop the society continuously.

IOP Conf. Series: Materials Science and Engineering **105** (2016) 011002 doi:10.1088/1757-899X/105/1/011002

Peer review statement

All papers published in this volume of *IOP Conference Series: Materials Science and Engineering* have been peer reviewed through processes administered by the proceedings Editors. Reviews were conducted by expert referees to the professional and scientific standards expected of a proceedings journal published by IOP Publishing.

IOP Conf. Series: Materials Science and Engineering **105** (2016) 012001 doi:10.1088/1757-899X/105/1/012001

Eco-friendly Fibre from Recycled Polypropylene of Bottle Cap Waste and Lignin

A S Soekoco[1] **, A Basuki** [1] **and Mardiyati** [1]
Fakultas Teknik Mesin dan Dirgantara, Institut Teknologi Bandung
Bandung, Indonesia

E-mail: asrilsenoaji@gmail.com, arifbasuki@material.itb.ac.id, mardiyati@material.itb.ac.id

Abstract. Ecofriendly fibre is one of potential alternatives to fulfill the rising demand in textile material supply which is limited due to the decreasing reserve of oil. Large amount of polypropylene waste from bottle cap and lignin as a byproduct from pulp industry are potential solutions. Grinded polypropylene bottle cap was blended with lignin powder in concentration of 5 wt. % processed by melt spinning at 170° C temperature. The fibres produced have an average diameter 170 and 250 micrometres . In view of the mechanical properties, the tensile strength is 11.9 MPa for fibre with 170 micrometres diameter and 14.7 MPa for fibre with 250 micrometres diameter. Fibre surface morphology was further studied using micron microscope, and the result shows black flocks spread in the fibre, indicating that the lignin does not blend evenly.

1. Introduction

Indonesia stands as the second biggest mismanage plastic waste country in the world after China [1]. This condition is potentially getting worse due to increment of plastic products demand as the population growth. On the other hand, world oil production as main material for plastic product including synthetic fibre has been decreasing. One largely applied solution for this issue is by reusing, reducing and recycling all plastic products including plastic bottle cap. However, recycling the bottle cap waste as fibre material has not been studied

Currently, many apparel producers are making eco-friendly fibre from recycled polyethylene terephthalate of plastic bottle waste. Study about synthetic fibre from plastic bottle cap waste is very interesting due to the abundant supply of this waste. According to Container Recycling Institute, at least there were 42.6 billions unit plastic bottle sold at 2010 in US [2]. The estimated waste weight is around 76,680 tons (average bottle cap weight is 1.8 grams). Moreover, Polypropylene also has interesting properties such as low melt processing temperature, low thermal conductivity (lower than wool) which is very useful for apparel in cold weather, good mechanical properties stability in wet condition, excellent chlorine resistance, etc. In addition, there is another material which is also abundant in the nature, the material is called lignin.

Lignin is the part of wood beside cellulose and other component. Lignin is the second most abundant biopolymer in the world after cellulose [3] readily available in the nature and also relatively inexpensive [4]. Lignin is also byproduct of pulp and paper industries which are using kraft process. Through this process, cellulose is isolated from wood and the other wood components including lignin are extracted into black liquor. Lignin can be gained by precipitating black liquor. This process well known as bio-refinery and currently available commercially.

In the past, there were some experiments about virgin polypropylene and lignin fibre. This paper focuses on producing and characterizing fibre from recycled polypropylene of bottle cap waste and lignin.

2. Experimental

2.1. Materials

Lignin for this research is obtained from Domtar Biochoice lignin. This lignin is byproduct of kraft process which only uses southern pine woods. The presentation of the solids is 66.31 %, presentation of the ash is 0.46% & pH is 3.45. Recycled polypropylene is taken from single plastic bottle cap waste. The bottle cap waste is cleaned prior blending process. The blending composition of polypropylene/lignin is 95/5 by weight.

2.1.1. Lignin analysis

Thermal analysis of the lignin was performed with differential scanning calorimetry (DSC) and thermogravimetric analysis (TGA) using a LINSEIS STA Platinum Series (simultaneous thermal analysis). DSC analyses were conducted with heating rate of 10 °C /min until the temperature of 310 °C . Sample weights were 5.3 ± 0.5 mg. Fourier Transform Infra Red (FTIR) analysis was run on Shimadzu.

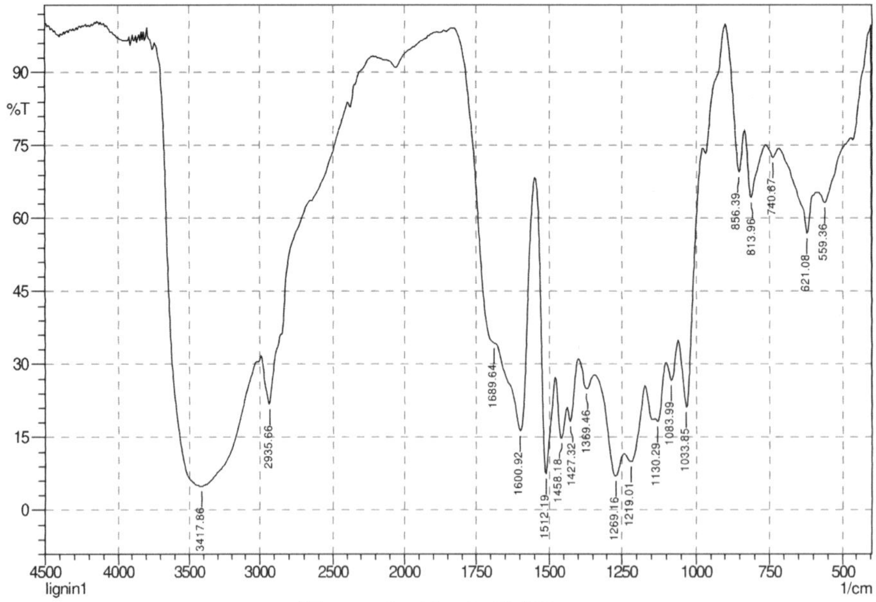

Figure 1. Lignin FTIR spectra

IOP Conf. Series: Materials Science and Engineering **105** (2016) 012001 doi:10.1088/1757-899X/105/1/012001

The FTIR spectra is in the line with the information from supplier information that lignin material was obtained from softwood (southern pine). The softwood FTIR spectra has more intensity absorption band at 1269 cm-1 than at 1230 cm and the intensive absorption existed on 1033 cm^{-1} (around 1030 cm^{-1}) is very typical for softwood lignin spectra [5]. The melting (softening) temperature of lignin was found around 187 °C.

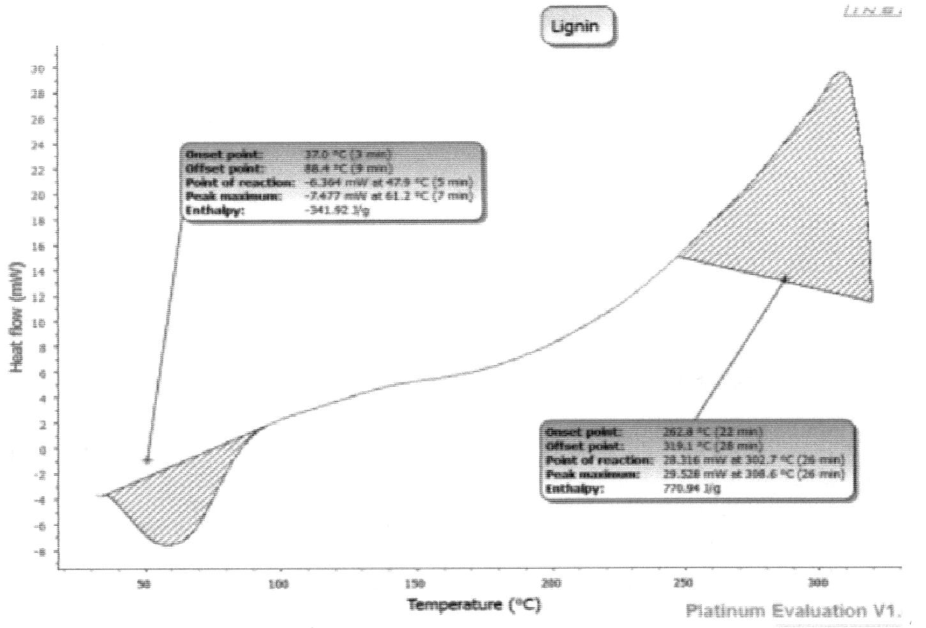

Figure 2. Lignin DSC analysis

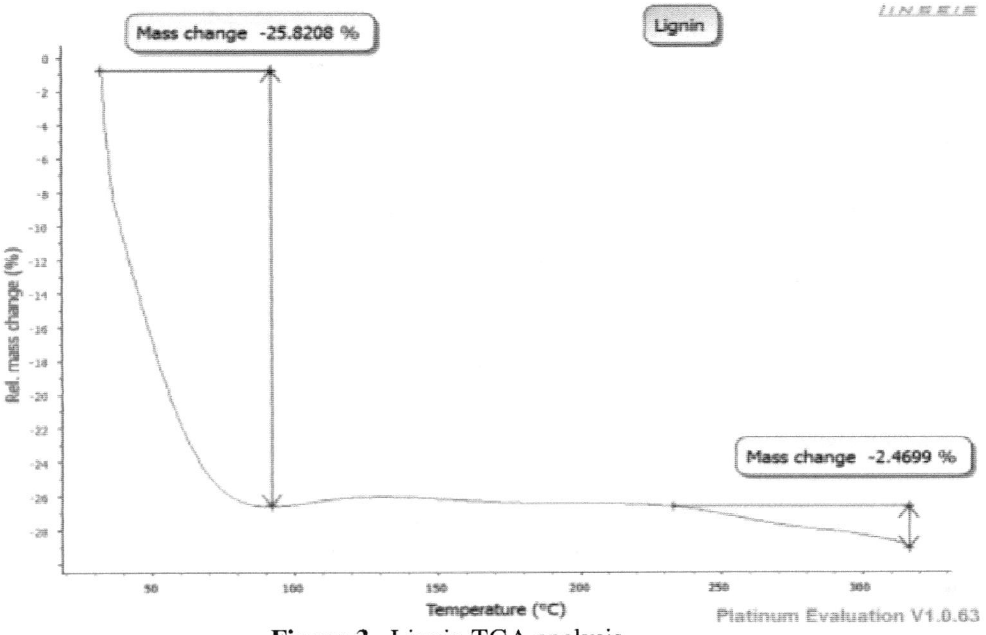

Figure 3. Lignin TGA analysis

IOP Conf. Series: Materials Science and Engineering **105** (2016) 012001 doi:10.1088/1757-899X/105/1/012001

2.1.2. Polypropylene waste analysis

Thermal analysis of the polypropylene was performed with differential scanning calorimetry (DSC) and thermogravimetric analysis (TGA) using same equipment that used for analysing lignin with same heating rate until the temperature of 310 °C Sample weights were 1.8 ± 0.5 mg. Fourier Transform Infra Red (FTIR) analysis was run on Shimadzu.

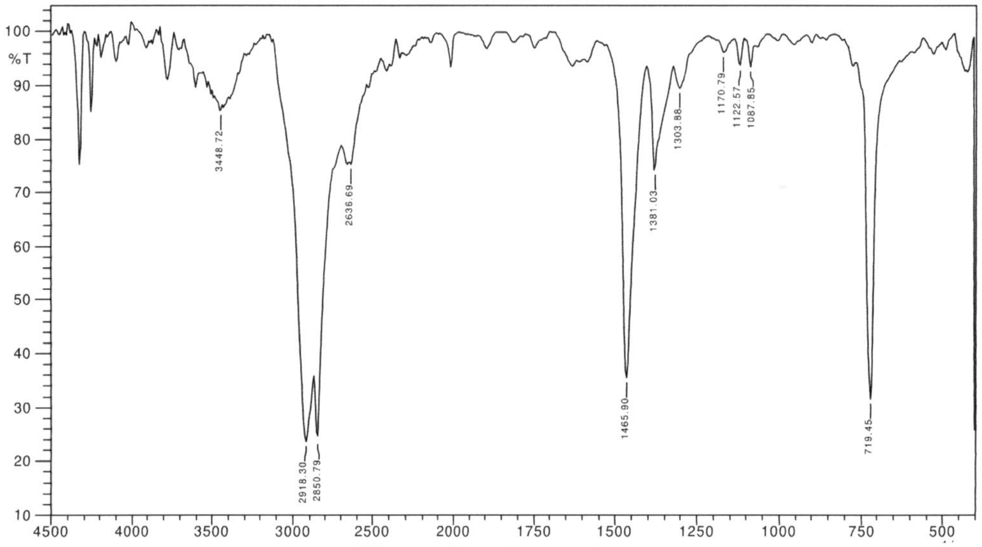

Figure 4. Polypropylene waste FTIR spectra

The FTIR spectra relatively is similar with the spectra in the handbook. FTIR spectra shows the intense absorption at 1087,1465, 1381, 2850 & 2918 cm [6]. The melting temperature of polypropylene was found around 125 c.

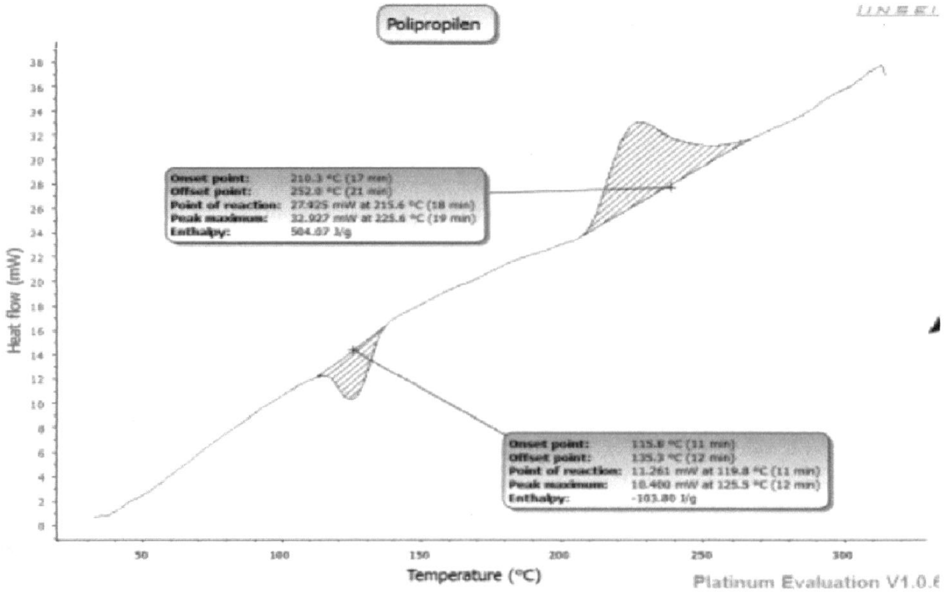

Figure 5. Polypropylene waste DSC analysis

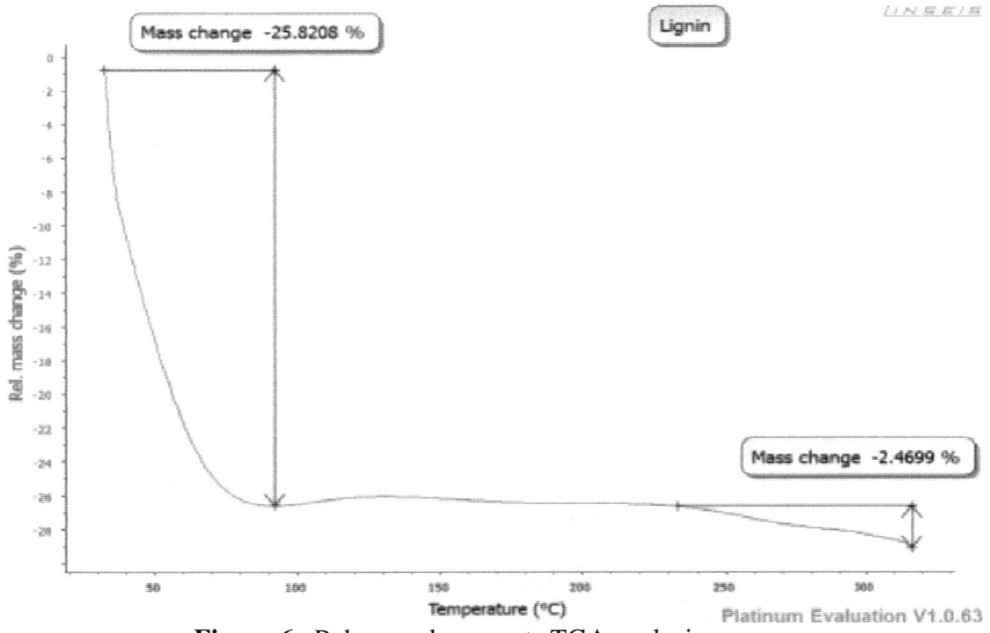

Figure 6. Polypropylene waste TGA analysis

2.2. Melt spinning machine

Melt spinning machine is built using plunger and dead weight to push molten polymer out through spinneret. The spinneret has single hole in 5 mm diameter. Autonics TZ4M is used as Temperature controller in this spinning machine. Its Digital temperature controllers offer quick setup and provide precise temperature management. This melt spinning machine uses Fotek Solid state relay with current load 40 Ampere. The maximum temperature can be reached with this melt spinning machine is 310 °C with heating rate is 7 °C / min. Fibre winder is using AC motor with capacity 480 – 600 RPM.

2.3. Fibre spinning

Thermal extrusion of the lignin fibers was performed at 170 ± 2 °C. After reaching the desired temperature, the heating temperature was maintained for 20 minutes to ensure the uniformity of molten polymer temperature. The molten polymer than quenched in open air at 24 °C temperature and drafted by winder. The fibres diameter were arranged by drafting setting. In this research the fibre diameter target was 170 micrometres (Fibre A) & 250 micrometres (Fibre B) concerning the broken fibre factor when the fibre spinning was conducted.

2.4. Mechanical strength of fibre

The mechanical strength of the fibers was measured in accordance with the ASTM standard (D 3379-75) with Mesdan Tensolab using a 1000 N load cell and a gauge length of 2,5 cm. Fibre diameters were examined using a calibrated Nikon optical microscope with Dinolite software assistance.

3. Result and discussion

3.1. Fibre diameter

The fibre diameter of these fibres varies with slubs (thick places which has bigger diameter in yarn/fibre) which are distributed unevenly. The slub places were distributed unevenly. This condition may cause broken fibre which will lead to production problem. For woven textile application, this condition will lead to production problem because it may cause broken fibre/yarn when the loom is running. The varying fibre diameter will also cause problem in machine setting such as the yarn tensioner at the knitting machine or gun hole at the loom. However, these slubs can create unique appearance in textile product, which are used by jeans maker for their product.

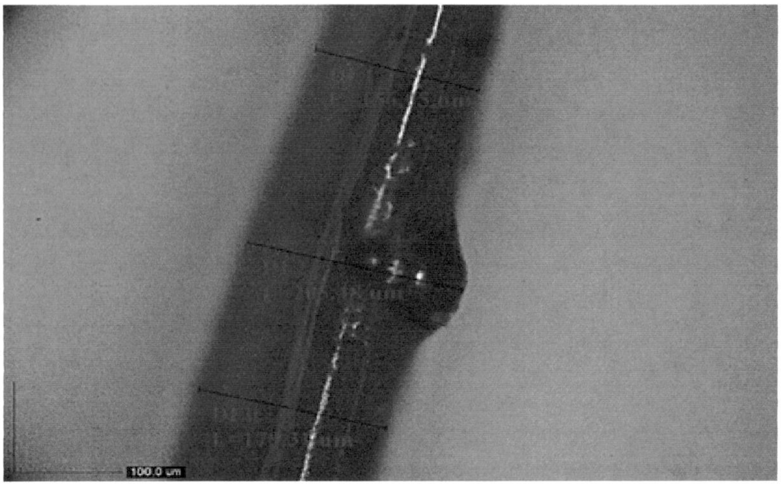

Figure 7. Fibre A with diameter target 170 micrometres

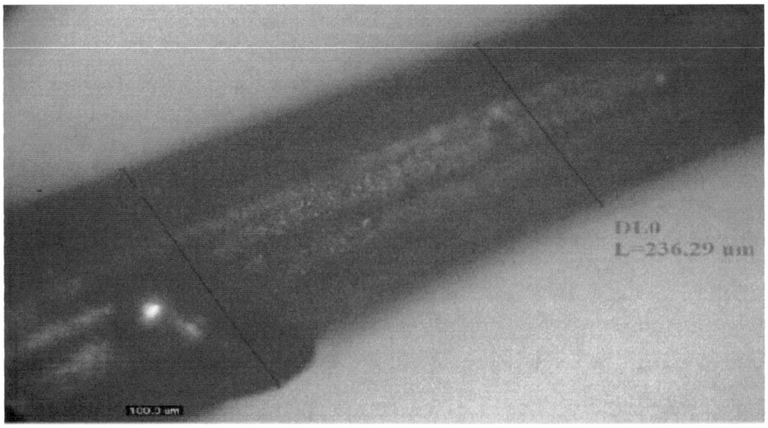

Figure 8. Fibre B with diameter target 250 micrometres

3.2. Fibre surface

The images from micron microscope shows black flocks within the fibre. The flocks are distributed unevenly and vary in size. These flocks are actually a group of lignin which is not well melted and blended with the polypropylene. Although the dimension of both materials have been grinded (to achieve minimum dimension), pounded and mixed, the molten polymer seems to be not well blended. The processing temperature and compatibility of both polymers need to be studied further to achieve better blends.

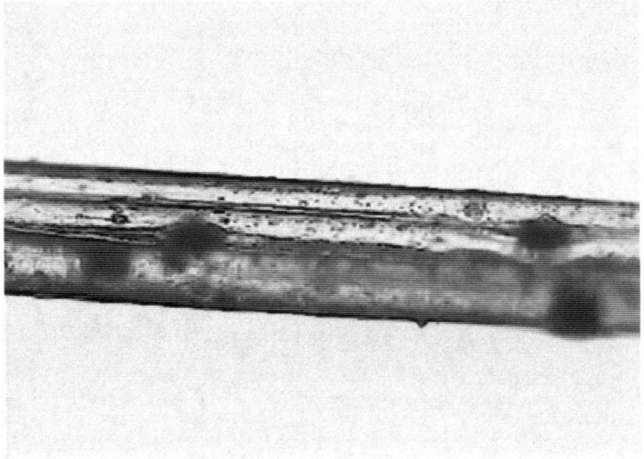

Figure 9. Black flock within fibre A surface (170 micrometres diameter)

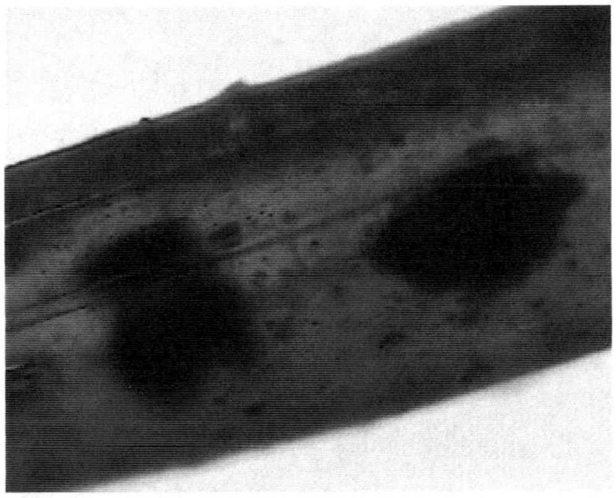

Figure 10. Black flock within fibre B surface (250 micrometres diameter)

3.3. Ultimate tensile strength of fibre

Ultimate tensile strength of fibre B with 0.25 mm diameter has higher ultimate tensile strength than fibre A. Both fibres were produced using same spinneret diameter and same plunger weight to push molten polymer out of spinneret, so the volume of polypropylene and lignin which passed through the spinneret were the same in both condition. The lignin is flocking and is not well blended with polypropylene and the dimensions range of flocking lignin for both condition are same. The difference of RPM winder results on the difference of fibre diameter due to draft resulted. The RPM winder is proportional to the draft resulted. The higher draft resulted, the smaller fibre diameter resulted. This condition will also affect to increment of polymer molecular orientation degree [7] as we can see in Figure 5. The increment of polymer molecular orientation may improve fibre strength.

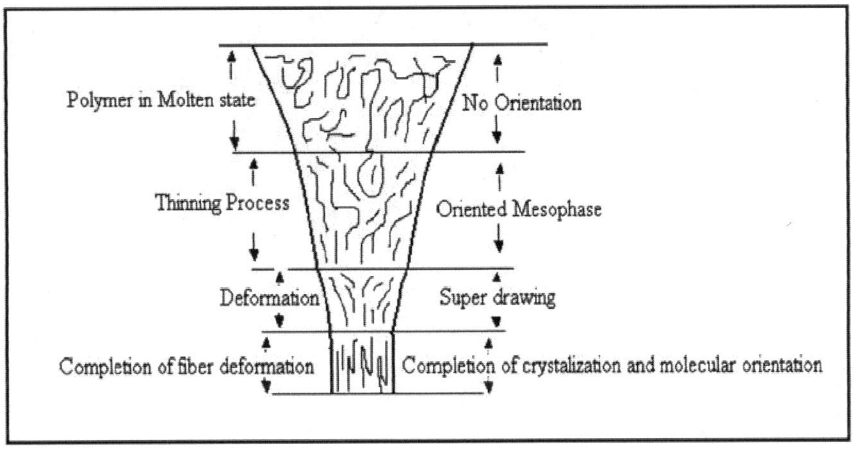

Figure 11. Polymer crystalinity and molecular orientation

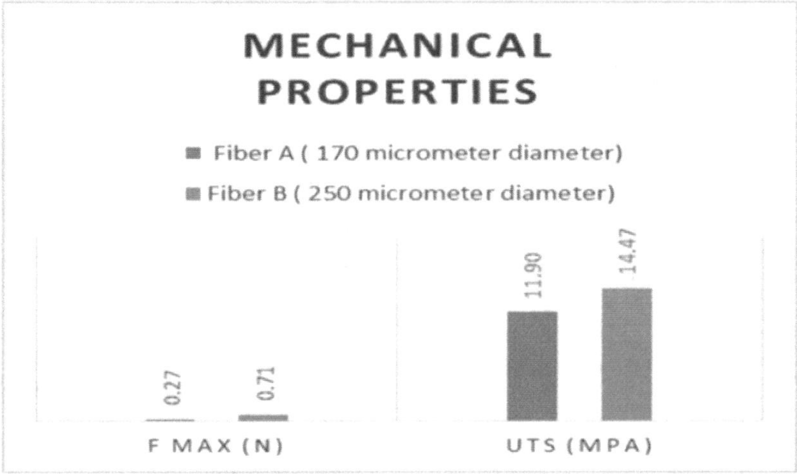

Figure 12. Maximum force and ultimate tensile strength of fibres

According to tensile testing result in Figure 12, fibre B has higher ultimate tensile strength than fibre A. This condition is not in line with relation of polymer molecular orientation with tensile strength but the result from fibre surface analysis shows that the fibre surface of fibre A is rougher than fibre B as we can see in Figure 13 and 14. This condition may cause increment of stress concentration and broken fibre in tensile strength test.

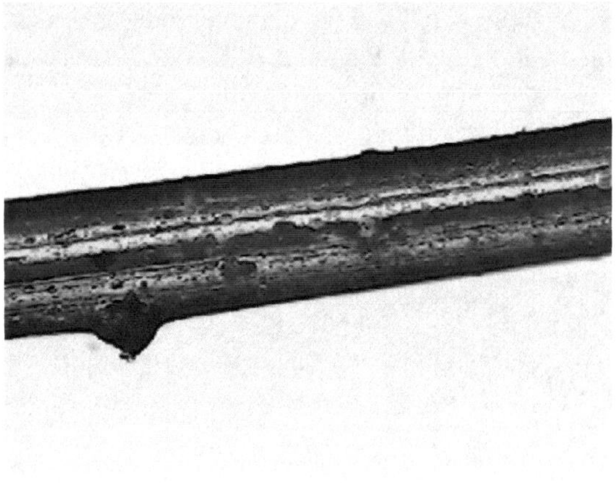

Figure 13. Surface of fibre A (170 micrometres diameter)

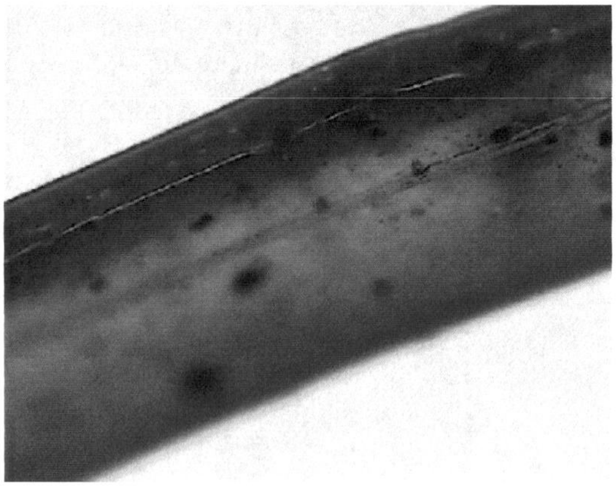

Figure 14. Surface fibre B (250 micrometres diameter)

4. Conclusions

Recycled polypropylene and lignin are potential materials for synthetic fibre, because both material has enough spinability to produce fibre by using melt spinning method. Further study is needed to produce high mechanical properties of recycled polypropylene lignin fibre, particularly to create good polymer blending. A good polymer blending will lead to even distribution of polypropylene and lignin in the fibre. The surface of fibre is also important to support mechanical properties of fibre resulted. All these conditions are very important especially for technical textile application. Poor polymer blending potentially leads to problem such as broken fibre.

5. References

[1] Jenna R. Jambeck, Miriam Perryman, Roland Geyer, Anthony Andrady, Chris Wilcox, Theodore R. Siegler, Ramani Narayan, Kara Lavender Law, 2015, Plastic waste inputs from land into the ocean

[2] Susan V. Collins, Jason Medbury, Fatemeh Bagheri, Chris Saprnicht, February 12, 2015, Container Recycling Institute, sciencemag.org

[3] Brett G. Diehl , Nicole R. Brown , Curtis W. Frantz , Matthew R. Lumadue Fred, Accepted 28 April 2013, Effects of pyrolysis temperature on the chemical composition of refined softwood and hardwood lignins, Elseviere, Cannon CARBON 60 (2013) 531– 537

[4] J.F. Kadla , S. Kubo , R.A. Venditti , R.D. Gilbert , A.L. Compere,W. Griffith C, accepted 8 July 2002, Lignin-based carbon fibers for composite fiber applications, Pergamon, Carbon 40 (2002) 2913–2920, 200

[5] Ivan Bykov, Characterization of natural and technical ligins using FTIR spectroscopy, 2008, Lulea University of Technology, Master Thesis, Department of Chemical Engineering dan Geosciences

[6] A.H. Kupstov & G.N. Zhinzhin, 1998, Handbook of Fourier transform raman and infrared spectra, Elsevier

[7] Raghavendra R. Hegde, M. G. Kamath, Atul Dahiya, Polymer crystallinity.

Acknowledgements

The authors would like to acknowledge Eddy Satriyo from Production Laboratory at ITB for supporting this research. We would also like to acknowledge Domtar for supplying Biochoice lignin for this research.

Calculating Contrast Stretching Variables in Order to Improve Dental Radiology Image Quality

Haris B Widodo[1], Arief Soelaiman[2], Yogi Ramadhani[2], Retno Supriyanti[2]
[1]Department of Dentistry, Jenderal Soedirman University
[2]Department Electrical Engineering, Jenderal Soedirman University
Jl. HR.Boenjamin 708, Purwokerto, Jawa Tengah, Indonesia

E-mail: retno_supriyanti@unsoed.ac.id

Abstract. Teeth are one of the body's digestive tract that serves as a softener food that can be digested easily. One branch of science that was instrumental in the treatment and diagnosis of teeth is Dental Radiology. However, in reality many dental radiology images has low resolution, thus inhibiting in making diagnosis of dental disease perfectly. This research aims to improve low resolution dental radiology image using image processing techniques. This paper discussed the use of contrast stretching method to improve the dental radiology image quality, especially relating to the calculation of the variable contrast stretching method. The results showed that contrast stretching method is promising for use in improving the image quality in a simple but efficient.

Keyword: dental radiology, low resolution, contrast stretching, diagnose, image processing

1. Introduction

Teeth are one of the body's digestive tract that serves as a softener food that can be digested easily. Histologically, the teeth and oral tissues derived from mesoderm and ectoderm, which has three main functions, namely, chewing (mastication), beauty (aesthetics), and speech (phonetic). In the mouth, the teeth embedded in the jawbone and are protected by the gums, with different forms according to function. One branch of science that was instrumental in the treatment and diagnosis of teeth is Dental Radiology. The use of X-ray has long been known as a tool in the field of general medicine and dentistry were very helpful in diagnosing the disease and to determine the treatment plan. Image generated x-rays of a patient to a dentist is very important especially for the presence of abnormalities that are not visible can be seen clearly, so that will really help a dentist in determining the diagnosis and treatment plan.

It has more than a century dentistry profession using radiographic examination as a means to obtain diagnostic information cannot be obtained from the clinical examination and other tests before. Until now, dental radiography became one of the important tools used in the treatment of modern dentistry. Good dental radiographic projections shooting intra oral and extra-oral almost a common procedure performed by a dentist in assisting the management of a case. Dental Radiography usefulness of which is to: (i) Radio diagnosis / Rontgen diagnose: used for diagnosis as apical or per apical abnormalities that are not detected clinically, abnormalities in the jaw, fracture of the jaw or tooth root and hidden caries (at the proximal or root caries) secondary caries, caries incipient, the depth of caries and others.

(ii) Treatment Plan: assist in making or determining a treatment plan, such as the determination of the location of the pins or implants, root canal condition, determination of the type and technique (iii) Supporting Treatment: help facilitate the conduct of a treatment, such as post-operative complications, endodontic treatment (iv) Evaluation of Care: used to evaluate the success or progress or control treatment (v) Radiography is one of the medical record that is very important and also for (vi) Interest forensic [1]. Based on usability, it can be said that the Dental Radiography plays an important role in various matters in the field of dentistry. One of the tools commonly used in dental radiography is panoramic machine. Figure 1 shows the dental panoramic machine.

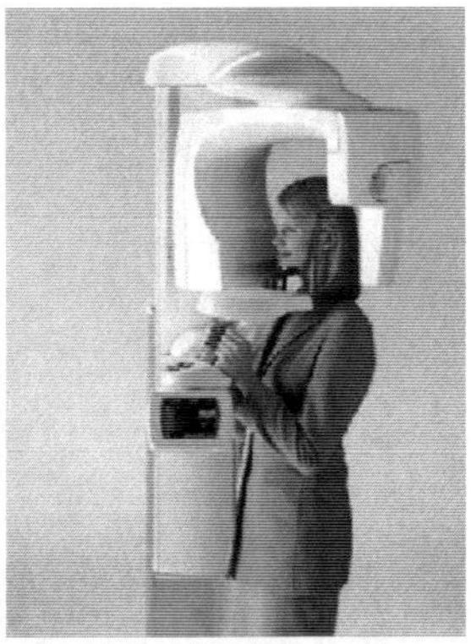

Figure 1. Dental panoramic machine
(Source: http://www.dentistryiq.com/ content/ dam/ etc 1)

However, given the prominence of panoramic machine then the price is directly proportional to its superiority, i.e. the better image quality produced by more expensive the price of its panoramic machine. In fact, especially in developing countries, existing panoramic machine is a machine that has a low resolution. This research aims to improve the quality of the image generated by the low quality panoramic machine.

On the other hand, the current technological developments especially in the field of image processing is growing rapidly. Digital image processing according to Gonzales [2] is manipulation and interpretation of digital images using computer assistance. Digital image processing aims to (i) improve the quality of the image, seen from the aspect of radiometric and geometric aspects. Radiometric aspects consists of contrast enhancement, image restoration, and color transformation while the geometric aspect consists of the rotation, scale, translation, geometric transformations. (ii) The process of withdrawal of information or description of the object or the introduction of objects contained in the image (iii) perform image selection characteristics for analysis purposes. (iv) Compression or data reduction for the purpose of data storage, data transmission, and processing time data. According to functions and benefits of digital image processing techniques especially for supporting technology in medical field, there are some researchers that worked in the medical field using image processing techniques. Supriyanti et al [3][4][5] using image processing techniques in the application of medical technology for developing countries, especially for the detection of cataracts [3][5] and improve the quality of the ultrasound image of the uterus [4]. Kronfeld [6] conducted research on three-dimensional modeling on dental image.[6] Wu Kan [7] conducted research on dental segmentation using morphological algorithm approach framework. Shah et al [8] conducted research on dental segmentation using active contour method without using edge detection, because this technique focuses on the intensity of the overall area of the teeth. Wanat [9] conducted research on dental radiology segmentation for forensic interest. Kustra [10] conducted a research on the analysis and segmentation of the tooth surface using Dental Cast Segmentation method. Dighe [11]. In this paper we emphasize our discussion about improving quality of low resolution dental radiology image using contrast stretching method especially for calculating variable R1, S1, R2 and S2. Point (r1, s1) and (r2, s2) will determine the shape transformation, and can be set to determine the level of the spread of gray level of the image produced. If r1 = r2 = s1 and s2, then the transformation will be a straight line, which means there is no change in gray level of the image produced. It is generally assumed r2 ≤r1 and s1 ≤s2 so that the function will return a single value and its value will always go up.

2. Methods

2.1. Image Acquisition

As discussed in above, in this research we will improve image quality of low resolution dental radiology. Input images are taken from low resolution panoramic machine from *Rumah Sakit Gigi dan Mulut Pendidikan Universitas Jenderal Soedirman.* However, because this is a manual panoramic machine, therefore in order to convert to digital image, we have to scan this images manually. Figure 2 shows examples of dental radiology images using in this research.

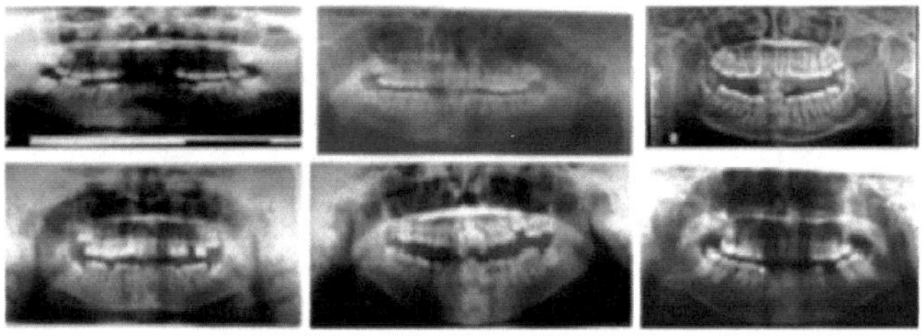

Figure 2. Examples of input images

2.2. Contrast Stretching

Contrast stretching method is part of the point processing method, which means that the process is only dependent on the value of the intensity of one pixel, independent of other pixels around it. Here it is assumed that the image has a grey level range of 0 to 255. In order to change contrast from original pixels is using condition as described in Figure 3.

If the grey level value a1 = a2 and b1 = b2, the transformation will perform a straight line, no change in the grey level image produced. However, if it is assumed grey level value a1 <a2 and b1 <b2 so that the function produces a value. To calculate it can be made the following three functions as described in Equation 1 to Equation 3.

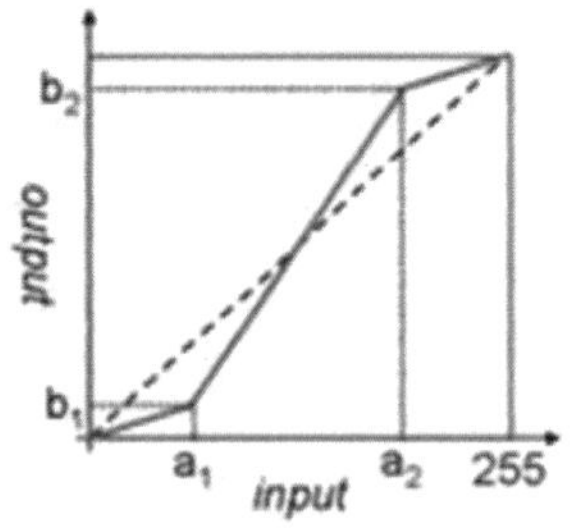

Figure 3. Contrast Stretching function
(Source: Gonzales, 2004)

For $0 \leq f_i(x,y) < a_1$, then :

$$f_o(x,y) = f_i(x,y)\left(\frac{b_1}{a_1}\right) \qquad \textbf{2.1}$$

For $a_1 \leq f_i(x,y) < a_2$, then :

$$f_0(x,y) = b_1 + (f_i(x,y) - a_1)\left(\frac{b_2 - b_1}{a_2 - a_1}\right) \ldots\ldots \ 2.2$$

For $a_2 \leq f_i(x,y) \leq 255$, then :

$$f_0(x,y) = b_2 + (f_i(x,y) - a_2)\left(\frac{255 - b_2}{255 - a_2}\right) \ldots\ldots\ldots \ 2.3$$

2.3. Image Histogram

Image histogram is a graph that illustrates the spread of pixel intensity values of an image or a certain part in the image. From a histogram can be determined relative frequency of occurrence of intensity in the image. The histogram can also show a lot of things about the brightness and contrast of an image. Therefore, the histogram is a valuable tool in the execution of image processing both qualitatively and quantitatively. In order to obtain a good image, then the spread of intensity values to be changed. The purpose of smoothing the histogram is to obtain a uniform spread of the histogram, such that every degree of grey has almost the same number of pixels. In the histogram, low-contrast image has a narrow shape which indicates that the image has a low dynamic range, which means low contrast. High-contrast image shows that the grey level histogram is widespread. Although the histogram only shows a general nature and does not provide information about the contents of the image, but the shape of the histogram can provide information about the possibility for improvement of the image contrast.

3. Results and Discussions

As has been discussed above, the input image has a color format with layer composition Red, Green and Blue (RGB). Formats such as these have a fairly large data size and can slow down the subsequent processing. To overcome this, the image must first convert grayscale type, which has a color type 8 bits (28 = 256), i.e. from 0 to 255 so that the processing can be performed with contrast stretching method. After making a histogram of the gray scale image, it can be seen the distribution of light and dark pixels through the histogram display. If the image looks too dark grayscale, then the pixels will be concentrated or dominate the left side, and if it is too light to be concentrated on the right side. Table 1 describes some examples of histogram analysis. Formatted as follows:

Table 1. Examples of Histogram Analysis

Input Image	Gray Image	Histogram	Histogram Region	
			Dark	Bright
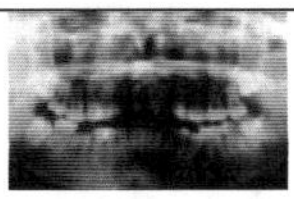	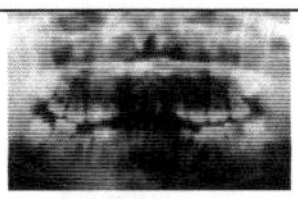	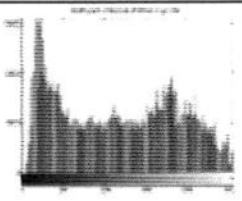	V	V
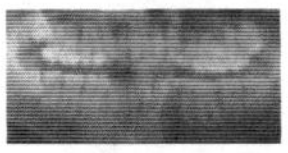	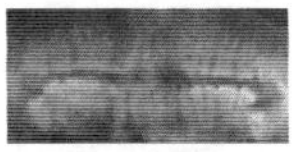	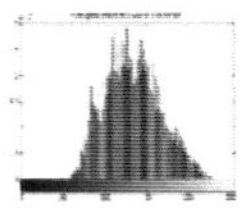	V	V
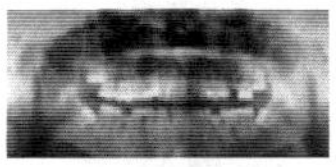	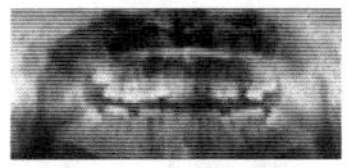	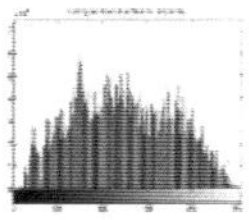	V	V
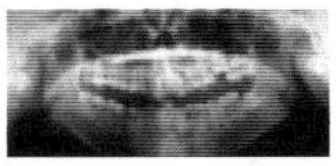	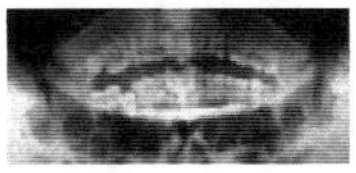	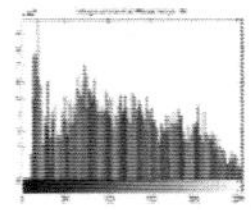	V	V

	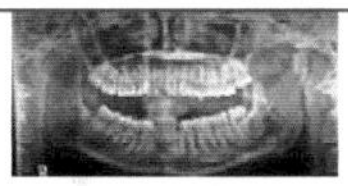	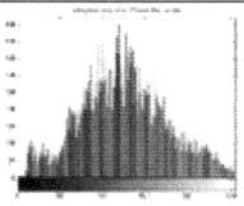	V	V
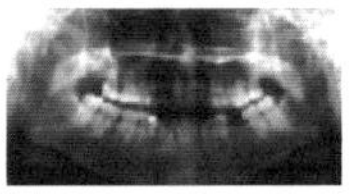	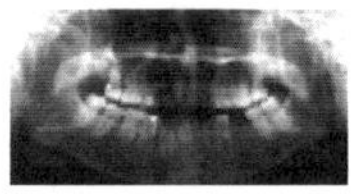	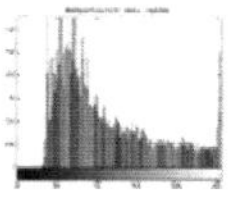	-	V

According to the results as described in Table 1, input images could be grouped into three classifications, namely: (1) concentration to the left (dark), (2) concentration to the right (bright), (3) Concentration evenly to the left (dark) and right (bright).

Contrast Stretching is the last stage of the process of improving the quality of dental radiology image by taking a grayscale image data. This method is conducted to get the maximum dental radiology image better than the initial conditions. This is done because some dental radiology images have poor contrast, therefore the image becomes blurred, and that makes it difficult to obtain clear information in determining the condition of the patient's teeth. It is necessary for contrast stretching the histogram of original image so that the image looks clearer teeth.

Furthermore, based on equations 1-3 then we did calculation variables R1, S1, R2 and S2 to determine the scale of contrast stretching. Table 2 describes examples of variables calculation results.

Table 2. Examples of Variables Calculation

Histogram	VARIABLES			
	R1	S1	R2	S2
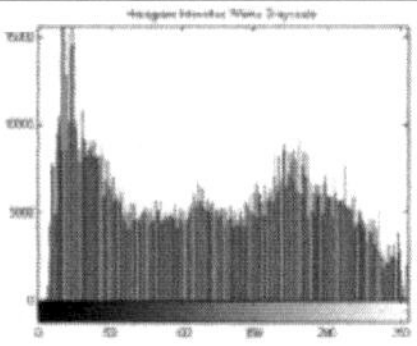	37	40	167	186
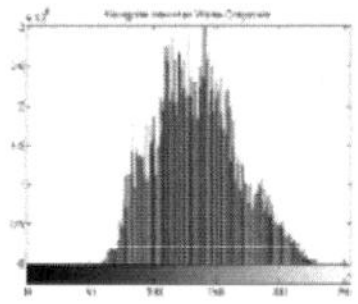	40	5	200	240
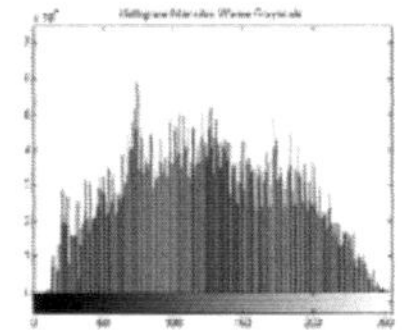	15	2	231	252
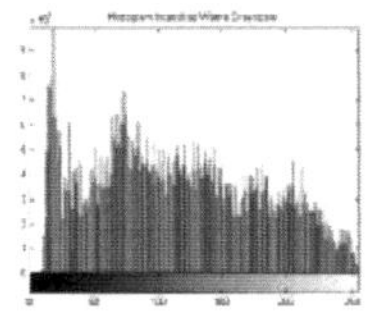	70	65	210	220

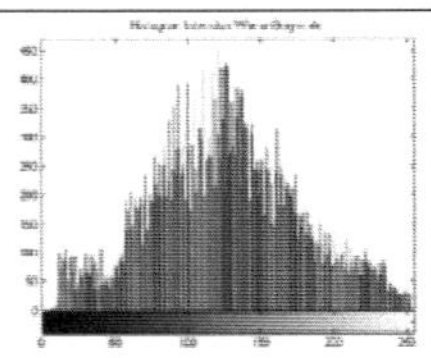	55	0	200	190
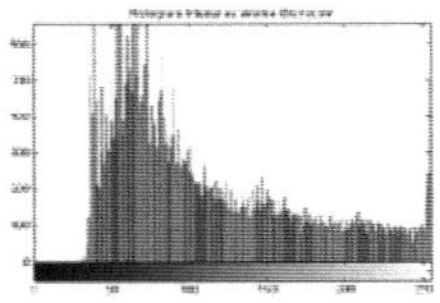	38	2	201	210

According to Table 2, Bright histogram has grey spread that dominates the right side. In order for the deployment of pixels evenly, the value of S1 is smaller than R1, This is because the smaller the value of S1 can encourage new pixel layout changes leading to the left. Dark histogram has grey spread that dominates the left. In order for the deployment of pixels evenly then the value of S1 is greater than the R1, This is because the larger the value of S1 can encourage changes in new pixels to the right. Evenly histogram has fairly grey spread from left to right. Input value almost the same type of dark histogram which distinguishes only determine the approximate value with reference seeking input grey pixel intensities (peak) at the centre of the histogram therefore it can be seen balance grey spread clearly of the image histogram. Furthermore, by using the input value R1, S1, R2 and S2, it can be determined scale of contrast stretching as described in Table 3.

Table 3. Example of Contrast Stretching Scale Area

Histogram	Scale Area		
	A1	A2	A3
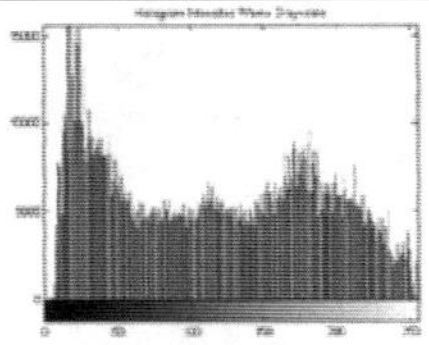	1,08108	1,12308	0,78409
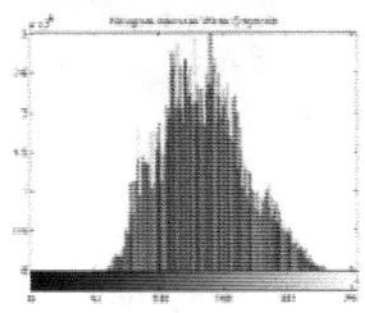	0,125	1.46875	0.272727
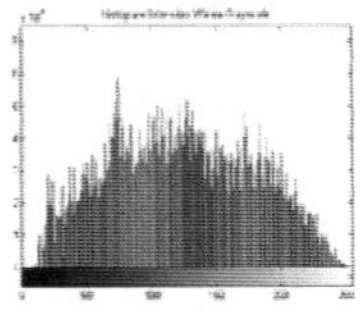	0,13333	1,15741	0,125
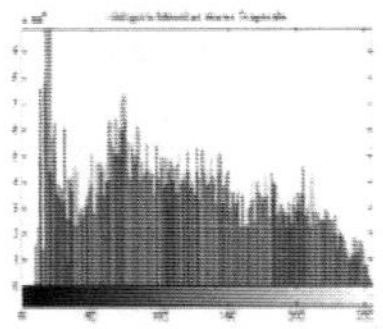	0,92857	1,10714	0,77778

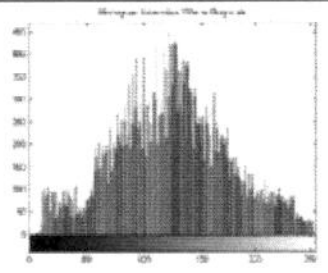	0,01818	1,30345	1,18182
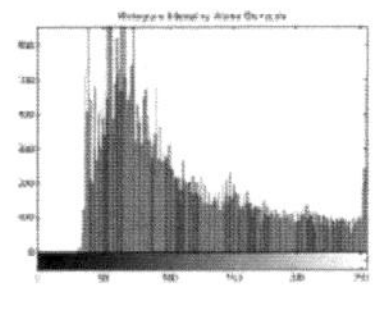	0,05263	1,27607	0,83333

According to Table 3, contrast stretching method has three areas, namely A1 (first area), A2 (second area) and A3 (third area). The area is an area stretching scaling based on the basic equations of contrast stretching. The area shows the value of a straight line gradient. If a straight line gradient is generating value <1, then the new pixel area will be narrowed, and the merger intensity values of the original pixels. Instead a straight line gradient > 1 there will be a widening or stretching the pixel area. Improved image quality using Contrast Stretching method does not change the intensity values of an image.

4. Conclusions
Contrast stretching method is promising for improving low resolution dental radiology images. However, Image processing algorithms can be improved so as to obtain a grey value R1, S1, R2 and S2 automatically. Extensional area adds to the original three, into four or more areas so that the areas to be improved to be more focused. Combine several methods both before and after contrast stretching process in order to obtain the result of improved dental radiology image quality better than the results now.

5. Acknowledgment
This work is supported by BLU Jenderal Soedirman University through Hibah Unggulan fiscal year 2015.

6. References
[1] D. Joen Iannucci and R. M. Laura Jansen Howerton, *Dental Radiography*, 4th ed. United States: Elsevier, 2012.
[2] R. C. Gonzales, R. E. Woods, and S. L. Eddins, *Digital Image Processing using Matlab*. United States: Pearson Prentice Hall, 2004.
[3] R. Supriyanti, H. Habe, M. Kidode, and S. Nagata, "Extracting appearance information inside the pupil for cataract screening," in *Proceedings of the 11th IAPR Conference on Machine Vision Applications*, pp. 342–345.
[4] R. Supriyanti, D. . Putri, E. Murdyantoro, and H. B. Widodo, "Comparing edge detection methods to localize uterus area on ultrasound image," in *Proc. of 2013 3rd Int. Conf. on Instrumentation, Communications, Information Technol., and Biomedical Engineering: Science and Technol. for Improvement of Health, Safety, and Environ., ICICI-BME 2013*, 2013, pp. 152–155.
[5] R. Supriyanti, E. Pranata, Y. Ramadhani, and T. I. Rosanti, "Separability filter for localizing abnormal pupil: Identification of input image," *Telkomnika*, vol. 11, no. 4, pp. 783–790, 2013.
[6] T. Kronfeld, D. Brunner, and G. Brunnett, "Snake-Based Segmentation of Teeth from Virtual Dental Casts, Computer-aided Diagnosis and Applications," *Comput. Diagnosis Appl.*, vol. 7, no. 7, 2010.
[7] K. Wu, L. Chen, J. Li, and Y. Zhou, "Tooth segmentation on dental meshes using morphologic skeleton," in *Computer and Graphics*, 2013.
[8] S. Shah, A. Abaza, A. Ross, and H. Ammar, "Automatic Tooth Segmentation Using Active Contour without Edges," in *Biometric Symposium*.
[9] R. Wanat, "A Problem of Automatic Segmentation of Digital Dental Panoramic X-Ray Images for Forensic Human Identification," in *Proceeding of the 15th Central European Seminar on Computer Graphics*, 2011.
[10] J. Kustra, A. Jalba, and A. Telea, "A Medial Point Cloud based Algorithm for Dental Cast Segmentation," in *014 IEEE International Conference on Consumer Electronics (ICCE)*.
[11] S. Dighe and R. Shriram, "Segmentation and Matching of Dental Radiographs used in Dental Biometrics," *Int. J. Sci. Appl. Inf. Technol.*, vol. 1, no. 2, 2012.

Performance analysis of cascaded h-bridge multilevel inverter using mixed switching frequency with various dc-link voltages

I B F Citarsa, I N W Satiawan, I K Wiryajati, and Supriono
Lecturer, Electrical Engineering Department,
Universitas Mataram, Jalan Majapahit Nomor 62, Mataram, Nusa Tenggara Barat Province, Indonesia

E-mail : ferycitarsa@unram.ac.id

Abstract. Multilevel inverters have been widely used in many applications since the technology is advantageous to increase the converter capability as well as to improve the output voltage quality. According to the applied switching frequency, multilevel modulations can be subdivided into three classes, i.e: fundamental switching frequency, high switching frequency and mixed switching frequency. This paper investigates the performance of cascaded H-bridge (CHB) multilevel inverter that is modulated using mixed switching frequency (MSF) PWM with various dc-link voltage ratios. The simulation results show the nearly sinusoidal load output voltages are successfully achieved. It is revealed that there is improvement in output voltages quality in terms of THD and low-order harmonics content. The CHB inverter that is modulated using MSF PWM with equal dc-link voltage ratio (½ Vdc: ½ Vdc) produces output voltage with the lowest low-order harmonics (less than 1% of fundamental) while the CHB inverter that is modulated using MSF PWM with un-equal dc-link voltage ratio (2/3 Vdc: 1/3 Vdc) produces a 7-level output voltage with the lowest THD (16.31%) compared to the other PWM methods. Improvement of the output voltage quality here is also in line with improvement of the number of available levels provided in the output voltage. Here only 2 cells H-bridge inverter (contain 8 switches) are needed to produce a 7- level output voltage, while in the conventional CHB inverter at least 3 cells of H-bridge inverter (contain 12 switches) are needed to produce a 7-level output voltage. Hence it is valuable in term of saving number of component.

1. Introduction

Multilevel inverters have been widely used in many applications since the technology is advantageous to increase the converter capability as well as to improve the output voltage quality. Multilevel inverters are able to produce a staircase output waveform, that is more approaching sinusoidal and produce less number of harmonics compare to the conventional inverter output voltage. There are mainly three types of multi level inverters, Diode clamped, Flying capacitor inverter and cascaded h- bridge multi-level inverter [1].

Cascaded h-bridge multilevel inverters have been developed for utility applications including utility interface of renewable energy, voltage regulation, VAr compensation, and harmonic filtering in power systems. [1]. A modified cascaded H-bridge multilevel inverter (MLI) is implemented for solar applications [2]. A new design of asymmetric cascaded H-bridge multilevel inverter is implemented in the grid side of the wind energy conversion system [3]. A Five level cascaded H-bridge multilevel Inverter is also designed for induction motor drive [4].

There are three classes of switching frequency for multilevel modulations: fundamental, high, and mixed switching frequency. Fundamental Switching Frequency PWM (FSF PWM) methods produce switch commutations at output fundamental frequency and can be aimed to cancel some particular low frequency harmonic [5-6]. High Switching Frequency PWM (HSF PWM) methods are the adaptation of standard PWM to multi levels and they are meant to switch at very high frequency, about 10 to 20 kHz [7]. In this class there are Carrier-Based PWM (CB PWM) or Sinusoidal PWM, and Space Vector PWM (SV PWM). CB PWM is mainly divided into two categories: Level-Shifted (LS PWM) [8-9] and Phase- Shifted (PS PWM) methods [9]. Finally the LS PWM can be divided into Phase Disposition (PD), Phase Opposition Disposition (POD) and Alternate Phase Opposition Disposition (APOD) modulations [10]. Mixed Switching Frequency PWM (MSF PWM) methods are those in which switches commutate at different frequency, like hybrid multilevel modulation, and are particularly suited for hybrid inverters where different cells can easily commutate at different frequencies [1].

A Hybrid / mixed PWM technique is a relatively new PWM strategy to control the CHB inverter. In the mixed switching frequency methods, one cell of H-bridge inverter is modulated in slow switching mode (fundamental frequency) and the other one is modulated in fast switching frequency mode, particularly using the level shifted carrier based PWM method and phase shifted carrier based PWM methods. The modulation technique offers possibility to apply various dc-link voltage ratios. This paper discusses the procedure to apply a mixed frequency PWM technique in the 2 cell CHB inverter with equal and un-equal dc link voltage ratios. The performance of cascaded 2-cell H-bridge (CHB) multilevel inverter using mixed switching frequency methods is investigated. The output voltages quality of the inverter in term of THD and the low order harmonics content are compared with the other PWM methods, namely fundamental frequency PWM, level shifted and phase shifted carrier based PWM methods.

2. Methodology

2.1 Cascaded h-bridge multilevel inverters

The structure of the cascaded H-bridge (CHB) may consist of two or more H-bridge inverters. The CHB inverter can be supplied by separated DC sources or a single DC source. The structure of the CHB inverter with separated DC-sources is shown in Figure 1. This type of inverter consists of two cells H-bridge inverter hence employs 8 power electronic switches. The modulation technique that applies to each cell of inverter may be the same or different. It varies from fundamental switching frequency PWM, carrier based PWM or combination of the two different PWM methods (known as mixed / hybrid PWM method).

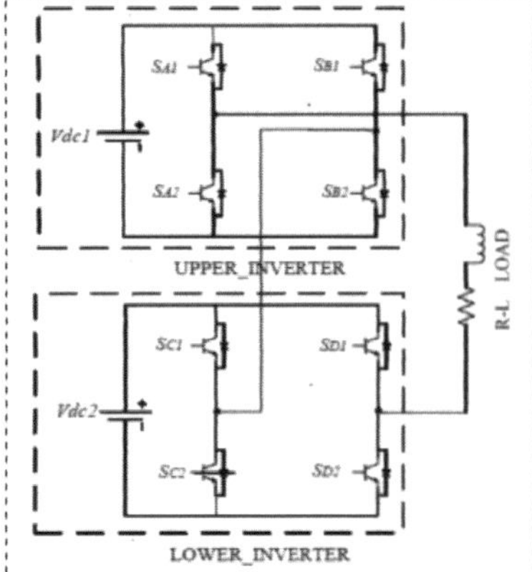

Figure 1. Five-level cascaded H-bridge inverter.

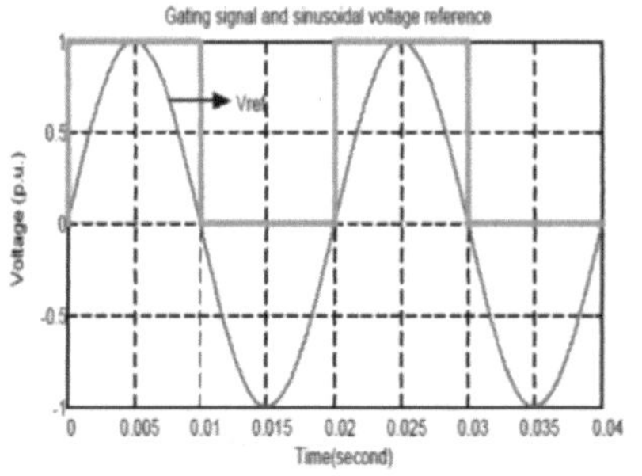

Figure 2. Gating signal using FSF PWM method

IOP Conf. Series: Materials Science and Engineering **105** (2016) 012003 doi:10.1088/1757-899X/105/1/012003

2.2 Fundamental Switching Frequency (FSF)PWM

Fundamental Switching Frequency PWM method controls the power electronic switches at fundamental frequency. The gating signals are simply generated by examining the sinusoidal voltage reference (Vref). When Vref has a positive (negative) value the upper (lower) switches are conducted (ON) otherwise the switches are dis-conducted (OFF).The operation of the FSF PWM method (for the upper switches) is governed using a simple logic as described in (1) and Figure 2.

$$\begin{aligned} if\ v_{ref} > 0\ then\, S_{(x=A,B,C,D)} = 1\ (High) \\ else\, S_{x1(x=A,B,C,D)} = 0\ (Low) \end{aligned} \tag{1}$$

Sx1 is the switching functions of the upper switches. The notation of switches refers to Figure 1. Since the switches operate in relatively slow switching, the output voltage contains high value of fundamental component as well as the low-order harmonics.

2.3 Phase Shifted PWM (PS PWM)

Phase Shifted PWM applies several triangular carriers that have the same frequency and same peak-peak amplitude, but there is a phase shift between any two adjacent carrier waves. For m voltage levels, (m-1) carrier signals are required and they are phase shifted with an angle of θ = (360°/m-1).

The gate signals are generated with proper comparison of carriers and a modulating signal. The comparison of voltage reference signal and triangular carriers and the respective gating signals for a five-level CHB inverter are shown in Figure 3 and 4.

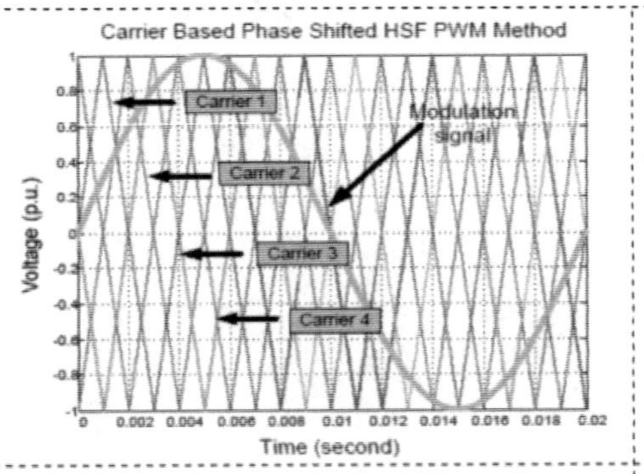

Figure 3. Comparison of modulation signal and carriers in PS PWM

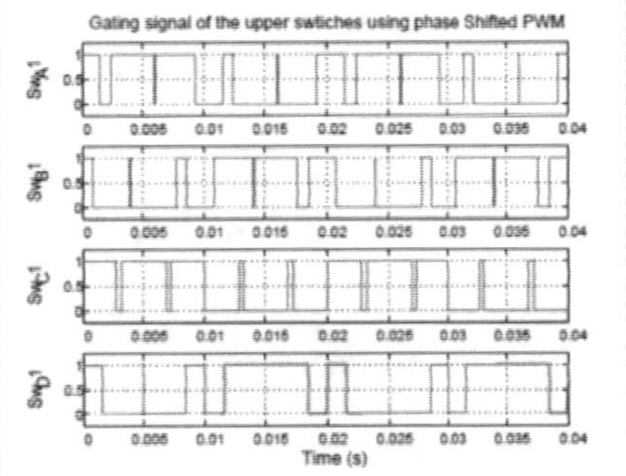

Figure 4. Gating signals using PS PWM of the CHB inverter upper switches

2.4 Level shifted PWM (LS PWM)

Level Shifted PWM also uses some carrier signals, but they are arranged in different levels among the carriers. According to the disposition of carrier waves, the LS PWM can be divided into three main types i.e (i) Phase disposition (PD), when all the carrier signals are in phase, (ii) Phase opposition disposition (POD), when all the carrier signals above zero reference are in phase but in opposition with those are below zero reference. (iii) Alternate phase opposition disposition (APOD): when the modulating signal of each phase is in opposition from each other.

The gating signals are generated by comparing the carrier waves and the modulating signal as shown in Figure 5 and 6.

IOP Conf. Series: Materials Science and Engineering **105** (2016) 012003 doi:10.1088/1757-899X/105/1/012003

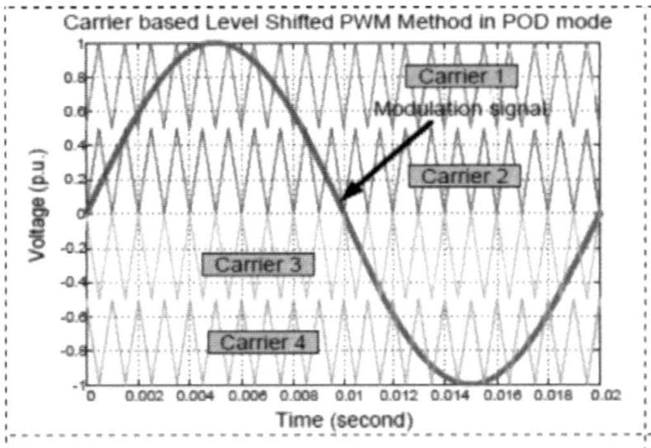

Figure 5. Comparison of modulation signal and carriers in LS PWM arranged in POD mode

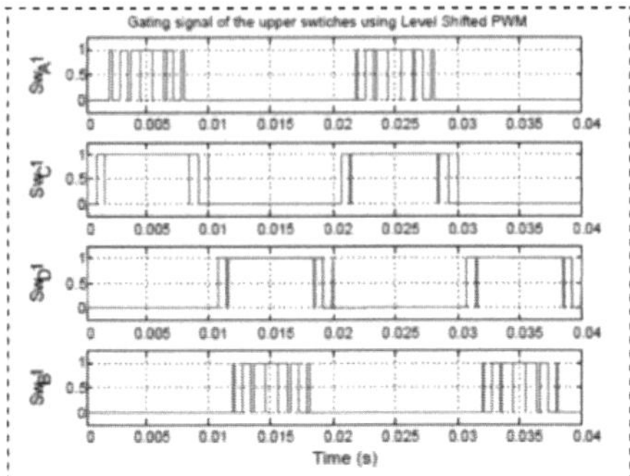

Figure 6. Gating signals using LS PWM for the CHB inverter upper switches

2.5 Mixed Switching Frequency (MSF) PWM

Mixed multilevel inverters may need different PWM strategies for the different stages composing them. For instance, a CHB inverter composed by one GTOs stage and one IGBTs stage may require two different switching frequencies, one for high-speed and one for low-speed devices. In this way, the output waveform, being the sum of the single stage outputs, presents the two frequencies in its harmonic content. The principle of a mixed switching frequency PWM of CHB inverter is shown in Figure 7.

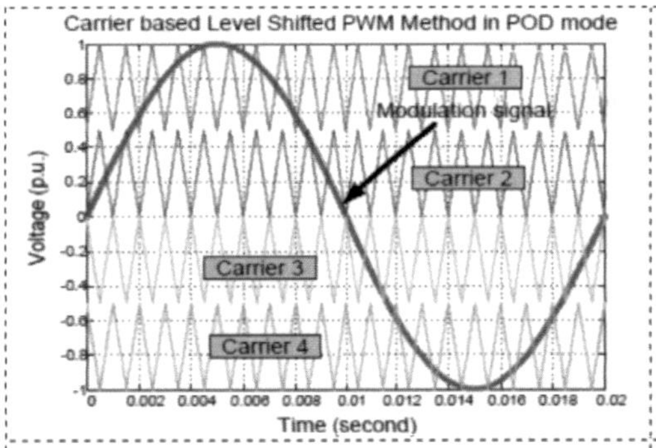

Figure 7. Schematic diagram of mixed switching frequency PWM of the CHB inverter

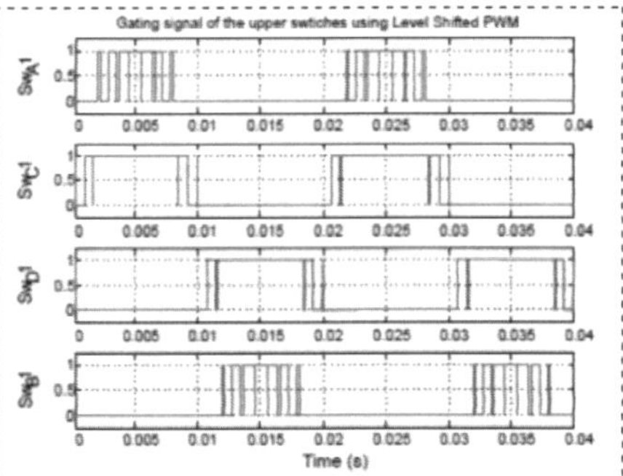

Figure 8. Method to generate the gating signals in MSF PWM (equal dc-link voltage)

The modulation signal of the MSF PWM is obtained by subtracting the sinusoidal reference signal and the output of the inverter that operates in FSF PWM. The gating signals for the inverter switches are generated by comparing the modulation signal and the triangular carriers. Figure 8 shows the modulation signal (bold line) along with the reference signal (dashed dot line) and output voltage of inverter operates in the FSF PWM (dashed line).

The gating signals of upper switches for the two H-bridge inverters are shown in the Figure 9. The gating signals of the lower switches are complementary to the gating signals of the upper switches. The gating signals of the MSF PWM with un-equal dc-link voltage ratio are generated in the same fashion as the equal dc-link voltage ratio.

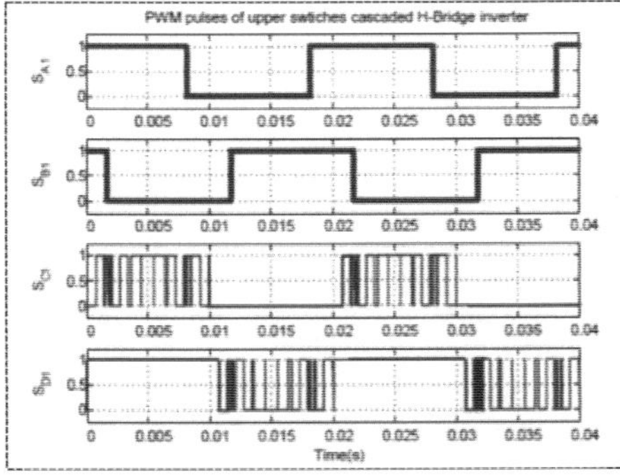

Figure 9. The gating signals of the CHB inverter with equal ratio of dc-link voltage

3. Simulations and Results

The developed PWM schemes were verified using computer simulations. The inverter is modelled using Sim-Power Blockset and the PWM pulses are generated using Simulink. In the simulation, the effect of inverter switches dead-times are neglected. Frequency switching of inverter operates in PWM mode is 1 kHz. The THD are calculated using (2) where the low order harmonics included in the calculation is up to 100.000 Hz (up to 2000th harmonics). The CHB inverter is supplied by separated dc-link voltages. Both inverters apply dc-link voltage ratio either 12 V : 12 V or 16 V : 8 V. In the two cases, the total of dc-link voltage is 24 V.

$$THD = \sqrt{\frac{\sum_{n=1,2,3,\dots}^{r} V_n^2 - V_1^2}{V_1^2}} \quad (2)$$

Vn is the rms value of nth harmonic, V1 is the value of the fundamental component, n is the low-order harmonics and r is the maximum harmonic includes in the THD calculation.

3.1 Output voltage for DC-link ratio of ½ Vdc : ½ Vdc

Figure 10 shows the output voltage of the multi-level CHB inverter using mixed switching frequency PWM technique with equal dc-link ratio and the respective harmonics spectrum. It is shown that the voltage output consist of 5 dc voltage levels {+24 V, +12 V, 0, -12 V, --24 V}. The fundamental component of the voltage output is 16.97 V (rms) and the value of THD is 25.39%. The harmonics spectrum indicates that the nearly sinusoidal waveform is achieved as the value of the low-order harmonics are significantly low. The zoom picture of the low order harmonics further shows the excellent quality of the output voltage. It reveals that the highest low order harmonics is the 15th harmonic (750 Hz) having the value of 2 V (rms) or 12% with respect to the fundamental, while the harmonics lower than the 13 harmonic are successfully eliminated by the proposed PWM method. All Switching combinations that produce 5 voltage levels in the output voltage are summarized in Table 1 [1].

Figure 11 and 12 show the output voltages of the individual inverter that are modulated using FSF PWM and HSF PWM methods and the respective harmonics spectrum. It is shown that the upper H-bridge inverter produces a three-level staircase waveform that contains high low-order harmonics and fundamental component. Meanwhile the lower H-bridge inverter also generates the same value of low-order harmonics. As a result the low-order harmonics do not appear in the output voltage. The lower inverter that operates in high switching mode has a role as a filter that prevents the low-order harmonics dismissed in the output voltage.

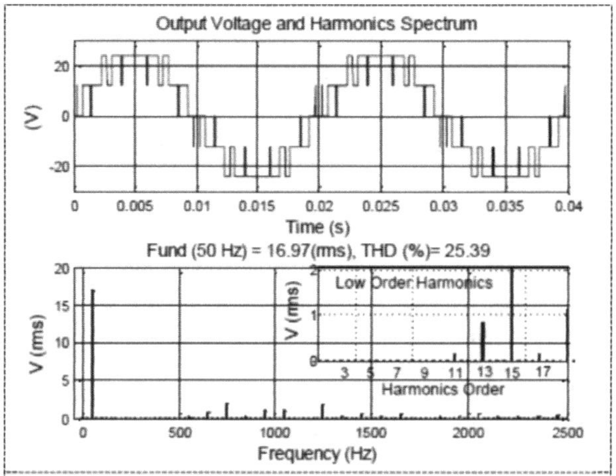

Figure 10. Output voltage and harmonics spectrum of MSF PWM method with dc-link ratio of ½ Vdc : ½ Vdc

Table 1. Operating switches with different voltage level for equal dc-link

SWITCHES ACTIVATED								VOLTAGE LEVEL
S A1	S B1	S C1	S D1	S A2	S B2	S C2	S D2	
1	0	1	0	0	1	0	1	24*V* (*Vdc*1+*Vdc*2)
1	1	1	0	0	0	0	1	12 *V* (*Vdc*2)
1	0	0	0	0	1	1	1	12 *V* (*Vdc*1)
1	0	1	1	0	1	0	0	12 *V* (*Vdc*1)
0	0	1	0	1	1	0	1	12 *V* (*Vdc*2)
1	1	1	1	0	0	0	0	0
1	1	0	0	0	0	1	1	0
1	0	0	1	0	1	1	0	0
0	1	1	0	1	0	0	1	0
0	0	1	1	1	1	0	0	0
0	0	0	0	1	1	1	1	0
0	1	1	1	1	0	0	0	-12 *V* (*Vdc*1)
0	0	0	1	1	1	1	0	-12 *V* (*Vdc*2)
0	1	0	0	1	0	1	1	-12 *V* (*Vdc*1)
1	1	0	1	0	0	1	0	-12 *V* (*Vdc*2)
0	1	0	1	1	0	1	0	-24*V* (-*Vdc*1-*Vdc*2)

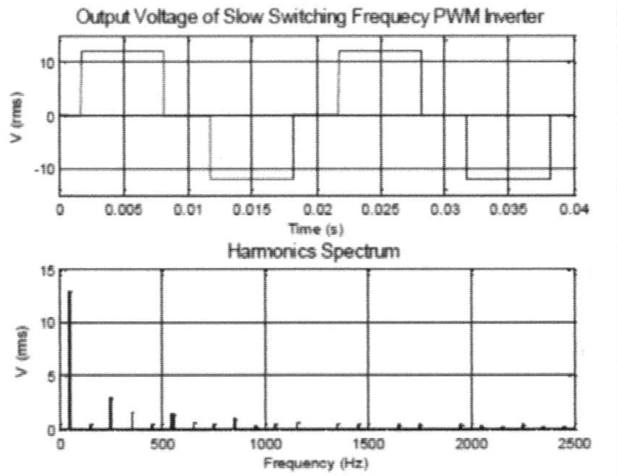

Figure 11. Output voltage of the FSF modulated inverter and harmonics spectrum

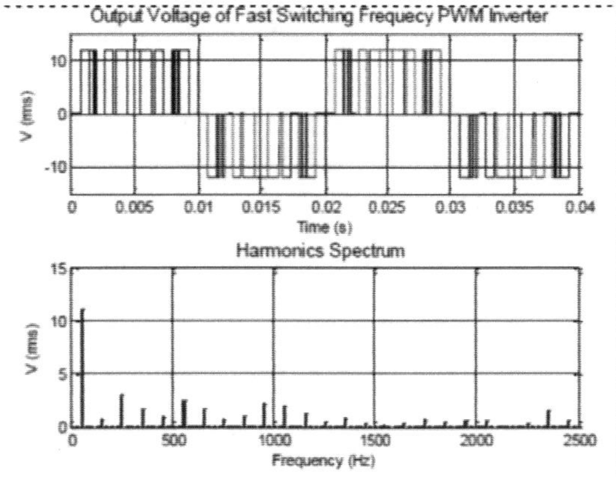

Figure 12. Output voltage of the HSF modulated inverter and harmonics spectrum

3.2 Output voltage for DC-link ratio of 1/3 Vdc: 2/3 Vdc

Figure 13 shows the output voltage of the multi-level CHB inverter using mixed switching frequency PWM technique with un-equal dc-link ratio of 2/3 Vdc :1/3 Vdc. It is clearly shown that the voltage output consist of 7 dc voltage levels {+24V, +16 V, +8 V, 0, -8 V, -16 V, -24 V}. The fundamental component is 16.97 V (rms) and the value of THD further improved to 16.41%. The zoom picture of the harmonics spectrum shows that the low-order harmonics are successfully suppressed into less than 5 % with respect to fundamental hence the output voltage is considered very close to sinusoidal. The improvement of the output voltage quality relates to the increased number of levels. This is the significant merit of the proposed PWM method. In the conventional CHB inverter, 3 cells H-bridge inverter (contains of 12 switches) are required to achieve seven levels in the output voltage. The number of level (l) is governed by, (l = 2n +1), where n is number of cell of H-bridge inverter [1]. In this method, only 2 cells H-bridge inverter (contains of 8 switches) are utilized to get a seven-level output voltage. All switching combinations that produce 7 voltage levels in the output voltage are summarized in Table 2 [1].

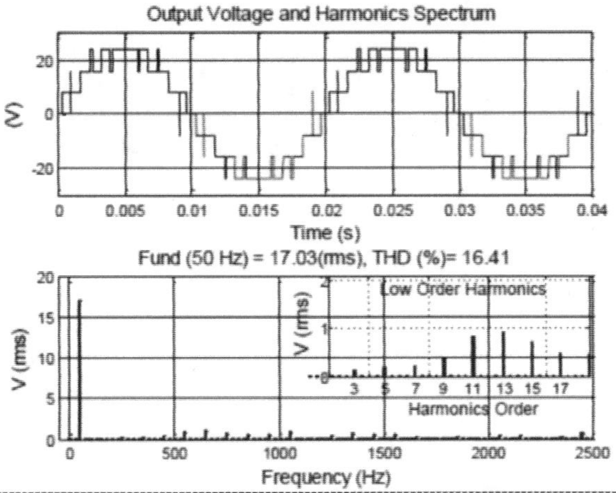

Figure 13. Output voltage and harmonics spectrum of MSF PWM method with dc-link ratio of 2/3 Vdc : 1/3 Vdc

IOP Conf. Series: Materials Science and Engineering **105** (2016) 012003 doi:10.1088/1757-899X/105/1/012003

Table 2. Operating Switches with different voltage level for unequal dc-link

SWITCHES ACTIVATED								VOLTAGE LEVEL
S A1	S B1	S C1	S D1	S A2	S B2	S C2	S D2	
1	0	1	0	0	1	0	1	24*V*(*Vdc*1+*Vdc*2)
1	1	1	0	0	0	0	1	8 *V* (*Vdc*2)
1	0	0	0	0	1	1	1	16 *V* (*Vdc*1)
1	0	1	1	0	1	0	0	16 *V* (*Vdc*1)
0	0	1	0	1	1	0	1	8 *V* (*Vdc*2)
1	1	1	1	0	0	0	0	0
1	1	0	0	0	0	1	1	0
1	0	0	1	0	1	1	0	0
0	1	1	0	1	0	0	1	0
0	0	1	1	1	1	0	0	0
0	0	0	0	1	1	1	1	0
0	1	1	1	1	0	0	0	-16 *V* (*Vdc*1)
0	0	0	1	1	1	1	0	-8 *V* (*Vdc*2)
0	1	0	0	1	0	1	1	-16 *V* (*Vdc*1)
1	1	0	1	0	0	1	0	-8 *V* (*Vdc*2)
0	1	0	1	1	0	1	0	-24*V* (-*Vdc*1-*Vdc*2)

The output voltages of the individual H-bridge inverters are shown in Figure 14 and 15. It is further confirmed that both inverters produce the respective low-order harmonics with almost the same magnitude hence avoid them to appear in the output voltage. In the ratio of 2/3 Vdc: 1/3 Vdc, the upper inverter generates relatively higher value of fundamental than the lower inverter. This implies that the upper inverter supplies more power to the load than the lower inverter. The benefit of this is that different power electronics switches may be used between two inverters. Yet the extra precaution should be taken to avoid the power transferred from upper inverter to lower inverter.

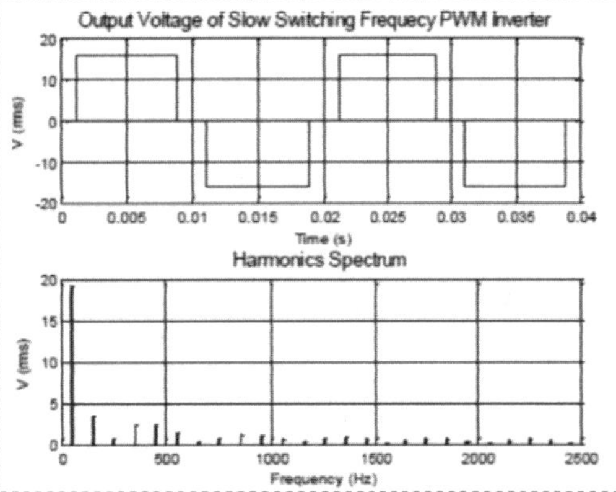

Figure 14. Output voltage of the FSF modulated inverter and harmonics spectrum

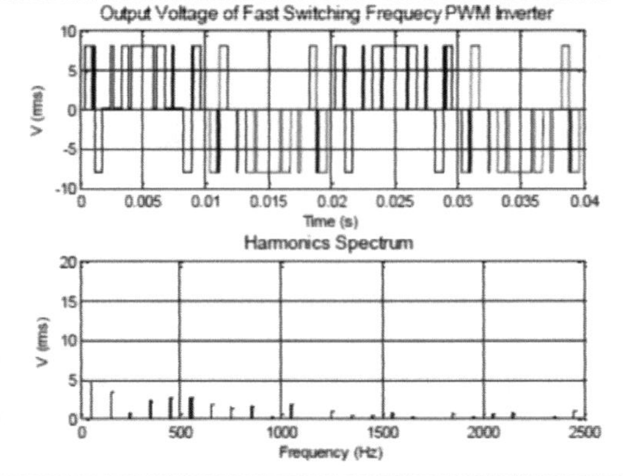

Figure 15. Output voltage of the HSF modulated inverter and harmonics spectrum

In order to further reveal the performance of the proposed PWM schemes, the THD and the content of the low-order harmonics is noted and compared with numerous PWM methods as presented in Figure 16. It is shown that the mixed frequency PWM with equal dc-link voltage (Mixed PWM-1) presents The lowest low-order harmonics among the other PWM methods while the mixed frequency PWM with un-equal dc-link voltage (Mixed PWM-2) shows the lowest THD (16.31%) among the other PWM methods.

Further investigation also found that the magnitude and position of the carrier signals in the mixed frequency PWM does not influence the quality of the output voltages. However, when phase shifted PWM method is applied in the mixed switching frequency PWM method, it is important to note that (based on the simulation process) the frequency switching have to be proportionally reduced twice as the frequency switching of the level shifted PWM in order to achieve a fair comparison.

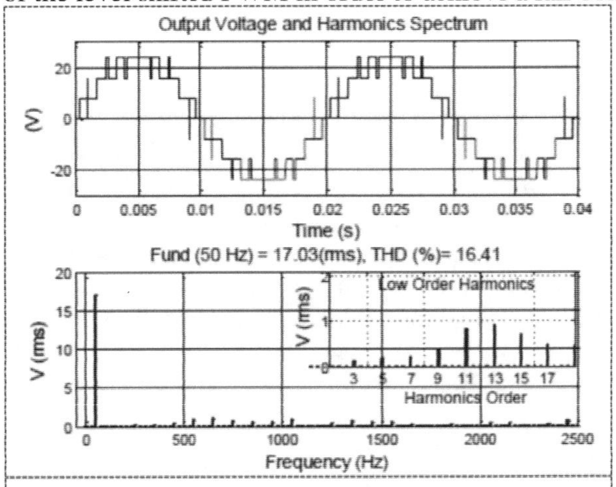

Figure 16. Comparison of low-order harmonics and THD of various PWM methods in CHB inverter

4. Conclusion

This paper investigates the performance of 2-cells CHB multilevel inverter that is operated in mixed switching frequency method with various dc-link voltage ratios. The developed PWM schemes produce output voltage with improved quality. The MSF PWM with equal dc-link produces output voltage with significant low low-order harmonics content (less than 1 % of fundamental) while the MSF PWM with un-equal dc-link voltage produces 7-levels in the output voltage in line with the lowest THD (16.31%) among several other PWM methods. Improvement of the voltage output quality is in line with number of available levels in the output voltage. Here only 2 cells H-bridge inverter (contain 8 switches) are needed to produce a 7-level output voltage, while in the conventional CHB inverter at least 3 cells of H-bridge inverter (contain 12 switches) are needed to produce a 7-level output voltage. Hence it is valuable in term of saving number of component.

5. References

[1] Lega A 2007 *Multilevel Converters: Dual Two-Level Inverter Scheme Final Dissertation Faculty of Engineering Department of Electrical Engineering University of Bologna*

[2] Ranjana M S B, Wankhade P S, Gondhalekar N D 2014 *A modified cascaded H-bridge multilevel inverter for solar applications International Conference on Green Computing Communication and Electrical Engineering (ICGCCEE) 1-7*

[3] John R C, Jude P S V, Mahalaksmi R 2013 Cascaded asymmetric multilevel inverter for wind energy conversion system *International Journal for Research and Development in Engineering (IJRDE),* 2 **24**-28.
[4] Venkadesan A, Panda P, Agrawal P, Puli V 2014 Cascaded H-Bridge multilevel inverter for induction motor drives *International Journal of Research in Engineering and Technology (IJRET)* 3 **260**-266
[5] Gomathi C, Navyanagath, Purnima S V, Veerakumar S 2013 Comparison of PWM methods for multilevel inverter *International Journal of Advanced Research in Electrical,Electronics and Instrumentation Engineering* 2 **6106**-6114
[6] Ismail B, Hassan S I S, Ismail R C, Haron A R, Azmi A 2014 Selective harmonic elimination of five-level cascaded inverter using particle swarm optimization *International Journal of Engineering and Technology (IJET)* 5 **5220**-5232
[7] Nho N V and Youn M J 2006 Comprehensive study on space vector PWM and carrier based PWM correlation in multilevel inverters *IEE Proceedings Electric Power Applications* 153 **149**-158
[8] Monicka J G and Jamuna V 2014 Optimal switching strategy of level shifted carrier based PWM technique for asymmetric multilevel inverter Proceeding of *International Conference on Interdisciplinary Research in Engineering and Technology (ICIDRET)* **146**-150
[9] Venkatakrishna A, Somanatham R, Reddy M S 2014 Phase shifted and level shifted PWM based cascaded multilevel inverter fed induction motor driveInternational *Journal of Current Engineering and Technology* 4 **350**-354
[10] Sambath E, Natarajan S P, Balamurugan C R 2012 Performance evaluation of multi carrier based PWM techniques for single phase five level H-bridge type FCMLI IOSR *Journal of Engineering (IOSRJEN)* 2 **82**-90

Acknowledgment
The Authors would like to thank the Ministry of Research Technology and Higher Education of Republic of Indonesia for the financial support in this research. The authors would also like to thank the Electrical Engineering Department of Universitas Mataram for the kind assistant in providing the laboratory equipment and other facilities.

IOP Conf. Series: Materials Science and Engineering **105** (2016) 012004 doi:10.1088/1757-899X/105/1/012004

Reaction Wheel Control Design Using Linear Quadratic Controller

Nur Nubli Muhamad, Erwin Susanto, Budi Syihabuddin, Ig. Prasetya Dwi Wibawa

School of Electrical Engineering
Telkom University
Jl Telekomunikasi 1 Terusan Buah Batu, Bandung, Indonesia 40257
nur.nubli.m@gmail.com

Abstract. This paper studies the design of active attitude control system of a nano-satellite in a single axis. In this paper, we consider dc motor based reaction wheel as an actuator, because of its pointing accuracy. However, the power consumption of the dc motor is often relatively large and needed to be optimized. Linear quadratic controller is supposed to have an ability to minimize power consumption and able to enhance the system performance. To show the advantage of this method, simulation result of attitude response, state trajectory, and trajectory of DC motor voltage are presented.

1. Introduction

Nano-satellite is a small sized satellite, weighted less than 10 Kg [1]. Nano-satellite has been widely used for various applications i.e mining exploration, disaster monitoring, educational purposes, remote sensing,etc [2]. One of previous researches on reaction wheel is reported in [3], in which Kalman Filter is implemented to enhance fault detection performance.

Active attitude control becomes necessary for nano-satellite to point its instrument (eg. camera) [4]. In this paper, reaction wheel is used as an actuator in active attitude control system. It is used as actuator due to its good pointing accuracy [5]. However, power consumption of DC motor in the reaction wheel is relatively large. Thus, optimal controller may become an answer for this circumstance.

Linear quadratic controller is an optimal control regulator, which poses quadratic cost function. The feedback gain of linear quadratic controller has to be determined, in order to minimized the cost function and optimizing system performance [6].The objective of this paper is to develop linear quadratic regulator for dc motor based reaction wheel. By using this controller, minimizing the power consumption of DC motor and reducing settling time of the system are done.

2. Controller design

The system consists of DC motor which is connected on the reaction wheel as its driver. As the reaction wheel begin to move, it controls the attitude of nano-satellite. The nano-satellite construction is illustrated by the following figure.

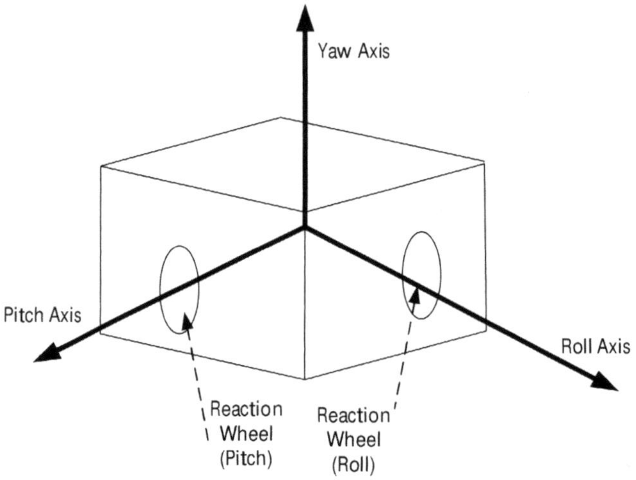

Fig. 1. Nano-satellite with reaction wheel construction

The used parameters to describe the dynamics of the reaction whell are as follows.

Table 1. Parameter description of Reaction wheel [5]

Parameter	Description	Value	Unit
I	Current of the dc motor on reaction wheel	-	A
ω	Angular speed of reaction wheel	-	Deg/s
θ	Angle of nano-satellite	-	Deg/s
R	Motor resistance	3.92	Ω
L	Motor inductance	$150\ 10^{-6}$	H
J	Inertia (including wheel)	$5.1181\ 10^{-5}$	Kg m^2
B	Damping ratio	$1{,}1261\ 10^{-6}$	Nm/(deg/s)
K_b	Back EMF constant	$2.7767\ 10^{-4}$	Volt/(deg/s)
K_t	Torque constant	$15.9\ 10^{-3}$	Nm/A
V	Voltage	-	Volt

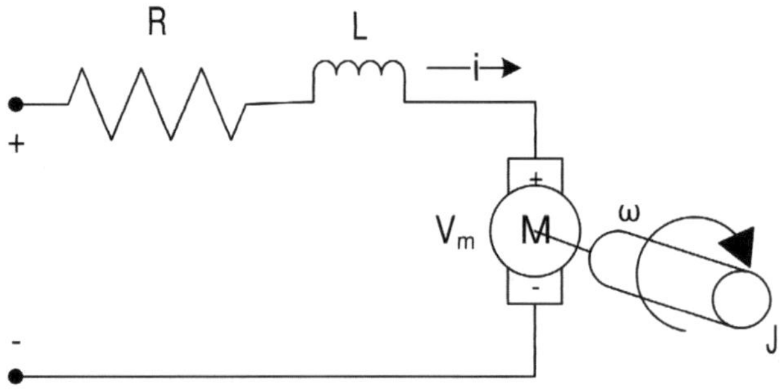

Fig. 2. DC motor schematic circuit [5]

The electrical dynamic equation of the DC motor above can be derived by applying Kirchoff's voltage law and written as bellow [5]

$$\sum V = -V + V_R + V_L + V_M = 0 \tag{1}$$

$$iRa + L\frac{di}{dt} + K_b\omega = V \tag{2}$$

$$\frac{di}{dt} = (-iRa - K_b\omega + V)/L \tag{3}$$

The mechanical dynamic equations of DC motor movement is

$$\sum \omega = J\frac{d\omega}{dt} + iK_t - b\omega = 0 \tag{4}$$

$$\frac{d\omega}{dt} = (iK_t - b\omega)/J \tag{5}$$

The angular rate movement of nano-satellite is assumed to be proportional with the movement of the DC motor, hence the equation becomes

$$\frac{d\theta}{dt} = K_a\omega \tag{6}$$

The state space equation for attitude of the nano-satellite is

$$\dot{x} = Ax + Bu, u = -kx \tag{7}$$

$$\dot{x} = Ax + B(-kx) = (A - BK)x \tag{8}$$

$$y = Cx \tag{9}$$

Refer to equation (3), (5), and (6); equation (7) and (9) become

$$\begin{bmatrix} \dot{x}1 \\ \dot{x}2 \\ \dot{x}3 \end{bmatrix} = \begin{bmatrix} \dot{i} \\ \dot{\omega} \\ \dot{\theta} \end{bmatrix} = \begin{bmatrix} -R/L & -Kb/L & 0 \\ Kt/J & -b/J & 0 \\ 0 & K_a & 0 \end{bmatrix} \begin{bmatrix} i \\ \omega \\ \theta \end{bmatrix} + \begin{bmatrix} 1/L \\ 0 \\ 0 \end{bmatrix} u \tag{10}$$

$$y = \begin{bmatrix} 0 & 0 & 1 \end{bmatrix} \begin{bmatrix} i \\ \omega \\ \theta \end{bmatrix} \tag{11}$$

The performance index, which needs to be minimized is

$$J = \int_0^\infty x^T Qx + u^T Ru\, dt, Q = C^T C, R = 1 \tag{12}$$

$$J = \int_0^\infty \theta^2 + u^T Ru\, dt, \tag{13}$$

The gain of linear quadratic controller could be determined after acquiring the state space equation and the performance objective for the attitude of the nano-satellite by using the following ricatti equation.

$$PA + A^TP - PBR^{-1}B^TP + Q = 0 \tag{14}$$

Where the feedback gain is

$$k = R^{-1}B^TP \tag{15}$$

Given ratio between angular speed of nano-satellite and angular speed of DC motor is 0.002, hence the state space of this system becomes

$$\begin{bmatrix} \dot{i} \\ \dot{\omega} \\ \dot{\theta} \end{bmatrix} = \begin{bmatrix} -26133.333 & -1.8444444 & 0 \\ 310.660642 & -0.0220027 & 0 \\ 0 & 0.002 & 0 \end{bmatrix} \begin{bmatrix} i \\ \omega \\ \theta \end{bmatrix} + \begin{bmatrix} 6666.7 \\ 0 \\ 0 \end{bmatrix} u \tag{16}$$

$$y = \begin{bmatrix} 0 & 0 & 1 \end{bmatrix} \begin{bmatrix} i \\ \omega \\ \theta \end{bmatrix} \tag{17}$$

By solving ricatti equation, we obtain feedback gain as bellow

$$k = [93.159\,10^{-6} \quad 0.0078 \quad 1] \tag{18}$$

As the feedback gain applied, the augmented matrix of A from state space matrix becomes

$$A = \begin{bmatrix} -26133.954 & -54{,}08968 & -6666.7 \\ 310.660642 & -0.0220027 & 0 \\ 0 & 0.00277 & 0 \end{bmatrix} \tag{19}$$

IOP Conf. Series: Materials Science and Engineering **105** (2016) 012004 doi:10.1088/1757-899X/105/1/012004

The eigen value of this system are -26133.3 , -0.33+0.33i, and -0.33 - 0.33 i. Since the real parts of eigen value are negatives, we can conclude that this system is asymptotically stable.

3. Simulation result and discussion

Attitude response of the nano-satellite is illustrated on Fig. 3. It shows that attitude of the nano-satellite converges to 0 degress, as it reaches its steady state condition. According to fig.2, it shows that system with linear quadratic controller reaches its steady state condition, faster than system without using state feedback controller.

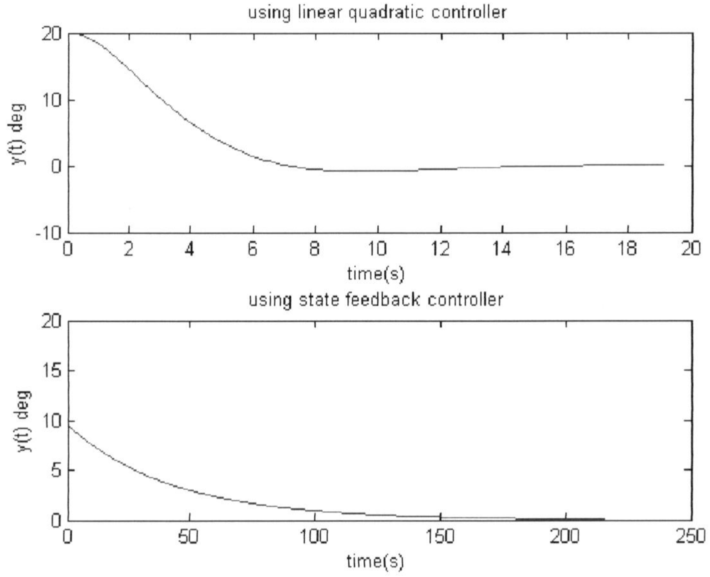

Fig. 3 Attitude response of nano-satellite

States trajectory of this system is illustrated in fig.4, x1 represents current of the dc motor on the reaction wheel, x2 represents angular speed of the reaction wheel, and x3 represents the angle of nano-satellite. It shown that this system achieves its steady state within 20 seconds, proving that linear quadratic controller has significantly enhance the performance of nano-satellite.

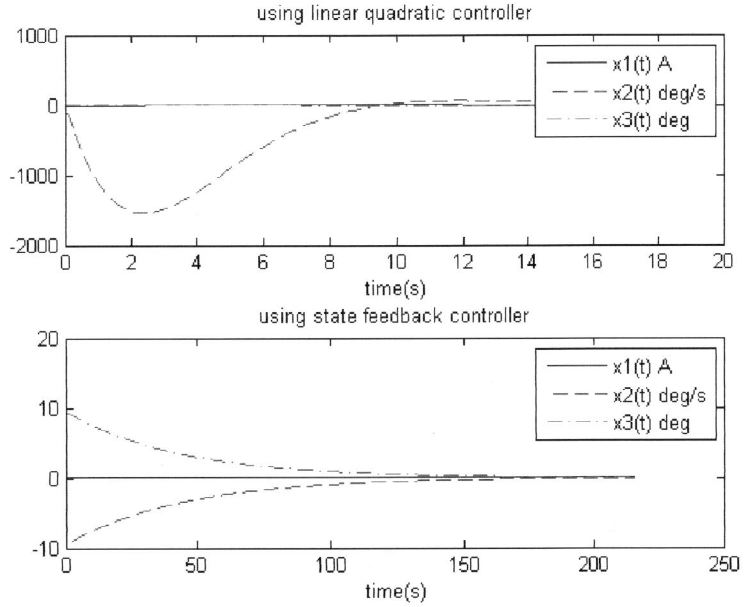

Fig. 4. States trajectory

Trajectory of control action of this system is shown in fig.5, illustrating used voltage to drive DC motor.

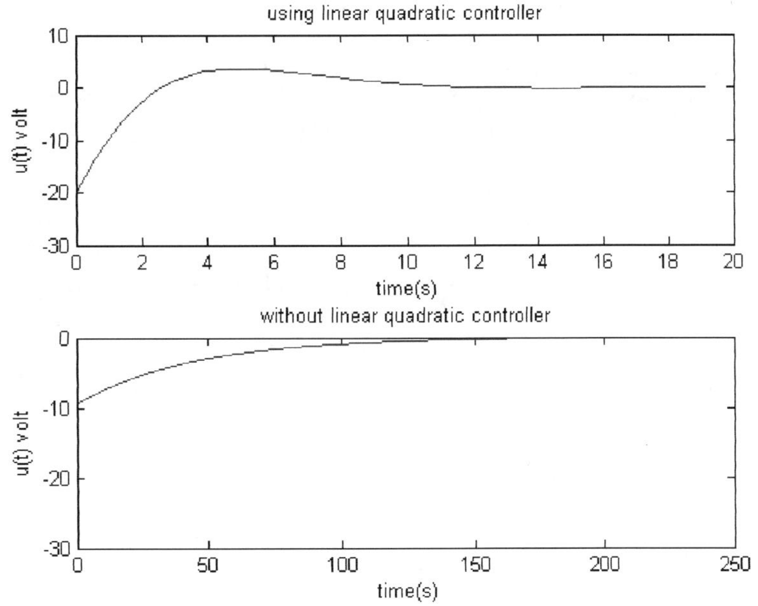

Fig. 5. Control action trajectory

Comparison of power consumption of DC motor between using and without using linear quadratic controller is illustrated in fig. 6. It shows that power consumption of DC motor is extremely larger, when state feedback controller is applied to this system. It can be infered that linear quadratic controller could minimize power consumption of DC motor.

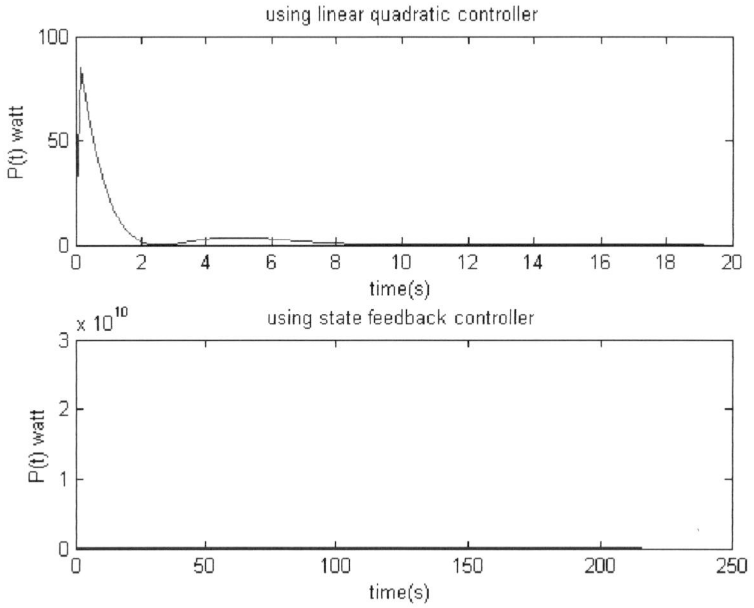

Fig. 6. Power consumption trajectory

4. Conclusion

We present controller design of nano-satellite attitude using linear quadratic controller. The simulation result shows that the attitude control can be obtained. Linear quadratic controller is designed to minimize the power consumption of the DC motor, which is given by performance index. It also shown that system with linear quadratic controller achieve its steady state condition, more than 180 seconds faster compared to system with state feedback controller. Simulation results are provided to demonstrate the effectiveness and efficiency of the controller.

References

[1] Tri Kuntoro 2011 Inusat-1: Satelit-Nano Perdana di Indonesia untuk Pendidikan dan Penelitian (Kursor)
[2] M Shadab Khan 2013 Smart and Effective Cost applications of Micro-Nano Satellite, (Japan Nano Satellite Symposium)
[3] J. Bouwmesteer and J. Guo 2010 Survey of Worldwide Pico-and Nano Satellite Missions, Distribution and Subsystem Technology (Acta Astronautica 67)
[4] Afshin Rahimi 2015 Enhanced Adaptive Kalman Filter for Reaction Wheel (IEEE Transactions on Aerospace and Electronics System)
[5] Helosman Figueiredo and Osamu Saotome 2013 Design of a Set of Reaction Wheels For Satellite Attitude Control Simulation (Cobem)
[6] Ashish Tewari 2002 Modern Control Design with Matlab and Simulink (John Willey & Sons)

IOP Conf. Series: Materials Science and Engineering **105** (2016) 012005 doi:10.1088/1757-899X/105/1/012005

Conceptual Model of Supply Chain Structure Mapping - A Case of Subsidized LPG Commodity in Yogyakarta

Joko Sulistio, Afifuddin Thoif, Aulia Fitri Alindira
Industrial Engineering Department, Universitas Islam Indonesia
Yogyakarta, Indonesia

E-mail: jokosulistio@uii.ac.id

Abstract. — In 2007, the government launched a conversion program of kerosene to LPG by issuing a Presidential Regulation No. 104/2007 on Supply, Distribution and Pricing LPG 3 Kg. Article 2 on the regulation says that setting the supply, distribution, and pricing of LPG 3 Kg include planning an annual sales volume of enterprises, the reference price and the retail price and conditions of export and import of LPG 3 Kg in order to reduce subsidies Kerosene especially to divert the use of kerosene according to government policy. In principle, the purpose of this policy is to reduce energy subsidies on commodities, especially Kerosene. Although the government claimed the conversion program is success, there are few problems arising from conversion program. In 2014, many scarcity and high price of LPG 3 Kg were reported. In this case, Pertamina was given full authority to manage all supply chain and distribution. Because the root of the problem of scarcity that occurred in the supply chain system has not been explained, the proposed solutions will also be partial and not comprehensive. Thus, this research will build a structural map of the causes of supply chain system LPG 3 Kg, as well as providing a comprehensive picture of system dynamics of LPG 3 Kg supply chain system which applied in Indonesia. And the result is expected as in form of Causal Loop Diagram of supply chain system.

1. Introduction

Demand for natural gas or liquefied petroleum gas (LPG) has increased significantly since the conversion program from kerosene to gas fuel in 2007. The conversion program refers to Presidential Regulation No. 104/2007 on Supply, Distribution and Pricing for LPG 3 Kg. Second article on Presidential Regulation says that the supply, distribution, and pricing for LPG 3 Kg include Enterprise annual sales volume, benchmark price and retail price as well as export and import of LPG 3 Kg in order to reduce subsidies on oil fuel particularly to divert the use of kerosene is rendering to government policy. Principally, the purpose of this policy is to reduce energy subsidies, especially in petroleum commodities. In 2006, kerosene subsidy is up to IDR 41.7 trillion. With the conversion of kerosene to LPG, the government is targeting to save IDR 15 trillion per year. After 8 years, the government can save 144.5 trillion in 2007-2014 [1]. PT Pertamina states can resolve the conversion program target of kerosene to LPG 3 Kg to 58 million household by the end of 2014. In this case, Pertamina was given full authority to manage all supply chain and distribution systems [2].

IOP Conf. Series: Materials Science and Engineering **105** (2016) 012005 doi:10.1088/1757-899X/105/1/012005

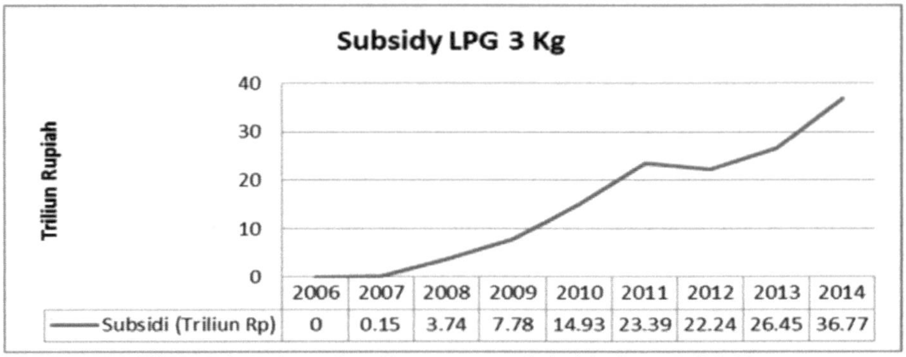

Figure 1. Indonesian Government Subsidy for LPG 3 Kg

Recently, the reformation spirit of Indonesian government for energy subsidy is very high. In 2015, Indonesian Proposed Budget agreed by DPR is allocated around IDR 275 Trillion for oil fuel subsidy. This subsidy includes premium for IDR 108,3 Trillion, diesel for IDR 80,2 Trillion, kerosene for IDR 6,1 trillion, LPG 3 Kg for IDR 55,11 trillion and electricity for IDR 68,7 trillion [3]. Figure 1 shows the behavior of subsidies for LPG 3 Kg by the government. Thus, the original purpose of energy conversion program to reduce energy subsidies, currently just shows the opposite direction.

Even though the government claimed success, there are problems arising from the conversion program. In 2014, scarcity and high price of LPG is widely reported. Early 2014, scarcity has reported in Tondano [4]. Followed by scarcity in Majalengka [5] and Batam in Mei [6] as well as Riau Provinsi in September [7] has been reported. Scarcity is also reported in 2008, causes by technical problem in Cilacap and Indramayu refuelling point [8]. Gas conversion policy in its development has been changing people's habits. Because the price is cheaper, people who previously using LPG 12 Kg switch to LPG 3 Kg [9]. In previous years, Pertamina has conducted market operation to overcome scarcity of LPG 3 Kg [10]. In addition, the distribution arrangement was also made to overcome the same problem [11]. In fact, agents and resellers are suspected of adding the complexity of distribution problem [12].

The cause of scarcity was allegedly from the mixing activity of transferring LPG 3 Kg to 12 Kg. This activity is one of criminal act with profit motif of price disparity between subsidies LPG 3 Kg with non-subsidies LPG 12 Kg [13]. Otherwise, besides Pertamina, the role of National Entrepreneur Accusation of Oil and Gas needs to be strengthened, as well as police controlling, prosecutors, legislators, and local government [14]. In addition, the increasing price of LPG 12Kg causes a lot of people switch to use LPG 3 Kg, it effects to high demand of LPG 3 Kg. These data shows that the root problem of scarcity has not well explained, this research will analyse the structural map of supply chain system of LPG 3 Kg as well as providing comprehensive picture of it system based on descriptive analysis method by utilizing Causal Loop Diagram.

2. Research Methodology

This research will use descriptive qualitative method. It is aim to gathering, enclose, and data interpreting which will then be processed to determine in detail the decision process making and distribution process on LPG 3 Kg distribution system. Gathered data will be used to develop Value Stream Map (VSM) and Causal Loop Diagram. VSM used to de-fine the whole distribution activity, whether they have value added or not, in delivering the product from Depot up to customers. While Causal Loop Diagram used as a tools to help researcher to mapping the system structure of LPG 3 Kg distribution. Both VSM and CLD used to help modeller to analyse the distribution system which focus on LPG 3 Kg distribution in Sleman regency, Daerah Istimewa Yogyakarta. The required data type for

IOP Conf. Series: Materials Science and Engineering **105** (2016) 012005 doi:10.1088/1757-899X/105/1/012005

this research is primary data collected by observation and interview. In this research, primary data will be collected from the Pertamina Yogyakarta, SPBE, agent, and station about the decision making process and order fulfilment order. Research methodology will adopted from the methodology offered by [15]. In general, model development will be done iteratively using following scheme.

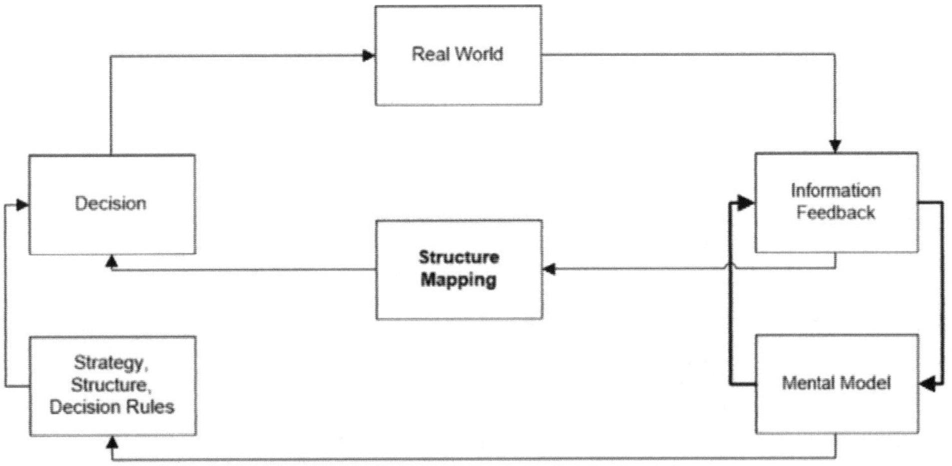

Figure 2. Scheme of Research Methodology

3. Result and Analysis

3.1. Identification of Supply Chain System LPG 3Kg

3 Kg LPG distribution channels in general is Depot → SPPBE → Agent → Base → Customer. Physical flow of 3 Kg LPG distribution starting from Depot heading SPPBE (Charging Station and Freight Bulk LPG), there are 2 different depots locations at Cilacap and Semarang. Delivery from Depot to SPPBE done every Monday to Saturday. Furthermore, after the truck arrived at SPPBE, the unloading process is carried then performed the process of updating inventory levels. After that, trucks from agent are going to SPPBE to fill the bulk LPG. When the trucks from agents comes to SPPBE, they will be checked for the administration and LO (Loading Order), after fulfilling all the administrative requirements then they will go to the bulk filling area if there is no queue then it will go directly to the truck unloading process that is loading the empty LPG to be filled. Then the empty bulk will go to the filling to the bulk process. The next process is the LPG which are already filled will be closed and given a protective plastic whose name matches the name of the agent in the Sealing process. After that is the quality control by dipping the LPG into the water to check for leaks tube. Then LPG directly inserted into the truck (truck loading). After the truck has finished loading, the next is the administrative process for issuing the street. Then LPG will be taken to the warehouse agent and will be distributed to Base Agent owned. Then the customer will come to the base to purchase LPG 3 Kg.

LPG distribution information flow is starting from Pertamina Yogyakarta which makes forecasting for daily quota Agent. Then the results of the daily quota Agents commonly called Kitir was proposed to Pertamina Regional IV, after the evaluation and approved then the Kitir published in the system information in Pertamina which can be accessed by the Agent and SPPBE. SPPBE will confirm to Pertamina Regional IV on the number of tons of LPG to be transported which later Pertamina Regional IV will confirm to the Depot. In the agency, after receiving Kitir containing daily quota Agent, the agent paid the amount of the quota to be able to fill up to SPPBE. Once it paid, a letter of Pertamina for charging to SPPBE would be issued.

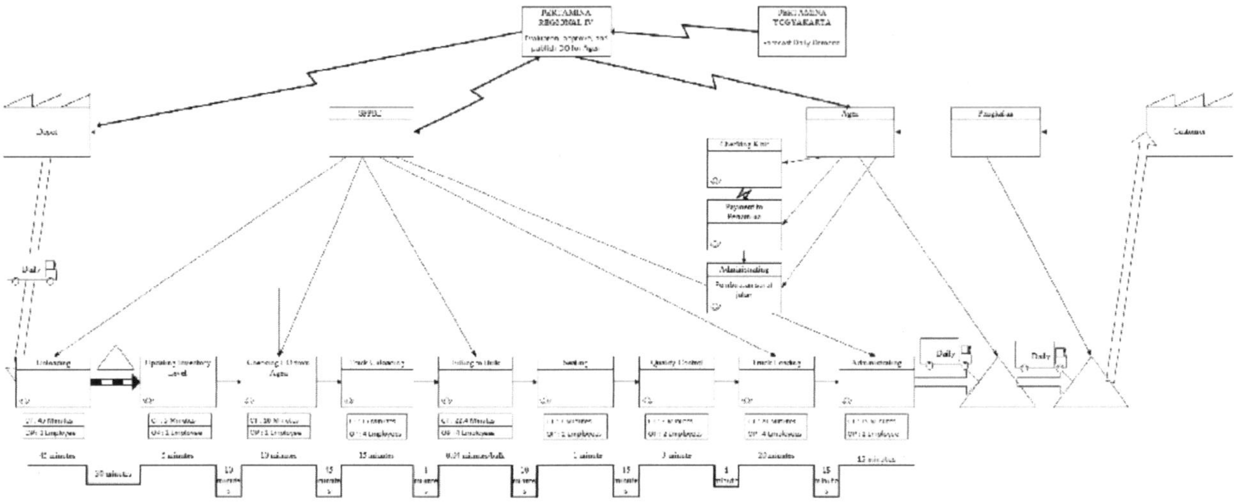

Figure 3. Value Stream Map of LPG 3 Kg Distribution

3.2. Causal Loop Diagram

Figure 4 shows the causal loop diagram of LPG 3 Kg distribution system. Causal loop diagram is consist of system elements which constituent the LPG 3 Kg distribution system. These elements have feedback relationship among others. The feedback could be positive or negative relationship. Positive relationship occurs when the value of an element increased, it led to an increase in other element value, or if the value of an element is decreased, it led to the decline of other element value. Otherwise, negative relationship among elements occurred if the increase of certain element value would cause other certain element value is decrease or contrarily.

The amount of LPG 3 Kg subsidy distribution is depend on the ability of each region subsidy budget. This budget will also affecting the price of LPG 3 Kg in certain region. The subsidy budget set by Sleman region government will then be distributed to LPG 3 Kg agent in form of quotas which done by Pertamina Yogyakarta. The quota become agent authority to take it in form of agent distribution based on quota. This quota is use to filling agent LPG bulk in SPBE which means decreasing SPBE stock. As the agent take the LPG quota, SPBE may supply the LPG from the Depot based on supply planning and considering safety stock and SPBE storage capacity. Once the agent take their quota, it means that the agent is adding their stock. These stock should delivered to the station subsidiary by the agent. Distribution quota for each station is depends on quota set by Pertamina, the number of station subsidiary by the agent, and the station capability to sell LPG 3 Kg bulk. It means that each of station may have different quota and delivery time. The station is responsible to sell the LPG 3 Kg bulk to the end customers. Quota given to station often create Gap between stock and demand, where the demand of LPG 3 Kg is more than the station stock. Subsidy which should delivered to micro business and middle and lower income people is adding with the others element which causes the unfulfilled demand. This unfulfilled demand may encountered by adding subsidy quota if the subsidy budget still sufficient enough. Otherwise, if there is no left subsidy budget, then there will be no additional subsidy quota.

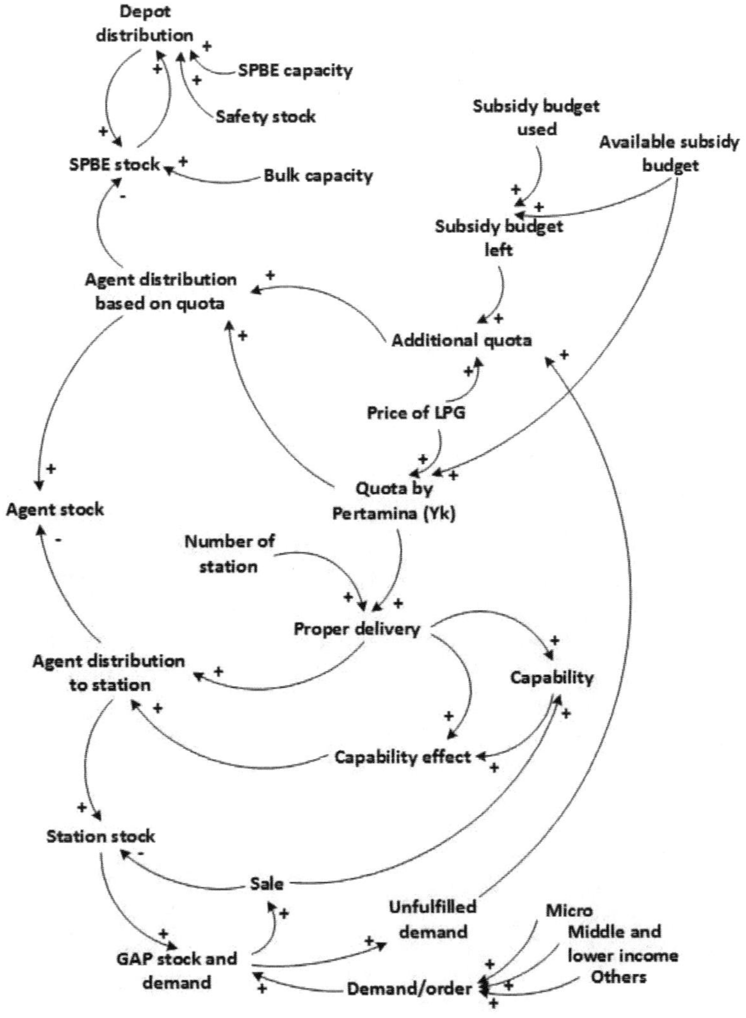

Figure 4. Causal loop diagram of LPG 3 Kg distribution

4. Discussion

In the beginning of conversion program, there are many step in conversion program which not proper with the regulation [16]. So, the conversion program target is not achieved as the purpose of LPG 3 Kg subsidy is originally targeting the micro business and middle and lower income. However, any parties exclude these parties can still buy LPG 3 Kg. As the growth of population, another problem occurred relates with the amount of subsidy. In this condition, a new facilities to support LPG 3 Kg distribution has to be built. Analysis to define new facilities has been done using P-Median method [17]. Optimization of supply chain configuration also has been studied previously using goal programming [18] as well as transportation route selection [19].

Many research about LPG 3 Kg has been previously done by researchers. However, there is no research done by utilizees CLD as a key to analyse the root cause of scarcity yet. As we know that causal loop diagram (CLD) or even called influence diagram is use to help modeller to understanding the system and to give the general image through cause and effect relationship in existing system

(system conceptualization). In this research, CLD help in providing the general supply chain system of LPG 3 Kg. Through CLD, modeller may easily construct the model based on assumptions. CLD provides simplicity in understanding the structure and behaviour of a system.

The purpose of LPG 3 Kg subsidy is originally targeting the micro business and middle and lower income. In the implementation, scarcity is often happen in many distribution area. Scarcity phenomenon itself is a condition where the amount of LPG 3 Kg in the market is lower than customers demand or order. In the development of causal loop diagram as an analysis of existing distribution system shows in Figure 3, it shows that there is unfulfilled demand element. The demand is come from micro business, middle and lower income and others. This condition shows that there is un-matched condition between the real and supposed condition especially for others element. Other element means the customers or end user that not supposed to buy LPG 3 Kg. In fact, there is middle-macro business and high income people still can buy LPG 3 Kg. In this condition, medium and high income people benefited most from the LPG pro-gram [20].

From the causal loop diagram, the station stock is causes by the distribution quota from agent to station. This condition means that station can determine their stock in order to meet customers demand. While the distribution quo-ta from agent to station is depend on the proper delivery or distribution planning which determine by Pertamina Yogyakarta. So, in order to fulfilled customer demand Pertamina Yogyakarta should increase the distribution quota. The ideal condition is LPG 3Kg can fulfil the demand in proper condition. But in this case, increasing quota of subsidy is not easy since it deal with government subsidy budget. A closed distribution system may become alternative solution for the government to distribute LPG 3Kg. An agent based model for closed distribution has been developed [21]. Simulation to compare open and closed distribution system also has been developed [22]. These research shows that closed loop distribution is more effective and efficient.

5. References

[1] Medanbisnis. 8 September 2014. Available: http://medanbisnisdaily.com/news/read/2014/09/08/116316/konversi_elpiji_3Kg_pemerintah_utang_IDR_27_triliun_ke_pertamina/#.VD4GgbCUdVU

[2] Metrosiantar. 22 April 2014. Available: http://www.metrosiantar.com/program-konversi-elpiji-3-Kg-bakal-selesai-pada-2014/

[3] Setyowati, D. 23 September 2014. Available: http://katadata.co.id/berita/2014/09/23/subsidi-bbm-2015-disepakati-IDR-276-triliun

[4] Tribunnews. 1 January 2014. Available: http://manado.tribunnews.com/2014/01/01/akhir-tahun-2013-kelangkaan-elpiji-3-Kg-jadi-masalah-di-tondano

[5] Republika. 4 May 2014. Available: http://www.republika.co.id/berita/nasional/jawa-barat-nasional/14/05/04/n510gk-kelangkaan-gas-elpiji-3-Kg-makin-parah-di-majalengka

[6] RRI. 30 May 2014. Available: http://www.rri.co.id/post/berita/81997/ekonomi/warga_batam_keluhkan_kelangkaan_elpiji_3_Kg.htm

[7] Riaupos. 9 September 2014. Available: http://www.riaupos.co/54015-berita-kelangkaan-elpiji-3-Kg-makin-parah.html#.VD32v7CUdVU

[8] Wahono, T. 12 Desember 2008. Kompas. Available: http://otomotif.kompas.com/read/2008/12/12/17305912/wapres.janji.atasi.kelangkaan.elpiji.dalam.dua.hari

[9] Republika. 15 Desember 2010. Republika Online. Available: http://www.republika.co.id/berita/breaking-news/ekonomi/10/12/15/152324-2014-distribusi-lpg-3-Kg-sistem-tertutup-se-indonesia

[10] Antara. 22 Agustus 2011. Antara News. Available: http://www.antaranews.com/berita/272891/pt-pertamina-segera-operasi-pasar-lpg

[11] Tempo. 15 Mei 2012. Tempo. Available: http://www.tempo.co/read/news/2012/05/15/058404049/Penyebab-Gas-Elpiji-3-Kilogram-Langka

[12] Oliez, M. 30 Januari 2013. Sindonews. Available: http://ekbis.sindonews.com/read/2013/01/30/34/712694/lpg-3-Kg-langka-akibat-agen-dan-pengecer-nakal.

[13] Lelyemin, P. P. 9 June 2014. Available: http://economy.okezone.com/read/2014/06/09/19/996055/atasi-kelangkaan-

elpiji-3-Kg-ct-minta-polri-razia

[14] Beritasatu. 11 Oktober 2014. Available: http://www.beritasatu.com/ekonomi/216561-selain-pertamina-kelangkaam-elpiji-3-Kg-di-medan-akibat-lemahnya-peran-distributor.html

[15] Sterman, J. D. (2000). *Business dynamics: systems thinking and modeling for a complex world.* Boston: Irwin/McGraw-Hill.

[16] Feriyanti, S. (2013). Evaluasi Implementasi Program Konversi Minyak Tanah Ke Gas Lpg 3 Kg Di Kelurahan Aren Jaya Kecamatan Bekasi Timur Kota Bekasi Tahun 2007. *Jurnal Administratio*, 3(6).

[17] Kawi, E. A., & Rusdiansyah, A. (2009). *Analisis Penentuan Lokasi Pembangunan Stasiun Pengisian Bulk Elpiji (SPBE) Untuk Program Konversi Minyak Tanah Ke Lpg 3 Kg Di Propinsi Jawa Timur Menggunakan Metode P-Median*. Teknik Industri ITS. ISO 690

[18] Anggraini, D., Pujawan, I.N., & Arvitrida, N.I. (2010). *Optimasi Konfigurasi Jaringan Supply Chain Hulu Gas LPG3 Kg di Indonesia.*

[19] Rahmawati, P. (2014). *Penentuan Rute Distribusi Gas Lpg Di Pt. Wina Putra Jaya Menggunakan Algoritma Clarke And Wright Savings.*

[20] Andadari, R. K., Mulder, P., & Rietveld, P. (2014). Energy poverty reduction by fuel switching. Impact evaluation of the LPG conversion program in Indonesia. *Energy Policy*, 66, 436-449.

[21] Cahyaningrum, A., & M. Simatupang, T. (2013). An agent based model for the subsidized LPG with a closed distribution system. *Business Process Management Journal*, 19(5), 841-860.

[22] Nariswari, N. P. A., & Pujawan, I. N. (2010). *Simulasi Penerapan Closed System Pada Distribusi Elpiji 3 Kg (Studi Kasus: Distribusi Elpiji 3 Kg Kec. Klojen-Malang.)*

Plastic Injection Quality Controlling Using the Lean Six Sigma and FMEA Method

A Mansur[1], Mu'alim[2], and Sunaryo[3]
[1,2,3] Industrial Engineering Department, Faculty of Industrial Technology, Universitas Islam Indonesia, Jl. Kaliurang km 14,5 Yogyakarta

E-mail: [1]agusmansur.am@gmail.com, [2]mu92alim@yahoo.com, [3]sunaryo@fti.uii.ac.id

Abstract. PT. Yogya Presisi Teknikatama Industri (PT. YPTI) is a mold, precision part, and plastic injection maker company. One of the obstacles faced by the company is the high level of nonconformity on its production results. The waste on production process can be identified and classified into four types, i.e.: a). during the process of injection molding machines, b). finishing and cutting processes, c). quality control process and d). the packaging process. The objectives of this research are minimizing the defective goods and reducing the waste using Lean Six Sigma and FMEA approaches, especially for Bush product. From the analysis result, defective types on Bush product can be classified into bubble, speckle, short shoot, sunken, sink mark, over-cut, flashing, and discolor. Based on the attributes data on Bush product, the DPMO score is 988.42 or the sigma level is 4.6, While the DPMO score on the variable data on each dimension i.e.: a). Slit width on the bottom side has DPMO score of 30119 (sigma level 3.37), b). Diameter of the circle on the top side has DPMO score of 392294 (sigma level 1.77), c). Product thickness on the top side has DPMO score of 70474 (sigma level 2.97), d). Product height has DPMO score of 82107 (sigma level 2.89), product thickness on the bottom side has DPMO score of 24448 (sigma level 3.47), and f). Diameter of the circle on the bottom side has DPMO score of 24448 (sigma level 3.47). The highest RPN score on the dominant types of product defects which needs improvement are the defective goods of bubble type has RPN score of 729, flashing and the molten material out on the heating channel has RPN score of 384, over cutting has RPN score of 324 and sink mark has RPN score of 270. The recommendations for improvement that can be given from this research are making checklist for maintenance and production monitoring, enhancing work supervision and inspection, as well as improving the environment and work stations.

1. Introduction
Development of manufacturing industries in Indonesia has increased by 6.8% in the last three years. For more than twenty years, the role of manufacturing industries to the economy of Indonesia has improved substantially [1]. In global market competition, products with good quality are absolutely required. Quality can be in the form of service to customers. To improve the products' quality, companies should recognize their processing capability. It is intended to determine the extent of the company's products to meet the customers' needs and furthermore it can be used as a basis for controlling and improving the quality of the product characteristics [2].

The key factor that will bring the manufacturing industries to success is quality. The quality of a product is defined as the degree/level where the product or service is able to satisfy the desires of customers (fitness for use). The quality of a product is its ability to provide appropriate performance or results, even in excess of what customers want [3]. Quality is the basic factor of the customers to get the product [4]. To maintain the consistency of the products produced and in accordance with the demands of the market, it is necessary to do a quality control on the production activities. [5] said that quality consists of some special products which can satisfy customers' need because using the product. [6] used the lean six sigma method to reduce rejected cover powder box and powder box products, [7] used the method to give improvisation at production line considering the environmental factor, while [8] gave internal customer satisfaction criteria beside cost and time in implantation the lean sigma.

PT. Yogya Presisi Teknikatama Industri (PT. YPTI) is one of the companies engaged in the service of making mold of a product, precision parts for the needs of industrial machines, as well as plastic injection. One of the products produced in the plastic injection division is Bush. It has some defects and there are also wastes such as short shoot, spots, sink marks, bubbles, over-cut, under-cut and scratch. As for the waste during the production process, there are material that comes out of the hopper dryer, reparation of broken machines and material rework results using crusher machines scattered around the production area.

Many types of defects and the waste that occurs during production process will cause more costs for companies. This research offers improvisation of quality control system using integration of the Lean Six Sigma and Failure Mode and Effect Analysis (FMEA) methods; not only for the products, but also for the production process. Therefore, by using the Lean approach it is expected to know how much the waste will increase the production costs and by using the Six Sigma concept, the company can improve the product quality level and minimize the defective product level. Improvement using the FMEA is expected to identify the root causes of defective products as well as priority improvement on the dominant factor which causes the defective products from the highest RPN score.

2. Literature Review

2.1 Quality

[9] interpreted quality in expanded definition, in which it is a dynamic condition related to products, services, people, processes, and environments that meet or exceed expectations. [10] explained that a product quality is an overall evaluation to customer for improved the performance of goods or services. Product quality is an important aspect to be noticed by customers before they decide to buy a product. The conventional definition of quality is usually directly describe the characteristics of a product such as performance, reliability, easy to use, aesthetics, etc.; while the definition of strategic illustrates that quality is everything that is able to meet the desires or needs of the customers [11].

2.2 Quality Control

[12] described Quality Control as a material control activity which aims to determine the actual material to comply with the conditions established during planning. Therefore, the quality control is an activity to maintain the standardization of the quality of a product or material from the process of preparation, storage, to production point of consumption by the consumer. There are four steps of quality control in general, i.e.:

- Setting the Standards
  It is to determine the cost of quality standards, quality standards of performance, quality standards of security and quality standards.
- Assessing Suitability
  It is to compare the problem of the products or services offered on the standards that have been set.
- Acting if necessary
  It is by correcting the problem and its cause through the factors that include marketing, design, engineering, production and maintenance that affect customer satisfaction.

- Planning for improvement
  It is by developing a continuous effort to improve the standards of cost, achievement, security, and reliability.

2.3 Six Sigma

Six Sigma is a quality improvement vision towards the target of 3.4 per million opportunities failures (DPMO) for each transaction of products (goods and services), vigorous effort towards perfection (zero defects) [13]. Six Sigma is a technique that allows objective assessment of process performance. The Sigma-metric quantifies the performance (and thus the risk) of an analytical testing process. The method decision chart is a visual tool that differentiates Sigma performance. The operating specifications chart is a visual tool that identifies the quality control (QC) required by a test, based on Sigma performance [14]. Six Sigma QC design tools can enhance FMEA, the risk assessment process, and design of quality. There are five basic stages in using the Six Sigma strategy, i.e. Define, Measure, Analyze, Improve, Control (DMAIC), in which the stages is continuous or creating a Six Sigma quality improvement cycle. The DMAIC can be described as follow:

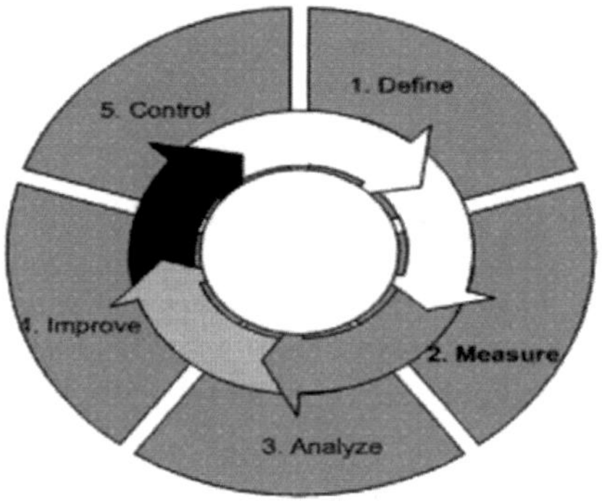

Figure 1. The DMAIC Cycle [15]

2.3.1. Define

It is an early operational stage in Six Sigma quality improvement program. In this stage, there are two important things to do, i.e. defining the core process of the company and defining the customers' needs specifically.

2.3.2. Measure

In the second stage of the Six Sigma quality improvement program, there are three main things [13], i.e. determining the key quality characteristics, developing the data collection plan, and measuring the performance baseline.

2.3.3. Analyze

In the third stage of the Six Sigma quality improvement program, there are three important things to do, i.e. determining the process stability and capability, determining the performance target of the key quality characteristics, and identifying the sources and root causes of the quality problems.

2.3.4. Improve

After identifying the sources and the root causes of the quality problems, we need to determine the action plan for the Six Sigma quality improvement. Basically the action plans will describe the allocation of resources and priorities and/or the alternatives made in the implementation of the plan.

2.3.5. Control

As a part of the Six Sigma approach, it needs controlling to ensure that the desired results are in the process of attainment. Results of the improve phase should be applied within a certain time to be able to see the effect on the quality of the product produced. At this stage the results of quality improvement are documented and disseminated, successful best practices in improving the process are standardized and disseminated, procedures are documented and used as working guideline standards, as well as the ownership or responsibility is transferred from the Six Sigma team to the owner or person in charge of the process.

2.4 Lean
Lean is a systematic methodology to reduce complexity and expedite the process by identifying the source of the waste in the process, because the waste can lead to the breakdown of the production flow. The concept of Lean is also defined as a systemic approach and systematic to identify and eliminate the waste or on-value-adding activities through continuous improvement radically by flowing the product (material, work-in-process, and output) and information using the pull system from internal and external customers for the pursuit of excellence and perfection [16].

2.5 Failure Mode and Effect Analysis (FMEA)
Another six sigma tools that are often used to identify the sources and root causes of the problem is the quality of FMEA. FMEA is a structured procedure to identify and prevent as much as possible failure modes. A failure mode is everything included in defects or failures in the design, condition beyond the limits established specifications or changes in the product that causes the disruption of the function of the product.

3. Research Methods
This research used quantitative methods, which can be interpreted as a research method based on the philosophy of positivism, used to examine the population or a particular sample, the sampling technique is generally done at random. Data are collected by using the research instruments, the analysis of quantitative data or statistics with the aim to test the hypothesis that has been established [17]. Field studies were conducted in this study including observation, interviews and identification to reveal the facts that there are deeper so as to provide solutions to existing problems.

4. Data Processing and Discussion
4.1 Stage 1: Define
In this stage, we did an identification of waste in the production process as well as identified the value added and non-value added activities in the company. The production process for Bush product in PT. YPTI was done in the plastic injection division which uses the SIPOC (Supplier, Input, Process, Output, and Customer) diagram flow system can be illustrated as follow:

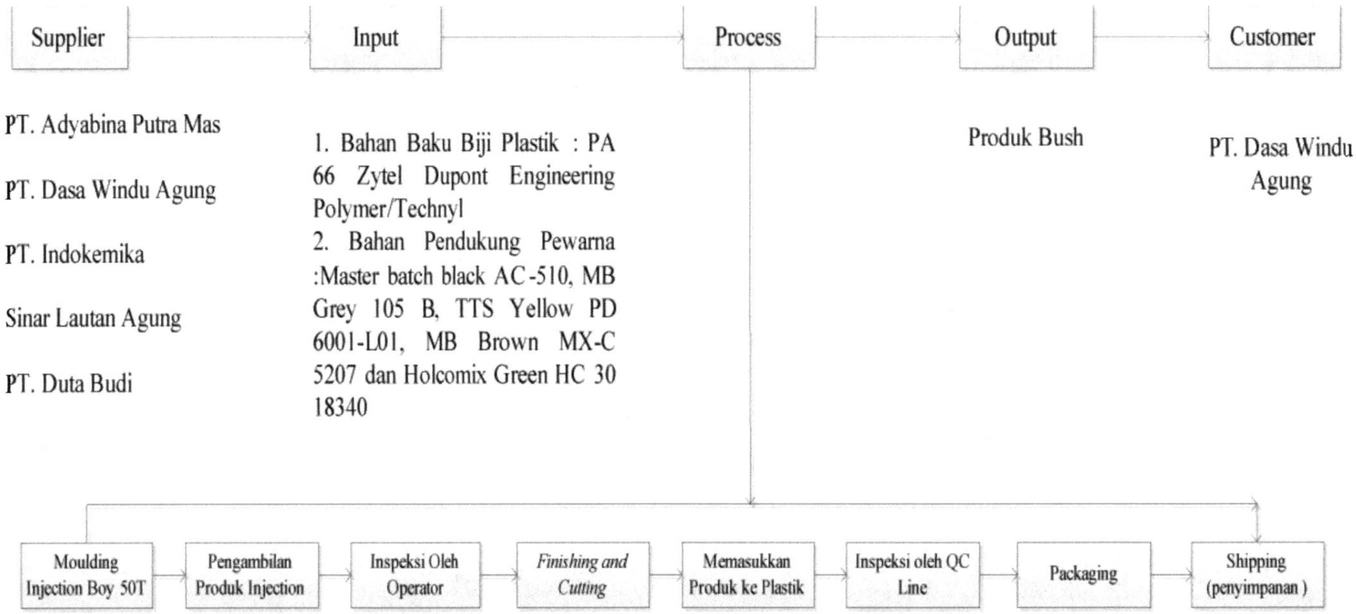

Figure 2. The SIPOC Diagram

And here is the technical figure for Bush product:

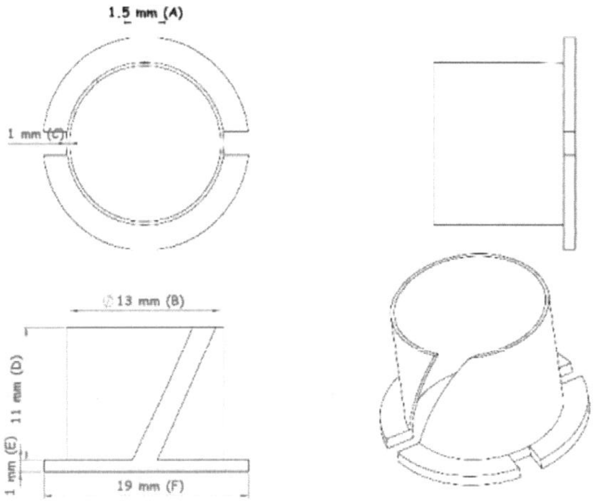

Figure 3. Dimension Specification of Bush product

Description of the product specifications are as follows:

No	Inspection Item	Standard Value			Inspection Instrument
		Nominal	Tolerance		
			Lwr	Upp	
I	**DIMENSION**				
	DIMENSI				Caliper
1	A	1.5	1.3	1.7	
2	B	Ø13	12.9	13.1	
3	C	1	0.95	1	
4	D	11	10.7	11.3	S/Ruller
5	E	1	0.9	1	
6	F	Ø19	18.9	19.1	

Figure 4. Bush Specifications

The types of waste in the Bush production process in PT.YPTI can be identified and classified into four processes, i.e. during the process of injection molding machines, finishing and cutting processes, quality control process and the packaging process. Besides, we found that there are several value added and non-value added activities for the company. Some of the non-value added activities will have a direct impact on the production process. For example the process of cutting products will be stopped when the operator fix the pedestal where the cuts and look for the products that fall. Activities to reprocess the defective products (rework) also add to production costs and make the production process to be longer. Efforts to prevent the recurrence of the non-value added activities can be done by increasing the discipline of production operators, setting the machine according to the procedures and minimizing the occurrence of defects in the product. Thus the waste and non-value added activities should receive serious attention from companies and carried out improvement efforts.

4.2 Stage 2: Measure

In this stage, we did measurement and deeper analysis of the defect types of Bush product, grouping the most potential defect types which affect the product quality and analyzing the stability and capability of the production process.

4.2.1. Critical to Quality (CTQ) Analysis

Based on the calculation result, we found the defective amounts and the percentage of each defect type, i.e. Bubble was 2,460 pcs (47.19%), Speckle was 1,521 pcs (29.18%), Short Shoot was 1,112 pcs (21.33%), Sink Mark was 48 pieces (0.92%), Discolor was 30 pcs (0.58%), Sunken was 20 pcs (0.38%), Flashing was 16 pcs (0.31%) and Over-cut was 6 pcs (0.12%).

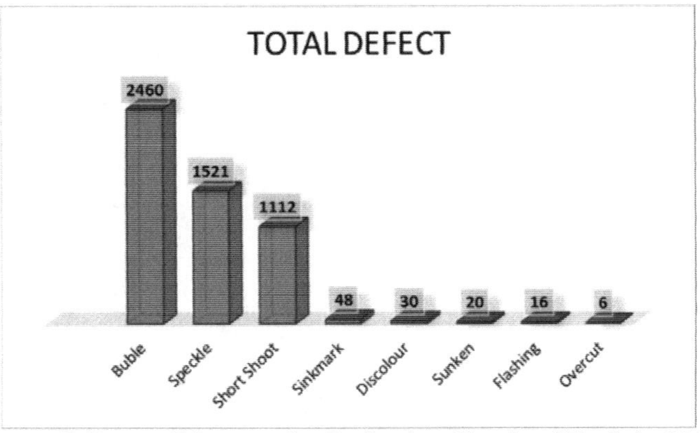

Figure 5. Pareto Diagram for Defective Goods of Bush in 2014

Based on the above figure, there are 3 out of 8 defect types which incurred the most on Bush production in 2014, i.e. Bubble, Short Shoot and Speckle; so those three will be considered as the Critical to Quality.

4.2.2. Analysis of Process Level and Output

Based on the total number of defects that is used as the attribute for the calculation of the whole products in 2014, the DPMO score is 988.42 or 4.6 sigma level, which means that the company is still producing defective products for 5213 pieces per one million products. For 2014 the overall process can be considered as quite competitive and are at the USA's average industry. As for the variable data, the DPMO score is quite high and sigma level is still far from the target. The variable of slit width on the bottom side has DPMO score of 30119 (3.37 sigma level), diameter of the circle on the top side has DPMO score of 392.294 DPMO (1.77 sigma level), the product thickness on the top side has DPMO value 70 474 (2.97 sigma level), high the product has DPMO score of 82107 (2.89 sigma level), product thickness on the bottom side has DPMO score of 24448 (3.47 sigma level) and the diameter of the circle on the bottom side has DPMO score of 24448 (sigma level 3.47). Based on the results of the data processing, the DPMO score of the Bush products' variable data is quite high and the sigma level is still far from the target, which is only 3.4 sigma. In other words, it can be said that the capability score of the process is still at the average level of Indonesia and less competitive, while at the present time many World Class companies have the process capability on quality control at the rate of 5-6 sigma, so they only produces the probability of failure per one million opportunities DPMO below 100.

4.3 Stage 3: Analyze

4.3.1. Process Stability Analysis

At this stage, we analyzed the process stability which was done towards the attribute data, i.e. the number of defects produced in January to December 2014.

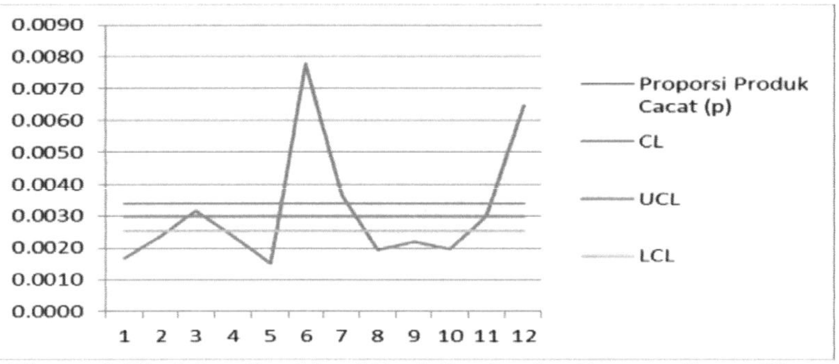

Figure 6. The Map Graphic of Bush Production Process Controlling

Based on the figure above, the controlling of Bush production process in 2014 is not stable. This is shown in the chart that in the 1, 2, 4, 5, 6, 7, 8, 9, 10, and 12 periods there were of control in the production process. In other words, there was proportion of defective products which passed the upper and lower limit control. Therefore, it needs some improvements and efforts to reduce losses due to the defective products.

4.3.2. Process Capability Analysis

Based on the results of data processing and capability analysis for the entire product process of Bush variable data, the index score of $C_{pm,}$ and C_{pmk} is less than 1 or C_{pm} and $C_{pmk}< 1$. It shows that the whole process were not able to meet the specifications set by the customers and not competitive to compete in the global market. Therefore it needs some improvement efforts to enhance the process capability by reducing and minimizing variation in each variable and conducting continuous improvement program in order to compete in the global market.

4.3.3. Source and Root Cause of Defective Products Analysis
The defective on Bush products are generally caused by several factors such as human, material, machines, methods and work environment. The main factor of the defective goods in PT. YPTI is that the machine's engine is old and the result of reconditioning. As for the human factor is because the production operators lack of discipline and fatigue due to positions and less ergonomic workplace. The material also became one of the causes of speckle defects, where the results of rework material mixed with another color material in the hopper dryer and there are still other waste materials, causing speckle defects in the product. Methods and the work environment factors were caused by operators who do not follow the Standard Operating Procedure of employment and rework processes. As for the environmental factors were caused by the noise generated from the crusher machine's engine sound during the rework process, the position of the air conditioners which are too far from the operators and the less clean workplace.

4.4 Stage 4: Improve

4.4.1. Improvement Analysis Using FMEA
From the calculation of FMEA, the highest RPN score on the dominant types of product defects which need improvement are the defective goods of bubble type with RPN score of 729, flashing and the molten material out on the heating channel with RPN score of 384, over cutting with RPN score of 324 and sink mark with RPN score of 270. Improvement recommendations to prevent recurrence of defects in the bush products are:

- For bubble defect :
  - Make sure the material passes through the proper drying.
  - Raise screw back pressure during plasticizing process to force the gas out of the cylinder.
  - Fix the gas ventilation system in the mold so that the gas is not trapped in the product.
  - Lower the speed of injection to allow time for the gas to come out through the ventilation system.
- For the molten material out on the heating channel problems:
  - Make sure the barrel is precise/fit the mold hole.
  - Clean the mold so the holes are not clogged and the material can flow smoothly into the mold.
- For the flashing and sink mark defect:
  - If the problem refers to the quality of the material, it can be replaced with a material with a grade which has a better flow rate and it can also reduce the temperature of the plastic.
  - If the problem refers to the wear of the mold, the mold can be repaired or replaced with a better mold.
  - Reduce product time cycle because it can cause flashing on the product if used for too long
- For over-cut defect
  - Improving the tools and production operator work station
  - Increase the rigor and discipline operator
  - Providing working facilities such as air conditioner or replace seats and a more ergonomic work desk.

4.4.2. Process Flow Analysis
The activities which have been identified and can be eliminated in the production process of a bush that is contained in the finishing and cutting process, the quality control process and the packaging process. Proposed improvement and simplification of the flow of the activities provided during the

process of finishing and cutting is to eliminate the activity of inserting the stem of products bush into the scoop before it is poured into the engine crusher, further improvement is to eliminate the activity of collecting and placing the product on the table of the operator before it packaged into size plastic 1kg/100pcs because such activity causes the table as a cutting area becomes narrower. Improvement of the process flow given during the process of quality control and packaging is to eliminate the activity, returns the defective product to the production station, perform the rework and re-count the number of products that have been packed into a plastic measuring 1 kg/100pcs. Those activities can be a waste for the company. Losses that could be caused by waste is that it can lead to a production process becomes longer, higher production costs and decreased worker productivity.

4.5 Stage 5: Control

This objective of this stage is to ensure that the improved conditions can be sustainable and maintained in the future. Some surveillance activities and the proposed improvements that can be made are as follows:

- Creating maintenance checklist sheet and monitoring the production process so that it could be a reference at the time of production in the future. The next control is to perform maintenance and checking the machine periodically intensively by Setter in order to minimize product defects caused by the machine factor.
- Improve supervision and work inspection. This control can be done by increasing oversight by the leader of the discipline of operators in order to minimize the production of defective products caused by human factors. Assessment of performance and provide training to the work force can be used as an attempt to improve the skills and abilities of workers, where the ultimate goal is to increase worker productivity.
- Fixing the environment and work stations. This form of control can be done by reducing the level of noise generated by the crusher machine's engine sound during the rework process and close the doors and installing a silencer in the Crusher room. Other controls can be done by improving the operator work station such as by replacing the chairs and tables more ergonomic so that workers can work safely, comfortably and can prevent the occurrence of occupational diseases.

5. Conclusion and Recommendation

5.1 Conclusion

Based on the results of the data collection and processing as well as the analysis that has been explained above, it can be concluded that:

- The types of waste in the Bush production process in PT. YPTI can be identified and classified into four processes, i.e. during the process of injection molding machines, finishing and cutting processes, quality control process and the packaging process. As for the types of defects found on the Bush product are bubble, speckle, short shoot, sunken, sink mark, over-cut, flashing and discolor.
- Based on the total number of defects that is used as the attribute for the calculation of the whole products in 2014, the DPMO score is 988.42 or 4.6 sigma level, which means that the company is still producing defective products for 5213 pieces per one million products. For 2014 the overall process can be considered as quite competitive and are at the USA's average industry. While the results of the data processing from the Bush products' variable data, the DPMO score is quite high and the sigma level is still far from the target, which is only 3,4 sigma. In other words, it can be said that the capability score of the process is still at the average level of Indonesia and less competitive.
- The highest RPN score on the dominant types of product defects which needs improvement are the defective goods of bubble type with RPN score of 729, flashing and the molten material out on the heating channel with RPN score of 384, over cutting with RPN score of 324 and sink mark with RPN score of 270.

5.2 Recommendation

The following are the recommendation from the researchers for the improvement of PT. YPTI:

- Creating maintenance checklist sheet and monitoring the production process.
- Enhancing the work supervision and inspection.
- Improving the environment and work stations.

6. References

[1] Kurniati Y and Yanfitri 2010 Dinamika Industri Manufaktur dan Respon Terhadap Siklus Bisnis *Buletin Ekonomi Moneter dan Perbankan* **October Volume**

[2] Susetyo J, Winarni, and Catur H 2011 Aplikasi Six Sigma DMAIC dan Kaizen sebagai Metode Pengendalian dan Perbaikan Kualitas Produk *Jurnal Teknologi* **Vol. 4**

[3] Kotler P and Kevin L K 2009 *Manajemen Pemasaran (translated by Bob Sabran) 13th edition* (Jakarta: Erlangga)

[4] Purnomo H 2004 *Pengantar Teknik Industri* (Yogyakarta: Graha Ilmu)

[5] Laksana F 2008 *Manajemen Pemasaran* (Yogyakarta: Graha Ilmu)

[6] Ahmad, et.all 2013 Analisa Penerapan Lean Six Sigma untuk MEngurangi Non Value Added Time dan Jumlah Produk Cacat pada Produksi Set Kotak Bedak *Jurnal-Pasti* **Vol. VII**

[7] Arifin M and Supriyanto H 2012 Aplikasi Metode Lean Six Sigma Untuk Usulan Improvisasi Lini Produksi Dengan Mempertimbangkan Faktor Lingkungan *JurnalTeknik ITS* **Vol. 1**

[8] Filardi F, et.all 2015 Implementation analysis of Lean Sigma in IT applications: a Multinational oil company experience in Brazil *Jurnal Procedia Computer Science* **55** 1221 – 1230

[9] Yamit Z 2010 *Manajemen Produksi dan Operasi 2nd Edition* (Yogyakarta: Ekonisia)

[10] Hansen D R, and Mowen M M 2005 *Akuntansi Manajemen 7th Edition* (Jakarta: Penerbit Salemba Empat)

[11] Gasperz V 2001 *Total Quality Management* (Jakarta: Gramedia Pustaka Utama)

[12] Suprihanto J 2000 *Penilaian Kinerja dan Pengembangan Sumber Daya Manusia* (Yogyakarta: BPFE)

[13] Gasperz V 2002 *Pedoman Implementasi Program Six Sigma Terintegrasi dengan ISO 90001:2000, MBNQA, dan HACCP* (Jakarta: Gramedia Pustaka Utama)

[14] Westgard S 2013 Prioritizing Risk Analysis Quality Control Plans Based on Sigmametrics *Journal of Clinics in Laboratory Medicine* **Vol. 33** 41 – 53

[15] Pande P, Neuman R P, and Cavanagh Ronald R 2002 *The Six Sigma Way* (Yogyakarta: Andi Offset)

[16] Gasperz V 2007 *Lean Six Sigma for Manufacturing and Service Industries* (Jakarta: Gramedia Pustaka Utama)

[17] Sugiyono 2010 *Metode Penelitian Pendidikan (Pendekatan Kuantitatif, Kualitatif, dan R&D)* (Bandung: Alfabeta)

IOP Conf. Series: Materials Science and Engineering **105** (2016) 012007 doi:10.1088/1757-899X/105/1/012007

Multimedia-Based Therapy Model for Non-Pharmacological Stroke with Decrease Impaired Muscle Strength

Rr Hajar Puji Sejati[1] *, Izzati Muhimmah[2] and Affan Mahtarami[3]

[1] Magister of Informatics, Faculty of Industrial Technology, Universitas Islam Indonesia
[2,3] Department of Informatics, Faculty of Industrial Technology, Universitas Islam Indonesi, Jl. Kaliurang Km14,5, Sleman, Yogyakarta

E-mail: hajarsejati@gmail.com, izzati@uii.ac.id, mahtarami@gmail.com

Abstract. Stroke patients who experience a decrease in muscle strength need to do exercises so that they can increase their muscle strength. In order to enable the patient does exercise independently the multimedia-based stroke therapy model is needed. These exercises can be done independently, with supervision of the family member at home. So, we develop prototype of the multimedia-based therapy for the family member so that they can assist patients performing exercises without attending therapy session in hospital. This model was built according to the advices from physiotherapist and a medical rehabilitation doctor. This model has been evaluated through focused group discussion by physiotherapists. And they gave positive responses to this proposed model.

1. Introduction

Definition of stroke according to the World Health Organization is prominent clinical signs of focal or global form of brain dysfunction that develops quickly, with symptoms lasting 24 hours or longer and can lead to death, with no apparent cause other than vascular [6] .Characteristics of all types of stroke is a neurological dysfunction with onset relatively sudden, involving one or all of the following signs: weakness, numbness, loss of sight, diplopic, dysarthria, gait abnormalities, aphasia, vertigo, or degrees of impaired consciousness [1]. Post-stroke rehabilitation is an effort to repair limb disability or decreased function. One motivation of this study is to help the families of stroke patients to perform muscle strength therapies independently at home. Post-treatment recovery and prevention of stroke risk factors can reduce mortality and increase post-stroke disabilities [2]. One part of post-stroke rehabilitation is therapy to improve muscle strength. The table 1 shows the muscle strength scale.

[1] Hajar is corresponding author

Table 1. Muscle strength scale table[a].

Findings	Grade
Normal motor power	5
Able to overcome gravity and significant resistance but strength not quite normal	4++
Able to overcome gravity and moderate resistance	4+
Able to overcome gravity and mild resistance	4
Able to overcome gravity but not resistance	3
Unable to overcome the force of gravity but able to move in the plane of the supported extremity	2
Flicker movements only	1
Total paralysis	0

[a] courtesy of Miller DW Hahn

Hemiparese stroke patients with decreased muscle strength is usually affect half of the body. Some stroke patients experience these because of having a stroke bihemiparese continued on different sides of a stroke before. Stroke patients, aided by family members, can do therapy to increase muscle strength independently at home. It is necessary to have and an exercises guidelines in hand to do so. Moreover, such multimedia-based guidelines were preferable because it could give more information and interactivity on stroke treatment guidelines for cases of decreased muscle strength.

Muscle strength 0 and 1 are suggested for therapy with a stimulant special tools. This research can illustrate some examples of multimedia visualization in the form of examples of movement therapy with patients with minimal muscle strength 2. If a patient has muscle power 2, it is expected to perform routine exercise therapy can be increased to have the muscle strength of 3 and so on until the maximum results obtained. Before doing movement therapy using this application, one is expected to consult with the patient's medical rehabilitation doctor and physiotherapist.

Multimedia-Based studies related to post-stroke rehabilitation have been studied: The book with the title active again after a stroke is equipped with a CD tutorial book is written by experts in the field of medical rehabilitation aimed to help stroke patients can perform therapeutic exercises independently with the help of a family member to do some post-rehabilitation movements [4]. Some of the videos available on YouTube which are published by hospitals in Singapore [8]. A product with a base pin board, namely designing plaything for post-stroke patients, has been investigated by Rosyada et al [7]. Identification System Based Rehabilitation Needs Affordable Technology For Stroke Patients In Indonesia has been investigated by Hariandja [9]. The interactive tools using smart phones that are connected between the patients with people who care has been investigated by Hariandja and Maitimo Robert [10]. However to our knowledge the study on animation therapy model for stroke patients has never been investigated.

This paper is outlined as follows. In section 2, explained the problem description. In section 3, the proposed method is presented. In section 4, we explained implementation about the software navigation, structured questions, and then therapy solution. In section 5, evaluation from focus group discussion of physiotherapist. Last but not least, in section 6 conclusion to further improvement.

2. Problem Description

How to develop an expert system for the treatment of stroke? How to integrate expert systems on multimedia-based applications? How to build an interface that is user friendly for patients with stroke? The difference with previous studies is that the proposed was built with animations and visualizations focus on therapy to improve muscle strength in stroke patients that experience decreased muscle strength. Multimedia animated visualization is more interactive therefore allow a user to use the navigation buttons on the menu available on the model.

3. Method

This study uses two methods in collecting the data, i.e.: secondary data collection methods (medical records from the hospital) and literature study method. There were101 medical record data collected to obtain examples of classification decreased muscle strength. A literature mainly focused on multimedia-based application development.

Expert systems are included in decision support systems. In general, expert systems is a system that is trying to adopt human knowledge into a computer, so that the computer can resolve the issue as it is commonly done by experts [3]. A general description of expert system on the model of stroke therapy tools are focused on patients with impaired muscle strength decrease. By doing exercises independently with the help of families the stroke patients were expected to their improve muscle strength until its maximum muscle strength. In general overview of the system can be seen in figure 1 below:

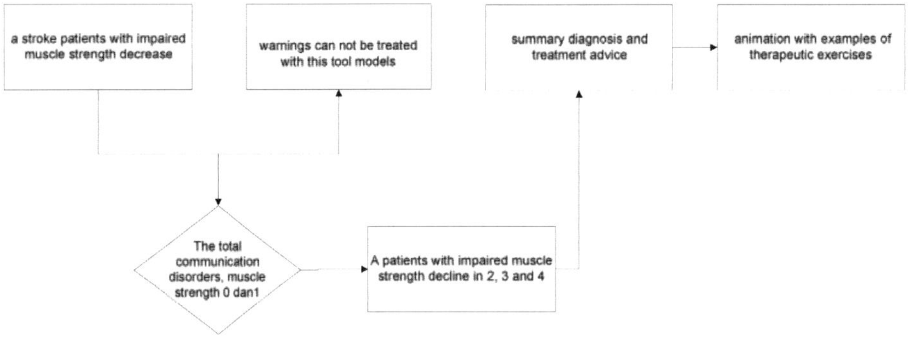

Figure 1. System overview.

On Figure.1 above can be defined when a patient experienced a total breakdown of communication or have very weak muscle strength cannot be treated with this tool. Patients who have the muscle strength 2, 3 muscle strength, and muscle strength 4 can be treated with this proposed tool.

The design of this application can be described with a Finite State Machine which can be seen in Figure 2 below:

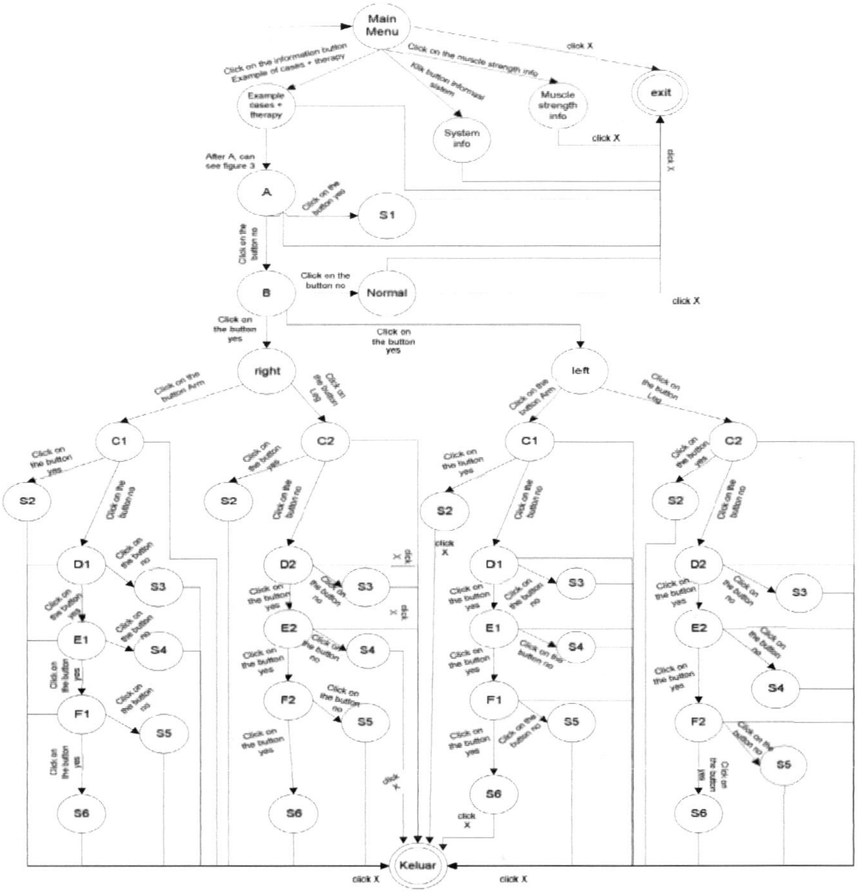

Figure 2. Finite State Machine

Finite state machine is the overall picture of the application, after caption "A" can see figure 3 on decision tree because this models was built according decision tree.

To diagnose muscle power stroke patients with impaired muscle strength decline in this study using structured questions. After going through the structured questions, the diagnosis of muscle strength of each part of the limb is shown so that it can be used to determine the therapeutic solution. The structured questions can be seen in the decision tree in Figure 3 below:

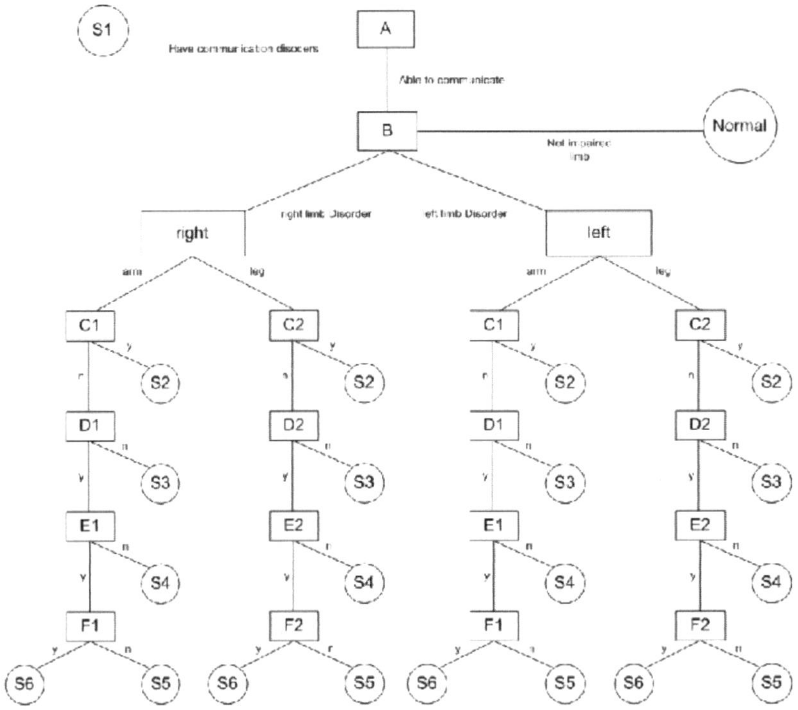

Figure 3. Diagnosis decision tree to determine muscle strength therapy solutions.

Figure 3 caption:

A	Does the patient have communication disorders?
B	Does the patient impaired limb?
C1, C2	Does the patient have a total muscle paralysis?
D1, D2	Does the patient can perform the movement without against gravity?
E1, E2	Does the patient able to do movement against gravity?
F1, F2	Does the patient can perform movement against gravity with mild loads.
S1	Cannot be treated with this application due to special handling needs in terms of communication.
N	Normal, does not need to do therapy.
S2	Cannot be treated with this application because paralyzed muscle strength (muscle strength 0).
S3	Cannot be treated with this application because muscle strength is flicker movements only (muscle strength 1).
S4	Solution therapy with a sliding movement to the right and to the left (muscle strength 2).
S5	Solution therapy with movement up and down without load (muscle strength 3).
S6	Solution therapy with movement up and down with the load (resistance) (muscle strength 4).

The answer of each limb in question will be used in the decision tree to produce a summary diagnosis. An example of this summary diagnosis is as follows: right hand muscle strength is normal, normal right leg muscle strength, muscle strength 3 left hand, left leg muscle strength 3. This tool would inform user that the patient has decreased muscle strength on both left hand and leg. When user clicking the button labelled "please click menu suggestions therapy", the example of animation of the therapy exercises will be played. There were sentences which describe how many times this movement should be done.

4. Implementation

The screenshot below describes the software implementation of the expert system, see figure 4. In this application, the "home" also serves as the main menu page that is used to select the modules that exist in the application. Graphical User Interface on this application can be seen in Figure 4 below:

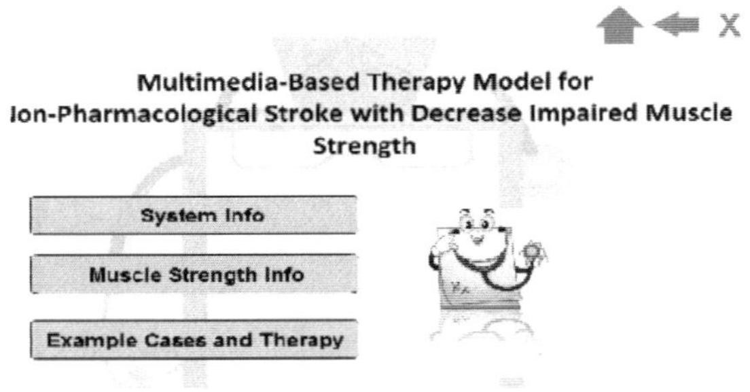

Figure 4. Graphical User Interface

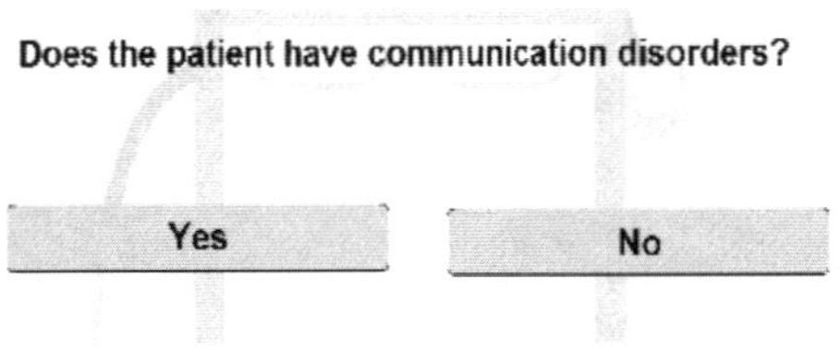

Figure 5. Structured question example about communication ability.

The system information menu is a general description of this application. The menu contains information about the scale of muscle strength and muscle strength information. Sample cases and therapy menu are some structured questions that lead to the diagnosis of muscle strength and therapy advice. The menu in this application is also equipped with the "home", "back", and "exit". Home button to return to the main menu, back button to go back into the previous menu, and out button to end the application program. When the example of button was clicked user will need to answer the structured questions on communication ability and muscle strength.

Examples of the structured questions on the model this tool can be seen in Figure 5 (question about communication ability) and Figure 6 below (questions referring to the diagnosis of muscle strength).

For communication disorders, muscle strength 0 and muscle strength 1, this tool cannot be used and the menu will appear warning that patient cannot be helped with this tool. One example of structured questions will refer to the third muscle strength can be seen in Figure 6 below:

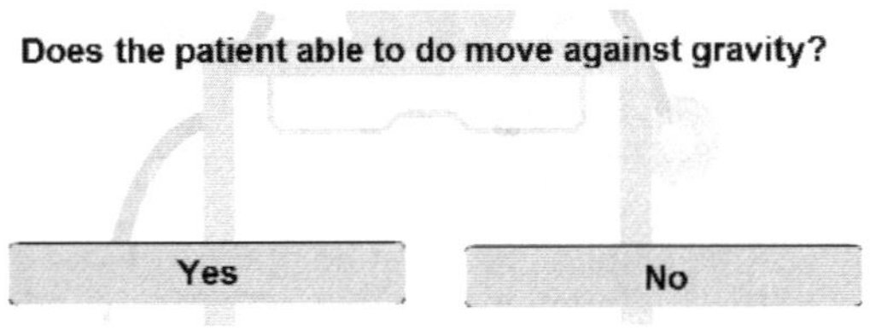

Figure 6. Structured question example to the diagnosis of muscle strength.

Figure 7. Animation example of exercise 1.

One example of the menu display muscle strength 3 therapy suggestions can be seen in Figure 7 above. Muscle strength 3 exercises for the wrist joints consist of two kinds of movement. If exercise1: selected hand movement joint parts lifted up his palms do against gravity. If exercise 2: selected hand movement wrist joints prone parts movement against gravity. For exercise 2 visualization can be seen in figure 8 below:

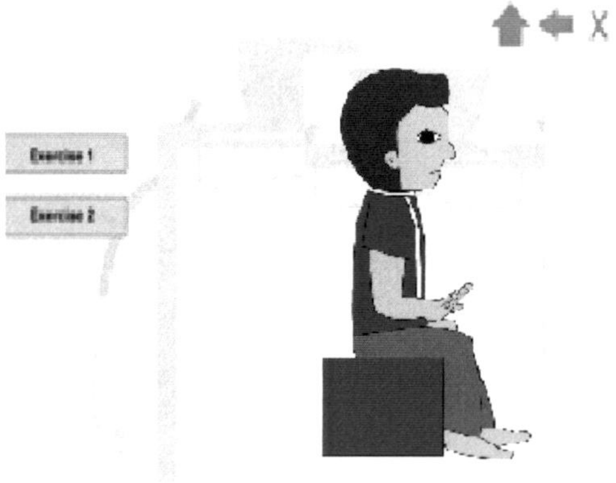

Figure 8. Animation example of exercise 2.

Therapy movement is repeated 8-10 times for each movement and exercise therapy is done 3 times a week, all movements to adjust the patient's condition. Once the patient can perform the movement with the optimal range of motion, there are advises to see the doctor to assist the muscle strength prior to go the next level therapy.

5. Evaluation

This model was shown to the group of 10 physiotherapist. They been asked about their opinion on some indicator, i.e.: application flexibility, application functionality, productivity applications and then usability according to ISO 9126. These indicators were then divided into 18 sub-indicators, i.e.: for application flexibility consists of 5 sub-indicators such as display, language, creativity, ease of use, and grouping action. For application functionality consists of 4 sub-indicators such as reliability, consistency, modelling material, and quality. For productivity applications consists of 5 sub-indicators such as the material suitability, submission info, level correctness, completeness of information, effectiveness of use. And the last indicator is usability theory according to ISO consists of 4 sub-indicators such as easy to understand, learn ability, operability, and attractiveness.

The FGD had been done by the group of physiotherapists. They evaluated the performance of the proposed model. The data is then tabulated to facilitate the testing process. The result given in this study are "expert system performance can be received by users" with a value of 75% acceptance. The total value of the average user acceptance is greater than the value of the hypothesis, the results obtained count value of 0.803 means that the application made resulting assessment data on the performance of the application being built was acceptable by physiotherapist. Moreover, they rate the application performance was “good enough". Result from focus group discussion is an average value of acceptance of more than 80%. However, they also pointed areas where the improvement need to be done, i.e. grouping of action, consistency, and operability. These indicators were rated 74%.

The groups also give some suggestions applications should be made more attractive by adding a better design to increase the interest of the user. This tool should be developed using mobile-based applications because mobile-based device is more user friendly. Some movements require a view from above in order to visualize the animation appear more clearly, but in this study only the view from the front and some animations with the view from the side. Model tools need to be developed to a wider scope, for example by adding exercise to the movement of the head. In anticipation of a situation which any family members who do not computers literate, it is necessary to add guidance and training before using this tool. Prior doing this therapy, patients are expected to check and consulate to a doctor because they need to assess their level of muscle rigidity possessed. Improvement to this model can be done by developing 3D animation.

6. Conclusion

Expert system for determining the level of muscle strength in stroke cases has been develop using combines decision tree successfully developed. Results of the evaluation models evaluated by physiotherapists have been considered as good and the application is expected to help the patient's family to do independent therapy. Further studies are needed to improve and extend coverage to the 3D animation.

ACKNOWLEDGEMENT

This work was supported in part by the Master of Informatics Faculty of Industrial Technology, Islamic University of Indonesia, and Hibah Pascasarjana DIKTI 2014/2015.

References

[1] Alway D & Cole J.2009.Esensial Primary Stroke For Service, translated by : Dwijayanthi Linda, Book Medical Publishers EGC.Jakarta.
[2] Gofir Abdul. 2009. Management Of Stroke. Pustaka Cendekia Press. Yogyakarta.
[3] Kusumadewi Sri. 2003. Artificial Intelligence (Techniques and Applications). Graha Ilmu. Yogyakarta.
[4] Lutfie Syarief Hasan. 2012. Post Stroke Recovery. Tiga Serangkai. Solo.
[5] Miller DW Hahn JF. Chapter 1: General methods of clinical examination. pags 31-32. IN: Youmans JR. Neurological Surgery 4 edition. W.B. Saunders Company. 1996.
[6] http://www.who.int/healthinfo/statistics/bod_cerebrovasculardiseasestroke.pdf diakses 22 Juni 2015.
[7] Rosyada, et al. 2010. Perancangan alat permainan untuk pasien pasca stroke. Jurnal Teknik Industri : Vol V. No 3. Universitas Diponegoro.
[8] http://www.youtube.com/watch?v=3GbpbuwcWDk
[9] Hariandja. 2013. Identifikasi Kebutuhan akan Sistem Rehabilitasi Berbasis Teknologi Terjangkau Untuk Penderita Stroke di Indonesia. Lembaga Penelitian dan Pengabdian kepada Masyarakat. Universitas Katolik Parahyangan. Bandung.
[10] Hariandja dan Robet Maitimo. 2014. Perancangan Alat Bantu Interaktif Penunjang Aktivitas Pendamping Insan Pasca Stroke. Lembaga Penelitian dan Pengabdian kepada Masyarakat. Universitas Katolik Parahyangan. Bandung.

IOP Conf. Series: Materials Science and Engineering **105** (2016) 012008 doi:10.1088/1757-899X/105/1/012008

PKK-man: A System to Manage PKK Activities in Indonesia

R N E Anggraini[1], S Rochimah and A R[2] Soedjono[3]

[1,2,3] Informatics Department, Institut Teknologi Sepuluh Nopember, Surabaya 60111, Indonesia

E-mail: ratih_nea@if.its.ac.id, siti@its-sby.edu, aranda.soedjono12@mhs.if.its.ac.id

Abstract. PKK community is a well-known women community in Indonesia. PKK was formed from national level to neighbourhood association level and usually has a regular event and several activities. This is the reason why PKK is expected to play an active role in national development, started from PKK member family. This research intends to develop a system to manage PKK activities and prepares information sharing in PKK community. The application was built using client server architecture. The web version of PKK-man was developed using PHP framework and was prepared to be accessed by PKK board member. Since they will deal with more data so a wide interface in web version will be more suitable. While the ordinary PKK member can access PKK-man through mobile application. The mobile version was built using PhoneGap framework that supports different mobile operating systems.

1. Introduction

Women play an important role for nation development as they are the key in establishing a strong family as the smallest part of a nation. A strong family will build a strong nation. Moreover, for developing country like Indonesia with around 50% of its population are women, women have a major role in alleviating poverty through community empowerment.

However, Indonesian social systems are still put women in a marginal position even though UUD 1945 (Indonesian constitution) mandate that all citizen have the same right in education. It means that there should not discriminate men and women education. In addition, all social aspects, such as politic and workplace, should not rule out women.

PKK is a community organization that empower women to participate in the development of Indonesia. This community was formed based on the need of activity container and efforts to increase family welfare. PKK groups were formed from the national level to neighborhood association level (rukun tetangga – RT). There are many PKK activities. Every month PKK used to have regular meetings. Other activities such as training, recitation, contest, etc. Somehow the activity of PKK sometimes are just a participation of women citizens in the society and are various based on its location and human resource quality [1]. However it can strengthen the bond between PKK members.

Many PKK member, especially in big cities, are a career women that are very busy. Some of them cannot always come to PKK regular meetings every month so they do not know the new update or information from their PKK community. Nevertheless they are willing to involve in PKK event once in a while and want to share information in their community.

The growth of information technology and internet usage facilitate information access anywhere and everywhere. Facilitating information access to PKK member become an important things. With more access to community information, it is expected PKK cadres will have stronger bond with PKK community and will increasingly play an active role in their community and further the success of national development.

Based on previous background, this research attempt to build a system to manage PKK activity and prepare information sharing in PKK community. The system employed client server architecture so the data will consistent between mobile and web application.

This paper consist of four sections. Firstly, the background of this research was explained in section 1. Next, the explanation about PKK and related work in detail will be given in section 2. Discussion about the proposed PKK-man system will be described in section 3. Finally, section 4 will give the conclusion of this research.

2. Literature Review

2.1. PKK (Pemberdayaan dan Kesejahteraan Keluarga)

PKK (Pemberdayaan dan Kesejahteraan Keluarga – Empowering and Family Welfare) is a community with women as a motor in supporting family welfare as the smallest unit in community that will affect the life of nation. This community was formed based on Decree of the Minister of Home Affairs and Regional Autonomy No. 53 Year 2000 on Empowerment and Family Welfare Movement (PKK). PKK groups has been formed from national level until neighborhood level [2].

PKK has a vision to realize family that faithful and devoted to god, noble and virtuous, healthy and prosperous, forward-independent, and gender equality as well as legal and environmental awareness. PKK has 10 principal program as follows [3].

1. Appreciation and practice of Pancasila
2. Mutual Cooperation (Gotong Royong)
3. Food
4. Clothing
5. Housing and Household Procedure
6. Education and Skills
7. Health
8. Life development of cooperatives
9. Environmental Sustainability
10. Planning Healthy

2.2. Client server architecture

Client server architecture consist of server system where the system processed or recorded data and client where user accessed the data. Web application usually used client server architecture. The system run when client request some services through the network using HTTP reques. Then server will reply with HTTP response and provide client's request (Saternos, 2014).

There are many application have been built using client server architecture. Most of them are web application for example BRAT application for text annotation using natural language processing (Stenetorp, et al., 2012). Others built mobile client server for authentication protocol used in elliptic curve cryptosystem (He, Chen, & Hu, 2012) and also for face recognition (Kremic, Subasi, & Hajdarevic, 2012)

2.3. Related Work

Previously several researchers have built applications for women. Kanno, Chen and Kitagawa built i-appli for women's health information management system. This application was built on mobile phone and are used to record women's health and give some advices. However this application was in Japanese so it can be used by Japanese only [4]. Other researches focused on women's health such as breast cancer applications [5] [6] and application for pregnant women [7] [8].

Indonesian researchers assessed ecommerce's benefits for Indonesian women entrepreneurs [9]. While Darmadi studied about the effects of women top management in firm performance. The result

IOP Conf. Series: Materials Science and Engineering **105** (2016) 012008 doi:10.1088/1757-899X/105/1/012008

showed that female top executives have more proportions in small firms or in another words, it is hard for women to hold seats on top management in large company [10].
From those studies, we have not found any researches related to women community in Indonesia especially PKK. Therefore in this study, we focus on building an application for PKK as a well-known women community in Indonesia.

3. PKK-man: A Proposed System

3.1. PKK-man System Architecture
The system consists of web application and mobile application. Web application is implemented to be used by board member because it will handle more business processes and has more data. While ordinary PKK members can access it through mobile apps. PKK mobile was built using PhoneGap framework that supported many mobile operating system. The architecture of e-PKK are shown in Fig 1.

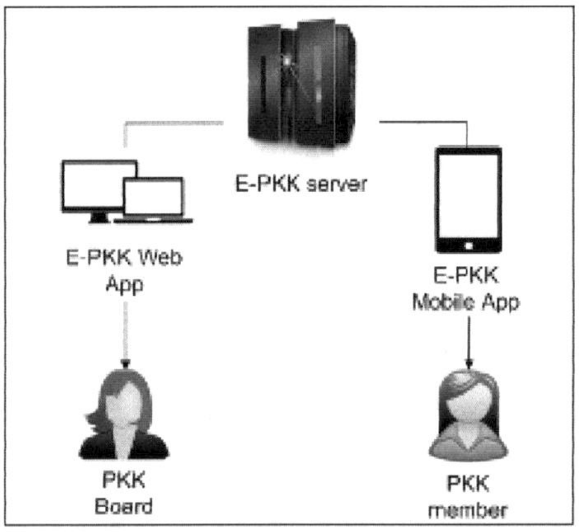

Figure 1. PKK-man Architecture

The system is built using client server architecture so that it can be accessed anywhere and anytime. PKK-man uses Apache as the web server and MySQL as the database server. In addition, the system is prepared to be used by PKK community in Indonesia which means it will record big data.

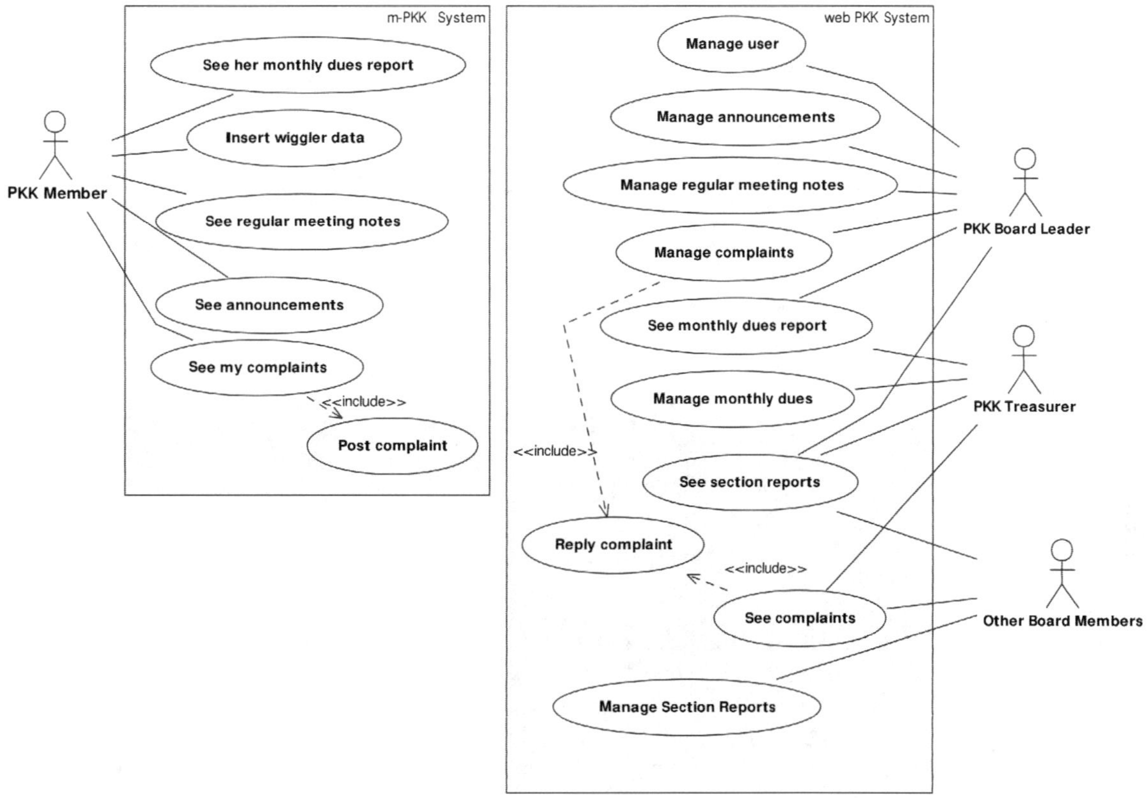

Figure 2. Use case of PKK-man application

3.2. Functional Requirements

Based on functional requirements of e-PKK in Fig 2, there are four user types to access this system. PKK member can accessed the system through mobile application. There are five features in the mobile PKK as shown in Fig 3 (a).

a. Monthly dues report

Every PKK community usually have monthly dues that have to be paid bay their member such us mandatory dues, garbage fees etc. Fig 3 (b) shows the interface of individual monthly dues report.

b. Wiggler data

PKK member need to report how many wiggler per scoop every month. This is used to monitor whether a location have a high possibility of dengue fever and malaria.

c. Regular meeting notes

By providing this access, it is expected that PKK member still be able to be aware of what was discussed in PKK meeting every month even they didn't come.

d. Announcement

This feature presented the announcement that was given by PKK board leader. The interface of announcement is shown in Fig 3 (c).

e. Complaint

PKK member can submit complaint to the board about their neighbourhood or PKK community and read replies for her complaint here.

IOP Conf. Series: Materials Science and Engineering **105** (2016) 012008 doi:10.1088/1757-899X/105/1/012008

(a) (b) (c)

Figure 3. Interface of PKK-man mobile

Web application implemented using PHP framework will be used by PKK board member. There are three types of board: Leader, Treasurer and Field Board. The leader will manage PKK user, announcements, regular meetings, and complaints. Monthly dues in a community will be managed by PKK treasurer. However the report can be seen by the leader. PKK treasurer will also be able to see and reply complaint. While field board will be able to manage section report and the report will can be read by leader and treasurer. Fig 4 shows the interface of PKK-man web version to manage regular meetings notes.

PKK-man system has been developed and all the functionalities has been tested successfully and works well. As feedback from the expected users are important, usability testing will be conducted to several PKK communities in Surabaya as the pilot project area. The evaluation and correction of the system will be done to improve PKK-man system for later use.

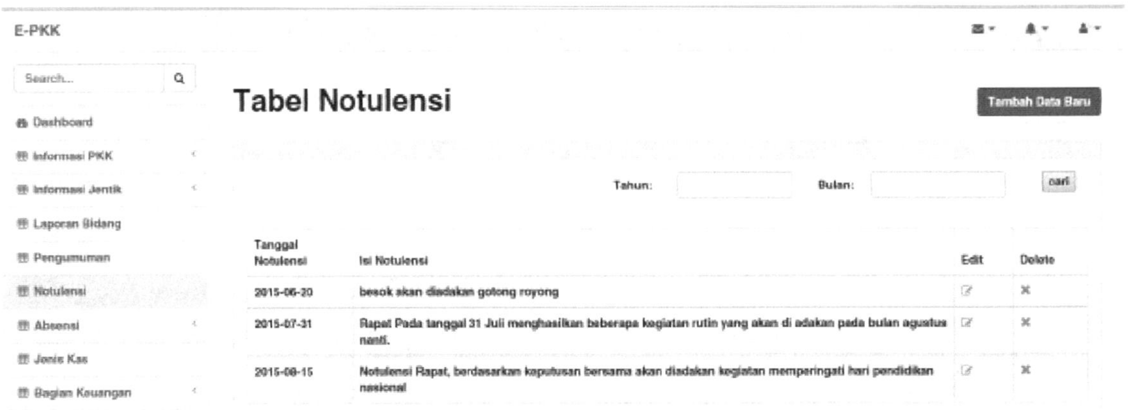

Figure 4. Interface of web PKK-man

4. Conclusion

PKK-man is a system that are built to manage PKK activities in Indonesia. The system are developed using client server architecture with Apache as the web server and MySQL as the database server. We prepared web version of PKK-man to be accessed by PKK board since a wide web interface is more suitable to manage a lot of data. PKK member can access the system through mobile application. PhoneGap framework was utilized to build PKK-man mobile version since it is support different mobile operating systems.

Acknowledgement
We would like to thanks to Dirjen Pendidikan Tinggi Indonesia (DIKTI) for funding our research through STRANAS Scheme.

References

[1] Jurusan Sosiologi UGM. *Pengabdian Masyarakat.* http://sosiologi.fisipol.ugm.ac.id/main/pengabdian-masyarakat/

[2] Sekretariat TP-PKK Pusat. *Sejarah Singkat Gerakan PKK.* http://tp-pkkpusat.org/index.php?option=com_content&view=article&id=83&Itemid=84

[3] Tim Penggerak PKK Kota Balikpapan. *10 Program Pokok.* http://pkk.balikpapan.go.id/index.php/10-program-pokok

[4] Kanno, Takehiro; Wenxi Chen; Kitazawa, M., *Development of i-appli for Women's Health Information Management System*, Computer and Information Technology, 2007. CIT 2007. 7th IEEE International Conference on , vol., no., pp.411,416, 16-19 Oct. 2007

[5] Cireşan, D. C., Giusti, A., Gambardella, L. M., & Schmidhuber, J. (2013). *Mitosis detection in breast cancer histology images with deep neural networks.* In Medical Image Computing and Computer-Assisted Intervention–MICCAI 2013 (pp. 411-418). Springer Berlin Heidelberg.

[6] Cardoso, J.S.; Domingues, I., *Max-Coupled Learning: Application to Breast Cancer*, Machine Learning and Applications and Workshops (ICMLA), 2011 10th International Conference on , vol.1, no., pp.13,18, 18-21 Dec. 2011

[7] An-Jim Long; Polun Chang, *The Use of Health Smart Card in Bridging Discontinuity of Care for Pregnant Woman*, Information Technology: New Generations, 2009. ITNG '09. Sixth International Conference on , vol., no., pp.1492,1497, 27-29 April 2009

[8] Megalingam, R.K.; Boopathi, K.; Sarathkumar, K.S.; Sreedevi, S.; Vishnu, G.B., *Assistive technology for pregnant women health care: Rural area, mobile ultrasound scan system (using ASTM E1384–07 standard)*, Global Humanitarian Technology Conference: South Asia Satellite (GHTC-SAS), 2013 IEEE , vol., no., pp.164,169, 23-24 Aug. 2013

[9] Kartiwi, M.; Gunawan, T.S., *Assessment of e-commerce adoption benefits by Indonesian women entrepreneurs*, Information and Communication Technology for the Muslim World (ICT4M), 2013 5th International Conference on , vol., no., pp.1,4, 26-27 March 2013

[10] Darmadi, S. (2013). *Do women in top management affect firm performance? Evidence from Indonesia.* Corporate Governance: The international journal of business in society, 13(3), 288-304

[11] He, D., Chen, J., & Hu, J. (2012). An ID-based client authentication with key agreement protocol for mobile client–server environment on ECC with provable security. *Information Fusion*, 223-230.

[12] Kremic, E., Subasi, A., & Hajdarevic, K. (2012). Face recognition implementation for client server mobile application using PCA. *34th International Conference on Information Technology Interfaces (ITI)* (pp. 435-440). Cavtat, Dubrovnik: IEEE.

[13] Saternos, C. (2014). Client Server Web Apps with JavaScript and Java . O'Reeilly.

IOP Conf. Series: Materials Science and Engineering **105** (2016) 012009 doi:10.1088/1757-899X/105/1/012009

Inventory Management for Irregular Shipment of Goods in Distribution Centre

Hitoshi Takeda [1], **Masatoshi Kitaoka** [2], **Jun Usuki** [3]

[1]Management Information Systems, Bunkyo University JAPAN
E-mail:takeda@shonan.bunkyo.ac.jp
[2]Industrial Engineering and Management, Kanagawa University JAPAN
E-mail: kita888da@gmail.com
[3]Communation Netwrok ,Kanagawa Institute of Technology JAPAN
E-mail:usuki@nw.kanagawa-it.ac.jp

Abstract. The shipping amount of commodity goods (Foods, confectionery, dairy products, such as public cosmetic pharmaceutical products) changes irregularly at the distribution center dealing with the general consumer goods. Because the shipment time and the amount of the shipment are irregular, the demand forecast becomes very difficult. For this, the inventory control becomes difficult, too. It cannot be applied to the shipment of the commodity by the conventional inventory control methods. This paper proposes the method for inventory control by cumulative flow curve method. It proposed the method of deciding the order quantity of the inventory control by the cumulative flow curve. Here, it proposes three methods. 1) Power method,2) Polynomial method and 3)Revised Holt's linear method that forecasts data with trends that is a kind of exponential smoothing method. This paper compares the economics of the conventional method, which is managed by the experienced and three new proposed methods. And, the effectiveness of the proposal method is verified from the numerical calculations.

1. Introduction

The distribution center only ship merchandise after they receive direct orders from customers, who may be wholesalers, supermarkets or large-scale stores. The wholesalers then ship merchandise to retailers, who in turn sell the merchandise to consumers. The supermarkets and other large-scale stores, on the other hand, sell the merchandise directly to consumers. Generally, the sale of merchandise directly to ultimate consumers fluctuates widely according to the characteristics of the merchandise. For this reason, the amounts shipped from distribution center to their customers and the times of shipment also vary greatly.The commodity goods(Foods, confectionery, dairy products, such as public cosmetic pharmaceutical products) does an irregular changes at the distribution center dealing with the general consumer goods. Therefore, the forecast of the shipment time and the shipment volume are difficult. It can not be applied to the shipment of the commodity by the

[1] 1100,Namegaya,chigasaki city, Kanagawa JAPAN
[2] 3-27-1,Rokkakubashi,Kanagawa-ku,Yokohama JAPAN
[3] 1030 Shimoogino Atsugi city, Kanagawa JAPAN

IOP Conf. Series: Materials Science and Engineering **105** (2016) 012009 doi:10.1088/1757-899X/105/1/012009

conventional inventory management techniques. For this reason, we have proposed a method of performing with the cumulative flow graph [1][2]. The irregular shipment characteristics of goods, we have proposed a theoretical formula that determines the order timing and order quantity. The cumulative inflow corresponds to the order quantity in inventory management. Here, it proposes three methods. 1) Power method ,2) Polynomial method[3] and 3)Revised Holt's linear method that forecasts data with trends that is a kind of exponential smoothing method. This paper compares the economics of the conventional method, which is managed by the experienced and three new proposed methods. And, the effectiveness of the proposal method is verified for the real distribution center.

2. Cumulative flow curve

Distribution center ship merchandise in direct response to orders from customers, be they wholesalers, supermarkets or other large-scale stores. The merchandise shipped from the distribution centers to wholesalers is then shipped by the wholesalers to retailers, as shown in Figure 1. Then the merchandise is sold by retailers to consumers. The merchandise shipped from distribution centers to supermarkets and other large-scale stores is sold directly to consumers at the supermarkets and those large-scale stores. Movement of warehouses and distribution center can be expressed by the cumulative flow curve. The cumulative flow curve is composed of the inflow, the outflow, and the processing line of the distribution center. Movement of the product can be represented by the inflow and outflow as shown in Figure 2[4][5]. Cumulative flow curve is a graph obtained by the cumulative value of the inflow and outflow of goods.Vertical interval of the cumulative line becomes the inventory volume and horizontal line becomes the lead time[6].

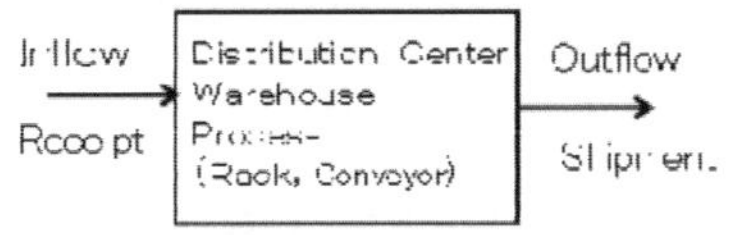

Figure 1 .Relation between inflow and outflow in distribution center

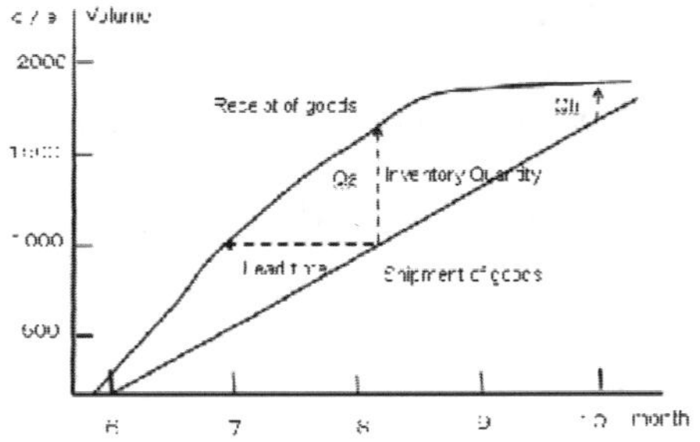

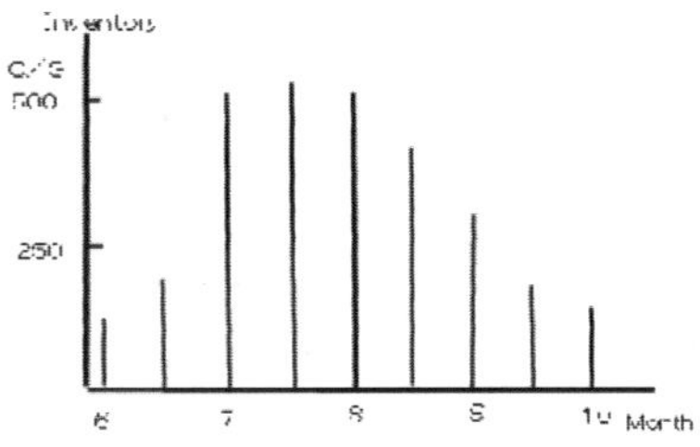

Figure2 . Cumulative flow curve at distribution center **Figure 3.** Amount of inventory to be changed every month

Figure 2 shows a state in which store them in stock products in distribution centers there. Figure 2 shows cumulative flow curve by orders from customers, and are shipping the goods from distribution centers. Figure 2 also shows that the inventory quantity Qa are increasingly July and August. And, inventory quantity Qb has become less in October. Figure 3 shows the inventory quantity calculated from the inflow and outflow of the cumulative flow curve. Inventory levels in Figure 3 is changed to ship the order from the customer on a monthly. In the distribution center, processing time of machine transport number of forklift, man-hours of loading and unloading work, such as conveyor come determined by the amount of stock[7].

3. Characteristic of shipment data

The amount and the time of the products shipped from the distribution center to the supermarket, the large-scale retail stores, and the wholesale make a big change. The reason for change is because the direct product to ship is intended for consumer food, confectionery, dairy products, cosmetics, and pharmaceuticals. In the distribution center, product management, inventory management, shipment planning, delivery vehicle planning, labor management is difficult as compared to the production plant for the change of the product sales of shipping to large. Production planning and inventory control in the production plant can easily control for planning and management since the fluctuation is little[3]. There are various patterns in shipping time by customer .In one pattern, the time of shipments is concentrated at the beginning of month ,whereas shipment time focuses at the end of a month in another pattern. In still another type of patterns, shipments are made almost every day. The shipment amount patterns also vary in a similar way to the shipping time patterns: large volume may be concentrated at the beginning of a month or at the end of month, or they can be almost same every day or differ every day. The shipment patterns The shipping pattern can be classified as follows.

1)There are shipments every day, shipments to the normal distribution. 2)Shipment shall be concentrated at the beginning of a month. 3)There is a ship at the end of the month. 4)Those to be shipped to day irregular. 5)There is a shipping amount on the day of at the beginning of a month, shipment shall disappear from the end of the month. 6)There are shipments amount at the end of the month, that there is no ship of a beginning month. 7)The various combinations of shipping these patterns exist. Inventory management of goods with shipping such properties is difficult and is not available, such as reorder point method in the past. Therefore, the method for inventory control, shipment management, storage management using the cumulative flow curves in the distribution center is proposed.

4. Relationship of cumulative inflow and cumulative outflow

Cumulative flow curve is made from the cumulative outflow and cumulative inflow as shown in Figure 4. In Figure 4, the top line of the graph represents the cumulative inflow, the lower line indicates the cumulative outflow. The vertical axis in figure 4 shows the amount of stock. The horizontal axis shows the date of arrival of goods. R1, R2, and R3 are the volume of inventories.

Q1, Q2, and Q3 are the amounts of arrival of goods that the distribution center ordered the supplier.

The t1 of the horizontal axis represents the day of order quantity Q1 is in-stock.

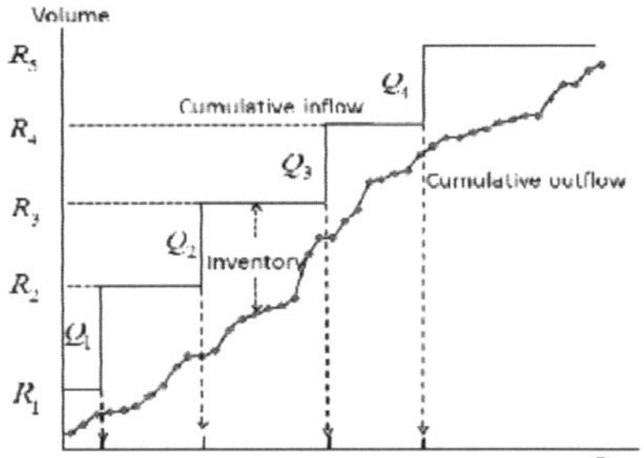

Figure 4. The cumulative inflow and outflow curve and the inventory

The following relations are approved between volume of inventories R and amount Q of the order.

$$R_2 = R_1 + Q_1 \;;\; R_3 = R_2 + Q_2 \qquad (1)$$
$$R_4 = R_3 + Q_3 \;;\; R_5 = R_4 + Q_4$$

The difference between the two graphs of the upper cumulative inflow and lower cumulative outflow is to the volume of inventories. The inventory control in the distribution center becomes a problem of deciding amount Q of the order.

5. Cumulative flow curve of shipping characteristics

5.1. Cumulative shipments characteristics shipment characteristics

The data must be cumulative in order to draw the cumulative flow curves of the product for the irregular shipment characteristics. Cumulative flow curve can be controlled by the amount of quantity and timing of arrival[8]. However, it is not possible to control the shipment quantity and shipment time. Because this is an order from the customer, it is not possible to control freely. It is necessary to examine the pattern at the shipment amount and the shipment time according to the product. Three shipments of the kind of product and patterns at the shipment time are shown here as a typical example.

1) Figure 5 is a product it ships at the beginning of the month, and without the shipment until the end of the month. The shipment of the cumulative flow curve is shown in figure 6[8][9][10].
2) Figure 7 is a product it ships at the end of the month, and without the shipment until the end of the month. The shipment of the cumulative flow curve is shown in figure 8.
3) Figure 9 is a product regularly shipped. T The shipment of the cumulative flow curve is shown in figure 10.

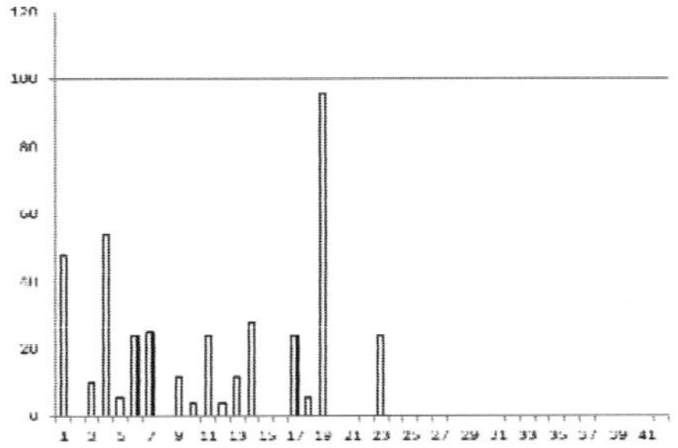

Figure 5 . Shipping is concentrated in the beginning of the month.(Item1)

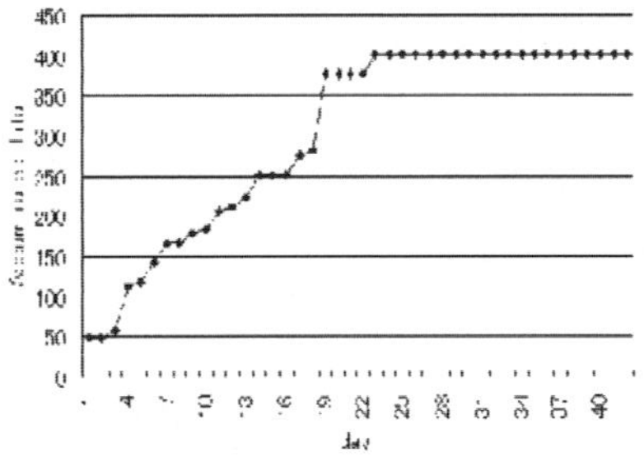

Figure 6. Cumulative flow curve for figure3

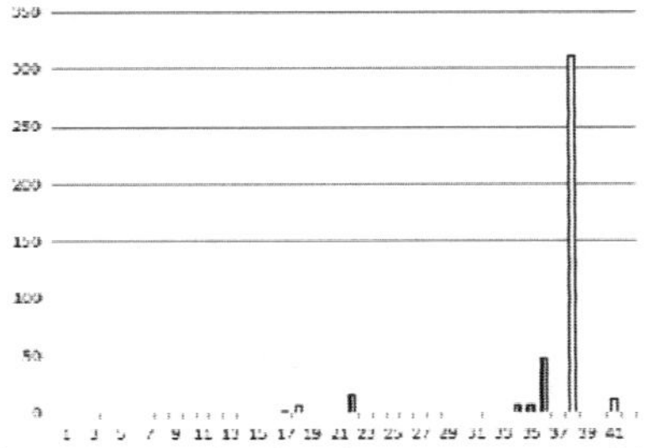

Figure 7. Shipping is concentrated in the end of the month.(Item2)

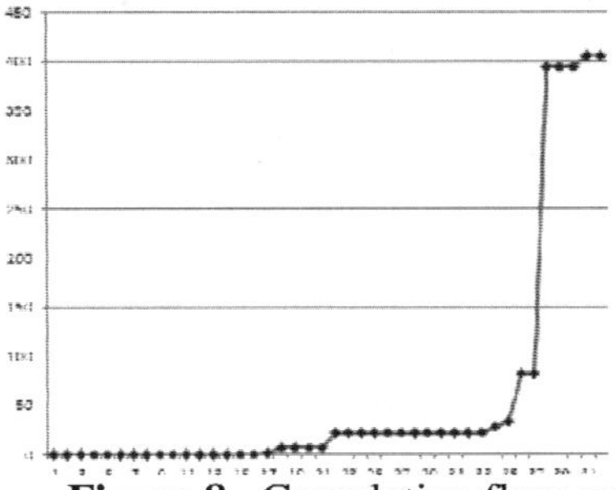

Figure 8. Cumulative flow curve for figure5

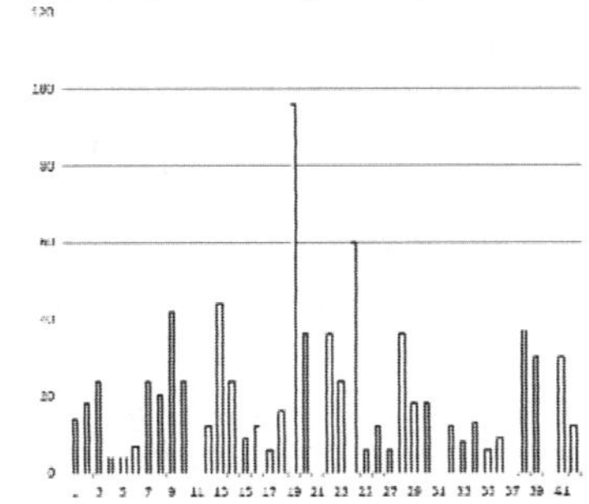

Figure 9. The product is evenly shipped (Item3)

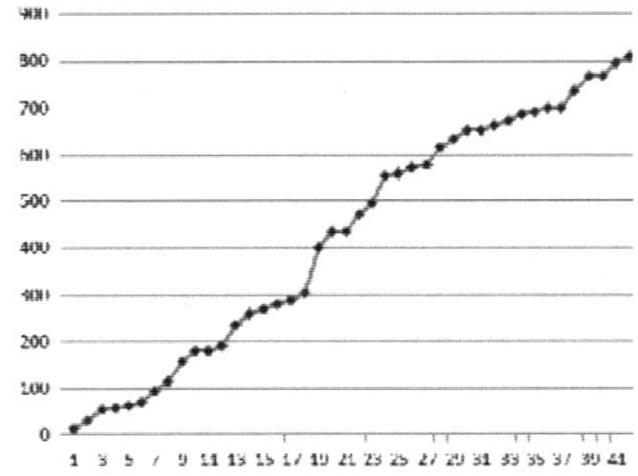

Figure 10.Cumulative flow curve for fig7

5.2. Inventory control that uses cumulative flow curve in distribution center

The current state of the inventory control that uses cumulative flow curve at the distribution center is examined here. Figure 11 is cumulative flow curve of item I4. The volume of inventories at this time is shown in figure 12, and the total of the volume of inventories becomes 12549. Cumulative flow curve of item 5 is similarly shown to figure 13. The volume of inventories is shown in figure 14. The total of the volume of inventories becomes 16507. The judgment whether this volume of inventories is a lot or is little cannot be judged because there is no technique of a scientific inventory management. However, because the shipment time and the shipment are irregular. It becomes possible to stock a lot of inventory in the distribution center [8][9][10].

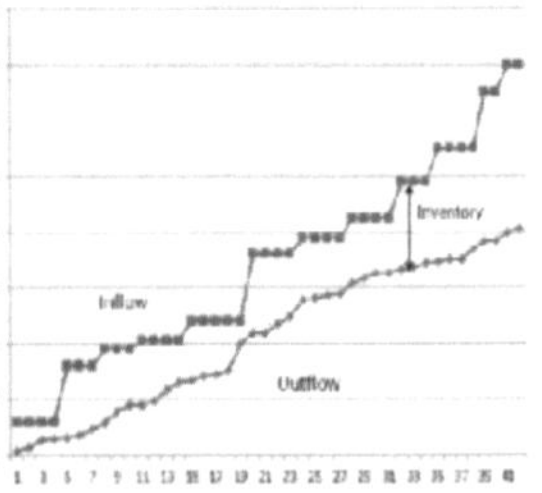

Figure 11. Cumulative flow curve of Item I4

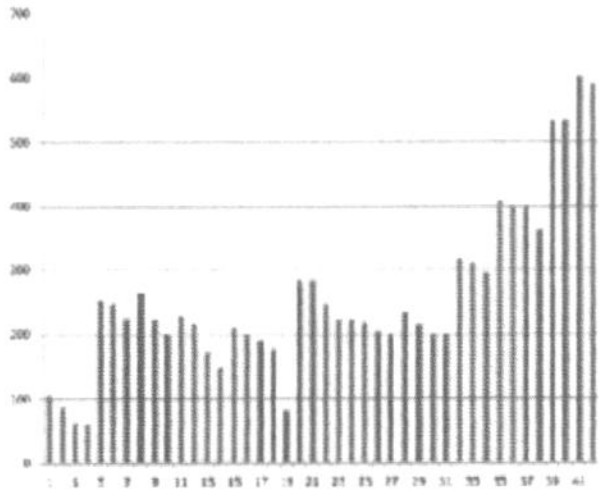

Figure 12. Volume of inventories of Item I4.

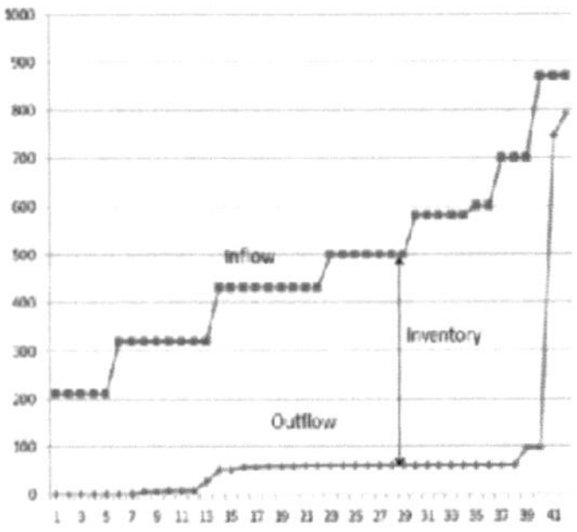

Figure 13. Cumulative flow curve of Item I5

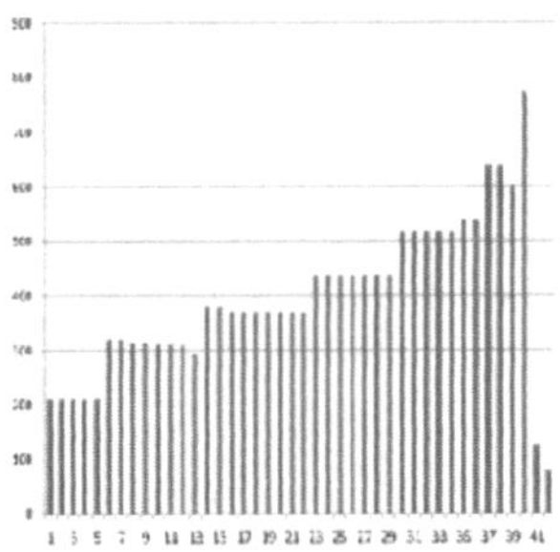

Figure 14. Inventory volume of Item I5

6. Cumulative inflow curve and volume of inventories

6.1. Decision of cumulative inflow curve

The inventory control is carried at the distribution center is a method by a past experience and intuition. In Figure 11 is the cumulative flow curve for item I4 and the Figure 12 is the volume of inventories. Figure 13 is the cumulative flow curve for item I5 and the Figure 14 is the volume of inventories. It is necessary to ship the amount (Outflow) of cumulative outflow by the order from the customer. Then, we can control at the distribution center is only an amount of the cumulative inflow from the vendor. The cumulative inflow (Inflow) Q shown in the upper row of figure 4shows the amount of received of goods from the vendor. The problem is the method of deciding amount (Inflow) of the cumulative inflow curve in figure 4.It is necessary to decide the method how the time of received of goods and the amount of cumulative inflow. The volume of inventories can be reduced by efficiently doing the amount of received of goods and time of received of goods. The relation of the cumulative inflow amount > amount of the cumulative outflow is approved from figure 4. The volume of inventories becomes the difference between the amount of cumulative inflow and the amount of the cumulative outflow. It proposes a new method of controlling the amount of the cumulative inflow to reduce the volume of inventories .When the difference is reduced, the volume of inventories is minimized. To reduce the volume of inventories, it proposes a new method here about time and inlet flow Q of the order to the amount of the accumulation inflow, that is, the vendor. When this difference is reduced, the volume of inventories is minimized. To reduce the volume of inventories, it proposes a new method here about time and inlet flow Q of the order to the amount of the accumulation inflow, that is, the vendor[11].

6.2. Inventory control by cumulative flow graph.

Cumulative flow graph puts out the order with t01 as shown in Figure 15. The commodity arrives to tR1. The lead time at this time becomes tR1-t01. However, the commodity handled here is a general consumption material. Therefore, it will be delivered the next day when ordering. Therefore, it is thought that there is no lead time. Amount Q of the order is assumed to take the periodic reordering method. The reorder cycle is assumed to be TC. Amount Q of the order will be decided with forecasting of the actual data. The forecast period becomes TC. The forecasting value is calculated for actual data of Figure 15. At this time, the safety stock SS is put in the forecasting value or the safety stock is separately installed.

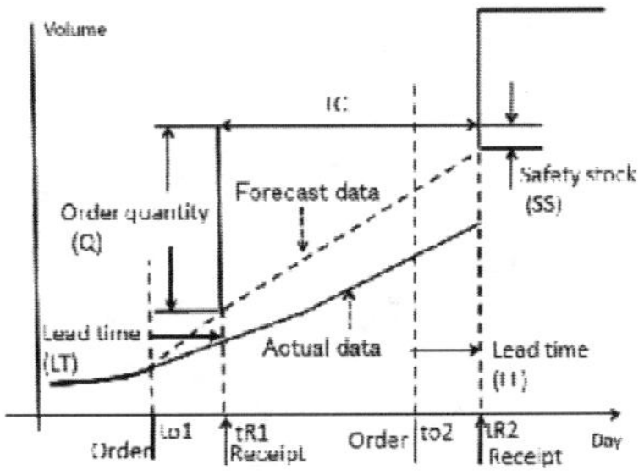

Figure 15. Order quantity of accumulative

In this research paper, the method of demand forecasting to validate the two methods according to the Holts exponential smoothing method and power approximation, polynomial approximation [11]

6.3. Order quantity of cumulative inflow by power and polynomial method

The forecast period is assumed to be TC. The forecast shows period p cycle TC by n. The predictive value becomes amount Q of the order. Amount Q of the order uses the forecast value by the exponential approximation and the polynomial approximation. The forecast time is assumed to be x. Stock shortage is caused when the forecast value is small. Therefore, the safety value Pσ of the upper bound is given to the forecast value. The expression of the forecast is calculated by using the past data. It forecasts p period ahead by the expression of the forecast of the past data of n period. As a result, the forecast by the power and the polynomial forecast becomes the next expression [8][9][10].

$$\hat{x}\ (t+p) = A(t+p)^B + p\sigma \qquad (2)$$

$$\hat{x}\ (t+p) = A_1 + A_2(t+p) + A_3(t+p)x^2 + p\sigma$$

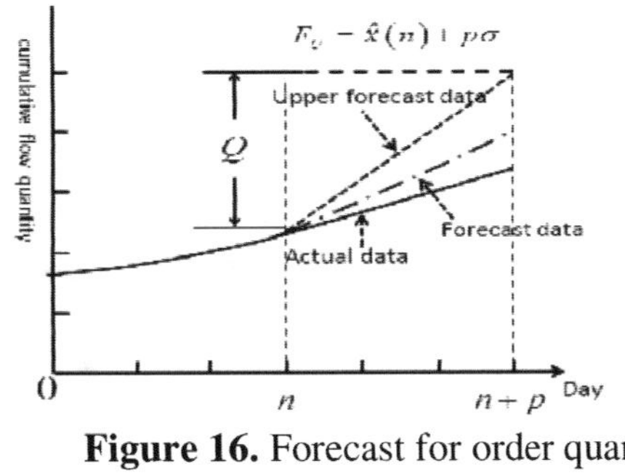

Figure 16. Forecast for order quantity

It applies to the commodity of Item3 shown in figure 9 to verify the proposed forecast method. The cumulative curve of Item3 becomes figure 10.

6.3.1 Numerical example 1

1） Power forecast method

It forecast to the shipment data of eight days in the past by the Power method. The expression of the forecast becomes as follows.

$$\hat{x}\ (t) = 15.843 \cdot t^{0.9282} + p\sigma \qquad (3)$$

Order quantity adds and calculates safety stock P σ in the expression of the forecast. As a result, order quantity became 82. Next, the prediction error occurs between a predictive value and actual value as shown in figure 17. Table 1 has shown the volume of inventories. The total of the volume of inventories becomes 158.

Table 1. Inventory and order quantity with power method

week	8	9	10
Inventory	82	40	16
week	11	12	Total
Inventory	16	4	158

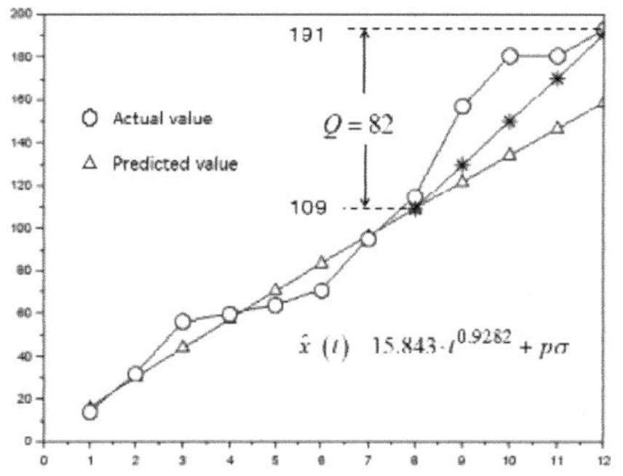

Figure 17. Power forecast of demand and order quantity

2） Polynomial forecast method

In the same way as the power method to calculate the forecast by polynomial to the shipment data of the past eight days. The expression of the forecast becomes as follows.

$$\hat{x}\ (t) = 8.5893 + 11.196 \cdot t + 0.1726 \cdot t^2 + p\sigma \qquad (4)$$

Order quantity is calculated by adding the safety stock P σ in the equation of prediction. As a result, order quantity became 108. Figure 17 has shown an expression of the forecast and actual data. Figure 17 has shown the relation between the safety stock and the volume of inventories when Order quantity is 100. Table 4 has shown the volume of inventories. The total volume of inventories becomes 218.

Table 2. Inventory and order quantity with polynomial method

week	8	9	10
Inventory	94	52	28
week	11	12	Total
Inventory	28	16	218

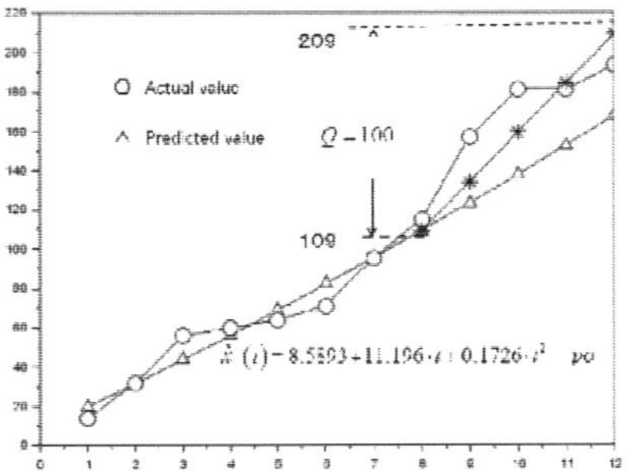

Figure 18. Polynomial forecast of demand and order quantity

6.4. Order quantity of cumulative inflow by Improved Holt's linear method

linear exponential smoothing is found using two constants,α and β,and three equations. The equation is given as follow[8][9][10].

$$L_t = \alpha \cdot Y_t + (1-\alpha)(L_{t-1} + b_{t-1})$$
$$b_t = \beta(L_t - L_{t-1}) + (1-\beta)b_{t-1} \qquad (5)$$
$$Q_{t+k} = L_t + b_t k + Z_t k$$

L_t denotes an estimate of the level and b_t denotes a estimate of slope of the time series at time t. L_t adjusts for the trend of previous period. b_{t-1} by adding it to the last smoothed value, L_{t-1}. b_t the updates the trend, which is expressed as the difference between the last two smoothed values. The trend is modified by smoothing with β the trend in the last period ($L_t - L_{t-1}$), and adding that to the previous estimate of the trend multiplied by $(1-\beta)$. Finally Q_{t+k} is used to forecast ahead. This equation different for Holt's linear method. The trend b_t is multiplied by the number of periods ahead to forecast k, and added to the base value of L_t [11].

6.4.1 numerical example 2

The predictor equation of Holt was calculated by using data here for eight weeks. Figure 13 shows the procedure for requesting the amount of the accumulation inflow from the forecast type of Holt of Item I1. The forecast value of t=8 and k=4 is calculated as α=0.8 and β=0.08. The forecast value of the seventh week becomes Q=67. In graph, triangle indicates the predicted value round is in the actual value. Table 7 shows the volume of inventories at this time.

Table 3. Inventory and order quantity with improved Holt's method

week	8	9	10
Inventory	78	36	12
week	11	12	Total
Inventory	12	0	138

IOP Conf. Series: Materials Science and Engineering **105** (2016) 012009 doi:10.1088/1757-899X/105/1/012009

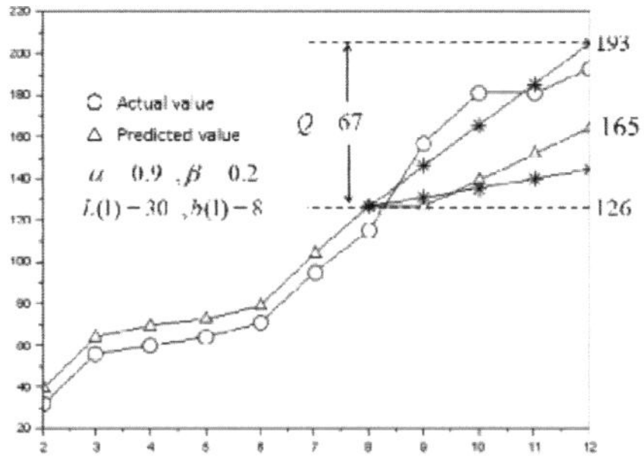

Figure 17. Forecast of demand and order quantity

7. Comparison between current method and improvement method

The volume of inventories of three proposed methods and current methods are compared.Table 4 is a method by the experience of the current method. The volume of inventories is examined for five week. The volume of inventories in the stock control managed in the distribution center experiencing becomes 1180.Compared to other methods of Table 5, the current management methods is clear that the amount of inventory is large. The improved Holt's method [7][8] is the most excellent clearly from Table 5.Because other forecasting methods are used power and straight line, the demand that fluctuates not adapted. For this, the prediction error grows. Power method is more excellent than the polynomial method because there are a lot of quantity safety stocks. The volume of inventories changes by setting the safety stock. Therefore, it is difficult to judge which of these two methods is excellent.

Table 4. Inventory quantity for current method

week	8	9	10
Inventory	275	233	209
week	11	12	Total
Inventory	239	227	1183

Table 5. Comparison between current method and proposed method

	Current method	Power method	Polynomial method	Improved Holt's method
Inventory quantity	1183	158	218	138

8. Conclusions

Up to now, the method of theoretically computing the order quantity in the cumulative flow curve has not been proposed. This paper proposed the method of deciding the order quantity of the inventory control by the cumulative flow curve. Three proposed methods are shown in the calculation of the order quantity. The time and volumes of the shipments from the distribution center to the customers are irregular. With the cumulative flow curve, it is possible to inventory control of products that the irregular shipping. To provide useful forecasts of such irregular shipping times and volumes, a method utilizing the power, polynomial method and improved Holt's method are presented. The volume of inventories of three proposed methods and current methods are compared. By applying the cumulative flow curves, it is clarified that the inventory control is possible for irregular shipment products.

IOP Conf. Series: Materials Science and Engineering **105** (2016) 012009 doi:10.1088/1757-899X/105/1/012009

References

[1] Kitaoka M, and Usuk J 2006 *Inventory Control of Distribution Center using Grey Theory and Accumulated Curve International Conference on Logistics and SCM Systems (ICLS2006), pp.29-38*

[2] Kitaoka M, Takeda H and Usuk J 2003 *Cumulative Curve Analysis in the supply chain using Grey, International Conference on Production Research, CD-ROM*

[3] Li J., Kitaoka M., and Suzuki S 2004 *Times series Analysis of Shipments and receipts for Distribution Center using GREY Theory and EIQ Table ICLS,246-252*

[4] Usuk J, Kitaoka M, andMatsuura H 2015 , GERT and Petri net Analysis for Performance Evaluation of Distribution Center *Information –An International Interdisciplinary Journal, Vol.18, No.2, pp.735-751*

[5] Kitaoka M, Iwase H, Usuki J and Nakamura R 2010 *Capability Measures and Estimation using Control Chart for Distribution Management Index Vol.3,No.1, J. Logistics and SCM Systems*

[6] Kitaoka M and Usuk J 2002 I*nventory Control of Distribution Center Using Grey Theory and Accumulated Curve ,Proc,18th,ICPR,Taiwan*

[7] Makridakis S,Wheelwright S and Hyndman V 1998 *Forecasting:Method and Applications* (John-Wiley and Son)

[8] Abraham B. and Ledolter J 2005 *Statistical Methods for Forecasting*.(Wiley Inter- Science)

[9]Vollmann T, Berry W, Whybark D and Jacobs F 2005 *Manufacturing Planning Systems for Supply Chain Management*(McGraw-Hill)

[10] Donald W. 2003, Inventory Control Management, (John Wiley&Sons)

[11] Simichi D.Levi P ,Kaminskty and Simichi E. 2008 *Designing and Managing the Supply Chain* (McGraw-Hill,I

IOP Conf. Series: Materials Science and Engineering **105** (2016) 012010 doi:10.1088/1757-899X/105/1/012010

1x2 Array of U-Slotted Rectangular Patch Antennas for High-Speed LTE Mobile Networks

P. Daud, Sri.H, D.Mahmudin , T.T.Estu, A.A.Fathnan, Y.N.Wijayanto, N. Armi
Research Centre for Electronics and Telecommunications, Indonesian Institute of Sciences (LIPI), Jl. Sangkuriang , Bandung 40135, Indonesia

E-mail: pmkdaud@gmail.com

Abstract. Microstrip antenna is a shaped thin board of antenna and capable for working at high frequencies. Microstrip antenna has a pattern strip shape in a various form, one of which is a rectangular shape. Microstrip antennas have some shortcomings which have narrow bandwidth and small gain, to cover the shortfall , the antenna is made using an array to increase the gain and u - slot to widen the bandwidth. In this paper will discuss the results of the antenna simulation using Ansoft HFSS software applications and their compliance with specifications designed antenna. In this issues we analysis the results of the design and simulation microstrip antenna at a frequency of 2.6 - 2.7 GHz for LTE applications

1. Introduction

LTE (Long Term Evolution) is a 4G wireless technology with have the ability to reach data transmission speeds of 100 Mbit / s for downlink and 50 Mbit / s for uplink. LTE can operate on the new spectrum band is 700 MHz and 2.6 GHz or on one of the cellular spectrum band has been allocated in the IMT -2000 standard (450 , 850 , 900 , 1800 , 1900, 2100 MHz) [1-7]. In this issues we analysis the results of the design and simulation microstrip antenna at a frequency of 2.6 - 2.7 GHz for LTE applications.

There is some considerations that should be take a good care and accurate with original and real data parameters of the substrate which will use for microstrip antenna to have the best calculations, and also need high precision for the advanced process or Etching process to make a layout which will be printed on the substrate surface.

2. Method of Design and Simulation

Specifications of the antenna to be design and simulated:

- Working frequency : 2.6 to 2.7 GHz
- Central Frequency : 2.65 GHz
- Bandwidth : 100 MH
- VSWR : < 2
- Return Loss : > 10 dB
- Impedance : 50 Ω
- Radiation pattern : Unidirectional
- Gain : > 5 dBi

- Polarization : Circular

Microstrip antenna to be simulated was rectangular shape, which will be arranged in arrays to increase the gain and U - shaped slot to widen the bandwidth. The substrate material will used is FR4 Epoxy with substrate characteristics as follows:

- Dielectric relative permittivity () : 4.4
- Dielectric thick (h): 1.6 mm
- Conductor thickness (t): 0.035 mm
- Tangent loss (α) : 0.024

The method to design and realization of the Patch Array Antennas can be determine using Flow chart which describe at Figure 1 below as follows

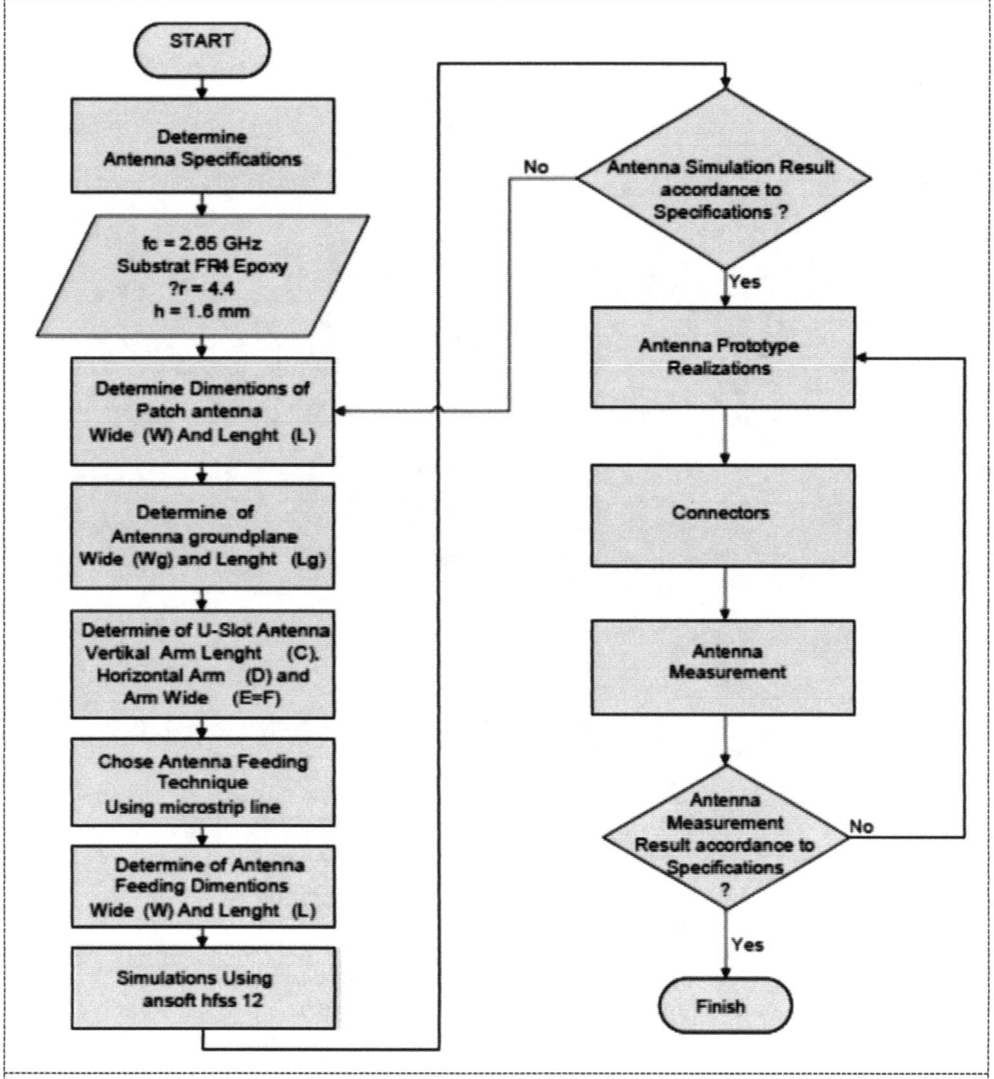

Figure 1. Flow Chart Method

To get and calculate the dimensions of the antenna used equation (1) to the equation (4) with the following equations [9.10]:

$$W = \frac{\lambda 0}{2}\left(\frac{\varepsilon_r+1}{2}\right)^{-\frac{1}{2}} \quad (1)$$

$$L = \frac{\lambda 0}{2\sqrt{\varepsilon_{eff}}} - 2\Delta L \quad (2)$$

$$\varepsilon_{eff} = \frac{\varepsilon_r+1}{2} + \frac{\varepsilon_r-1}{2}\left[1+\frac{12h}{W}\right]^{-\frac{1}{2}} \quad (3)$$

$$\Delta L = 0.412h\frac{\varepsilon_{eff}+0.3\left(\frac{W}{h}+0.264\right)}{\varepsilon_{eff}-0.258\left(\frac{W}{h}+0.813\right)} \quad (4)$$

To get and calculate the dimensions of u - slot antenna using equation (5) to the equation (7), with the following:

$$\frac{C_1}{W} \geq 0.3 \quad (5)$$

$$D = \frac{c}{f_{low}\sqrt{\varepsilon_{eff}}} - 2(L + 2\Delta L - E) \quad (6)$$

$$E = F = \frac{\lambda}{60} \quad (7)$$

$$A = \frac{Z_0}{60}\sqrt{\frac{\varepsilon_r+1}{2}} + \frac{\varepsilon_r-1}{\varepsilon_r+1}\left(0.23+\frac{0.11}{\varepsilon_r}\right) \quad (8)$$

$$\frac{W_c}{h} = \frac{8e^A}{e^{2A}-2} \quad (9)$$

$$\lambda_g = \frac{c}{f_0\sqrt{\varepsilon_{eff}}} \quad (10)$$

$$L_c = \frac{\lambda_g}{4} \quad (11)$$

The result of calculations can be seen at Table 1 below:

Table 1. Calculation dimensions result of the antenna

Component	Dimensions
Length of the patch (L)	29,49879 mm
Wide of the patch (W)	34,44871 mm
Length of the ground plane (Lg)	39,09879 mm
Wide of ground plane (Wg)	44,04871 mm
Length of the slot (C)	10,33461 mm
Wide of the slot (D)	4,96523 mm
Wide of the Arm slot (E=F)	1,88679 mm
Length of the Feeder Line ()	3,05894 mm
Wide of the Feeder Line ()	15,50886 mm

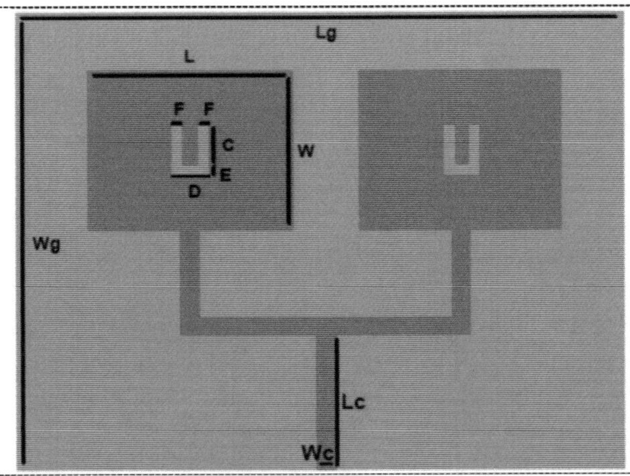

Figure 2. Dimensions and Shape Form of the antenna

3. Result and Discussion

The result of simulation from several parameter to frequency response such as VSWR, Return Loss (L), Impedance and Gain, also working frequency and Bandwidth, given by charts below. Display antenna modeling simulation results with Ansoft HFSS v.12 as follows:

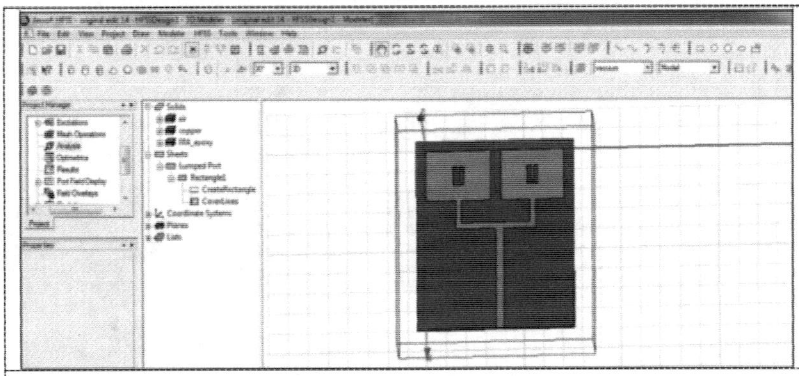

Figure 3. Dimensions of the antenna simulation results.

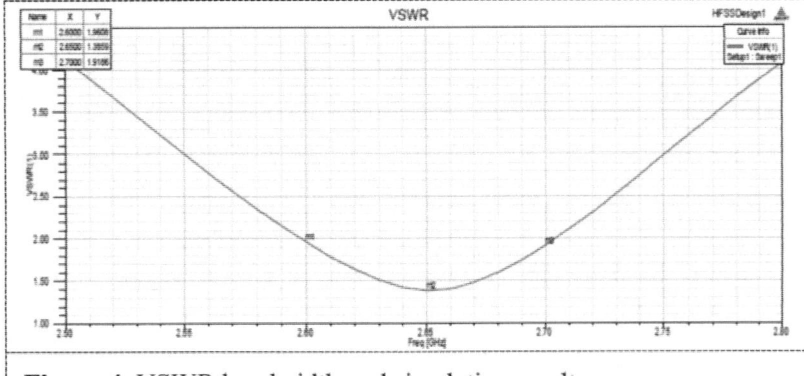

Figure 4. VSWR bandwidth and simulation results.

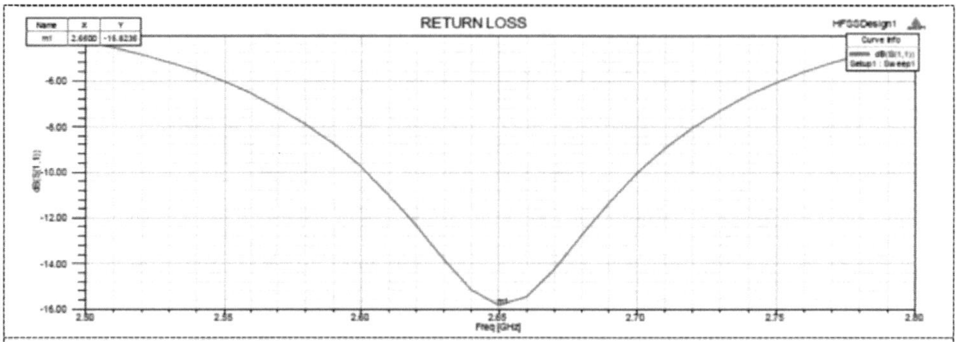

Figure 5. Return Loss simulation results.

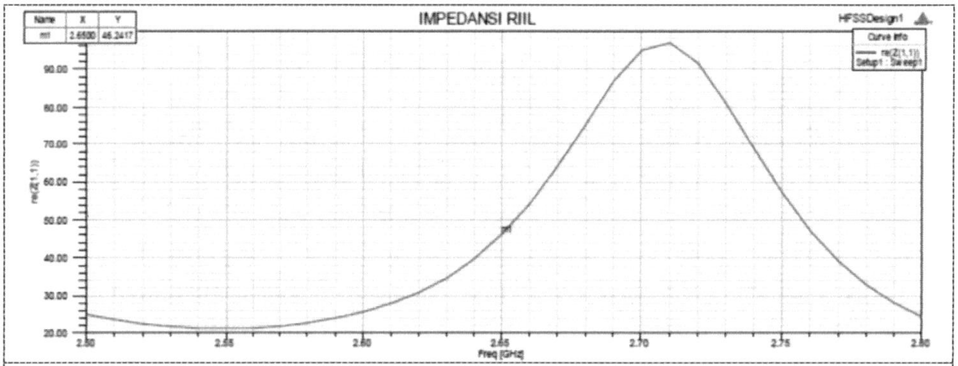

Figure 6. Real impedance simulation results.

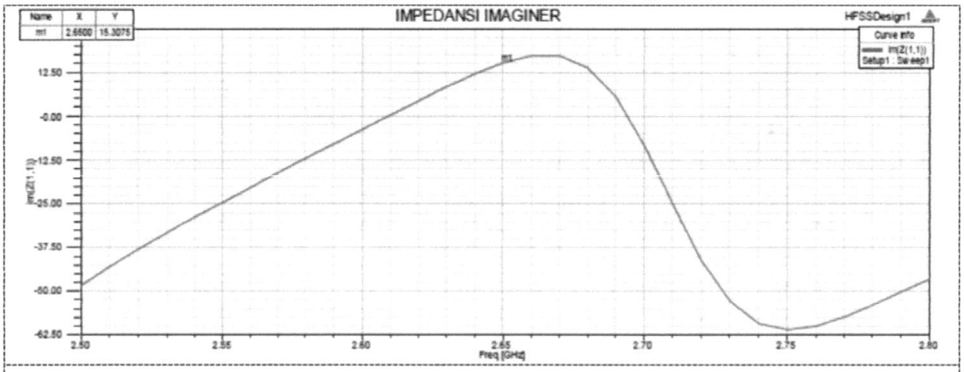

Figure 7. Imajiner impedance simulation results.

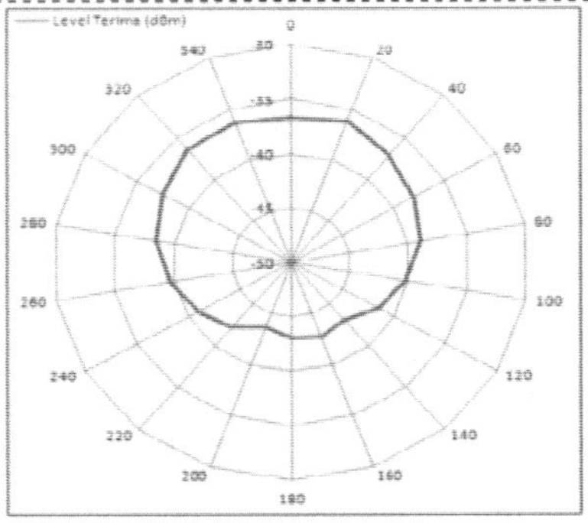

Figure 8. Antenna Polarization Measurement Result.

Figure 9. Microstrip Antennas Realization On FR4 Epoxy Substrat.

Table 2. Design Result and Simulation

Parameter	Spesifications	Simulation Result
Frequency Range (MHz)	2600 - 2700	2600 - 2700
Central Frequency (MHz)	2650	2650
Return Loss (dB)	> 10	15.82
Bandwidth (MHz)	100	100
VSWR	< 2	1.39
Impedance (Ω)	50	46.24 + j15.31
Gain (dBi)	> 5 dBi	-
Radiations Pattern	Unidirectional	-
Polarization	Circular	-

Here we used Rationing techniques electromagnetically coupled (EMC) where In this technique , there is no metallic contact between the patch with other elements , rationing the patch is done by means of electromagnetic coupling of elements feeding . This technique is commonly used in microstrip antenna that uses a stacked configuration. Using Slot antennas configurations in the center of Patch, made the frequency bandwidth more wider and support the need of bandwidth LTE applications that we need, about 100MHz. From the simulation results obtained and compared with the design specifications requested seen almost all requirements are match then the antenna can be realized and fit to work at a frequency of 2.6 - 2.7 GHz for LTE, R factor is very influential on the dimensions of the antenna.

The antenna can work at frequencies from 2.6 to 2.7 GHz with Bandwidth 100 MHz achieved disposals with U-Slot. The simulation of VSWR is 1.048 < 2, Return loss is 15.82 dB and Impedance 46.24 + j15.31Ω close to 50 Ω. The antenna Polarization is circular, Gain antenna simulation results is >5 dBi. All the Results already accordance with the specifications is desirable.

4. Conclusion

The antenna can work at frequencies from 2.6 to 2.7 GHz and useful for High-Speed LTE Mobile Communications. Suggestion to get the antenna performance Better in realization, strips of the microstrip antenna required accuracy and high precision, installation and soldering connectors should be arranged as good as possible in order to match strip channel conditions. To have the best measurement result, the measurements should carried out in the room such as anechoic chamber.

5. References

[1] From the Network Migration Process Analysis Technology towards UMTS LTE technology. [On Line]. Available: http://www.scribd.com/doc/39596509/Proposal [January 7, 2014]

[2] Balanis, Constantine A. 1982. *Antenna Theory - Analysis and Design*. United States of America: John Wiley and Sons, Inc.

[3] The definition and types of antennas. [On Line] Available: http://adibdevc.blogspot.com/2012/02/definisi-dan-jenis-jenis-antena-dalam.html [February 2, 2014]

[4] Hapsari, Natalia Rina. 2013. *Design and Realization of Rectangular Microstrip Antenna Wideband at 2.3 to 2.39 GHz frequency*. IT Telkom Bandung: Unpublished.

[5] Kraus, John. D. 1997. *Antennas Second Edition*. New York: McGraw-Hill.

[6] Maula, Akbar Muhammad Rusdi.2012. *Design and Realization of Microstrip Antenna Using Array Methods*. IT Telkom Bandung: Unpublished.

[7] The development of communication technology antenna [OnLine]. Available: http://id.wikipedia.org/wiki/Perkembangan_teknologi_komunikasi_antena [January 30, 2014]

[8] Tobing, Artios Lemuel L. 2010. *Characteristics Analysis of Microstrip Transmission Line. University of North Sumatra*: Unpublished.

[9] *Class Module Antennas and Propagation* 2008. IT Telkom Bandung: Unpublished.

[10] Daud, Pamungkas 2007 Coupled Patch Array Antenna For Survelliance Radar, *Proceeding of Telematics System Services,and Applications (TSSA)* 2007, Bandung 6-7 December 2007, Bandung Institute of Technology, Indonesia, ISSN:1693-993X

Acknowledgement

The authors would like to thank to Research Centre for Electronics and Telecommunication, Indonesian Institute of Sciences for their supports during analysis and design of the system. Also the authors thank to Research Center for Electronics and Telecommunications, Indonesian Institute of Sciences (LIPI) to support in experiment facilities and administration procedure.

IOP Conf. Series: Materials Science and Engineering **105** (2016) 012011 doi:10.1088/1757-899X/105/1/012011

Improving Reliability of Service Operation Using FMEA: Review and New Opportunity for Investigations

Agung Sutrisno[1], **Indra Gunawan**[2]

[1]Department of Mechanical Engineering,
Sam Ratulangi University, Kampus Bahu, Manado, Sulawesi Utara, Indonesia 95115
E-mail: agungsutrisno@unsrat.ac.id

[2]School of Engineering and Information Technology
Federation University, Churchill, VIC 3842,Australia
E-mail: indra.gunawan@federation.edu.au

Abstract. Despite its growing contribution to the global economy, investigation on the application status of service FMEA study to support realization of reliable service operation is very limited in literature. Motivated by such situation, the paper presented an initial survey on the status and research gaps in developing and applying FMEA in service sectors. Systematic preliminary survey using specific criteria are undertaken. Our study indicated that development and application of service FMEA are partially addressing the characteristics of service operations and it is still applied into the good deed and profit oriented operations. Opportunities for further investigation pertaining to advancement of its decision supporting tool for service risk appraisal, its modification to cope with sustainability related requirements and application of service FMEA in not for profit oriented operations are presented as new avenues for further investigation

1. Introduction

In line with growing contribution service sector to the global GDP nowadays, maintaining reliable service operation is undoubtedly important for sustaining world economy [1] . Failed service provision will yield into customers' dissatisfaction and incur service quality loss. In industrial practice, practitioners using service FMEA as means to assess reliability by identify the service failure modes, appraise the risk of faulty service occurrences and finding solutions to remedy or prevent its re-occurrence in future. Due to its beneficial effects, application of FMEA methodology is widely spreading in various service operation fields. However, in spite of its versatility in its application, studies discussing opportunities intended to improve reliability by using FMEA in the literature such as the studies of [2], [3], [4] and [5] are mostly still focusing on improving product design and manufacturing process related - reliability and paying very little attention on service operation reliability.

Motivated by scarcity on the availability on references on improving reliability of service operation, this study is intended to review and proposed new perspectives on improving reliability of service operations by using Service FMEA approach by extending to the study of [6] based on time period of 2011 onward. The goal of this study are as follows, first we intend to consolidate contemporary papers pertaining to the application of service FMEA in service operations, second; to classify service FMEA papers based on criteria such as advancement of the methodology in service risk appraisal, application of service FMEA in supply chain context, modification of FMEA, and its integration with other engineering tools. Third, to investigate the most important ideas based on the service FMEA survey and proposed new prospects on service FMEA studies for further investigations.

2. Research Methodology

In an attempt to reach our research goal, literature survey is conducted. Studies pertaining to service FMEA in the field of quality and reliability management in various publisher databases between time span 2011 onward is accomplished. The papers pertaining to the FMEA in service operations are retrieved from various journals publishers such as Elsevier Science Direct, Emerald, Taylor and Francis, IGI Global Publishers, Springer, Hindawi, Sage, and Directory of Open Access Journals (DOAJ), Cambridge University press and Wiley Interscience. Only periodicals written in English are retrieved. The keyword "FMEA", "FMECA" and "service FMEA" and "Service FMECA" are used to consult the aforementioned literature database. Reference classification criteria used in literature search are service failure risk reprioritization methodology, supply chain FMEA and integration of service FMEA with other engineering methodology are used for categorizing service FMEA literature. The result of our initial literature survey then classified into various classification criteria as depicted in Table 1, Table 2, and Table 3.

Table1. SERVICE FMEA REFERENCES ON ADVANCING SERVICE FAILURE RISK REPRIORITIZATION METHODOLOGY

Service FMEA Aspect	Methodological basis	References	Service Application Area
Failure Risk Reprioritization	1-10 Ordinal Scale	[7]	Knowledge Management
	AHP (Analytical Hierachy Process)	[8]	Healthcare
	Decomposition Theory	[9]	Trade Service
	Bayesian Network Causal Model	[10]	Healthcare
	Fuzzy Logic	[11]	Healthcare
	Fuzzy Rule and Bayesian Network	[12]	Security
	Cloud Theory	[13]	Maintenance
	Logarithmic scale	[14]	Healthcare
	Cost Oriented Fuzzy Theory, GRA and Profitability Theory	[15]	Healthcare
	2-Tuple Linguistic Representation Model and GRA	[16]	Healthcare
	Grey Relational Analysis	[17]	Healthcare
	Data Envelopment Analysis	[18]	Tourism Service

Table2. Classification Of Supply Chain FMEA Studies And Its Application Settings

Supply Chain FMEA Scope	References	Methodology	Application Setting
Failure Risk Reprioritization	[19]	Fuzzy AHP and Fuzzy TOPSIS	Manufacturing Supply Chain
	[20]	MICMAC Method	Food supply Chain
	[21]	Bayesian Belief Networks	Aerospace Mfg Supply Chain
	[22]	Bow-Tie Diagram and Fuzzy Inference system	Manufacturing Supply Chain
Risk Mitigation	[23]	MICMAC	Horticultural Supply Chain
	[24]	Grey Theory and Digraph Approach	Electronics Supply Chain

Table3. Integration Of Service Fmea With Other Engineering Improvement Tool

Integration of Quality Tools into Service FMEA	References	Application Area	Application Context
FMEA-IDEFO	[25]	Healthcare Service	Companywide
FMEA-Control Chart	[26]	Manufacturing	Companywide
FMEA-QFD	[27]	Healthcare	Organizational wide
FMEA-SWOT Analysis	[28]	Gas Distribution Service	Company wide
	[29]	Manufacturing	Companywide
FMEA and Design of Experiments	[30]	Manufacturing	Companywide
FMEA and Non Homogeneous Poisson Process	[31]	Maintenance	Companywide
FMEA and Process Activity Mapping	[32]	Manufacturing	Companywide

In spite of the fact that service FMEA studies is continuously growing and becoming more versatile, driven by the escalation of global disastrous events [33], the Inseparability, Heterogeneousity, Intangibility and Perishability (IHIP) characteristics of service operations [34] and the rising of the grey and black swan type risk events' occurrences [35], in our point of view, much unanswered questions are still waiting for further investigations. Considering that enhancement and application of service FMEA is having a vast scope, we only limit our ideas into the few aspects which covering enhancement of FMEA's Risk Appraisal, Modification of Service FMEA, Application to Non Profit Oriented Service and integration with other engineering tools. Some new research opportunities for advancing and expanding application of service FMEA studies are discussed in brief in the followings.

3. New Opportunities of FMEA Studies in Service Operations

3.1. Advancement of Failure Risk Assessment, Its Root Causes Re-prioritization, and Mitigation in Service Operations.

3.1.1. Prioritizing Root Causes of Failure for Risk Reprioritization

In an attempt to determine appropriate risk-based risk recovery measures, determination of root causes of the service failure occurrences is important. Contrary to the abundance of methodology to rank failure risk criticality in FMEA, the existence of methodology to prioritise root cause of failure occurrence is very limited in literature. Extending on the work of [36] and [37] in appraising the weight of failure root causes and ranking their hierarchy using other type of multi criteria decision making methodology and probability theory is a potential research direction.

3.1.2. Revision of the failure risk Evaluation by considering failure time occurrence and Interdependencies of Failure Event occurrences

In FMEA methodology, the magnitude of the risk is measured by its RPN. Due its simplicity, the use of RPN based on the multiplication among FMEA indexes on failure occurrence scale, detectability and severity using 1-10 ordinal like scale is very versatile. However, it has limitation such as ignorant to the failure time occurrence, which in real situations may escalate loss over time. Furthermore, although many studies have been proposed to improve determination on the component of probability of failure occurrences in FMEA as exemplified by [38], and [39], their previous works is based on the idea that probability of failure occurrences is independent among others. In real situation, however, the occurrence of failure modes is sometimes dependent among others. Reformulation of failure risk assessment quantitatively by considering dependencies among others is important and to our knowledge, never been investigated in previous Service FMEA studies.

3.1.3. Development of FMEA Method as Risk Distribution Tools in Supply Chain Application Framework

In supply chain risk assessment framework, quantifying the magnitude of the SCM risk is only half way to battle the risk. In an attempt to mitigate the adverse effect of risk event occurrences in supply chain context, decision makers must also consider the affected parties due to the risk exposure, which may endanger sustainability of multiple tiers in SCM framework. While many studies focus on appraising the magnitude of the risk by using many advanced mathematical tools such as Graph Theory, endeavour to design capability of FMEA as risk distribution tool for risk distribution to the affected parties in the supply chain context is vacant in references.

3.1.4. Modification of FMEA for Waste Appraisal Tool in Maintenance as Service Supporting Operation.

Driven by sustainability issues nowadays, currently global manufacturers face the challenge of creating green and sustainable manufacturing practice [40]. Sustainable manufacturing demands the efficient resource utilization and minimization of any activities which do not add value. From this point of view, application of lean manufacturing principle into maintenance is obviously relevant in supporting sustainable manufacturing as the goal of lean philosophy is on curbing the root cause of resource wastage. Nevertheless, studies which discuss on the advancement of maintenance methodology such as [41] indicate to the escape on the focus of developing maintenance model based on lean principle. Narrowing down the gaps, opportunities in this directions may commence by development of modified service FMEA model dealing with reprioritization of maintenance waste, which to our knowledge, is never been investigated in previous service FMEA references.

3.2. Application of Service FMEA in Not for Profit Oriented Operations.

According to the study of [42], the escalation of global disastrous events is coming at an alarming rate. In order to cope with above situation, designing a reliable service recovery operations is believe to be able to mitigate the negative impact of disastrous events occurrence. In this situation, investigation of the utilisation of FMEA as Risk Appraisal tool in disaster recovery actions can be basis for a start. An opportunity in this direction for instance is the modification of Service FMEA to appraisal the criticality of customers' needs in humanitarian situation. Another opportunity to apply FMEA in not for profit oriented operations is in the area of educational service provision in a supply chain framework. An investigation of [43] indicated that attention of previous studies in supply chain operations are almost neglect to pay attention on investigation of risk appraisal in educational service supply chain settings.

3.2. Integrating Service Life Cycle Analysis and Anticipatory Failure Determination Engineering Tools in Designing Predictive Service FMEA Model

In an attempt to enhance capability of FMEA, many studies integrating it with other engineering tools has already been proposed as given in table 4. Nevertheless, above mentioned studies are not sufficient since those still uncover limitation of service FMEA in dealing with service defectiveness at second and sub sequent limitation. And also, the birth of grey and black swan risk event types indicated to the need to design a predictive oriented model beyond the utilization of integrated FMEA and NHPP model such as the work [31]. In this regards, integration of service FMEA with Service Life Cycle Analysis and Anticipatory Failure Determination (AFD) methodology can be chosen as an opportunity for further studies.

4. Conclusions

In this paper, a preliminary survey and classification scheme of current service FMEA studies from current periodical - type references using specific criteria are proposed. In spite of the facts that applications of service FMEA is continuously growing and becoming more advanced, driven by nowadays business situations getting more complex due to uncertainties and characteristics of the service process itself, many challenges are still awaiting. In our paper, some directions of investigating service FMEA for future studies are proposed. Obviously, by using a very limited literature, it is clear that this study is not an exhaustive investigation and future research could reiterate our initial findings by using more databases. Furthermore, discussions on opportunities for development of service FMEA model could include some issues such as its application in virtual settings, its integration with performance measurement tools, inclusion of people's belief in apprising the criticality of defective services and establishment of time -considering multi criteria decision support model for service failure risk appraisal.

5. Acknowledgements

The authors wishing to acknowledge financial support from Indonesian Ministry of R&D and Higher Education under Research Scheme *International Collaborative Research and Joint Publication 2015.*

6. References

[1] Abu Soleh Z and Anjalin U 2011 *J S.Science and Man.* **4** 253-60.
[2] S.S.Bhamare, Om P. Yadav and Ajay Rathore 2007 *I J Rel Saf.* **1** 377-10
[3] Om P Yadav and Nanua S 2008 *I J Prod Dev* **5** 4-16.
[4] Enrico Z 2009 *Rel.Eng Sys Saf.* **94** 125-41.
[5] Ram M 2013 *I J Syst. Assu. And Eng and Man.* **4** 101-17.
[6] Sutrisno A. and Lee, T. R. 2011 *I J Eng Sci and Tech* **7** 27-38.
[7] Luo and G. G. Lee 2015 *TQM and Bus Ex* **26** 62-75.
[8] Ravid, Bhaskany and Dror 2011 *IIE Trans Heath Sys Eng* **1** 91-00
[9] Sant'Anna 2013 *Int J Qual Rel Man* **29** 349-62.
[10] Song B C Lee and Y Park *Int J Prod. Econ.* **141** 493-504.
[11] C Kahraman I Kaya and O Senvar *Hum Eco Risk Ass* **19** 535-52
[12] H Alyami P.T.W Lee J Yang Riahi S Bonsal J Wang 2014 *Mar Pol and Man* **41** 634-50.
[13] R Liao J Bian L Yang S Grzybowski 2014 *Int Trans Elec En Sys* **23** 1172-90.
[14] K.C Chu and L.P Hung 2014 J App Math 1-17.
[15] Y Geum Y Cho and Y Park *Int J. Prod. Econ.* **141** 493-04
[16] S. A Rahimi A Jamshidi D Ait Kadi and A Ruiz 2014 *Qual. Rel.Eng.Int.* **31** 601-15.
[17] H C Liu L Liu and P Li 2014 *Int J. Sys Sci* **45** 2012-30.
[18] D.S Chang and J.H. Chung *J Trav and Tour Mkt* **29** 817-834.

[19] Samvedhi A. Vipul J and F.T.S Chan 2013 *Int J Prod. Res.* **51** 2433-42.
[20] Elleuch W Hachica and H Chabcoub *J Risk Res* **17** 641-33.
[21] Badurdeen et al 2013 *J Man Tech Man* **25** 631-54.
[22] F Aqlan and S.S Lam 2015 Int. J Prod. Econ. **161** 54-63.
[23] R Astuti, Marimin,Machfud, Arkerman, Purwanto and Meuwessen 2013 Op.Supp. Ch Man Int J. **6** 11-25
[24] R. Rajesh V Ravi and V Rao 2015 Int. J.Prod. Res. **53** 238-257.
[25] M Bevilacqua G Mazzuto and C Paciarotti 2015 I.J .Proc Man. **8** 25-43.
[26] X Zhao 2011 Int J. Ind Eng. The. App. Prac. **18** 444-51.
[27] Chen S.H. 2014 *Qual and Quan* 1-6
[28] Agung S H.M.Kwon I.Gunawan S Eldridge and T.R. Lee 2015 *Int J. Qual. Prod.* Man. in print.
[29] A.P Punasvervaran N Jamibollah and N Zorazin 2014 *Am. J. App Sci.* **11** 1332-42.
[30] H.Y Jen and Y.L Liu 2012 *TQM and Bus Excel.* **23** 1171-89.
[31] Srivastawa and Mondal 2015 *Int J. Serv Op.Man* **19** 319-37.
[32] Marriot B. Reyes J.A.G. Meyer H.S. and Anthony J 2013 *J Man.Tech Man* **24** 197-217.
[34] Whybark 2015 *Int. J. Qual.In.* **1** 1-10.
[35] He and Goh T.N 2015 *Qual.Tech and Quan Man.***12** 83-92
[36] Aven T 2015 *Rel Eng Sys Saf* **134** 83-91.
[37] M. Braglia 2000 *Int.J.Qual.Rel.Man.* **17** 1017-33.
[38] J.H. Ae, S. H. Hong, M. K. Lee, and H.M. Kwon 2015 ***ICIC Ex.Lett Part B: App***. 6 383-390.
[39] Senol S 2007 *Qual. Man.* J **14** 29-40
[40] Braaksma A.A.J Meesters A.J. Klingenberg W. and Hicks C 2012 Int. J. Prod. Res**.** **50** 6904-17.
[41] Garreti M.and M. Taisch 2011, *Prod. Plan Con*t. **23** 83-04.
[42] Garg H. and S. G. Deskmukh 2006, *J. Qual. Main.Eng.* **12**, 205-38.
[43] Abu Nahleh Y.L.A 2014, *PhD Dissertation*, RMIT, Australia.

[44] Habib Md M. 2010 Int J Bus Man and Soc Sci. **1** 79-87.

IOP Conf. Series: Materials Science and Engineering **105** (2016) 012012 doi:10.1088/1757-899X/105/1/012012

The effect of filler addition and oven temperature to the antioxidant quality in the drying of *Physalis angulata* fruit extract obtained by subcritical water extraction

R F Susanti[1,2] and G Christianto[1]
[1]Chemical Engineering Department, Industrial Engineering Faculty, Parahyangan Catholic University, Bandung, West Java, Indonesia
[2] Korea Institute of Science and Technology, Hwarangno 14-gil 5, Seongbuk-gu, Seoul, 136-791, Republic of Korea

E-mail: santi@unpar.ac.id

Abstract. *Physalis angulata* or *ceplukan* is medicinal herb, which grows naturally in Indonesia. It has been used in traditional medicine to treat several diseases. It is also reported to have antimycobacterial, antileukemic, antipyretic. In this research, *Pysalis angulata* fruit was investigated for its antioxidant capacity. In order to avoid the toxic organic solvent commonly used in conventional extraction, subcritical water extraction method was used. During drying, filler which is inert was added to the extract. It can absorb water and change the oily and sticky form of extract to powder form. The effects of filler types, concentrations and drying temperatures were investigated to the antioxidant quality covering total phenol, flavonoid and antioxidant activity. The results showed that total phenol, flavonoid and antioxidant activity were improved by addition of filler because the drying time was shorter compared to extract without filler. Filler absorbs water and protects extract from exposure to heat during drying. The combination between high temperature and shorter drying time are beneficial to protect the antioxidant in extract. The type of fillers investigation showed that aerosil gave better performance compared to Microcrystalline Celullose (MCC).

1. Introduction

Physalis angulata, well known as *ceplukan,*is a wild medicinal herb found in shaded fields, gardens, wastelands, fallow fields, along roadsides, in open forests, and other places in Indonesia [1]. In folk medicine, all of the parts of *Physalis angulata* plant such as roots, flowers, leaves, stems and fruits have been used to treat some diseases including asthma, diuretic, malaria, hepatitis and cancer treatment [2,3]. There has been significant interest in evaluating the phytochemical and pharmaceutical properties of this plant, including the potential of *Physalis angulata* plant as antimycobacterial, antileukemic, antiseptic, etc [4]. However there were less reports regarding the antioxidant potential of *Physalis angulata*. Our group is the first who investigated the antioxidant extraction from *Physalis angulata* by subcritical water extraction [5,6,7].

[1] To whom any correspondence should be addressed.

IOP Conf. Series: Materials Science and Engineering **105** (2016) 012012 doi:10.1088/1757-899X/105/1/012012

Antioxidant itself is chemical compounds which can prevent oxidation inside our body and protect our body from free radical by neutralizing it. Free radical is very dangerous which can weaken our immune system, damage and kill our cells, also generate diseases in our body [8]. *Physalis angulata* is commonly extracted by organic solvents such as methanol, ethanol, and hexane. However, organic solvents are also well known for its toxicity, carcinogenic, and hazards for environment. It will not simply disappear during extract evaporation and will still remain in the extract. Therefore, pharmaceutical industries tried to exchange the uses of these organic solvents with the more friendly solvents, such as water and supercritical CO_2. Water has an unique characteristic. When water is in normal (room) condition, it behaves as a very polar solvent. However, its polarity can be adjusted by changing its temperature and pressure. The condition of water which its temperature is above boiling point and below its critical point, meanwhile its pressure maintains its liquid form is called subcritical water. Subcritical water is chosen as the solvent of this experiment because water is non-toxic, cheap, friendly to environment with adjustable polarity [9,10].

In this experiment, subcritical water was used as a solvent. In previous report, high temperature about 200-250 °C had shown better result [6]. In this research, high temperature of 250 °C, pressure of 100 barand 15 min of extraction time was utilied to extract *Physalis angulata* fruit. Crude extract which contains oleoresin (oily and solid form at once) often causes difficulty in the determination of drug dosage. The treatment of crude extract into dry extract can be obtained by drying in dryer. However, the bioactive compound in crude extract which contains mostly antioxidant may damage during exposure in heat.In addition just simply drying resulted in a sticky form of extract. The addition of filler was intended to make the drying process faster, to keep the flavor, to add the quantity, and to make the powder form of extract [11]. It is inert, added typically to make consistency of drug. In this research, extract qualities were determined based on total phenol, flavonoid and antioxidant activity. The effects of filler types (Microcrystalline Cellulose (MCC) and aerosil), filler concentrations and drying temperatures were investigated to the extract obtained from *Physalis angulata* fruits which were dried in vacuum drying oven.

2. Experimental

2.1. Subcritical Water Extraction

The extraction was conducted in Subcritical Water Extraction batch apparatus (Hanyang Accuracy, Korea). The batch reactor system has volume of 150 mL, is equipped with electric heating for heating the reaction to the desired temperature and cooling jacket for decreasing the reactor temperature once the reaction finished. The temperature and pressure of the system are controlled using control box. Ten grams of the sample was put in the reactor basket. The reactor then was closed and purged with nitrogen for about 15 minutes to remove the oxygen in the reactor and lines. The distilled and deionized water (DDI) water was also purged to remove the dissolved oxygen. Then, DDI water was delivered to the reactor using high pressure metering pump (HKS-600, Hanyang Accuracy, Korea) and bought to the desired pressure. The solution then was heated to the desired extraction temperature and kept for specific time, called as static time. After the time was fulfilled, the reactor then was cooled by opening the insulation and flowing cooling water. The extract was then filltered out with fine grade filter paper (No 393, Sartorius, Germany) to make sure no residue left in the extract. For drying experiment, water was evaporated partially from the extract using rotavapor (RE-3000A, Xi an HEB Biotechnology Co. Ltd, Shanghai, China) until extract volume of 30 ml.

2.2. Drying Experiment

The extract with volume about 30 mL was dried further by vacuum drying oven. Prior to the drying, an amount of filler was added. The drying temperature was set and the extract was dried and weighted every hour. The drying was stopped if the constant weight of extract was observed. The water content then was checked by moisture analyzer (Mettler Toledo, HR83, USA). The dry extracts were stored at refrigerator prior used.

2.3. Analysis
Prior used in analysis, extract powder was dissolved in methanol p.a to separate the extract from filler. The solution then was filtered using filter paper Sartorius 388 (Sartorius Stedim Biotech S.A.), the filtrate was used for analysis while the filler was dried and weighted for calculation purpose.

2.4. Total phenol
The total phenols of the extracts were analyzed by Follin-Ciocalteu method, which used gallic acid as the standard solution. Accurately, 0.5 mL extracts 500 ppm (5 mg in 10 ml DDI water) were mixed with 2.5 mg Follin Ciocalteu reagent (1:10) and 2 mL of Na_2CO_3 solution (75 g/L). After added with reagent and Na_2CO_3, the mixture solution was incubated at 40°C for 30 minutes, and the absorbance was measured at 735 nm.

2.5. Total flavonoid
The total flavanoids content of extracts were analyzed using calorimetric method, which used quercetin as the standard solution. As much as 0.25 mL extracts were mixed with 1.25 mL DDI water and 0.075 mL $NaNO_2$ 5%-wt. After 5 minutes, 0.15 mL $AlCl_3$ 10% was added into the solution. 6 minutes later, 0.5 mL NaOH 4%-wt was added into the solution. Immediately after 10 minutes, water was added to the mixture to bring the final volume of 4 ml. The mixture was thoroughly mixed and allowed to stand for another 10 minutes. Then the absorbance was measured at 510 nm.

2.6. Antioxidant activity
The stock DPPH solution was prepared by dissolving 7.9 mg DPPH in 100 ml methanol p.a. (0.2 mM). The solution was obtained by diluting a specific amount of dry extracts with 100 ml methanol p.a. The different concentration of solutions (20-400 ppm) were made by diluting the extract mother liquor. Then, 2 ml of this solution was allowed to react with 2 ml of the DPPH solution for 24 h and the condition was kept away from any sources of light. Then the absorbance was measured at 517 nm.

3. Result and Discussion

3.1. Total Flavanoid
The total flavanoid analysis was conducted by calorimetric method. The total flavonoid was quantified by quercetine as a standard. The effect of MCC addition to the total flavonoid at different drying temperature is illustrated in figure 1 while the effect of aerosil addition is shown in figure 2. It can be shown clearly that the higher amountof filler addition, the better the total flavonoid in extract. It is due to the higher concentration of filler the shorter the drying time, as shown in table 1 for MCC and table 2 for aerosil. For example, by increasing the MCC concentration from 0 wt% to 20 wt%, the length of drying time at 50°C decreased by half and total flavonoid increased by 7 folds. Filler may protect the antioxidant from exposure to heat. It might be also the reason why an . increase in drying temperature from 40 to 60 °C also increased the total flavonoid, it was due to the time needed for drying became shorter at high temperature.

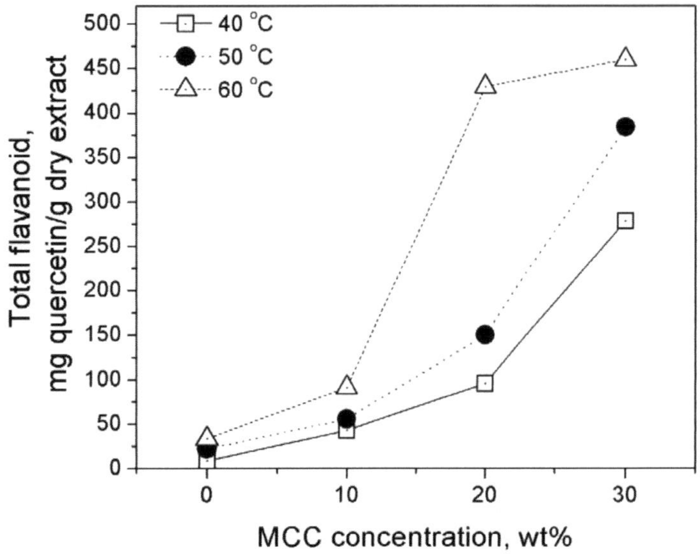

Figure 1. Total flavonoid as a function of drying temperature and MCC concentration

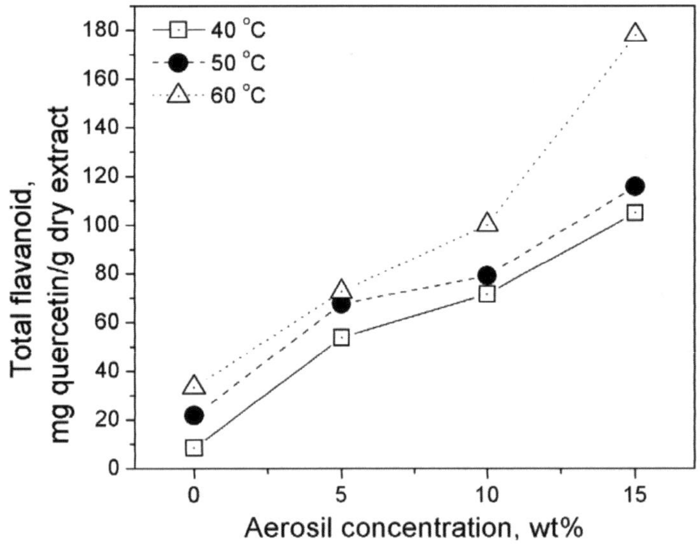

Figure 2. Total flavonoid as a function of drying temperature and aerosil concentration

IOP Conf. Series: Materials Science and Engineering **105** (2016) 012012 doi:10.1088/1757-899X/105/1/012012

Table 1. The length of drying time and water content in the drying of *Physalis angulata* fruit extract in addition of MCC filler

MCC Concentration (wt%)	Drying temperature (°C)	Drying time (min)	Water content (wt%)
0	40	4330	2.65
	50	3645	2.52
	60	2680	2.44
10	40	3330	3.67
	50	2350	2.55
	60	1830	3.23
20	40	3005	2.66
	50	1820	2.32
	60	1360	1.90
30	40	1990	2.52
	50	1340	2.84
	60	1050	2.35

Table 2. The length of drying time and water content in the drying of *Physalis angulata* fruit extract in addition of Aerosil filler

Aerosil Concentration (wt%)	Drying temperature (°C)	Drying time (min)	Water content (wt%)
0	40	4480	2.76
	50	3770	2.59
	60	2800	2.49
5	40	3830	2.33
	50	2755	1.77
	60	2450	1.44
10	40	3240	1.27
	50	2105	1.42
	60	1810	1.22
15	40	2670	1.08
	50	1610	1.28
	60	1440	3.67

3.2. Total phenol

The effect of MCC addition to the total phenol at different drying temperatures is illustrated in figure 3 while the effect of aerosil addition is shown in figure 4. The lower the drying temperature, the longer the drying time. For instance, in MCC case, by increasing the temperature from 50 to 60 °C can shorten the drying time till 2 times so that long exposure in heat can be avoided. An increase in filler concentration (0-30% for MCC and 0-15% for aerosil) increases the total phenol in all drying temperatures. It was because the higher the concentration of filler, the shorter the drying time as shown in table 1. For instance, the addition of 30 wt% MCC to sample which dried at 60°C can shorten drying time till 50%. Filler may absorb water and protect the antioxidant from deteriorations because of thermal heat.

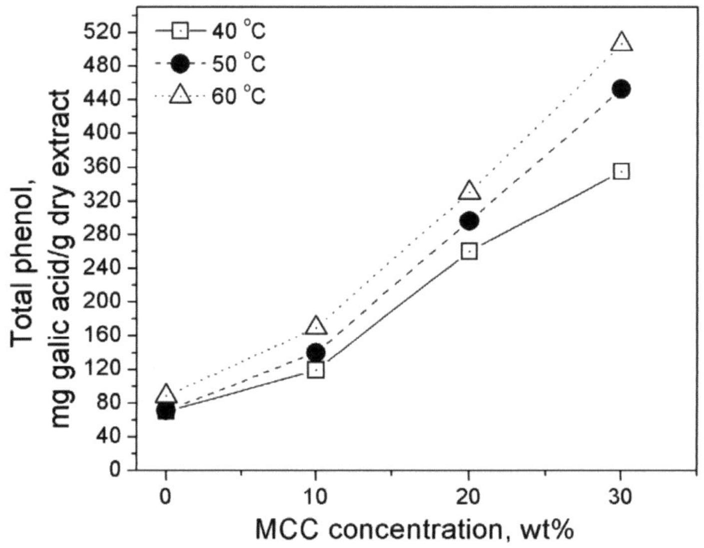

Figure 3. Total phenol as a function of drying temperature and MCC concentration

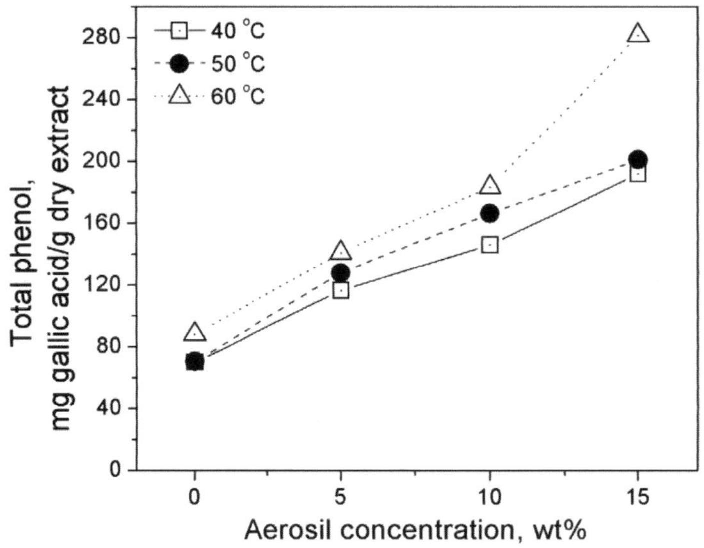

Figure 4. Total phenol as a function of drying temperature and aerosil concentration

IOP Conf. Series: Materials Science and Engineering **105** (2016) 012012 doi:10.1088/1757-899X/105/1/012012

3.3. Antioxidant Activity

Antioxidant activity is shown by IC50 value. IC50 represents the concentration of extract that is required for 50% inhibition of oxidation. It means the lower the IC50 value, the smallest extract needed for inhibiting the oxidation, means the strongest the antioxidant activity. The IC50 value at different filler concentrations and drying temperatures are shown in figure 5 for MCC and figure 6 for aerosil. For both of filler types, the higher concentration of filler resulted in the lower IC 50 or better antioxidant activity. The addition of more filler helps to absorb water, so exposure in high temperature can be avoided. An increase in drying temperature increased the antioxidant activity, shown by a decrease in IC50 value. The higher the drying temperature shortened the drying time as shown in table 1 and 2, so reduced exposure of extract to the heat. The results were confirmed by total phenol and flavonoid which showed the same trends. Started from the additional of 10 wt% filler, antioxidant in *Phyalis angulata* are categorized as strong antioxidant (IC50<50).

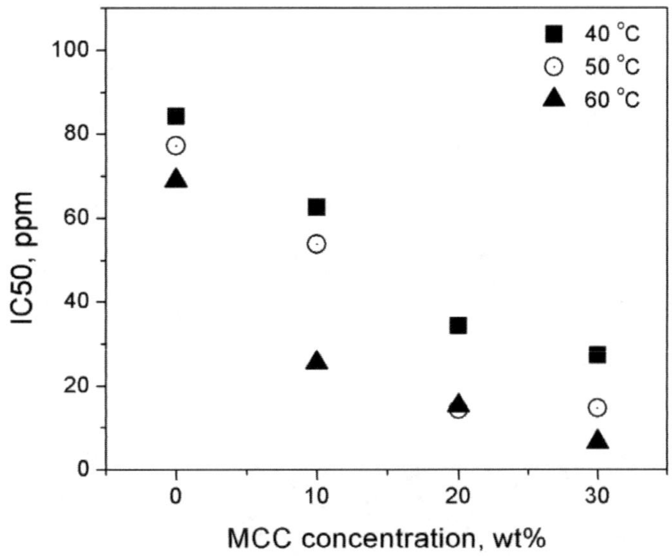

Figure 5. IC50 as a function of drying temperature and MCC concentration

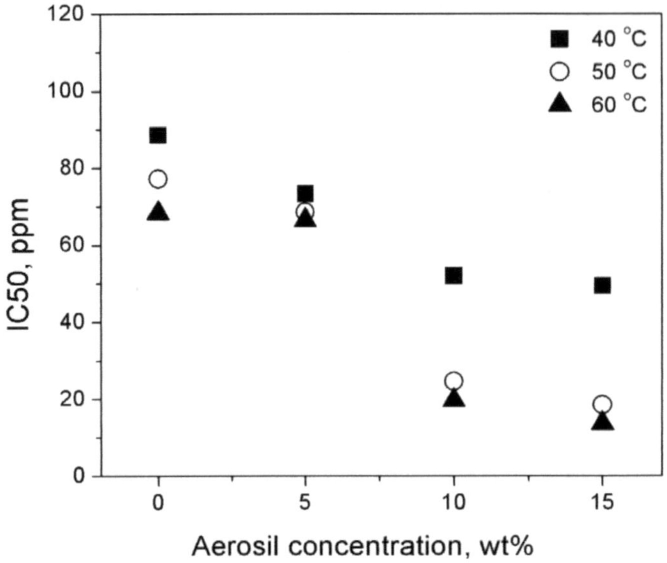

Figure 6. IC50 as a function of drying temperature and aerosil concentration

3.4. The effect of filler types to total phenol, flavonoid and antioxidant activity

Table 3. Comparison between MCC and aerosil at the same concentration (10 wt%)

	Drying temperature (°C)	MCC	Aerosil
Total flavanoid	40	42.70	71.52
	50	55.36	79.13
	60	91.05	100.14
Total phenol	40	119.02	146.11
	50	139.47	166.35
	60	169.43	183.31
IC50	40	62.54	51.99
	50	53.83	26.27
	60	25.50	19.62

The comparison between performance of MCC and aerosil can be shown in table 3. The effect of filler types was observed at drying temperature of 40 - 60°C using 10 wt% of filler. Aerosil showed better total phenol and flavonoid results in all temperatures observed. It may be because aerosil (50 g/L) has much lower density compared to MCC (200-500 g/L) [12, 13], so that for the same weight, the volume of aerosil was more than MCC. The impact would be in a better ability of aerosil to absorb water compared to MCC for the same quantity, giving faster drying time.

4. Conclusion

The addition of filler has been proven to improve extract quality in the term of antioxidant capacity. Both of MCC and aerosil gave the same trend for total phenol, flavonoid and antioxidant activity. An increase of filler concentration could enhance the total phenol, flavonoid and antioxidant activity. Prolong exposure in heat could deteriorate the extract, however addition of filler could shorten the drying time so the antioxidant in the extract could be maintained. An increase in filler concentration from 0 wt% to 30 wt% for MCC and 0 wt% to 15 wt% for aerosil and oven drying temperature from

40-60°C has been proven to improve the antioxidant quality because of reduction in drying timeExtract was protected from long exposure to heat. Comparing aerosil and MCC at the same concentration showed that aerosil gave better performance.

Acknowledgments
The author gratefully appreciate the financial support from Korea Institute of Science and Technology (KIST), Korea through KIST Collaboration Alumni Project and Lembaga Penelitian dan Pengabdian Masyarakat (LPPM) UNPAR through Internal Funding

References

[1] Nahid S, M A Hassan, Momtaz B, Mahbuba S 2008 *Bangladesh J.* **37** 195-198
[2] Chang J C 2008 *Pharmaceutical Biology* **46** 724-731.
[3] Hieronymus B S 2008 (Tangerang: PT Agromedia Pustaka) pp 142
[4] Diones K *Biota Neotropica* **13**
[5] Ratna F S, Ignatius J R and S Garmi 2013 *Conf. Regional Symporium on Chemical Engineering* (Philippines: Bohol)
[6] Ratna F S, Kennedy, Ashianti S W 2014 *Conf. International Seminar of Chemical Engineering in Minerals & Materials Processing*
[7] Ratna F S, Kevin K, Amadea V, Ignatius J R 2015 *Modern Applied Science* **9** 190-198
[8] Kosasih E N, Tony S, Hendro H 2006 (Jakarta: Pusat Kajian Nasional Masalah Lanjut Usia)
Cut F Z, Julianti Br Tarigan, Herlince S 2008 *Jurnal Biologi Sumatera* **3** 7-10.
[9] Ali K, Meredith A W, Thomas H R 2013 *Molecules* **18** 2328-75.
[10] Miguel De La Guardia, Salvador G 2012 *Handbook of Green Analytical Chemistry* (Chichester: John Wiley & Sons, Ltd.)
[11] Michalczyk M, Macura R and Matuszak I 2009 *J. Food Process* **33** 11-21
Nadia D M, Nourhene B, Nabil K, Francis C, Catherine B 2012 *Food and Bioproducts Processing* **90** 433-441.
[12] Jonat S, Hasenzahl S, Drechsler M, Albers P, Wagner KW and Schmidt PC 2004 Powder Technology **141** 31-43
[13] Gregory T, Fabrice K, Bruno L, Brian C, Brigitte E 2014 Int. J. Pharmaceutics **473** 64-72

Mobile application for guidance and provision of toddler's nutrition to support e-PKK

S Rochimah[1]**, F Y Sianipar**[2] **and R N E Anggraini**[3]

Department of Informatics , Institut Teknologi Sepuluh Nopember, Surabaya, Indonesia

Email : [1]siti@its-sby.edu, [2]febryanyeremi@gmail.com and[3]ratih_nea@if.its.ac.id

Abstract. *Pembinaan Kesejahteraan Keluarga*(PKK)is an Indonesian community with women as its member, especially housewife. It has many purposes, such ascollaborating the knowledges among members, monitoring children's health, supporting healthy life style in the family. This article is part of our research in building e-PKK, an integrated application to support many activities in PKK's business processes. In this paper we build a module to guide and provise toddler's nutrition to be used by mother. This application is very useful since baby's growth phase is an important phase to be noticed by mother.Using this application, mother can easily obtain baby's growth information whenever and wherever they are via their smartphone. This mobile device applications using backward chaining and forward chaining method. Backward chaining method is a method that uses a goal-based approach, while the forward chaining conducting a set of data for later inference process to find optimal conclusion. Moreover, this apllication provides recommendations of groceries, recipes, as well as the suitability of foodstuffs according to the age of early baby's growth and the type of baby's allergic. In addition, it also provide information about baby's nutrition, growth benchmarks, and first aid. Besides, the application can be used to monthly baby growth record like *Kartu Menuju Sehat* (KMS) or Growth Monitoring Card, storing toddlers weighing, immunization and provision of vitamin A. An additional feature of this application is the complaints system, where other can ask directly to health care center about toddlers' growth.

1. Introduction

Baby's growth phase is an important phase to be noticed by every mother. Experts argued about golden period time, which is between 0-2 years, 0-3 years, 0-5 years or 0-8 years. But all of them agreed that early age of the baby's growth is the golden period. In this period, baby's brain ability to absorb every kind of information is very high, so parent's role is needed in order to educate and optimize children intelligence intellectually, emotionally and spiritually. However, food allergies are common in this period than in adulthooddue to incomplete baby's digestive tract.

Today's technology developed very fast compared to previous decades. Along with it, mobile technology has indirectly become a part of people's daily lives, for example smart phones. It is undeniable that more people used smartphones regardless of their status or age restrictions.

We have had many applications on smartphone mobile device that is offering information about nutrition and development of the baby, so the baby's mother can easily obtain every kind of information about the growth of the baby whenever and wherever they are via their smartphones devices. However, each of this application still have their shortcomings. For example, the application for baby food recipes are not given information about solutions to problems that often occur at the

IOP Conf. Series: Materials Science and Engineering **105** (2016) 012013 doi:10.1088/1757-899X/105/1/012013

time of feeding, whileapplications for baby's growth are not given information about how to stimulate the growth of the baby[1].
Therefore, this article demonstrates the development of an application using the PhoneGap platform framework that can run on mobile devices. Thisapplication use backward chaining and forward chaining methods to provide recommendations of groceries, recipes, as well as the suitability of foodstuffs according to the age of early baby's growth and the types of baby's allergy. In addition this application also can provide information about baby's nutrition, baby's growth benchmarks, and baby's first aid. This application can also be used to record information like *Kartu Menuju Sehat* (KMS) or Growth Monitoring Card used in some baby's health care center, as recording information about toddlers weighing, baby's immunization and provision of vitamin A. An additional feature of this application is given the complaints system, where mothers can ask directly to the health care center about the growth of toddlers through this application.
With the making of this application, it is expected that this application will be able to help parents to get useful information about baby's nutrition, baby's growth, and appropriate baby's first aid in case of accident in the baby or baby's disease, and also help parents to monitor the baby's growth.

2. Literature Study

2.1. Existing Application
Recently there are so many applications that can be used to help providing information about nutrition as well as providing information about the baby's growth. Some of the applications are as follows.

1. Baby Solid Food
   This mobile device application contains information about some kind of food that can be consumed by the baby aged 0-12 months. This application also provides recipes that can be made in accordance with the age of the baby[2].
2. Baby Development Track Growth
   This mobile device application contains information about the benchmarks of toddler's growth divided into four parts, namely motor skills, cognitive development, language skills, and social and emotional skill[3].

2.2. Forward Chaining
The application of operators to those structures in the database that describe the task-domain situation –to produce a modified situation–is often called reasoning forward. The object is to bring the situation, or problem state, forward from its initial configuration to one satisfying a goal condition [4]. The advantages of forward chaining method compared to backward chaining method is that a new data can be inserted into the database table inference and there is a possibility to make changes to the rules of inference.

2.3. Backward Chaining
Reasoning backward involves another type of operator than forward chaining, which is applied, not to a current task-domain situation, but to the goal. The goal statement, or problem statement, is converted to one or more subgoals that are (one hopes) easier to solve and whose solutions are sufficient to solve the original problem. These subgoals may in turn be reduced to sub-subgoals, and so on, until each of them is accepted to be atrivial problem or its subproblems have been solved[4].

2.4. Growth Monitoring Card
Growth Monitoring Card is a card contains the child's normal growth curve based on the index of the anthropometric weight in accordance to the baby's age. By using the Growth Monitoring Card, the growth disorders or the risk of excess nutrients can be known in advance, so preventive action can be carried out more quickly and precisely before the problem become more severe[5].

2.5. PhoneGap
PhoneGap is an open source-based framework released by a company in America called Nitobi. PhoneGap framework allows everybody to develop a mobile device application using the expertise of HTML, CSS, and JavaScript. An application created by using the PhoneGap framework can be developed into a variety of different platforms[6].

2.6. jQuery
jQuery is a fast, small, and feature-rich JavaScript library. It makes things like HTML document traversal and manipulation, event handling, animation, and Ajax much simpler with an easy-to-use API that works across a multitude of browsers. With a combination of versatility and extensibility, jQuery has changed the way that millions of people write JavaScript[7].

2.7. jQuery Mobile
jQuery Mobile is a HTML5-based user interface system designed to make responsive web sites and apps that are accessible on all smartphone, tablet and desktop devices[8].

2.8. jqPlot
jqPlot is a plotting and charting plugin for the jQuery Javascript framework. jqPlot produces beautiful line, bar and pie charts[9].

3. Analysis and Design

3.1. Requirement Specification
System requirements is shown in use case diagram, showing relationship between actorsand system. Based on use case diagram in Figure 1, user can see the recommendations for baby's nutrients according to the age and the type of baby's allergy, learning the baby's growth benchmarks, as well as read first aid of common diseases that often affects the baby. Moreover, users can record information of baby's growth just like the original KMS card, and can ask for the help of health care centre about the development of the baby.

3.2. Recommendation Rules Design
Some rules were built to recommend food for baby based on age and allergies. Age category along with the type of allergy can be seen inTable **1**. This table is used in advanced methods of backward and forward chaining to obtain the rules of nutritional recommendations by category of age and the type of allergy.

Table 1. Age category and allergy type.

Age Category	Allergy			
	Intolerent Lactose	Cow Milk Casein	Egg	Sea Fish
0-6 months	✓	✓		
6 months	✓	✓		
7-9 months	✓	✓		
9-12 months	✓	✓	✓	
12-18 months	✓	✓	✓	✓
18-24 months	✓	✓	✓	✓

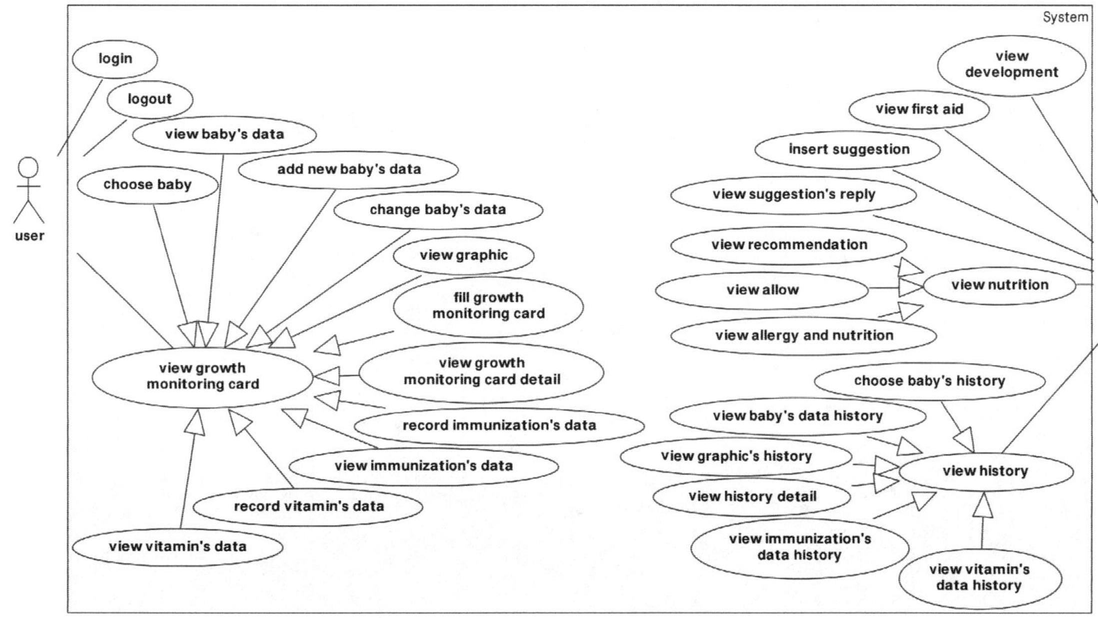

Figure 1. Use case diagram.

3.3. Database Design

The database used in this application is SQL and SQLite. The logic of this database is stored in the form of javascript and sql file. SQLite database can be accessed using Local Storage APIs provided by PhoneGap framework. As for the SQL database, it can be accessed using application that has been stored in server.

4. Implementation

4.1. Interface Implementation

This section discusses the implementation of the user interface. This interface was developed using HTML5, CSS, JavaScript, and jQuery Mobile. All pages are included in the file named index.html.

4.1.1. Login Page.

This page providing a form that can be used to login. Users need to enter a phone number and password to log into the system. This page is shown inFigure 2.

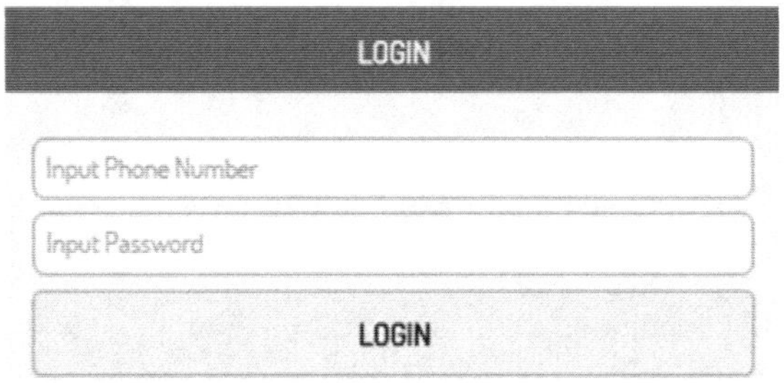

Figure 2. Login page.

4.1.2. Main Menu Page.

This page showedsix main menusused as a navigation to move to other pages, namely Nutrition page, Development page, First Aid page, Growth Monitoring Card page, History page, and Suggestion and Complaints page.The interface is shown in Figure *3*.

4.1.3. Nutrition Page.
This page showing three submenu pages, namely Recommendation page, Allow page, and Information page. This page shown inFigure 4.

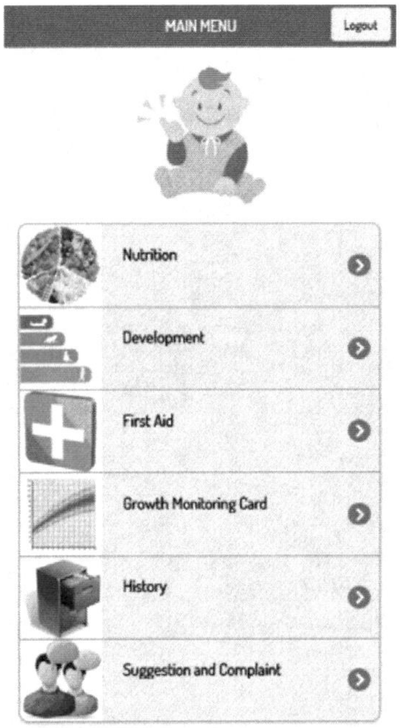

Figure 3. Main menu page.

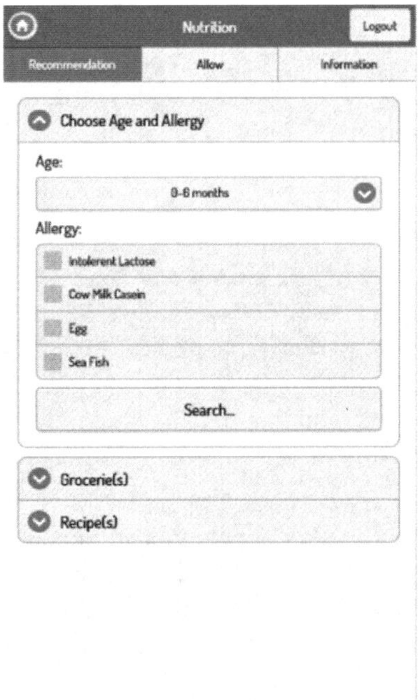

Figure 4. Nutrition page.

*4.1.4. Development Page.*This page showing a list of growth benchmarks information according to the age of the baby. This page shown in Figure *5*.

*4.1.5. First Aid Page.*This page showing a list of diseases information and their treatment. This page shown in Figure *6*.

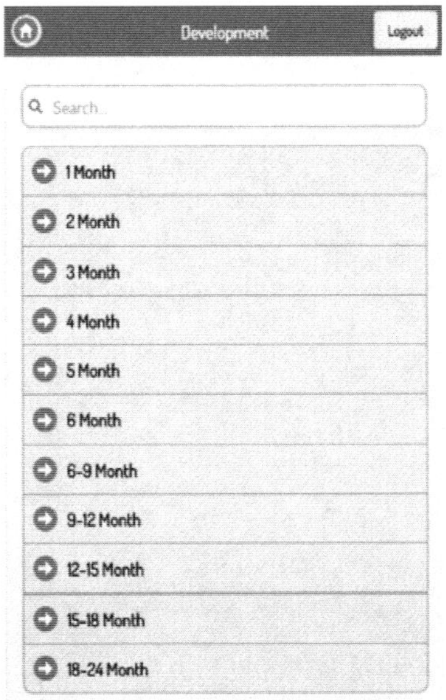

Figure 5. Development page.

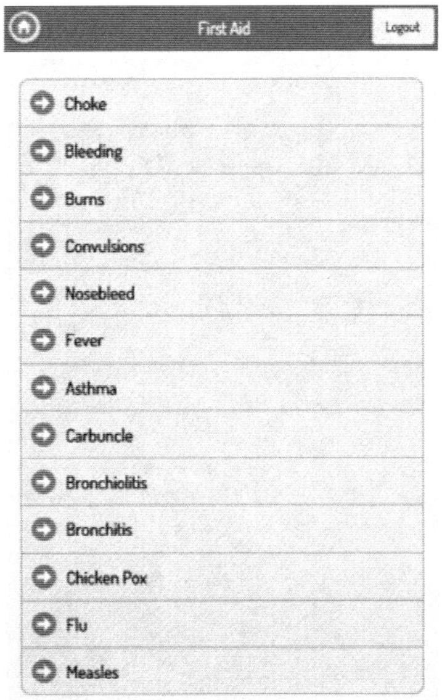

Figure 6. First aid page.

4.1.6. Growth Monitoring Card Page.
On this page there are four submenu pages, namely Baby's Data page, Graphic page, Immunization page, and Vitamin A page. This page shown in Figure *7*.

4.1.7. History Page. On this page there are four submenu pages, namely Baby's Data page, Graphic page, Immunization page, and Vitamin A page. This page shown in Figure *8*.

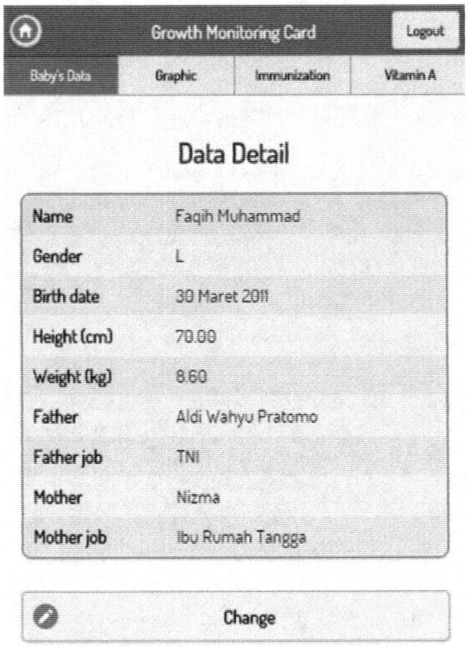

Figure 7. Growth monitoringcard page.

Figure 8. History page.

*4.1.8. Suggestion and Complaint Page.*On this page there are two submenu pages, namely Complaints Replies page and Input Complaints page. This page shown in Figure ***9***.

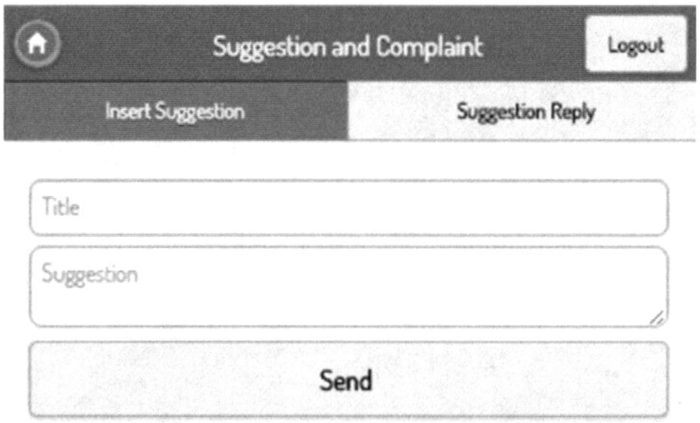

Figure 9. Suggestion and complaint page.

4.2. Database Implementation

Implementation of the database is performed using SQLite Database as well as SQL Database which are shown in Figure **10** and Figure **11**below. SQLite database used to implement database in the mobile device, whilst SQL database used for web application.

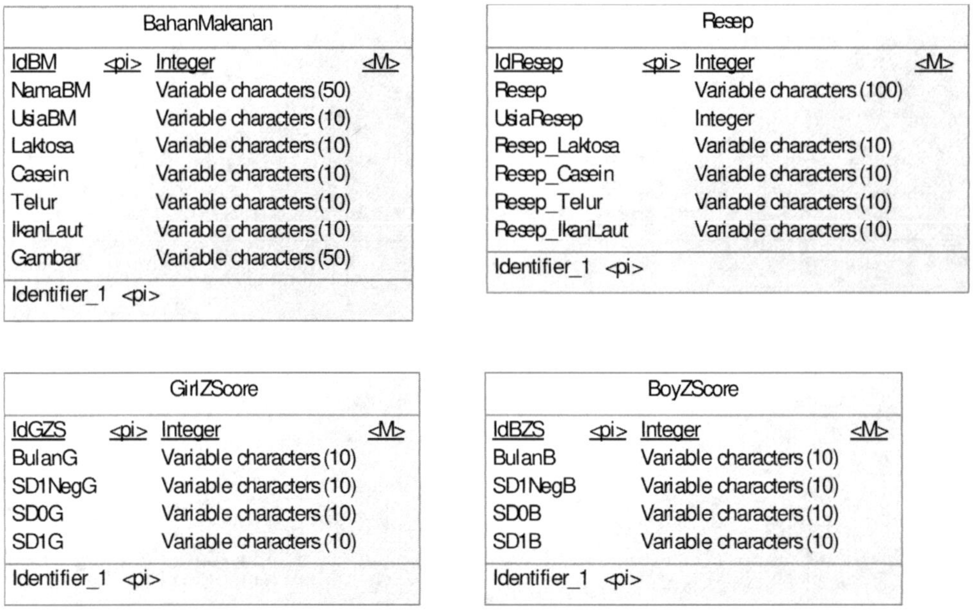

Figure 10. SQLite database implementation.

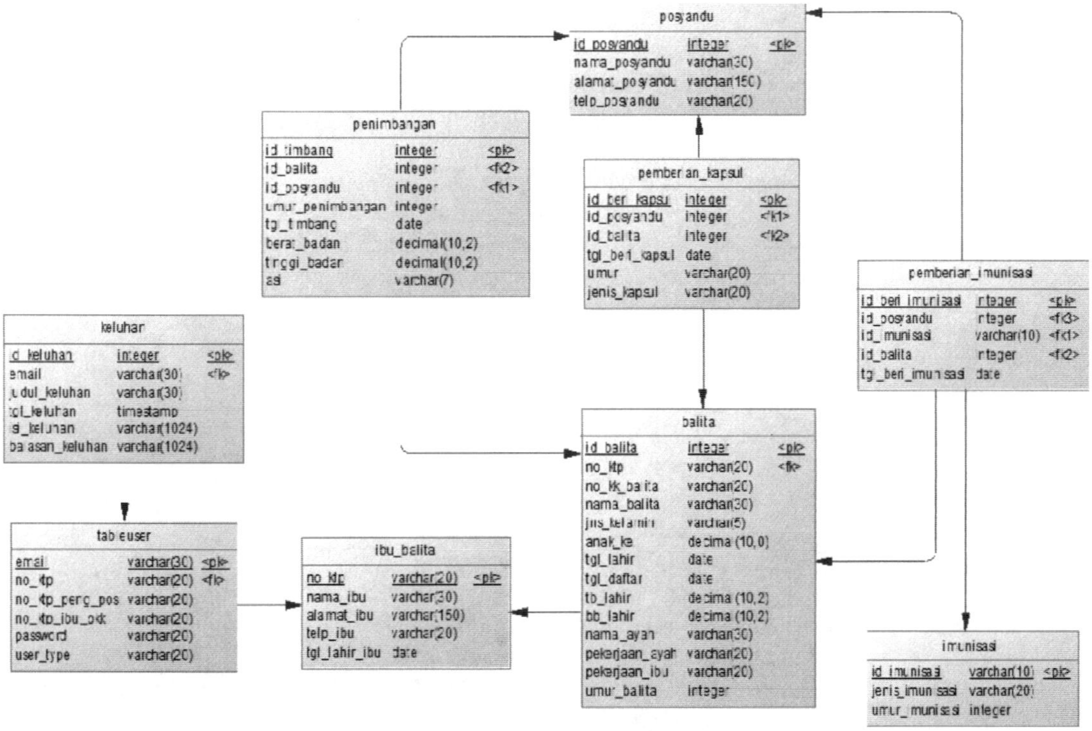

Figure 11. SQL database implementation.

5. Testing and Evaluation

5.1. *Functionality Testing*

The test which were conducted is functionality requirement testing. Functionality testing was using a black box testing method, wherein the method is carried out to ascertain the suitability of the output of the system. Testing the functionality of the system is done by setting up a number of scenarios to measure the success rate of the testing.

5.2. *Evaluation of Functionality Testing Result*

Based on functionality testing using the scenario that has been provided, obtained test results can be seen inTable 2.

Table 2. Evaluation of functionality testing result.

Testing Code	Testing Name	Result
UT-001-1	Testing of login 1 feature	Success
UT-001-2	Testing of login 2 feature	Success
UT-002	Testing of logout feature	Success
UT-003-1	Testing of see recommendarion 1 feature	Success
UT-003-2	Testing of see recommendarion 2 feature	Success
UT-004-1	Testing of see allow 1 feature	Success
UT-004-2	Testing of see allow 2 feature	Success
UT-005-1	Testing of see allergy and nutrition information 1 feature	Success
UT-005-2	Testing of see allergy and nutrition information 2 feature	Success
UT-006-1	Testing of see growth 1 feature	Success
UT-006-2	Testing of see growth 2 feature	Success
UT-007	Testing of see first aid feature	Success

Testing Code	Testing Name	Result
UT-008	Testing of choose a baby feature	Success
UT-009	Testing of see child data feature	Success
UT-010	Testing of add new baby data feature	Success
UT-011	Testing of change baby data feature	Success
UT-012	Testing of see graphic feature feature	Success
UT-013	Testing of fill growth monitoring card feature	Success
UT-014-1	Testing of see growth monitoring card detail 1 feature	Success
UT-014-2	Testing of see growth monitoring card detail 2 feature	Success
UT-014-3	Testing of see growth monitoring card detail 3 feature	Success
UT-014-4	Testing of see growth monitoring card detail 4 feature	Success
UT-015	Testing of immunization record feature	Success
UT-016	Testing of see immunization data feature	Success
UT-017	Testing of vitamin record feature	Success
UT-018	Testing of see vitamin data feature	Success
UT-019	Testing of choose a baby history feature	Success
UT-020	Testing of see baby data history feature	Success
UT-021	Testing of see graphic history feature	Success
UT-022	Testing of see growth monitoring card detail history feature	Success
UT-023	Testing of see immunization data history feature	Success
UT-024	Testing of see vitamin data history feature	Success
UT-025	Testing of inputting suggestion feature	Success
UT-026	Testing of see suggestion replies feature	Success

6. Conclusion

Out of all processes that have been passed, – design process, implementation process, and testing process – can be concluded that it is possibe to built an application that contains information about nutrition, growth, and treatment of diseases of babies aged 0-2 years old, it is possible to built an application for recording information about the baby's development aged 0-5 years old like a Growth Monitoring Card, and also it is possible to built an application to be used for communication with the health care center.Moreover, this application can provide the correct output results according to the purpose of the application shown by the test results which were success.

7. Future Work

Here are some future works for the development and improvement of the system in the future. Among them are the following.

1. Growth benchmark included with illustrations.
2. Complaints and suggestion feature can be given comments by the mother and the health care center (currently only frequently asked questions).
3. Applications can be run on a platform other than Android.

References

[1] Anggraini R N E, Rochimah S and Dalmi K D 2014 *Mobile Nutrition Recommendation System For 0-2 Year Infant*

[2] Google Play Store 2014 *Baby Food Recipe*

[3] Google Play Store 2014 *Baby Development Track Growth*

[4] Barr A and Feigenbaum E A 1981 *The Handbook of Artificial Intelligence*

[5] Kementrian Kesehatan Republik Indonesia 2010 *Penggunaan Kartu Menuju Sehat (KMS) Bagi Balita*

[6] Yudistira Y 2011 *Membuat Aplikasi iPhone Android & BlackBerry Itu Gampang*
[7] The jQuery Foundation2014*jQuery*
[8] The jQuery Foundation2014*jQuery Mobile*
[9] Leonello C 2009 *jqPlot*

Application Of Decision Tree Approach To Student Selection Model– A Case Study

Harwati[1] and Amby Sudiya[2]

[1]Industrial Engineering, Islamic University of Indonesia, Yogyakarta, Indonesia
E-mail: harwati82@gmail.com

[2]Industrial Engineering, Islamic University of Indonesia, Yogyakarta, Indonesia
E-mail: ambysudiya@gmail.com

Abstract. The main purpose of the institution is to provide quality education to the students and to improve the quality of managerial decisions. One of the ways to improve the quality of students is to arrange the selection of new students with a more selective. This research takes the case in the selection of new students at Islamic University of Indonesia, Yogyakarta, Indonesia. One of the university's selection is through filtering administrative selection based on the records of prospective students at the high school without paper testing. Currently, that kind of selection does not yet has a standard model and criteria. Selection is only done by comparing candidate application file, so the subjectivity of assessment is very possible to happen because of the lack standard criteria that can differentiate the quality of students from one another. By applying data mining techniques classification, can be built a model selection for new students which includes criteria to certain standards such as the area of origin, the status of the school, the average value and so on. These criteria are determined by using rules that appear based on the classification of the academic achievement (GPA) of the students in previous years who entered the university through the same way. The decision tree method with C4.5 algorithm is used here. The results show that students are given priority for admission is that meet the following criteria: came from the island of Java, public school, majoring in science, an average value above 75, and have at least one achievement during their study in high school.

Keywords: Classification, Decision tree, Selection, Criteria, Student

1. Introduction

Data mining is the process of searching for a pattern or interesting information in the selected data by using techniques or methods. With the increasing sophistication of technology, databases are now able to store large capacity data. Data mining is useful whenever the system is dealing with large data sets (Agarwal, et al, 2012). In a data set which is very much present, hidden information stored is an important strategy. Rich data without analysis just makes the database only as a repository for data

[1] To whom any correspondence should be addressed.

that rarely visited. Consequently, important decisions are often made not by the rich data information stored in the database but only based on the intuition of decision makers. Data mining has been widely applied in various fields such as healthcare, manufacturing, banking, marketing, business, security, entertainment and so on. Today has emerged a renewed interest in research that uses data mining in the field of education. This is a relatively new field appears, called Educational Data Mining (EDM). EDM is emerging discipline, concerned with developing methods to explore patterns in the data that is unique to find the knowledge derived from the education domain (Abu Tair and El Halees, 2012). That knowledge is then used to better understand the students and the learning process. In the education system, the amount of hidden information in data stored in the database of institution increased rapidly. This database can contain student records such as registration details, profiles of students and academic achievement.

One data mining techniques that most useful in the field of education is a classification. Classification is the pro-cess of finding a model that describes and differentiates the data classes or concepts, in order to able to use the model to predict the class of the object whose class label is unknown (Umamaheswari and Niraimathi, 2013). Not only in education, but the classification has also been widely applied in various fields such as healthcare, manufacturing, CRM, even in the field of intelligence (Padhy, et al, 2012). The use of this technique on several studies in the field of education have provided benefits to educational institutions. The main objective of the institution is to prepare a quality education for students and to improve the quality of managerial decisions. One way to improve the quality of students is by select the new students more selective. At the Islamic University of Indonesia, admission new students consists of a wide variety of selection paths, one of them is Searching Student with Achievement (SSA). SSA is a pattern of admissions by selection for Academic achievement and interest talent. All the department admit students through the SSA by setting quotas and the selection is determined by the Faculty. Students for Academic Achievement, Student Achievement Division of Sports and Arts, Seed Excellence Scholarship, and Hafiz / Hafidzah Qur'an (www.pmb.uii.ac.id). Industrial Engineering Department at this time, especially not having a standard model selection criteria for SSA. Selection is only done by comparing the application file of prospective students in every registration schedule by management assessment, so that subjectivity is possible to occur because of the lack comparison criteria that can discriminate the quality of prospective students with each other. Certainly this issue demands to be addressed immediately considering SSA selection takes place every year in eight period of admission. If there is no standard model selection, it is highly possible to the injustice in accepting or rejecting new students where prospective students have the same relative performance.

By utilizing the technique of classification, it can be constructed a model selection for new SSA students that contains criteria to certain standards such as the area of origin, the status of the school, the average value of high school grade and so on. Such criteria are determined by using classification rules that appear based on academic achievement, in this case represented by GPA of SSA student in previous years. Classification will be implemented using Decision Tree algorithm C.4.5 with consideration of decision tree models preferred among others because the classification algorithm is an algorithm that easily developed and implemented (Anyanwu and Shiva, 2010). Selection models built is expected to become the standard for the selection of new students in the future, especially for the SSA track. Thus, subjectivity in determining the acceptance of SSA could be avoided and could ultimately homogenize the quality of new students accepted in the Industrial Engineering Department.

2. Aim of the Research

Referring to the problem above, the purpose of this study was to determine the criteria in the admission model of new student from SSA track based on classification academic achievement previous SSA student with decision tree technique.

3. Research Methodology

Classification is a process to find a model that describes or differentiate the concept or class of data, in order to be able to predict the class of an object whose class is not known (Umamaheswari and Niraimathi, 2013). There are several classification techniques that can be used, among other: Decision Tree, rule-based, neural networks, support vector machine, Naive Bayes and nearest neighbour. One method commonly used Data Mining is a decision tree. Decision tree is a flowchart structure that resembles a tree (the tree), where each internal node denotes a test in an attribute, each branch represents a test result, and a leaf node represents a class or class distribution (Sivaram and Ramar 2010). Groove on the decision tree traced from the root node to the leaf node that holds the class prediction for the example. Decision tree is easy to convert to classification rules. The concept of Decision Tree is essentially converts the data into a decision tree and decision rules.

Much of the research on the application of data mining in particular classification techniques in various fields such as the medical field, marketing, manufacturing and in the academic field. Chaurasia and Pal (2013) predict the possibility of cardiac patients by using three classification techniques well as the classification algorithm Naïve Bayes, J48 Decision Tree, and Bagging algorithm. By using 11 attributes such as gender, age, blood pressure, heart rate and other physical conditions, this study managed to classify patients with heart disease by about 85% and the accuracy of the pre-diction time is only about 0.05 seconds. It is very useful to predict the possibility that a person will develop heart dis-ease simply by looking at the previous data record in a short time and accuracy are relatively high

Classification can also be used to predict the possibility of a customer of a bank, insurance and retail (Karim and Rahman, 2013). By using individual data such as status, occupation, education, home ownership status of these studies can predict a person's tendency to be open a savings account, a credit card or become a member in a place of shop-ping. The algorithm used is a C4.5 decision tree algorithm and Naïve Bayes classification as easily applied and under-stood by the relatively high degree of accuracy. Manufacturing also take advantage of this classification technique in decision making. As well as a study in Taiwan conducted on 66 of data to predict the failure of where 43 the data used as the training data set and the data 23 is used as a data testing (Yeh, et al, 2011). With the level of accuracy of 97.6% for the training set and 86.9% for data testing, the study resulted in rules of classification failure moulding process is based on several attributes that occur during the manufacturing process are: Temperature of injection, velocity, time packaging, injection pressure, injection time and so on.

In the academic field, Thomas and Galambos (2004) using a decision tree classification techniques to predict the satisfaction of students in the school. It is done to predict whether the student concerned shall continue education to the next level at the same school or not, as well as to determine what are the attributes that make the students decide to continue their education in the same school. Decision tree can also predict and evaluate the performance of students in a university (Baradwaj and Pal, 2011). The results can help university earlier to identify students were threatened by dropouts so that it can be given special treatment and provide an opportunity for the supervisor to provide guidance and counseling. Osmanbegović and Suljić (2012) have utilized the technique decision tree is compared with the two techniques other classification that Naïve Bayes (NB) and the Multilayer Perceptron (MLP) for predicting the performance of students based on 12 attributes were: gender, origin of high school, the number of families, distance home, scholarship, learning time, the value of entrance exams and so on.

In this study classification techniques will be used to predict the performance of students who entered the university through the SSA where the rules obtained can be input for decision-makers in determining the criteria for acceptance of next new student who entering through the same track.

3.1 Attribute

Attribute functions as the object to be observed and the effect of the changes. In the new student admission system SSA track, some attributes are taken from the database that was available so it is not necessary anymore distributing questionnaires to obtain data. Some of these attributes include:

- Origin
- The origin of school (public or private)
- Department of high school (science or social)
- Average value
- Achievement
- GPA

Data cleansing is done to remove data that has entries imperfect as missing data, invalid data or also just a typo. Cleaning this data will affect the performance of the system because the data will decrease in amount and complexity.

3.2 Profile of Respondent

Data were collected from 179 secondary data of active students who entered through SSA track in the space of 3 years. The profile of the respondents is presented in the following table:

Table 1. Profil of Respondent.

No	Attributes	Profile
1	Gender	Men 61%, Women 39%
2	Origin	West Java 22%, Middle Java 40%, East Java 7,9%, Non Java 29%
3	High School Status	Public 72%, Private 27%, Islamic 6,2%, others 5%
4	Concentration/Majoring	Science 97%, Social 3%
5	Parents Occupation	Gouverment Staff 39%, Private Staff 15,6%, Entrepreneur 30%, Others 15%)
6	Average Value in High School	Excellent (above 75) 51%, Good 49%

Meanwhile, distribution of achievements obtained by the SSA student during his education at the university are presented in the following figure:

Figure 1. Distribution of Student SSA Performance by GPA

4. Result and Discussion

In the process of data classification, C4.5 Decision Tree algorithm is used. This technique is very commonly used to construct a decision tree because it is easy to understand and flexible compared with other algorithms such as Naïve Bayes, CART, J48, and so on. The stages in the classification consists of: Develop tree, tree transform into rules, simplify and test rules, determines the final rules, counting accuracy and accuracy level

4.1. Information gain Calculation

Information gain (IG) is used as a reference for the formation of a branch in the decision tree. IG value indicates the effectiveness of an attribute in classifying data. The value of the highest IG will serve as the root node in the decision tree. IG is calculated using the following formula:

$$IG(S,A) = Entropy\ (S) - \sum_{v \in values\ (A)} \frac{|S_v|}{|S|} Entropy\ (S_v) \qquad \text{.......................} (1)$$

Information Gain calculation for all attributes used is shown in the following table:

Table 1. Information Gain.

No	Attributes	Information Gain
1	Gender	0,001308678
2	Origin	0,016357926
3	High School Status	0,06366159
4	Concentration/Majoring	0,004976685
5	Parents Occupation	0,003094282
6	Average Value in High School	0,005293175
7	Achievement	0,004571812
8	Type of SSA (Academic/Art)	0,025250454

From the result above, it was found that the greatest value of information gain is school status. It means that attribute is the root node that is elected and for other attributes automatically become a certain level of internal node according with the calculation some iterations

4.2. Decision Tree Model

Classification calculations is assisted by application software Rapid Miner. One part of the resulting decision tree model presented in the figure below:

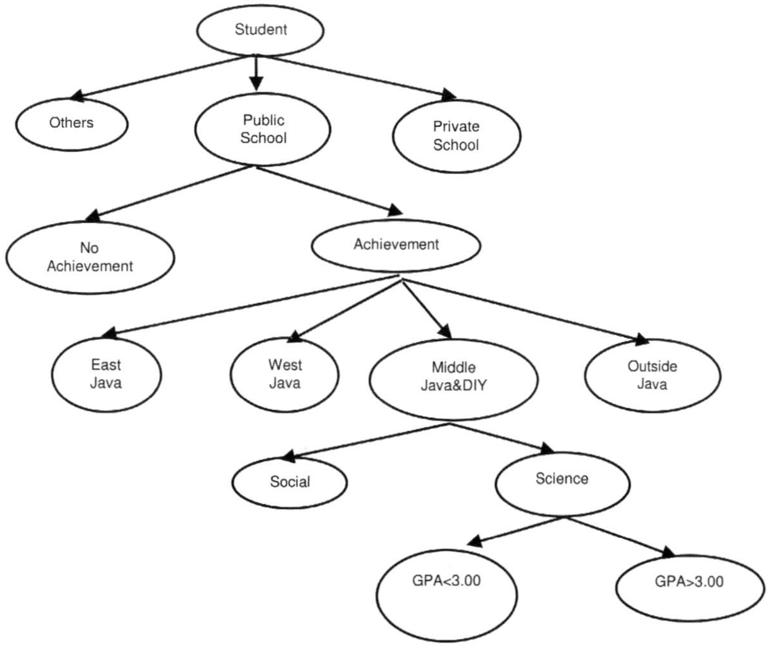

Figure 2. Decision Tree Example

4.3. Rules Construction

There are 50 rules that are formed from the whole decision tree model. Some of the main rules involving all the attributes and characteristics of students with satisfaction GPA are:

- IF the status of the school is the State, the achievement is yes, origin is from Central Java, majoring in school is science: THEN GPA is Satisfactory
- IF the status of the school is the State, the achievements are not, origin is from Central Java Value SSA is Good , the work of parents are civil servants, majoring in school is science, SSA is from Academic Track , gender is Female: THEN GPA is Satisfactory
- IF the status of the school is the State, the achievements are not, origin is from outside Java, the job of parents is the business man , gender is Male, majoring in school is science, Line SSA is Academic: THEN GPA is Not Satisfactory.
- IF status is a State school, the achievement is yes, origin is from outside Java, the work of parents are civil servants, Value SSA is Good: THEN GPA is satisfactory.
- IF the status of the school is private, gender is Male, origin regions are East Java, Value SSA is Very Good parents work is businessman, majoring in school is science, Line SSA is Academic achievements are not: THEN THEN GPA is satisfactory
- IF is a private school status, gender is female, the region of origin is Central Java, the job of parents is businessman: THEN GPA is satisfactory

Table 1. Attributes for Satisfactory GPA.

No	Attributes	Category
1	Gender	Men, Women
2	Origin	Java
3	High School Status	Public
4	Concentration/Majoring	Science

5	Parents Occupation	Every Category
6	Average Value in High School	> 75
7	Achievement	Yes
8	Type of SSA (Academic/Art)	Academic

4.4. Rules Construction

Accuracy level the proximity of measurement of the quantity of the actual values. In decision tree accuracy level is calculated by this following formula:

$$\text{Accuracy Level} = \frac{\text{Number of correct Prediction}}{\text{Total Prediction}} \text{x } 100\%$$

$$\text{Accuracy Level} = \frac{15}{90} \text{x } 100\%$$

Accuracy Level = 22%

The result shows that the prediction accuracy rate is above 22%. This means the possibility of failure prediction is 78%. Low levels of accuracy can be caused by several things, including the amount of data that is too few, too many variables, or even may be due to the random pattern of the respondents that are difficult to predict based on existing data.

5. Conclusion

From all the analysis above, it can be concluded that although the accuracy rate is very low but it can be predicted the characteristics of students form SSA admission that may be able to produce a good performance. These characteristics are came from the island of Java, public school, majoring in science, an average value above 75, and have at least one achievement during their study in high school.

6. Acknowledgement

This research was supported by a grant of Industrial Engineering Department of Universitas Islam Indonesia. The data, statement, and the view expressed are solely the responsibility of the authors.

References

Abu Tair Mohammed M., El-Halees M Alaa, 2012, Mining Educational Data to Improve Students' Performance: A Case Study , ISSN 2223-4985 Volume 2 No. 2.

Agarwal Sonali, Pandey G. N., Tiwari M. D, 2012, Data Mining in Education: Data Classification and Decision Tree Approach International Journal of e-Education, e-Business, e-Management and e-Learning, Vol. 2, No. 2.

Anyanwu N Matthew, Shiva G Sajjan G, 2010, Comparative Analysis of Serial Decision Tree Classification Algorithm International Journal of Computer Science and Security, (IJCSS) Volume (3): Issue (3)

Ayinde A., Dr Adetunji., Bello and Odeniyi, 2013 Performance Evaluation of Naive Bayes and Decision Stump Algo-rithms in Mining Students' Educational Data, IJCSI International Journal of Computer Science Issues, Vol. 10, Issue 4, No 1

Baradwaj K Brijesh, Pal Saurabh, 2011, Mining Educational Data to Analyze Students" Performance, (IJACSA) In-ternational Journal of Advanced Computer Science and Applications, Vol. 2, No. 6, page 63

Baradwaj K Brijesh, Pal Saurabh, 2011, Data Mining: A Prediction For Performance Improvement Using Classifica-tion, (IJCSIS) International Journal of Computer Science and Information Security, Vol. 9, No. 4

Chaurasia Vikas, Pal Saurabh, 2013, Data Mining Approach to Detect Heart Dieses, International Journal of Ad-vanced Computer Science and Information Technology (IJACSIT) Vol. 2, No. 4, Page: 56-66, ISSN: 2296-1739

Gubosh Soumi , Kumal K Sanjay, 2013, Comparative Analysis of KMeans and Fuzzy C Means Algorithms, IJACSA) International Journal of Advanced Computer Science and Applications,Vol.4, No.4

Kalyani M Raval, 2012, Data Mining Techniques, International Journal Of Advanced Research In Computer Science And Software Engineering Volume 2, Issue 10.

Karim Masud, Rahman M Rashedur , 2013, Decision Tree and Naïve Bayes Algorithm for Classification and Genera-tion of Actionable Knowledge for Direct Marketing, Journal of Software Engineering and Applications, 6, 196-206.

Keim A Daniel, 2002, Information Visualization and Visual Data Mining, IEEE Transactions On Visualization And Computer Graphics, Vol. 7, No. 1

Kotsiantis Sotiris, Kanellopoulos Dimitris, Association Rules Mining: A Recent Overview, GESTS International Trans-actions on Computer Science and Engineering, Vol.32 (1), pp. 71-82

N Sivaram, K Ramar, 2010, Applicability of Clustering and Classification Algorithms for Recruitment Data Mining, International Journal of Computer Applications (0975 – 8887) Volume 4 – No.5.

Osmanbegović Edin, Suljić Mirza, 2012, Data Mining Approach For Predicting Student Performance, Economic Re-view – Journal of Economics and Business, Vol. X, Issue 1.

Padhy Neelamadhab, Mishra Pragnyaban , Panigrahi Rasmita, 2012, The Survey of Data Mining Applications And Feature Scope, International Journal of Computer Science, Engineering and Information Technology (IJCSEIT), Vol.2, No.3

Pujari Pushpalata, Gupta Jyoti Bala, 2012, Exploiting Data Mining Techniques For Improving The Efficiency Of Time Series Data Using Spss-Clementine, Journal of Arts, Science & Commerce E-ISSN 2229-4686ISSN 2231-4172, ol.– III, Issue 2(3).

Sahu Hemlata, Shrma Shalini, Gondhalakar Seema, 2012, A Brief Overview on Data Mining Survey, International Journal of Computer Technology and Electronics Engineering(IJCTEE) Volume 1, Issue 3:114

Thomas H Emily, Galambos Nora, 2004, What Satisfies Students? Mining Student-Opinion Data with Regression and Decision Tree Analysis Research in Higher Education, Vol. 45, No. 3

Umamaheswari K, S. Niraimathi, 2013, A Study on Student Data Analysis Using Data Mining Techniques, Interna-tional Journal of Advanced Research in Computer Science and Software Engineering, Volume 3, ISSN: 2277 128X Issue 8.

Yeh Duen-Yian, Cheng Ching-Hsue, Hsiao Shih-Chuan, 2011, Classification Knowledge Discovery In Mold Tooling Test Using Decision Tree Algorithm,) 22:585–595 DOI 10.1007/s10845-009-0321-7

Performance Test of Various Types of Antenna Arrays in Real Propagation Environment

SetiyoBudiyanto[1], BenyNugraha[2] and Dian WidiAstuti[3]

[1,2,3] Department of Electrical Engineering,
Universitas Mercu Buana, Jakarta, Indonesia

E-mail: budiys1@gmail.com, beny.nugraha@gmail.com, dian_widia1@yahoo.com

Abstract. The research was conducted on various types of antenna arrays namely Uniform Array, Binomial Array, Dolph-Chebyshev Array, and Taylor Array. This research is done in the real propagation environment in order to define precisely the number of antenna elements, the distance between the elements, the angle of the antenna arrays, the side lobe level and the n-bar array distribution. The testing process is done by using Matlab and the Non-Uniform Array Simulation Program. The results obtained for various types of antenna arrays are as follows: On Uniform Array produces Half Power Beam Width (HPBW) of 10.152° and directivity of 10 dB, on Binomial Array generates Half Power Beam Width (HPBW) of 20.245° and directivity of 7.47 dB, on Dolph-Chebyshev Arrayproduces Half Power Beam Width (HPBW) of 20.304° and directivity of 4.0185 dB, and on Taylor Arrayproduces Half Power Beam Width (HPBW) of 12.78° and directivity of 8.9 dB.
Keywords—Binomial Array; Dolph-Chebyshev Array; Real Propagation; Taylor Array; Uniform Array

1. Introduction

In communication world, antennas are considered as the most important component of a fully-functional communication process. An antenna is defined as a tool that is used to transmit and receive electromagnetic wave, this wave is transmitted from the transmitter to the receiver. A good antenna design will lead to the increase of the whole communication system's performance. One example is the design of an array antenna. Several antennas are arrayed according to certain geometric and electrical configuration to form an antenna array. In order to simplify the analysis, synthesis and fabrication process, the antennas are arrayed with the same type, such as array dipole, array waveguide and array microstrip.[1-3]

In this paper, antenna arrays with four different types are designed, these types are Uniform Array, Binomial Array, Dolph-Chebyshev Array, and Taylor Array. Each type has different characteristic, thus, the design between each type will vary. The simulation process is done in the real propagation environment, thus, the number of antenna elements, the distance between the elements, the angle of the antenna arrays, the side lobe level and the n-bar array distribution can be defined beforehand. From the simulation results, it can be defined that there is a relation between the number of antenna elements

[1] To whom any correspondence should be addressed.

and the directivity, when the number of antenna elements increased then the electromagnetic wave that is generated will be more direct [4]. In [5], research on the characteristic of the non-uniform antennas using firefly algorithm is conducted; in [6], research regarding the integration between antenna array and solar cell multicrystal silicon is done; in [7], the multicoupling of the antenna array is researched; in [8], research of antenna array using the stepped and half bow-tie slotted is conducted; in [9], genetic algorithm is used to research the non-uniform antennas; in [10], research regarding the non-uniform antennas concentric circular is done; in [11], research regarding the non-uniform antenna array spacing which focused on the travelling rectangular design is conducted. Based on all of the above references, there is a research opportunity regarding the Half Power Beam Width (HPBW) comparison between the uniform array antennas and the non-uniform array antennas, such as Binomial array, Dolph-Chebysev array, and Taylor array. HPWB in an antenna is an angle between the mainlobe and other lobes, which can be vertical or horizontal beam.

The paper is organized as follows: in Section 2, the theory of antenna arrays in the real propagation environment is discussed, the simulation process is described in Section 3, in Section 4, the simulation results as well as its analysis are discussed, finally, the conclusion is presented in Section 5.

2. Antenna Arrays in Real Propagation Environment

As mentioned in Section I, the simulation process is done in the real propagation environment so that the value of the number of antenna elements, the distance between the elements, the angle of the antenna arrays, the side lobe level and the n-bar array distribution can be defined first and also can be changed to differentiate the results between each value.There are four antenna arrays to be designed, and they can be divided in to two types which are the Uniform Array and the Non-uniform Arrays (Binomial array, Dolph-Chebyshev Array, and Taylor Array.

2.1. Uniform Array

For the Uniform Array, there are two alignment types that are tested in this paper, they are the Broadside Array and Endfire Array.

The Broadside Array alignment has a maximum radiation angle value, ϕ, of 90° [4]. The radiation pattern for Broadside Array alignment can be seen in Figure. 1

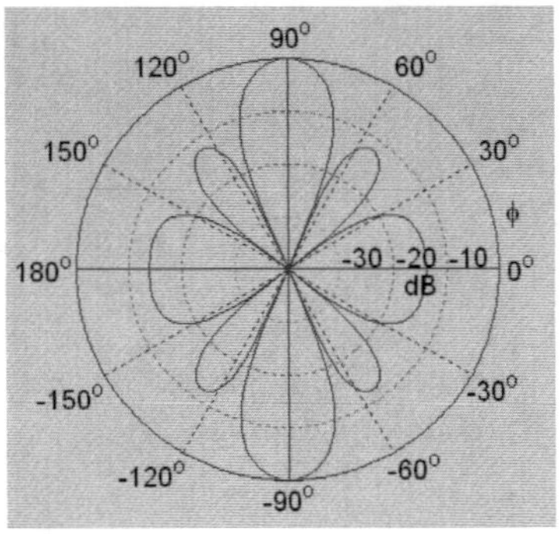

Figure1. Radiation Pattern for Broadside Array

The following equations can be used in order to calculate the value of HPBW and the value of Directivity for Broadside Array:

$$HPBW \approx 50.76 \times \frac{\lambda}{Nd} \tag{1}$$

$$D_{dB} = 10\log\left[2N\left(\frac{d}{\lambda}\right)\right] \tag{2}$$

The Endfire Array alignment has a maximum radiation angle value, ϕ, of 0° or 180° [4]. The radiation pattern for Endfire Array alignment can be seen in Figure. 2.

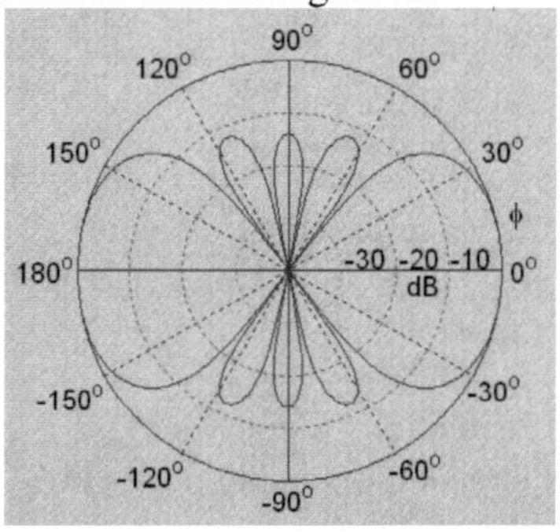

Figure2. Radiation Pattern for Endfire Array

The following equations can be used in order to calculate the value of HPBW and the value of Directivity for Endfire Array:

$$HPBW \approx \left(2\sqrt{0.886\frac{\lambda}{Nd}}\right) \times \frac{180}{\pi} \tag{3}$$

$$D_{dB} = 10\log\left[4N\left(\frac{d}{\lambda}\right)\right] \tag{4}$$

The simulation process is done in the real propagation environment so that the value of the number of antenna elements, the distance between the elements, and the angle of the antenna arrays can be defined before the simulation and also can be changed to differentiate the results between each value.

2.2. Non-Uniform Array

A non-uniform array antenna is an antenna array that has the same distance between elements but has different amplitude. The non-uniform arrays that are used in this research are Binomial Array, Dolph-Chebyshev Array, and Taylor Array.

2.2.1. Binomial Array

Binomial Array generates a really small sidelobe, even for the parameter d (distance between elements) value smaller than $\lambda/2$ the sidelobe is not visible, however Binomial Array generates a wide mainlobe.

Equations can be used in order to calculate the value of HPBW and the value of Directivity of Binomial Array are as follows:

$$HPBW \approx \frac{1.06}{\sqrt{N-1}} \tag{5}$$

$$D_{dB} = 10\log\left(1.77\sqrt{N}\right) \tag{6}$$

Due to its characteristic, in order to find an optimum result, the distance between elements should be set to $d \leq \lambda/2$.

2.2.2. Dolph-Chebyshev Array

Dolph-Chebyshev Array is more flexible than Uniform Array and Binomial Array because the sidelobe can be adjusted freely. In order to obtain an optimum result, the sidelobe should be adjusted to the uniform sidelobe.

In order to calculate the value of HPBW and the value of Directivity for Dolph-Chebyshev Array, the following equations can be used:

$HPBW = f\left(hpbw_{uniform-array}\right)$, where f is the Bordening Factor: $f = 1 + 0.632\left[\frac{2}{R}\cosh\sqrt{\left(\cosh^{-1} R\right)^2 - \pi^2}\right]^2$,

and R is the Sidelobe ratio.

$$D \approx \frac{101.5}{HPBW}$$

2.2.3. Taylor Array

Taylor Array has similar characteristic to theDolph-Chebyshev Array, its sidelobe can be adjusted to the uniform sidelobe to achieve optimum result, however, only some of the first sidelobe that can be adjusted in the Taylor Array.

In order to obtain the value of HPBW and the value of Directivity for Taylor Array, the following equations can be used:

$$\Theta_0 \simeq 2\sin^{-1}\left\{\frac{\lambda\sigma}{\pi l}\left[(\cosh^{-1} R_0)^2 - \left(\cosh^{-1}\frac{R_0}{\sqrt{2}}\right)^2\right]^{1/2}\right\}$$

$$\sigma = \frac{\bar{n}}{\sqrt{A^2 + \left(\bar{n} - \frac{1}{2}\right)^2}}$$

$$D \approx \frac{101.5}{HPBW}$$

3. Simulation Process

The whole simulation process usedwith Mathlab sofware, the flowchart for the simulation process can be seen in Figure. 3.

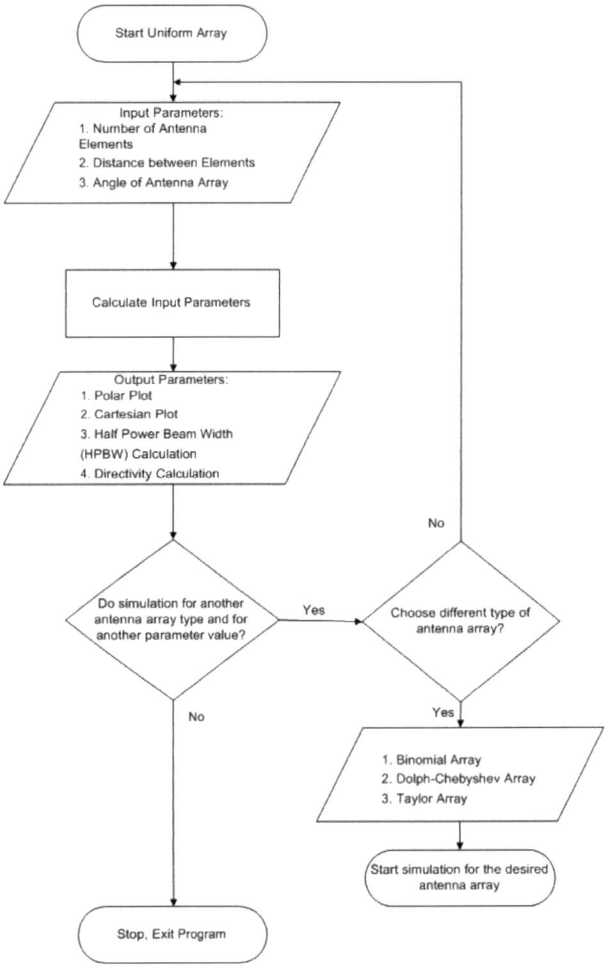

Figure3. Simulation Process Flowchart

The initial process of the simulation is by starting the simulation for the Uniform Array. As this process is done in the real propagation environment, the next step is, the value of the parameters such as number of antenna elements, distance between elements and the angle of the antenna array is defined. The third and the fourth step calculate the input value to get the Polar and Cartesian plot and also calculate the value of HPBW and directivity.

Once the simulation for the Uniform Array is done, the next simulation can be run. For the next simulation, another type of antenna array can be selected and the value of the input parameters can be changed as well.

The input parameters that can be changed are number of antenna elements (N), distance between elements (d), and angle of antenna array (ϕ).

IOP Conf. Series: Materials Science and Engineering **105** (2016) 012015 doi:10.1088/1757-899X/105/1/012015

4. Result and Analysis

This Section presents the results from the simulation process as well as the analysis for all of the results.

4.1. Uniform Array

For the Uniform Array, there are two alignment types that are tested in this research namely the Broadside Array and Endfire Array.

For Broadside Array alignment, the parameters that are defined first are: $N = 5$, $d = 1/2\lambda$, and $\phi = 90^{o}$. The result of this simulation can be seen in Figure. 4.

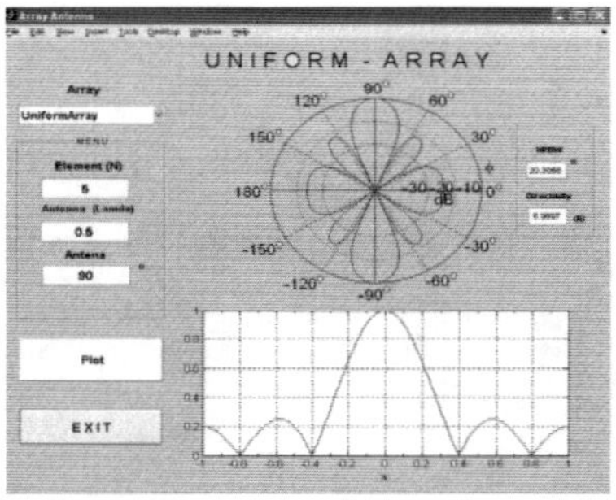

Figure4. Broadside Array Simulation Result for$N = 5$

For the next simulation, the parameters are changed as follows: $N = 10$, $d = 1/2\lambda$, and $\phi = 90^{o}$. The result can be seen in Figure. 5.

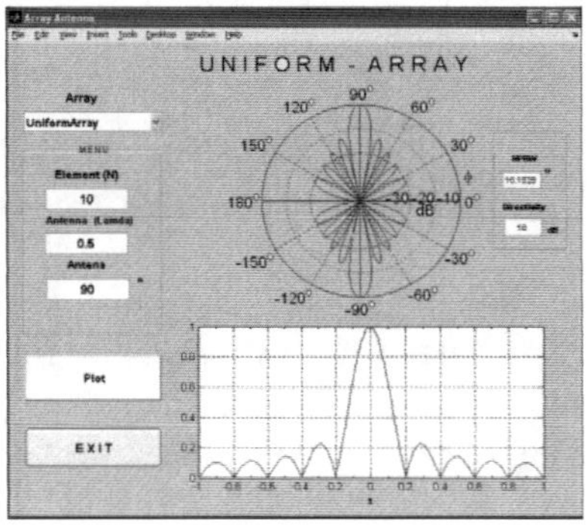

Figure5. Broadside Array Simulation Result for $N = 10$

It can be seen from the above simulations, for Broadside Array, the more the number of the elements, the larger the Directivity value but the sidelobe (value of HPBW) became smaller.

For Broadside Array alignment, the parameters that are defined first are: $N = 5$, $d = 1/2\lambda$, and $\phi = 0^{o}$ and 180°. The result of this simulation can be seen in Figure. 6.

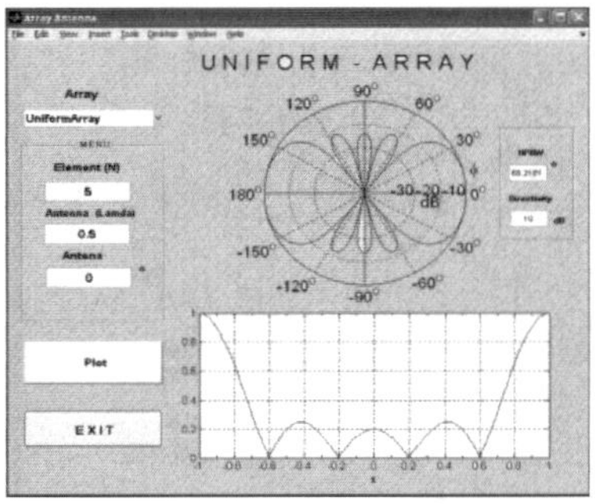

Figure6. The Endfire ArraySimulation Resultfor $N = 5$

For the next simulation, the parameters are changed as similar to the second simulation of Broadside Array: $N = 10$, $d = 1/2\lambda$, and $\phi = 90^{\circ}$. The result can be seen in Figure. 7.

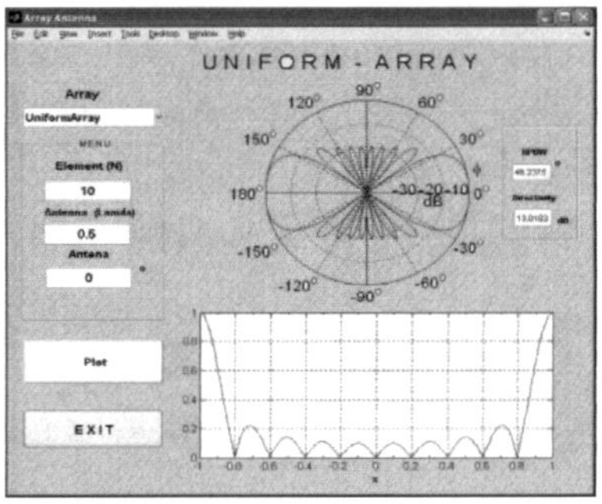

Figure7. The Endfire Array Simulation Resultfor $N = 10$

From the above simulations, it can be concluded that the larger the number of the elements, the smaller the HPBW value and the larger the Directivity Value.

4.2. Binomial Array

For Binomial Array, the parameters that are defined first are the following: $N = 5$, $d = 1/2\lambda$, and $\phi = 90^{\circ}$. The result of this simulation can be seen in Figure. 8.

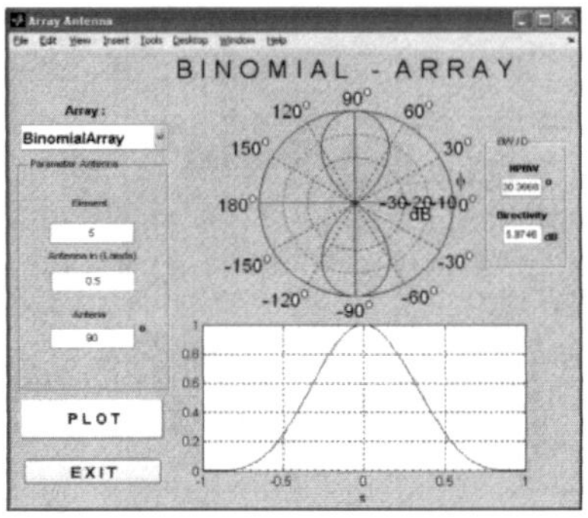

Figure8. The Binomial ArraySimulation Resultfor$N = 5$

For the next simulation, the parameters are changed as follows: $N = 10$, $d = 1/2\lambda$, and $\phi = 90^{\circ}$. The result can be seen in Figure. 9.

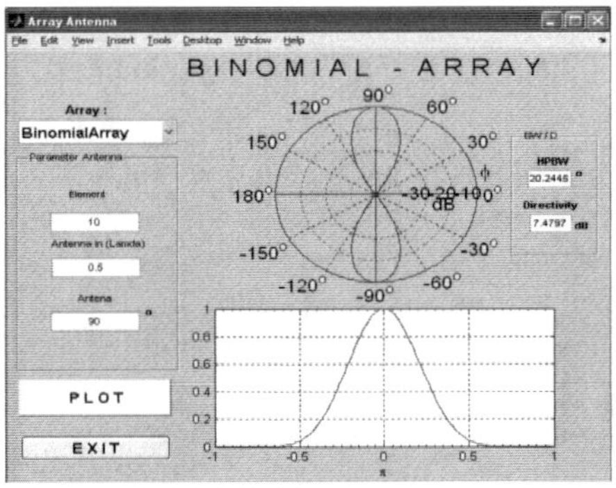

Figure9. Binomial ArraySimulation ResultforN = 10.

From the above simulations, it can be concluded that Binomial Array has the same characteristic as the Uniform Array, which is the larger the number of the elements, the smaller the HPBW value and the larger the Directivity Value.

4.3. Dolph-Chebyshev

For Dolph-Chebyshev Array, the sidelobe is adjusted first to get an optimum result. The parameters are as follows: $N = 5$, $d = 1/2\lambda$, and sidelobe, $R = 26$ dB and $R = 30$ dB. For $R = 26$ dB, the result can be seen in Figure 10.

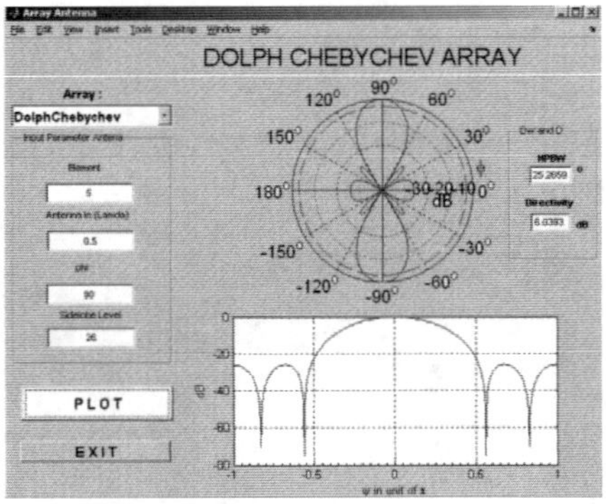

Figure10. Dolph-Chebyshev Array Simulation Result for$R$ = 26 dB

For R = 30 dB, the result can be seen in Figure. 11.

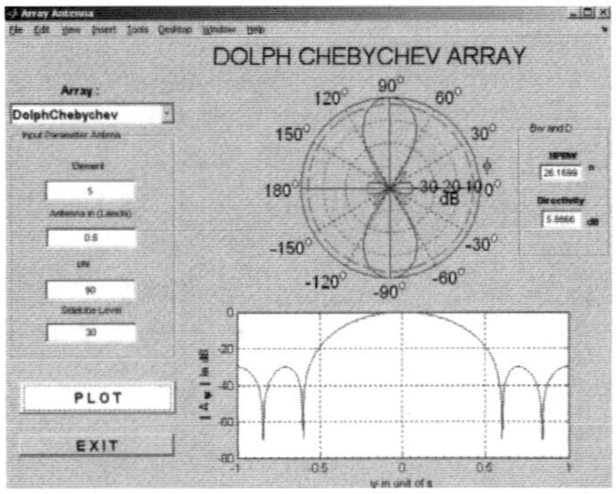

Figure11.Dolph-Chebyshev Array Simulation Result for R = 30 dB

From the above simulations, it can be concluded that with the same value of number of elements, the larger the value of the sidelobe will result to the larger the value of the HPBW and the smaller the value of the Directivity.

4.4. Dolph-Chebyshev

It has been mentioned in Section II that Taylor Array has similar characteristic to the Dolph-Chebyshev Array, its sidelobe can be adjusted as desired. Thus, for Taylor Array, it is the number of the element that is changed, the sidelobe value remains unchanged.

For the first simulation, the parameters are the following: N = 10, d = $1/2\lambda$, and R = 25 dB. The result is presented in Figure 12.

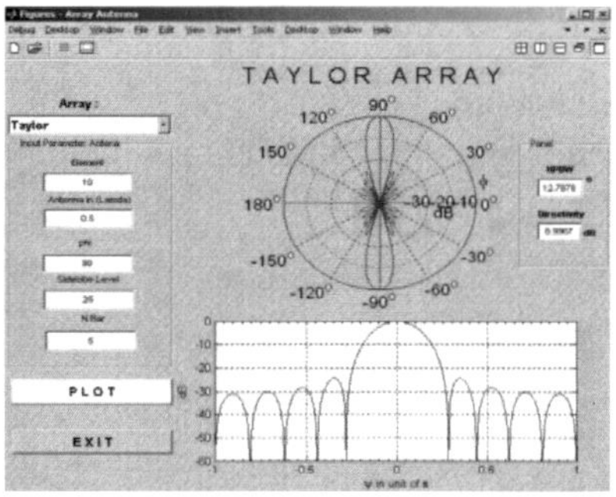

Figure12. The Taylor Array Simulation Result for$N$ = 10

For the second simulation, the parameter that is changed is only the number of elements, it is changed to N = 15. The result can be seen in Figure 13.

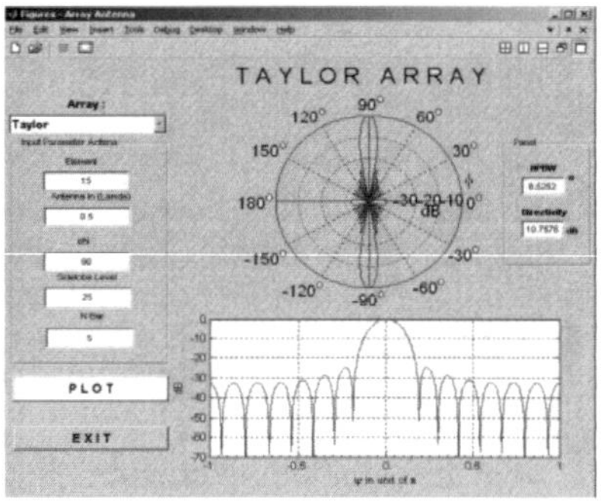

Figure13. The Taylor Array Simulation Result for N = 15

It can be concluded from the above simulations that the larger the number of the elements, the larger the Directivity value and the smaller the HPBW value.

The simulation results for all types of array and for all scenarios can be seen in Table1.

TABLE 1. SIMULATION RESULTS

Array Type	Output Value	
	Half Power Beam Width (HPBW)	***Directivity***
Uniform Array – Broadside Array Alignment (N=5, d=1/2λ)	20.5036	6.9897 dB
Uniform Array – Broadside Array Alignment (N=10, d=1/2λ)	10.152°	10 dB
Uniform Array – Endfire Array Alignment (N=5, d=1/2λ)	68.2181°	10 dB
Uniform Array – Endfire Array Alignment (N=10, d=1/2λ)	48.2375°	13.01 dB
Binomial Array (N=5, d=1/2λ)	30.367°	5.97 dB
Binomial Array (N=10, d=1/2λ)	20.245°	7.47 dB
Dolph-Chebyshev Array (N=5, d=1/2λ, R = 26 dB)	25.2581°	6.041 dB
Dolph-Chebyshev Array (N=5, d=1/2λ, R = 30 dB)	26.1699°	5.8666 dB
Taylor Array (N=10, d=1/2λ)	12.78°	8.9 dB
Taylor Array (N=15, d=1/2λ)	8.5252°	10.7576 dB

5. Conclusion

There are two alignments for the Uniform Array, it can be concluded thatfor Broadside Array, the more the number of the elements, the larger the directivity value, but the sidelobe (value of HPBW) became smaller, meanwhile for the Endfire Array, it can be concluded that the larger the number of the elements, the smaller the HPBW value and the larger the Directivity Value.From the simulation results, it can be concluded that theBinomial Array has the same characteristic as the Uniform Array, which is the larger the number of the elements, the smaller the HPBW value and the larger the Directivity value.Fromthe Dolph-Chebyshev Array simulations it can be concluded that with the same value of number of the elements, the larger the value of the sidelobe will result to the larger the value of the HPBW and the smaller the value of the Directivity. The last simulation for the Taylor Array, itcan be concluded that the larger the number of the elements, the larger the Directivity value and the smaller the HPBW value.

6. References

[1] Budiyanto, S., Asvial, M., Gunawan, D.,"Implementation Dedicated Sensing Receiver (DSR) in 3G - WiFi Offload", Proceedings - 2014 International Conference on Smart Green Technology in Electrical and Information Systems: Towards Greener Globe Through Smart Technology, ICSGTEIS 2014

[2] Daniel J. Ludick, Rob Maaskant, David B. Davidson, Ulrich Jakobus, Raj Mittra, and Dirk de Villiers,"Efficient Analysis of Large Aperiodic Antenna Arrays Using the Domain Green's Function Method", IEEE Transactions on Antennas and Propagation, VOL. 62, NO. 4, APRIL 2014, pp. 1579 – 1588

[3] P. Brahma, P. Nandi, A. Senapati and J. S. Roy, "Reduction of Side Lobe Level of Thinned Phased Array Antenna using Genetic Algorithm", International Journal of Computer Applications Volume 112 – No. 8, February 2015, pp. 13 – 15

[4] Amer Tawfeeq Abed,"Study of Radiation Properties in Taylor Distribution Uniform Spaced Backfire Antenna Arrays", American Journal of Electromagnetics and Applications, 2014, pp. 23-26

[5] Mohammad Asif Zaman and Md. Abdul Matin, " Nonuniformly Spaced Linear Antenna Array Design Using Firefly Algorithm", International Journal of Microwave Science and Technology Volume 20 No. 12, 2012

[6] O. O'Conchubhair, P. McEvoy, and M. J. Ammann,"Integration of Antenna Array with Multicrystalline Silicon Solar Cell", IEEE Antennas and Wireless Propagation Letters, Vol. 14, 2015 pp. 1231 - 1234

[7] Hoi Shun Lui, Hon Tat Hui, and Marek Bialkowski,"Mutual Coupling in Antenna Arrays 2011",Editorial of International Journal of Antennas and Propagation Volume 2012 (2012)

[8] Ankit Sharma and Ravi Mohan,"Antenna Array with Stepped & Half Bow-Tie Slotted Microstrip Rectangular Patch Elements", International Journal of Communication Engineering and Technology.Volume 4, Number 1 (2014), pp. 1-6

[9] Ridwan, M. ; Abdo, M. ; Jorswieck, E." Design of non-uniform antenna arrays using genetic algorithm",13th International Conference on Advanced Communication Technology (ICACT), 2011 , Seoul 13-16 Feb. 2011, pp. 422 – 427

[10] Sharaqa, A.H.; Dib, N.I.,"On the optimal design of non-uniform concentric circular antenna arrays", IEEE Antennas and Propagation Society International Symposium (APSURSI), 2012 pp. 1 – 2

[11] Oraizi, H.; Behbahani, A.K.; Noghani, M.T.; Sharafimasouleh, M.,"Optimum design of travelling rectangular waveguide edge slot array with non-uniform spacing", IET Microwaves, Antennas & Propagation, Year: 2013, Volume: 7, Issue: 7 Pages: 575 - 581

Integrated Model for E-Learning Acceptance

Ramadiani[1], A Rodziah[2.1], S M Hasan[2.2], A Rusli[2.3], and C Noraini[2.4]
[1]Computer Science, Natural Resources and Mathematic Faculty, Mulawarman University, East Kalimantan, Indonesia.
[2]Information System, Computer Science and Information Technology Faculty, Universiti Putra Malaysia, Malaysia.
E-mail: mmi_ugm04@yahoo.com, rodziah@upm.edu.my, mhbselamat@gmail.com, rusli@upm.edu.my, norainip@upm.edu.my

Abstract E-learning is not going to work if the system is not used in accordance with user needs. User Interface is very important to encourage using the application. Many theories had discuss about user interface usability evaluation and technology acceptance separately, actually why we do not make it correlation between interface usability evaluation and user acceptance to enhance e-learning process. Therefore, the evaluation model for e-learning interface acceptance is considered important to investigate. The aim of this study is to propose the integrated e-learning user interface acceptance evaluation model. This model was combined some theories of e-learning interface measurement such as, user learning style, usability evaluation, and the user benefit. We formulated in constructive questionnaires which were shared at 125 English Language School (ELS) students. This research statistics used Structural Equation Model using LISREL v8.80 and MANOVA analysis.

1. Introduction

E-learning is a method of learning that is offered by many universities and educational institutions. Using e- learning is expected to improve services and educational outreach to the many remote areas. Especially for companies and institutions that have branches all over the world. They need e-learning to train employees and serve customers. E-learning is a distance learning system which offers training courses, and custom tailored to the needs of learners [1]. E-learning will not work if the system is not used in accordance with user needs. It is a measure allowing users to progress at their own speed [2].

E-learning is expected to facilitate the weakness that existed or found in the conventional education. Through e-learning is hoped education become more accessible, cheaper, more fun and easier to share and to learn [3]. But in reality, the e-learning was not entirely successful as expected. The success of e- learning process, like any other product depends greatly on learners' satisfaction and some other factors that will encourage learners' intention for continuous use [4]. The e-learning system implementation at universities has encouraged a number of studies. Yet, the influence of various students' experiences relating to their satisfaction and continuance intention is not clearly understood [5].

Why e-learning does not function properly because it is not in accordance with the user need (perceived usefulness and perceived ease of use) [2]. The purpose of application is to provide ease of use, not to answer the difficulties were faced by students [7]. The important to determine whether application useful or not is usability.

In many projects, the problem with the user interface is not visible until the system is ready for use and the users complete their work. At that time, may be very costly to make changes. There are many theories that discuss the interface evaluation design, but they are not integrated into a single conceptual framework that facilitates their usage by developer [6]. Various problems will arise in the design of the user interface to suit the needs of e-learning users.

The effectiveness of learning and design of the interface will be significantly related. Interface is used to communicate with the user in an interactive system. E-learning will be less than optimal if the system is not effectively used in accordance with user needs. Well-designed interface, as good educators, and instructional materials, must make a "teacher-student" relationship that guides the user to learn and enjoy what they do [2]. Good interface can be challenging for users to expand their understanding of user interfaces and computer systems. Thus, the user interface is very important for designers to understand and be aware of user habits, physiological, and the ability of the user.

This study aims to identify and measure the variables to develop e-learning users interface model based on these reasons;

a) Many theories had discuss about the interface usability evaluation and user technology acceptance separately, actually why we do not integrate between interface usability evaluation and user acceptance to enhance e-learning process. Therefore, the evaluation model for e-learning interface acceptance is considered important to investigate.
b) How to develop the integrated e-learning user interface evaluation model, based on the user interface acceptance measurement and the structural model. Is there any influence of user learning style, usability function and user benefits to success e-learning interface?

2. Phases of research

In generally, there are three phases in this research activities; first phase is information gathering and formulation, the second phase is estimation fit model and the last phase is model prototyping and evaluation. When designing an e-learning interface, instructors are faced with many considerations and decisions that consequently affect how students experience instruction construct and process knowledge. These decisions related to the didactic design of e-learning course may refer to one of five fields of instruction [8].

2.1. Model Design

Model design includes organization of the research activity in model proposing which are most likely to achieve this study objective. It can be described as a blue print to perform this model, which involves the description of model approach, information gathering, user needs identification, envisioning and evaluation, requirement specification, estimation, assessment, modification, fit model, prototype developing, testing, evaluation and model contribution, also sampling size, tool and methods of data collection to answer specific research questions or for testing research hypothesis. The purpose of model design is to provide answers to model research questions and to control model research variance.

These are the steps for developing UIA model in this research;

- Developing good theoretical Model and some hypotheses to represent that model
- Case screening, missing data and unengaged responses

- Missing variable, Skewness, and Kurtosis
- Exploratory factor analysis
- Clean pattern matrix, convergent validity, discriminate validity, reliability
- Confirmatory factor analysis and Structural model

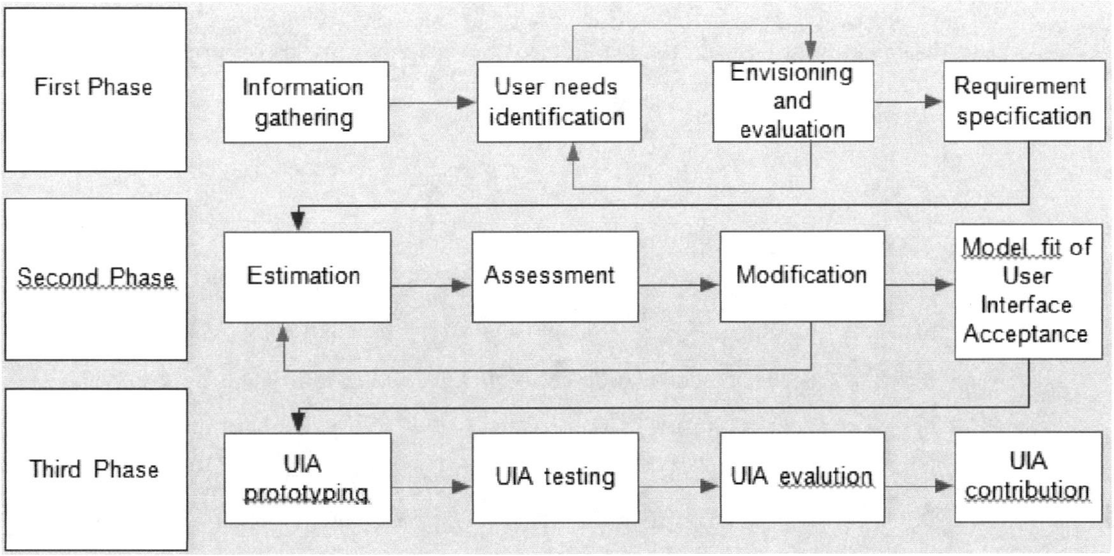

Figure1. Research Activities

We can illustrate this research activities in Figure 1 which each process has some input and output data. The researcher collected the learner problem in e-learning interface acceptance. The research activities such as; information gathering, user needs identification, envisioning and evaluation, and formulate the user requirement specification for research hypothesis. User needs identification in this research was collected from observation and case study at Language Technology Centre in ELS in UPM.

2.2. First Phase of Research

Understanding user requirement is an integral part of information system design and is critical to the success of interactive system. The problem of this study is how to develop integrated model in e-learning interface system. The system is benefits if it was developed with understanding the user needs and requirement. It can include; increase productivity, enhance quality of work, reduction in support and training costs, to encourage their knowledge and to help them finishing their task, and improve user satisfaction.

2.2.1. Information Gathering. This step is related to the initial formation of the structural equation model, prior to estimation. Initial model was formulated based on a previous theory or research. In this study, the previous researches were used as references, for instance, user learning style and usability evaluation. Especially user benefit is the original theory for these research hypotheses. There are some factors that need to be prepared in order to test hypotheses to represent the issues contained in this research. This research model consists of:

- User Learning Style [9]; Learning style, motivation and user knowledge
- Usability [3]; Know-ability, operability, efficiency, robustness, safety and subjective satisfaction
- *User Benefits*; Media elements, communicativeness and user expectation

In a confirmatory technique, the model must be specified correctly based on the type of analysis that the researcher is attempting to confirm. When building the correct model, we use two kinds of variables, exogenous and endogenous variables. In this model, a variable can act as both independent and dependent variable. Two main components of models are distinguished in SEM [10].

2.2.2. User Need Identification.

Based on the previous evaluations and theories about the interface acceptance, we developed hypotheses and designed new model then followed by making some in Table 1 there are fifteen hypotheses in this research, twelve hypotheses for measurement model and three hypotheses for structural model.

Table 1 Research Hypothesis

Hypothesis	Research Hypothesis
*H*1 10	There is no correlation between User Learning style and User's Style
H 2 10	There is no correlation between Motivation and User's Style
H 3 10	There is no correlation between Knowledge ability and User's style
H 4 20	There is no correlation between Know-ability and Usability
H 520	There is no correlation between Operability and Usability
H 620	There is no correlation between Efficiency and Usability
H 720	There is no correlation between Robustness and Usability
H 820	There is no correlation between Safety and Usability
H 920	There is no correlation between Subjective Satisfaction and Usability
*H*10 30	There is no correlation between Media elements and User Benefits
*H*11 30	There is no correlation between Communicativeness and User Benefit
*H*12 30	There is no correlation between User expectation and User Benefits
*H*13 40	There is no correlation between User's Style and UI Acceptance
*H*14 40	There is no correlation between Usability and UI Acceptance
*H*15 40	There is no correlation between User Benefit and UI Acceptance

2.2.4. Research Question Design.

This research has twelve indicator variables and each indicator variable has three criteria. Totally the variables in this research are 36 attributes as shown in Table 2

Table 2 Research Attributes

Variables	Goals	Attributes
Y1	Learning style	CIEP Level, Study habits
Y2	Motivation	frequency of learning time, finishing study,
Y3	Knowledge ability	grades
Y4	Know-ability	learnability, understandability, memorability
Y5	Operability	ease of use, effectiveness, flexibility
Y6	Efficiency	user workload, efficiency, productivity

Y7	Robustness	error management, trustfulness, errors
Y8	Safety	safety, secure, comforTable
Y9	Subjective Satisfaction	Attractiveness, Compliance, Satisfaction
Y10	Media element	Usefulness, Completeness, Increase
Y11	Communicativeness	Simple, Intuitive, Perceptive
Y12	User Expectation	User need, Capability, Expectation

2.2.5. Data Collection Method. The research was held in July 2013 at English Language School language Center Course in Putra Malaysia University. There are nine level of English class in English Language School; each level consists of around ten until 20 students. We involved all students in each level to become the research respondents. The questionnaires were distributed when the students finished their exercises in language laboratory (LTC). The minimum sample size recommended [11] for the sample in this study, depending on the number of variables to be studied. The formula is as follows: k (k+1) / 2, where k is the number of variables. It needs at least samples to calculate the minimum model of in this research is: 12(12+1)/2=78 respondents in this research.

In data research collection, the authors distributed 125 questionnaires and only 120 were returned. After disability questionnaires were eliminated, the total samples were used 116 respondents.

2.3. Second Phase of Research
In this phase, the researcher did the model estimation and measurement in UIA model. The research activities in these phases such as; estimated the research hypothesis model, model assessment and produced the last fit model for the UIA to answer the research questions and hypothesis.

2.3.1 User Acceptance Model Estimation. Data collected in this study is ordinal data that has changed to continuous data, so the estimation method we use is the method of maximum likelihood. After it was estimated then we prepared the suitability test of the model. We compare the estimation results of the early models to the latest models. Then we calculate the average value of user interface acceptance. Model fit test was performed in order to assess whether the construct model suitable and consistent with the data collected.

Figure 3 The First Model Estimation

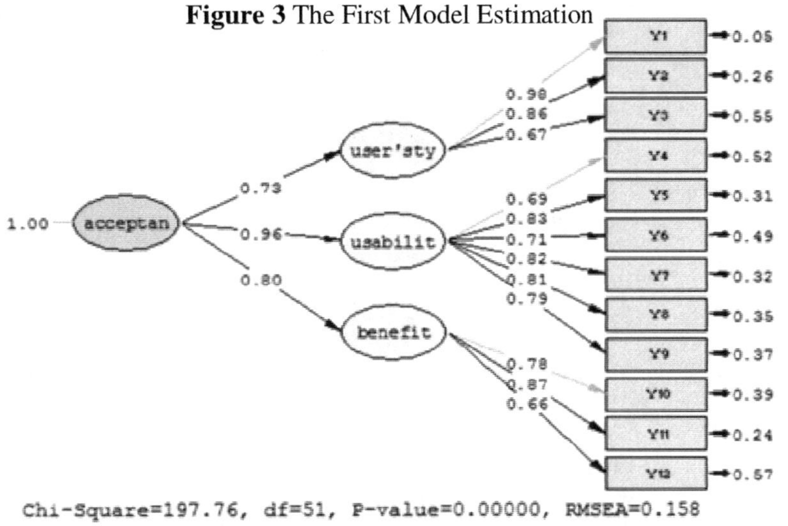

Figure 3 shows that all hypothesis model (measurement and structural model) estimation are significant (p= .000), we can see that γ1= 0.73, γ2= 0.96 and γ3= 0.80. Chi-Square= 197.76 and RMSEA= .158. This evaluation is done to measure the reliability and validity of the concept model of UIA. T-value ≥ 1.96 of mean relevant variables significantly related to the concept of related design, and is a verification of the relationship between variables and constructs that have been defined. The high factor loading value is greater than 0.70.

2.3.2 User Interface Acceptance Measurement. Reliability is used to see the consistency of a measurement. High reliability suggests that the indicators have consistently high in measuring latent constructs. The evaluation of the reliability can be done by using two types of measurements that measure reliability or construct composite reliability and variance extracted measure. Measure of the reliability of a construct can be said to be good if the value of its 0.70 construct reliability and variance extracted value of its 0.50. Examination of the structural model is done by examining the significance of the estimated coefficients. If the value of t is 1.96 then the coefficient is significant.

2.3.3 UIA Model Modification. The model may require to be customized in order to get better the model fit, so estimating the relationships between variables to be expected. Modification indices report the change in χ^2 that result from freeing fixed parameters: usually, as a result adding a path to a model which is presently set to zero. Modifications to a model, especially the structural model, are changes to the theory claimed to be true. Modifications must make sense in terms of the theory being tested, or be approved as limitations of that theory [10].

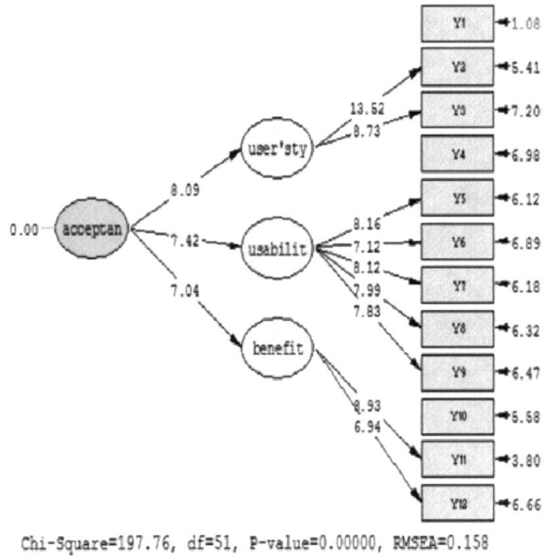

Figure4. T value estimation

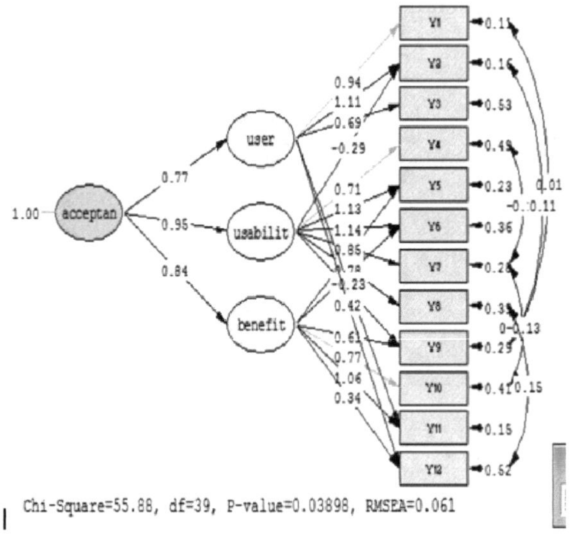

Figure5. Model Modification

Figure 4 shows that all our hypothesis model (measurement and structural model) estimation are significant (p= .039), we can see that γ1= 0.77, γ2= 0.95 and γ3= 0.84. Chi-Square=55.88 and RMSEA=0.061. T values are above than 1.96, it means that the model is significant. Figure 5 shows that all hypothesis model (measurement and structural model) estimation are significant (p=0.039), we can see that γ1=0.77,γ2= 0.95 and γ3= 0.84. Chi-Square = 55.88 and RMSEA= 0.061.

Table 3 Research Hypothesis Results

Name	Lambda	t value	H_0	Research
Knowledge Ability	2.90	*		
Motivation	1.18	5.41	Rejected	H21 Accepted
Learning Style	0.66	7.20	Rejected	H31 Accepted
Knowing-Ability	1.21	6.98	Rejected	H42 Accepted
Operability	0.84	6.12	Rejected	H52 Accepted
Efficiency	0.66	6.89	Rejected	H62 Accepted
Robustness	1.36	6.18	Rejected	H72 Accepted
Safety	0.84	6.32	Rejected	H82 Accepted
Subjective Satisfaction	1.74	6.47	Rejected	H92 Accepted
Media Element	0.93	5.58	Rejected	H103 Accepted
Communicative	0.97	3.80	Rejected	H113 Accepted
User Expectation	5.54	6.66	Rejected	H123 Accepted
User Style	0.73	8.09	Rejected	H134 Accepted
Usability	0.96	7.42	Rejected	H144 Accepted
User Benefit	0.80	7.04	Rejected	H154 Accepted

Table3 is about the hypotheses measurement model result estimation. It values were confirmation from LISREL output. All attributes in our hypotheses are significant as indicator variables where all t-values are above 3.80.

Table 4 Variance Extracted and Construct Reliability of model

Variables	Construct Reliability (>0.70)	Variance Extracted (>0.50)
User style	0.88	0.71
Usability	0.90	0.61
User Benefit	0.82	0.60
Acceptance	0.73	0.69

Based on Table 4 model of e-learning user interface has a highly significant correlation values and strong construction between variables, which is evidenced by the size of the construct reliability values above 0.70 and the value of its variance extracted 0.50.

2.4. Third Phase of Research

2.4.1. User Interface Acceptance Prototype. Beside UIA model, this study also developed a prototype for e-learning acceptance evaluation. The application was built using Learning Management System Moodle v1.9. Figure 6 describes e-learning prototype for user engaged in learning process.

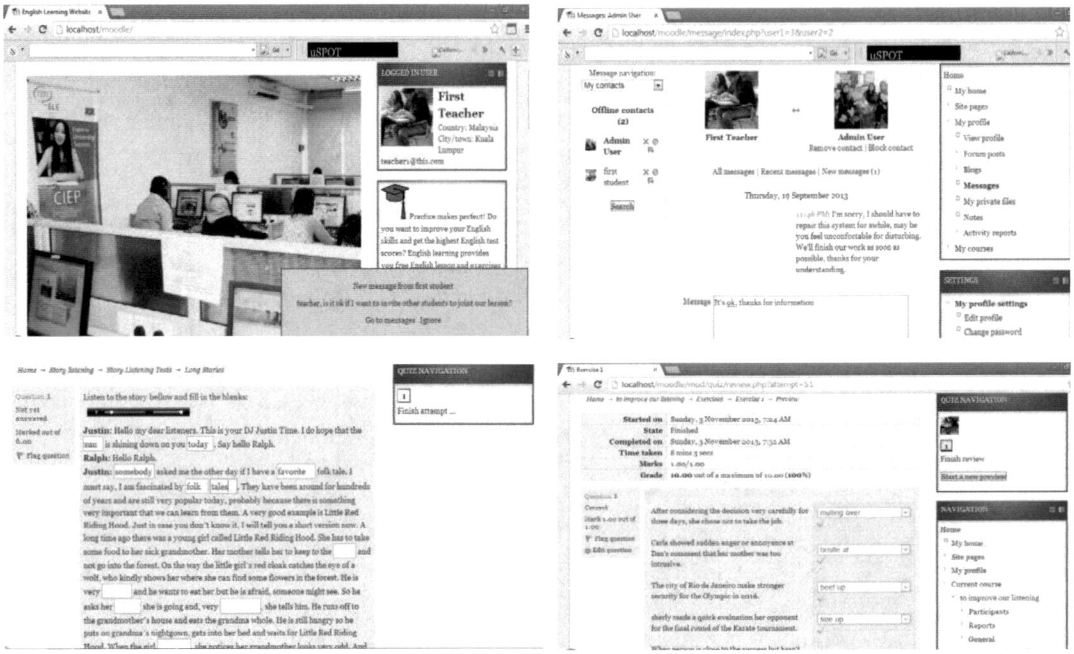

Figur 6. Prototype for E-learning Acceptance

2.4.2. User Acceptance Testing.
In model testing we tried the user to evaluate the application program. There are two different student groups who evaluated e-learning interface. Each group has experiment using Learning Technology Center. In this phase, the researcher shared some questioners about the e-learning interface Acceptance. The questioners were collected and measured using MANOVA. The objective of MANOVA is to determine whether there are differences among group members on dependent variables. The first group based e- learning application and the second groups based on grades.

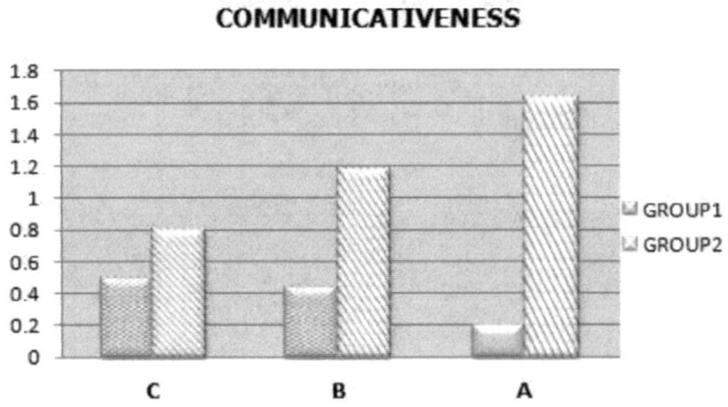

Figure 7 Communicativeness Acceptances

Figure 7 describe about respondent perception about communicativeness in both groups; group 1 (dark bar) is ELS without UIA model and group 2 (light bar) is ELS with UIA model. The dark bar is group 1 (C=0.50, B=0.44, A=0.20) and the light bar is group 2 (C=0.80, B= 1.18, A= 1.63). From the profile plot we can see communicativeness has significant different between two groups (H0 is rejected; p=0.000).

IOP Conf. Series: Materials Science and Engineering **105** (2016) 012016 doi:10.1088/1757-899X/105/1/012016

3. Conclusion

User Interface is very important to encourage the users to use the application. Failure in using e-learning interface could be caused by application development which does not provide ease of use; neither answers the difficult task faced by its users. Therefore, user interface is required to be evaluated in term of user interface usability and technology acceptance. Yet, many theories discussed user interface usability evaluation and technology acceptance separately. Thus, it is necessary to correlate between interface usability evaluation and user acceptance. Hopefully this model can be considered in developing an e-learning application in the future.

4. References

[1] Soufiane B. 2009 Learning style appropriate to the personal character of a learner: "*Pedagogical Indexing Learning Object*". IEEE.

[2] Guralnick, D. 2006 "*How to Design Effective, Motivating User Interfaces*". American Society for Training & Development TechKnowledge Conference, Denver, Co.

[3] K Marçal de Oliveira 2010 *New research challenges for user interface quality evaluation* Interaction Homme-Machine ACM New York: 287–294.

[4] Chiu, C.-M., Sun, S.-Y., et.al. 2007 An empirical analysis of the antecedents of web-based learning continuance. Computers & Education, 49 (4), 1224–1245.

[5] Wing S. Chow and Si Shi 2014 Investigating Students' Satisfaction And Continuance Intention Toward E-Learning: An Extension Of The Expectation–Confirmation Model. Procedia – Social and Behavioral Sciences 141: 1145 – 1149

[6] Ríos A, García A and Bonillo. 2010 *Usability: A Critical Analysis and a Taxonomy*. Intl journal of Human–Computer Interaction 26 (1): 53–74.

[7] Olga.B.A. 2004 *The metrics for quantitative evaluation of user interface usability construction methodology*. SPECOM: 9th Conference Speech and Computer St. Petersburg, Russia, September 20-22.

[8] Manuela P., Brigitte M., and Daniel M., 2010. Students' expectations of and experiences in e-learning: Their relation to learning achievements and course satisfaction. Computers & Education 54:222–229

[9] Sfenrianto, Hasibuan ZA, Suhartanto H 2011 The Influence Factors of Inherent Structure in e-Learning Process. *International Journal of e-Education, e-Business, e-Management and e-Learning –IJEEEE*, 1(3), 217-222.

[10] Von Oertzen, T. 2010 "Power equivalence in structural equation modeling". *British Journal of a Mathematical and Statistical Psychology* 62 (2): 257–272.

[11] Joreskog, Karl.G.dan D.Sorbom 1996 LISREL 8: *Us er's Reference Guide*. Chicago: SSI, Inc.

The Influence of Variation in Time and HCl Concentration to the Glucose Produced from Kepok Banana

Rohman Widodo M [1], Denny Noviyanto [2] and Faisal RM [3]

[1,2,3] Department of Chemical Engineering, Faculty of Industrial Technology, Islamic University of Indonesia, Yogyakarta, Indonesia

Email Address: faisal.rm@uii.ac.id

Abstract - Kepok banana (*Musa paradisiaca*) is a plant that has many advantagesfrom its fruit, stems, leaves, flowers and cob. However, we just tend to take benefit from the fruit. We grow and harvest the fruit without taking advantages from other parts. So they would be a waste or detrimental to animal nest if not used. The idea to take the benefit from the banana crop yields, especially cob is rarely explored. This study is an introduction to the use of banana weevil especially from the glucose it contains. This study uses current methods of hydrolysis using HCl as a catalyst with the concentration variation of 0.4 N, 0.6 N and 0.8 N and hydrolysis times variation of 20 minutes, 25 minutes and 30 minutes. The stages in the hydrolysis include preparation of materials, the process of hydrolysis and analysis of test results using Fehling and titrate with standard glucose solution. HCl is used as a catalyst because it is cheaper than the enzyme that has the same function. NaOH 60% is used for neutralizing the pH of the filtrate result of hydrolysis. From the results of analysis, known thatthe biggest yield of glucose is at concentration 0.8 N and at 30 minutes reaction, it contains 6.25 gram glucose / 20 gram dry sampel, and the convertion is 27.22 % at 20 gram dry sampel.

Keywords: Kepok Banana, Time and HCl Concentration, Glucose

1. INTRODUCTION

Banana plants are soft-stemmed plants with high trunk and capable of up to 4 m, wide leaves and root fibers. Almost every part of this plant can be used. Stems, leaves, fruit, flowers, and even the cob. Banana fruit has sweet and soft texture. Suitable to be consumed by variety of ages. Can be directly eaten or cooked into other foods. wide leaves can be used as food wrap, or as decoration. Soft stems can be used as decoration, planting media, fodder and others. Before bananas are really old, bud resembles a heart, called cardiac bananas. Some varieties of banana can be used as processed food and has a good taste. cob is immersed parts of the plants in the ground where the discharge of roots. cob can be used as the banana weevil chips, shredded banana cob, and flour banana weevil.

Many advantages can be exploited from every part of the banana plant, but we tend to get benefit from the fruit only. We grow and harvest the fruit without taking the advantages from other parts. So they would be a waste or detrimental to animal nest if not used. Therefore, it takes about a week to maximize yields of banana plants, especially for the most rarely used cob.

Problem Formulation

a. What is the effect of reaction time for glucose resulting from the hydrolysis of banana weevil?
b. What is the effect of the solvent to the concentration of glucose produced by the hydrolysis of banana weevil?

Limitation Problem

a. Researchers only compare the effect of reaction time and solvents the concentration of the glucose content in the banana weevil.
b. Weevil banana plant tuber crops studied is kepok banana (Musa Paradisiaca)
c. Flour banana weevil is not determined its scale mesh
d. The solvent used is a solvent HCL The concentration variation is 0.4 N, 0.6 N and 0.8 N
e. Variations on the hydrolysis time is 20 minutes, 25 minutes and 30 minutes after boiling
f. Hydrolysis process used did not consider the speed stirrerrotation
g. Room temperature practicum ignored
h. Not using the average volume of filtrate

Research Objectives

The aim of this study is:

a. Analyzing the glucose content in the banana weevil with hydrolysis
b. Knowing the influence of reaction time and the concentration of solvent to the content of glucose obtained from hydrolysis of banana weevil.
c. Utilizing the banana weevil crop waste into products that have a commercial value

Benefits Research

The benefits of this research are:

a. Making waste banana weevil as economically valuable goods
b. Contributing as a useful knowledge to be reference for further research

2. RESEARCH METHODOLOGY

Secondary data and reference are collectedfrom the internet, books or college textbooks related to the problem being studied. Primary data collected by direct experiments on samples that were analyzed and tested in the laboratory.

Data Sampling

Sample conducted in this study is harvesting of banana weevil starch hydrolysis with hydrochloric acid (HCl) by varying of concentration and reaction time, using a three-neck flask and magnetic stirrer. This research was conducted in the laboratory Introduction to Chemical Engineering, Department of Chemical Engineering, Faculty of Industrial Technology, Islamic University of Indonesia.

Equipment and Materials Used

The materials used

a. Flour tuber Bananas
b. HCl 2 N
c. NaOH 60%
d. Aquadest
e. Fehling A
f. Fehling B
g. Methyl Blue
h. Glucose Standard

IOP Conf. Series: Materials Science and Engineering **105** (2016) 012017 doi:10.1088/1757-899X/105/1/012017

Tools used

a. Buret
b. Stoves
c. Scissors
d. Mixer
e. Glass Tubes
f. The universal pH indicator
g. The glass beaker (600 ml).
h. The glass beaker (250 ml)
i. The glass beaker (100 ml)
j. Erlenmeyer (250 ml)
k. Erlenmeyer (50 ml)
l. GlassesFunnel
m. Flask 1 L
n. Flask (500 ml)
o. Heating mantle and stipper
p. Thermometer

Research Procedure
Hydrolysis Process

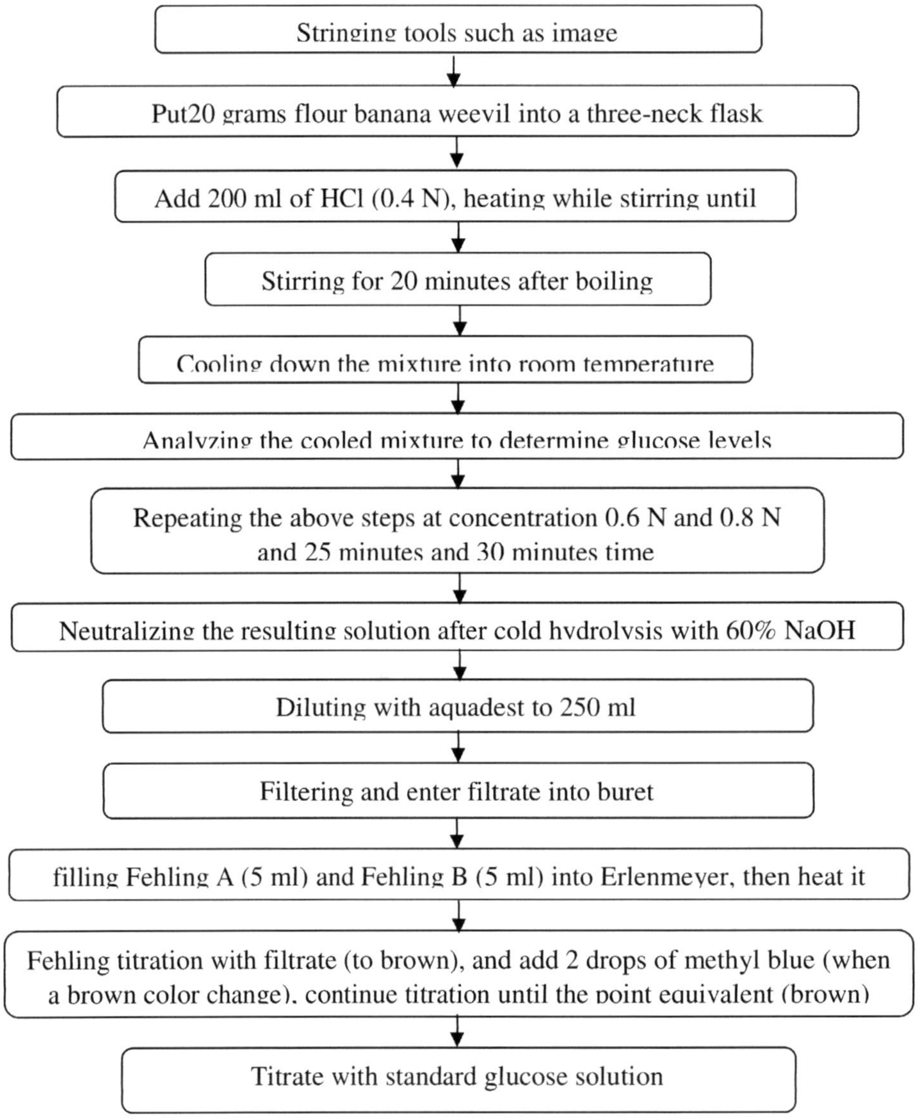

Figure 2.1 Flowchart of hydrolysis process

3. EXPERIMENT RESULTS

This experiment uses materials derived from banana weevil is made such that intangibles such as flour, but not the same mesh uniformity. Observation temperature is 105^{0}C, but is not so considered. Results from this study is shown in tabular form as shown below.the average temperature of the experiment: 105^{0}C

sample weight: 20 g

Total volume: 250

Table 3.1 Results of experiment

waktu (menit)	V NaOH (ml)	V filtrat (ml)	V larutan glukosa standart (ml)	jumlah glukosa total C (gr)
untuk konsenterasi 0.8 N				
20	6.5	13	12	1.1538
25	6.4	13.9	13	1.1691
30	6.7	15	15	1.2500
untuk konsenterasi 0.6 N				
20	5.3	11	9	1.0227
25	5.5	12.8	10.8	1.0547
30	5.5	13	11.2	1.0769
untuk konsenterasi 0.4 N				
20	3.9	11	7	0.7955
25	4.1	11.2	7.6	0.8482
30	4	11.6	8	0.8621

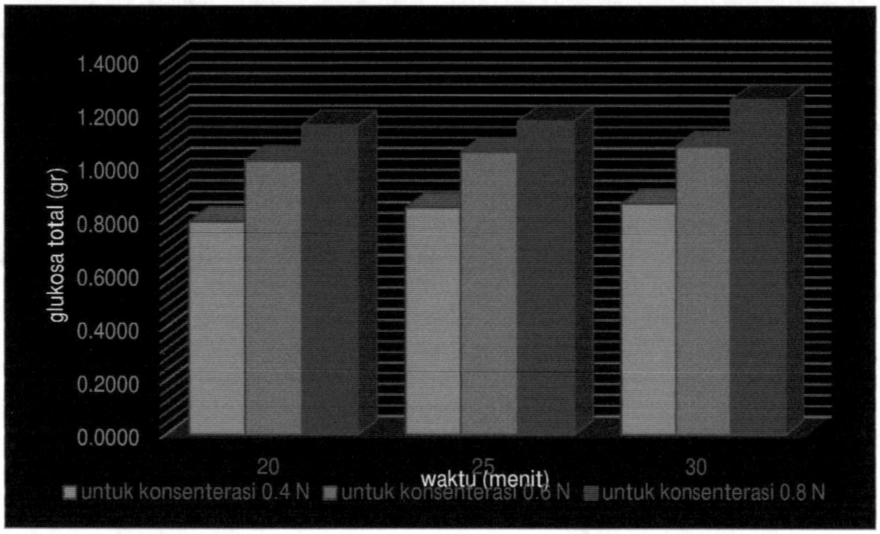

Figure 3.1 Graph relationship variation concentration and reaction time for glucose produce

4. DISCUSSION

This hydrolysis reaction is very slow. Catalyst should be added to accelerate the reaction. Catalysts which can be used are acids and enzymes. Banana weevil hydrolysis into glucose in stirred tank is used as an acid catalyst, because the acid is more economical in terms of time and cost than the use of enzymes as catalysts. Used catalyst HCl grading 0.4 N, 0.6 N and 0.8 N. In addition to the catalyst, the assumption that reducing the size can speed up the process of hydrolysis, because the reaction will be distributed evenly throughout the surface of the banana weevil. But in this study does not emphasize the mesh size of the banana weevil hydrolyzed.

Broadly speaking, this study include raw material preparation, hydrolysis and analysis of the results of hydrolysis. At the stage of raw material preparation, banana weevil cleaned and diced, washed and the sap is removed. The sap is removed because it may become caramel when hydrolysis takes place. In this case, the removal of sap by using boiling method within 1 hour. Results boiling in the form of tubers which are more tender texture and water are turned into purple, then hump is aerated to cool. After the hump cool, then cut into small, after it is inserted into the juicer, the aim is downsizing, resulted in the slurry, a mixture of smooth hump of water, so the water must be separated to get the weight of dry matter. Separation done with extortion using a cloth. Once separated, the devastated solids banana weevil dried manually using sunlight to avoid the formation of caramel when using a drying oven.

Once dried, the tubers then destroyed again in order to obtain a finer texture. However, at the time of destruction, not all materials can be smoothed with a same level of refinement. This is because the results of the first destruction is chewy. Once crushed, tubers with 20 mg weight were 9 samples.

Once the raw material preparation banana weevil, researchers prepare research tools and other materials. Among create a solution HCl 0.4 N, 0.6 N and 0.8 N using 2 N HCl using the dilution method, namely by dilution using distilled water and making mix Fehling Fehling A and B each 5 ml.

Once the tool is installed hydrolysis, rare-step hydrolysis is carried out at the time of entering material 20 mg banana weevil and 0.4 N HCl solution into the 200 ml three neck flask, heating and stirring using a magnetic stirrer. After boiling record the time and let the reaction continues until 20 minutes. After 20 minutes, the hydrolysis process is stopped. Then dirolisa result is cooled to room temperature. Hydrolysis results in the form of orange-red colored solution. Then the focus in this research is on a variety of HCl and hydrolysis time.

After hydrolysis is complete, the resulting solution is analyzed for levels of glucose hydrolysis. Hydrolysis resulting solution pH neutralized using 60% NaOH solution. NaOH alkaline is added to make this solution becomes neutral. Selection of neutralization using NaOH as the conditions under which it is definitely acidic hydrolysis, likely due to the influence of a catalyst HCl. It can be seen from the table, the greater HClconcentration given, the more also NaOH used to neutralize the pH of the solution. From the table and graph shows that the moreHCL concentration of acid used as a catalyst, the more the volume of sodium hydroxide used as a neutralizing acid. This shows that the acidic hydrolysis process is strongly influenced by HClconcentration.

Parameters that this solution is at neutral pH universal indicator shows the number 7. Once the neutral solution was filtered using filter paper. Once filtered, the filtrate was added to the burette. The aim of filtering by filter paper is to filter the solution from the solid particles, so it does not make burette clogged.

At the stage of the analysis of glucose, hydrolysis resulting solution which has been neutralized, diluted and filtered and then used to titrate Fehling Fehling A and B which have been mixed and heated to boiling using the Erlenmeyer. Dilution goal is uniformity filtrate thus simplifying the calculation of the analysis. The purpose of the test is to determine the carbohydrate Fehling reduced. Fehling titrated using the filtrate to brown, after titration continued until the chocolate brown color equivalent (no color change), but if the color changes, add 2 drops of methyl blue and titration was continued until completely brown equivalent. The amount of volume of filtrate that is used to titrate Fehling, The greater HClconcentration, the greater the volume needed to titrate Fehling, nor with the hydrolysis time. The longer the time of hydrolysis, the more volume of filtrate needed to titrate Fehling until it reaches a equivalent point. The relationship between the volume and concentration of HCl of the filtrate to variations and temperature hydrolysis can be seen in the table results of the research.

The next step is to get the glucose Fehling titrate titration results with the filtrate using a standard glucose solution. This is almost the same as the Fehling test. However, to determine more specific results from the monosaccharides resulting from the hydrolysis of carbohydrates. Results of analysis attached to the table. At the concentration variation is given in this study, the higher the HCl concentration, the higher the glucose produced. This is because acid serves as a catalyst for the destroyer of carbohydrates, making them easier to hydrolyzed. This is because destructed carbohydrates will have a large cross-sectional area and the more experience a reaction, so the more glucose is generated.
Variations in the reaction time of this study also affect the results of glucose produced. Shown in the table, the longer the reaction time, the greater the glucose results generated from the hydrolysis process of the banana weevil.

5. CONCLUSION

The conclusions that can be drawn from this study is that the banana weevil hydrolysis into glucose in stirred tank can be done by using HCl as a catalyst. In addition, the results of this study are influenced by variables of concentration and duration of hydrolysis catalyst. The greater concentration of the catalyst, the greater the glucose levels resulted. This is due to the increasingly sour nature of the catalyst, the higher The concentration, it would be destructive to the composition of the polysaccharide carbohydrate. Thus, the same treatment to the higher levels of catalyst, then the sooner the destruction of carbohydrates into sugar. So that within the same hydrolysis, rise more quickly obtain a catalyst The concentration of glucose.
In addition to the catalyst concentration, duration of hydrolysis also affects glucose levels generated. The longer hydrolysis time, until the optimum point, the higher the glucose produced. In this study, the determination of the time variation of hydrolysis is still in optimal time range, i.e. from 1 minute to 1 hour. So that changes can still be seen hydrolysis time effects on glucose levels generated.
In this study, the largest yield obtained catalyst 0.8 N The concentration and reaction time of 30 minutes with a yield of 6:25 grams of glucose / 20 grams of material, as well as the conversion of 27.22% at 20 grams of dry matter.

Suggestions

This research of glucose levels of banana weevil is conducted to get a comparison of glucose levels with variable concentrations of catalyst and duration of hydrolysis. So it suggests further research using different variables like the use of different varieties of banana weevil, the use of different catalysts, variable of temperature and variable of different catalysts concentration.

REFERENCES

[1] Deman J.M 1997 *Kimia Makanan* Ed.2 Bandung : ITB,.
[2] Groggins P.H 1958 *Unit Processes in Organic Synthesis* 3rded Mc.Graw Hill Book Company Inc New York
[3] Kerr R.W 1987 *Chemistry and Industry of Starch* 2rd ed. Academic Press Inc. New York
[4] Kirk, R.E and Othmer D.F 1983 *Encyclopedia of Chemical Technology* 3rd ed John Wiley and Sons Inc New York
[5] Levenspiel O 1972 *Chemical Reaction Engineering* 2rd ed John Wiley and Son Inc New York
[6] Matz S.A 1970 *Sereal Technology* The Avi Publishing Co. inc West Port, Connecti-cut,.
[7] Supranto 1998 *Proses Industri Kimia II* Teknik Kimia FT UGM, Yogyakarta,
[8] Artati, and Enny Kriswiyanti, et.al 2012 *Pengaruh Jenis dan Konsenterasi Asam Terhadap Kinetika Reaksi Hidrolisis Pelepah Pisang*. Jurusan Teknik Kimia Fakultas Teknik Universitas Sebelas Maret, Surakarta,.
[9] Nyimas Chairunisa 2010 *Tinjauan Hidrolisis Pati Bonggol Pisang untuk Pembuatan Bio-ethanol* JurusanTeknik Kimia, FakultasTeknik, Politeknik Negeri Sriwijaya

[10] Sukmawati R.F, and Milati, S 2009 *Pembuatan Bioethanol dari Kulit Singkong* Program Studi D III Teknik Kimia FakultasTeknik, Universitas Sebelas Maret Surakarta
[11] Hikmiyati N and Yanie N.S 2009 *Pembuatan Bioetanol dari Limbah Kulit Singkong Melalui Proses Hidrolisa Asam dan Enzimatis* JurusanTeknik Kimia, FakultasTeknik, Politeknik Negeri Sriwijaya,.
[12] *Modul Petunjuk Praktikum Pengantar Teknik Kimia II.* Laboratorium Pengantar Teknik Kimia. JurusanTeknik Kimia, FakultasTeknologi Industri, Universitas Islam Indonesia, Yogyakarta.
[13] Rivai, H 1995 *Asas Pemeriksaan Kimia* Jakarta :UI, Jakarta
[14] Sari D.P and. Eka J.P 2013 *Proses Hiodrolisa Tepung Jagung menjadi Glukosa dengan Tanki Berpengaduk* Jurusan Teknik Kimia, Fakultas Teknologi Industri, Universitas Islam Indonesia, Yogyakarta,.

Compressed Air System Optimization: Case Study Food Industry in Indonesia

Endang Widayati, [1] **Hasril Nuzahar**[2]

[1] Education and Training Center for Electricity, New, Renewable Energy and Energy Conservation, Jalan Poncol Raya No. 39, Ciracas, Jakarta Timur, Indonesia

[2] Ministry of Energy and Mineral Resources Republic of Indonesia, Jalan Medan Merdeka Selatan No.18, Jakarta Pusat, Indonesia

E-mail: ris4_ed@yahoo.com; nuzahar@yahoo.com

Abstract. Compressors and compressed air systems was one of the most important utilities in industries or factories. Approximately 10% of the cost of electricity in the industry was used to produce compressed air. Therefore the potential for energy savings in the compressors and compressed air systems had a big challenge. This field was conducted especially in Indonesia food industry or factory. Compressed air system optimization was a technique approach to determine the optimal conditions for the operation of compressors and compressed air systems that included evaluation of the energy needs, supply adjustment, eliminating or reconfiguring the use and operation of inefficient, changing and complementing some equipment and improving operating efficiencies. This technique gave the significant impact for energy saving and costs. The potential savings based on this study through measurement and optimization e.g. system that lowers the pressure of 7.5 barg to 6.8 barg would reduce energy consumption and running costs approximately 4.2 %, switch off the compressor GA110 and GA75 was obtained annual savings of USD 52,947 ≈455 714 kWh, running GA75 light load or unloaded then obtained annual savings of USD 31,841 ≈ 270,685 kWh, install new compressor 2x132 kW and 1x 132 kW VSD obtained annual savings of USD 108,325 ≈ 928,500 kWh. Furthermore it was needed to conduct study of technical aspect of energy saving potential (Investment Grade Audit) and performed Cost Benefit Analysis. This study was one of best practice solutions how to save energy and improve energy performance in compressors and compressed air system.

Keywords: compressor, compressed air system optimization, efficiency, energy saving, running cost, light load, unloaded

1. Introduction

1.1. Background
Energy has a very important role and becomes a basic necessity in sustainable national economic development. Energy demand grew at an average of 7% per year (DESDM, 2009). On the supply side, the growth rate of energy supply could not keep up the pace of growth in energy demand. This condition was increasingly gap between demand and supply. Therefore, energy conservation was an importance throughout the world, including Indonesia. There were strong indications that the use of energy in all sectors which includes households, industry, buildings (public and private) and trasportation was inefficient (Warent, Ernest et.al, 2010).
Almost all industries use compressed air system/compressed air (Thumman & Younger, 2007). Compressed air systems were often as significant energy users in industry/ factory. Industrial/plant used compressed air for process or production. Generally the compressed air generated by the compressed air unit with ranged from 5 to 50,000 Hp (Harding & Moore, 2012). Approximately 70-90% of compressed air was lost in the form of heat that could not be used, friction, misuse and noise. This was why efforts to increase the energy efficiency of the compressors and compressed air systems become an important target area in industry/plant (Warent, Ernest et.al, 2010). Based on these facts it was necessary to investigate the energy efficiency improvement in compressors and compressed air system through this study.
The results of this study were expected to be used as reference/ benchmark for industry/factory to increase the efficiency of the compressors and compressed air systems and also provided reference support for industrial energy audit training in our training center.

1.2. Problem Identification
Generally about 10% of the cost of electricity in industry is used to produce compressed air. So far, the management considers that the compressed air was cheap because air or fluida to be compressed was available and free of charge. Otherwise indicated that the compressed air was an expensive resource therefore that must be used efficiently. Besides the energy losses - in the compressed air system was large enough so that the potential savings of about 30% was high enough (Kane & Medaris, 2003). How does the solution to the problem so that the compressor and compressed air systems could be optimized in order to save energy and costs?
Compressed air system optimization was the method how to approach the compressed air system as a whole system in order to save energy and cost. Energy savings in the compressors and compressed air systems had considerable potential (Kazuhide Konitoku, 2014). However, the techniques and the study of it is not widely practiced, especially in the industry/factory in Indonesia. In industry/factory compressed air is generally provided based on users request. If the compressed air demand of users are fullfilled it mean that compressor and compressed air system in a good condition although the users request or compressed air demand over, waste energy and cost.

1.3. Study Purpose
This study aimed is to obtain the potential energy savings in compressed air systems in industry/ factory so that it could provide a solution to the problems of energy supply, especially in the energy performance improvement. The results of this research could be used for industry/factory as benchmark for energy performance of the compressor and compressed air systems.

2. Literature Review

2.1. Compressor

Compressor was a utility to increase the air pressure by compressing the gas or air. Compressor worked by sucking in atmospheric air (Abels & Kisscook, 2011). If the compressor worked at a higher pressure than the atmospheric pressure then called a booster compressor (booster), and if the compressor was working under atmospheric pressure, it is called a vacuum compressor (Abels & Kisscook, 2011). Low-powered compressor used single phase electric motor or gasoline engine while large power compressors required 3 phase electric motors, diesel engines or gas turbines.

2.2. Compressed Air System

Compressed air system consisted of supply and demand side. Supply side included compressors and air treatment while the demand side such as distribution, storage and end-use equipments (Harding & Moore, 2012). Main component compressed air system consisted of intake air filters, inter-stage coolers, after-coolers, air-dryers, moisture drain traps, receivers, piping network, filters, regulators and lubricators. It was described in Figure 1.

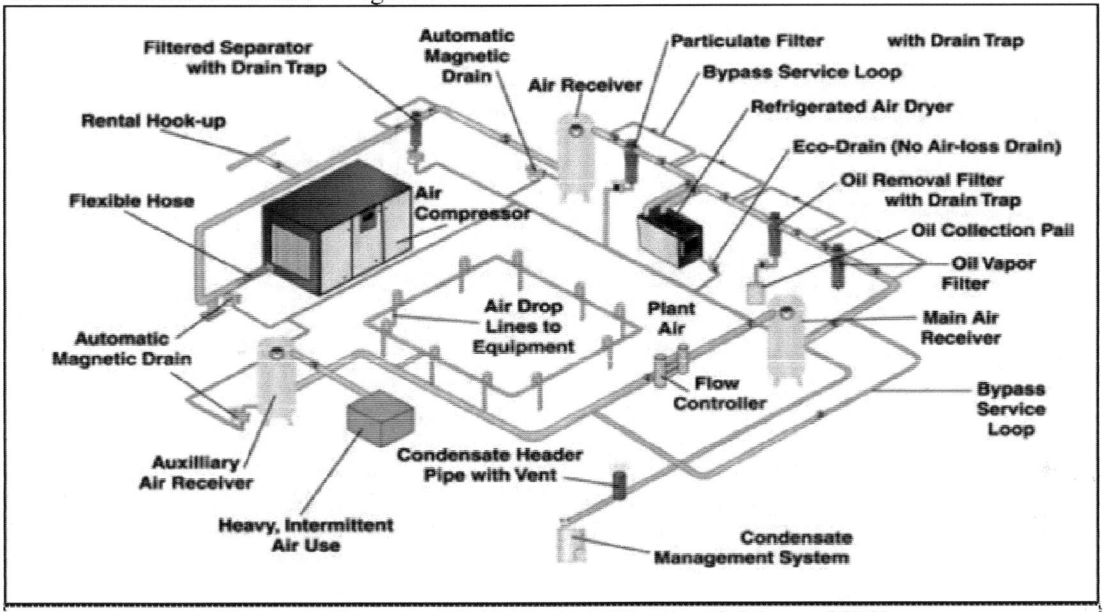

Figure 1. Compressed Air System (Harding & Moore, 2012)

2.3. Compressed Air

Compressed air was a gas fluid/compressed air. Compressed air had a great ability to store energy per unit volume. At the time of the expansion of the stored energy was released as work (Beals, CE, 2009). In these cases applied equation:

$$\mathrm{W}P = \frac{F}{A} \quad (1)$$

$$\frac{Po\ Vo}{To} = \frac{P1V1}{T1} = R = constant \quad (2)$$

2.4. Compressed Air System Optimization

Compressed air system optimization is an engineering approach to determine the potential for energy savings in the compressed air system which included evaluation of energy needs in industrial plant/factory, adjustment of the supply, eliminating or reconfiguring the use and operation of inefficient, converting and equipping equipment to fit the needs and improve the efficiency of operations. In this case, boundary need to be set in advance as an initial step in the study of air compressed system optimization (Harding & Moore, 2012).

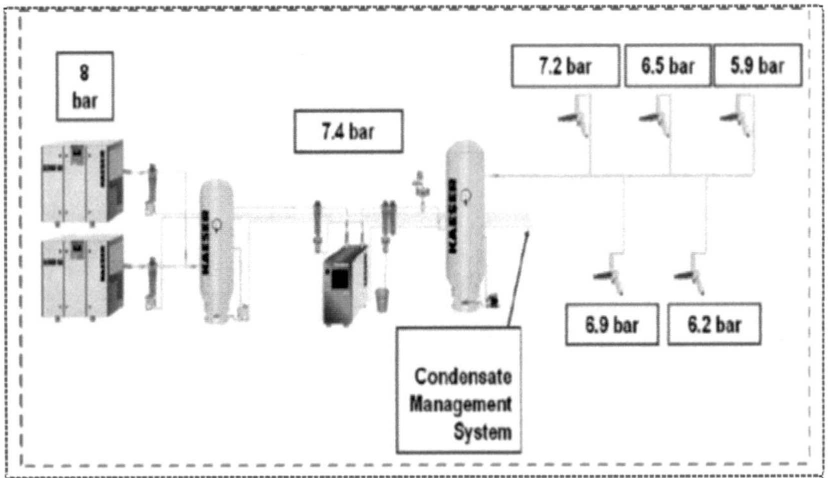

Figure 2. Boundary (Harding & Moore, 2012)

2.5. Energy Consumption in Compressor

It was only about 10-20% from total energy consumption in compressed air system had been using to compress the air (Kazuhide Konitoku,2014) as described in Figure 3.

IOP Conf. Series: Materials Science and Engineering **105** (2016) 012018 doi:10.1088/1757-899X/105/1/012018

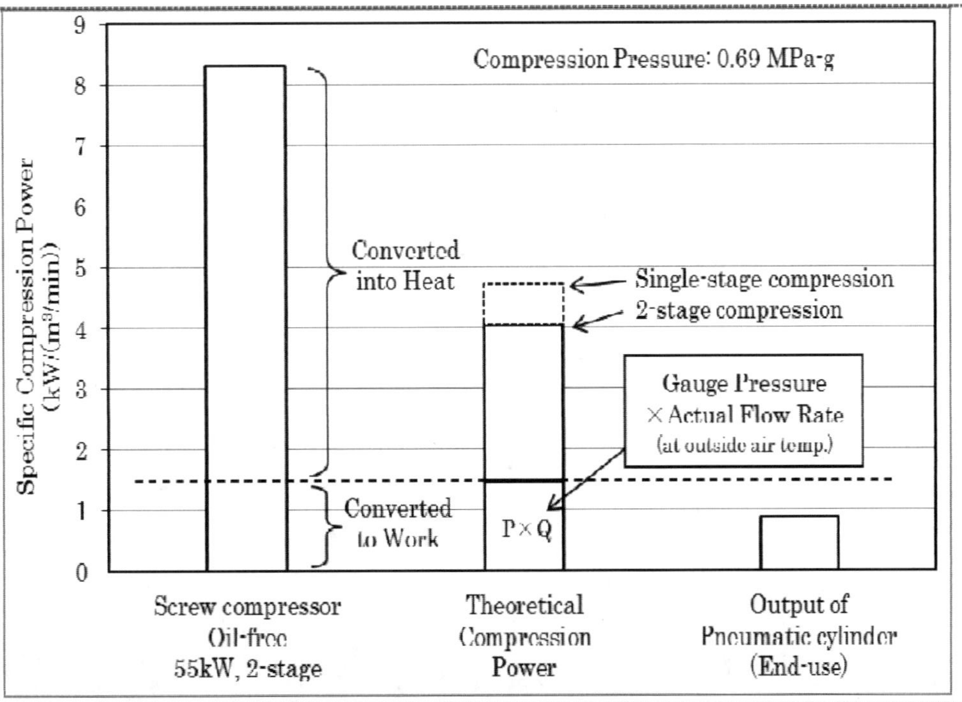

Figure 3. Energy Consumption Pattern in Compressed Air System (Kazuhide Konitoku, 2014)

3. Research Methodology

3.1. Methodology

The methodology used in this study was a qualitative and quantitative methods with steps as follows:

- Preparation,
- Collecting secondary data,
- Field observation and measurement,
- Data verification and
- Data analysis

4. Result of Study

4.1. Overview of Energy Use

Types of energy and enery use in this plant (locus of study) was as follow:

- Electricity;
- Coal (steam
- Liquid Petroleum Gas
- Diesel oil

The electrical energy generated by the Power Plant with coal fuel. The type of power plant was co-generation with 70% of the steam produced to generate electricity and 30% steam for process.

4.2. Investigation and Compressed Air System Optimization

The compressor was one of the most important utilities and compressed air (CA) was needed in the production process as well as operations in this study. CA system used to meet the demand site consists of 8 (eight) compressors with 1 (one) main system and local instrument air system.

4.2.1 Compressor Performance

Compressors performance had been described in Figure 4.

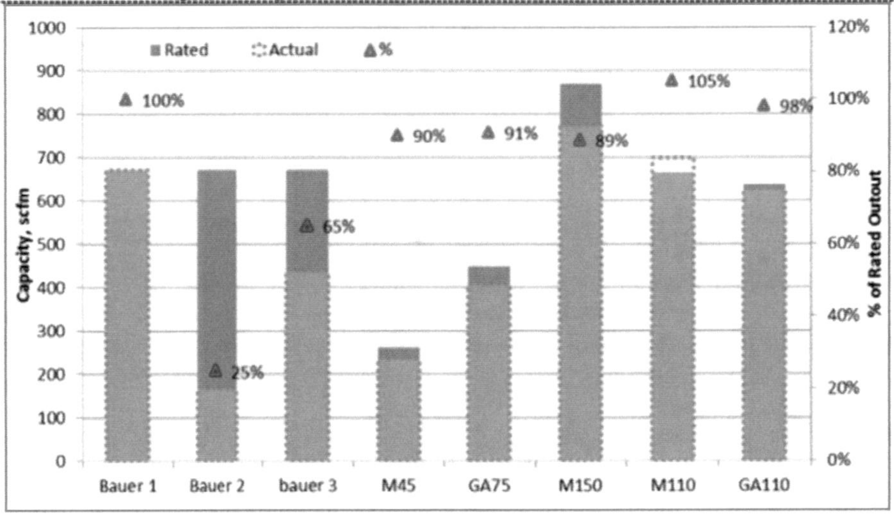

Figure 4. Compressor Performance

The performance in the figure 4. based on specific energy consumption (kwh/cfm). It was necessary able to compare the performance of each compressor, as shown in Table 1.

Table 1. Compressor Performance and Specifc Energy Consumption

Compressor	Bauer 1	Bauer2	Bauer3	M45	GA75	M160	M110	GA110
Rated (scfm)	671	671	671	261	449	869	665	637
Actual (scfm)	670.6	167.5	437.2	235.1	408.1	772.1	699.7	626.3
% Rated	100	25	65	90	91	89	105	98
SEC (kWh/100 cfm)	0.3	4.0	0.6	0.4	0.4	0.5	0.3	0.3

Based on the figure 4 and table 1 as well as visual observation in the field, it could be compared compressor performance for each compressor was as follows:

- Bauer 1, very good (efficient): 0.3 kWh/100 cfm
- Bauer 2, very bad (very energy wasteful): 4.0 kWh/100 cfm
- Bauer 3, bad (energy wasteful): 0.6 kWh/100 cfm
- M45, good (efficient): 0.4 kWh/100 cfm
- GA75, good (efficient): 0.4 kWh/100 cfm

4.2.2 Compressed Air Demand Profile

IOP Conf. Series: Materials Science and Engineering **105** (2016) 012018 doi:10.1088/1757-899X/105/1/012018

To obtain the compressed air demand profile was conducted by measuring the flow rate (flowmetering) in-GA75 Line 1, Line 2 and Line1-Main. Measurement results could be seen in the graph Figure 5.

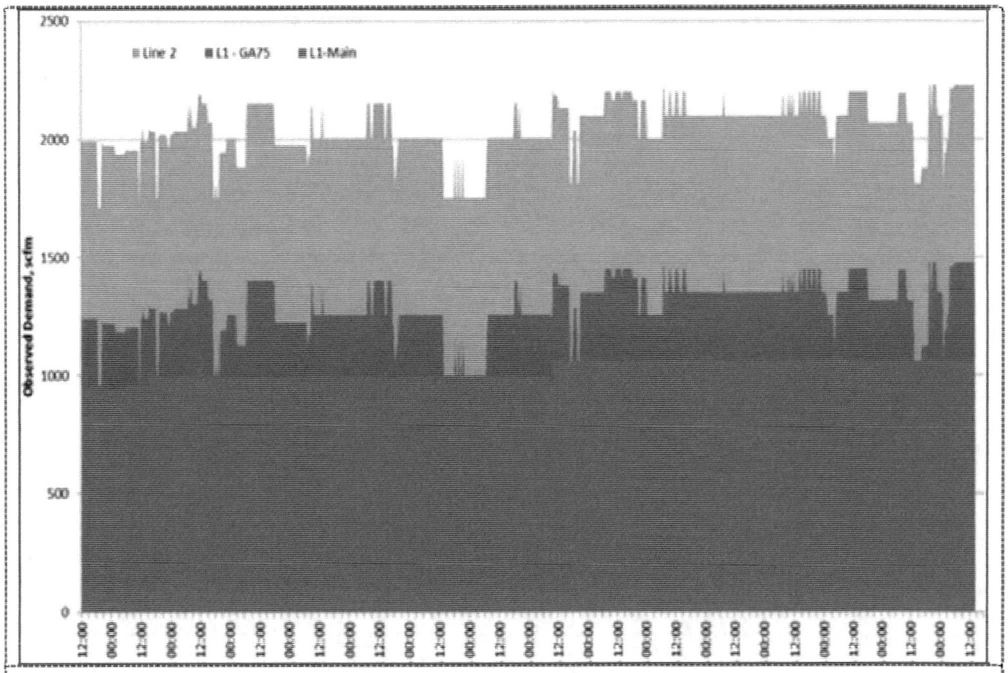

Figure 5. Compressed Air Demand Profile

From the figure 5, it was obtained the average demand in 2100 scfm.

4.2.3 System Pressure

Measurement of the compressed air system pressure was conducted by using a pressure transducer and the results could be seen from the graph Figure 6.Based on Figure 6 ilustrated that the system pressure of 7.2 barg with a range between 6.5 – 7.5 barg. This indicated that the pressure could be reduced up to 6.8 barg. It meant that with a pressure of 6.8 barg pressure system had been insufficient and had also considered the margin for the safety factor. Reducing the pressure of 1 barg would reduce energy costs 6-7% (Harding & Moore, 2012).

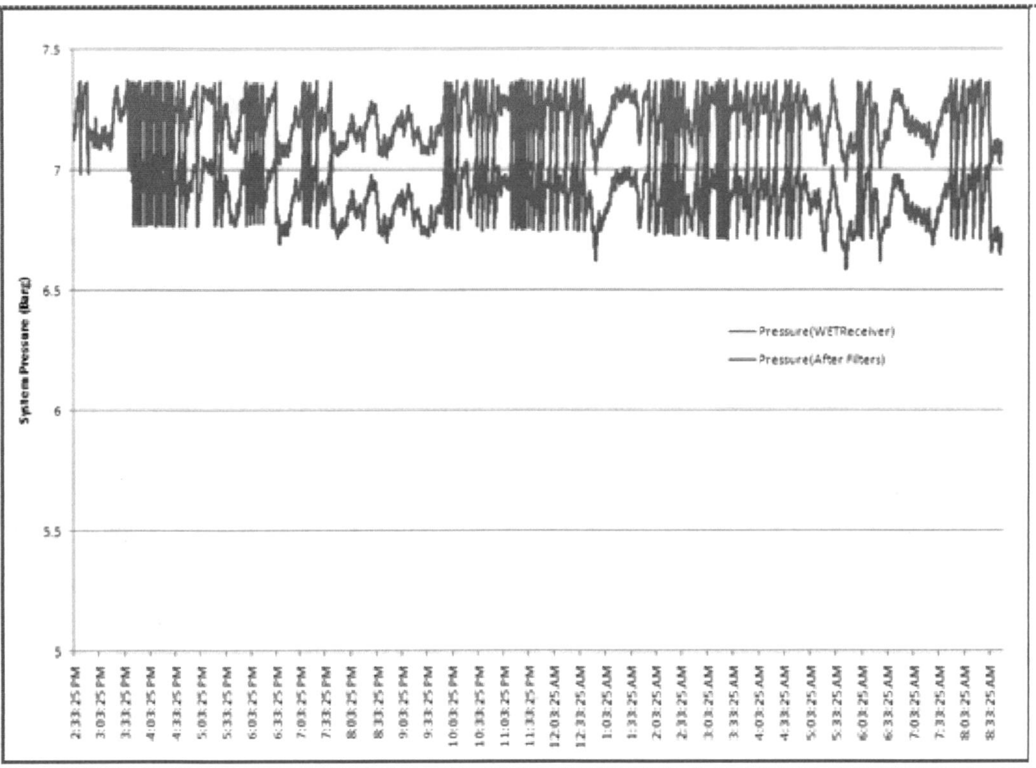

Figure 6. System Pressure

4.2.4 Annual Generation Cost

Compressed air generation costs per year was estimated based on the average demand by simulation. The simulation results verified based on secondary data, energy costs for compressors and compressed air systems. Estimation of annual costs generation could be seen in Figure 7. From Figure 7, obtained cost estimation for the annual generation average demand (2100 scfm) was 396,666 USD which consist of 366,400 USD Main System and 30,266 USD Class Screen System. The result was verified with the real costs 400,712 USD; which consisted of 4,310,600 kWh Main System and 356,300 kWh class screen System. The difference between estimation and real cost was about 4,046 USD or about 1.02%, so that the annual cost estimation compressed air generation could be accepted.

IOP Conf. Series: Materials Science and Engineering **105** (2016) 012018 doi:10.1088/1757-899X/105/1/012018

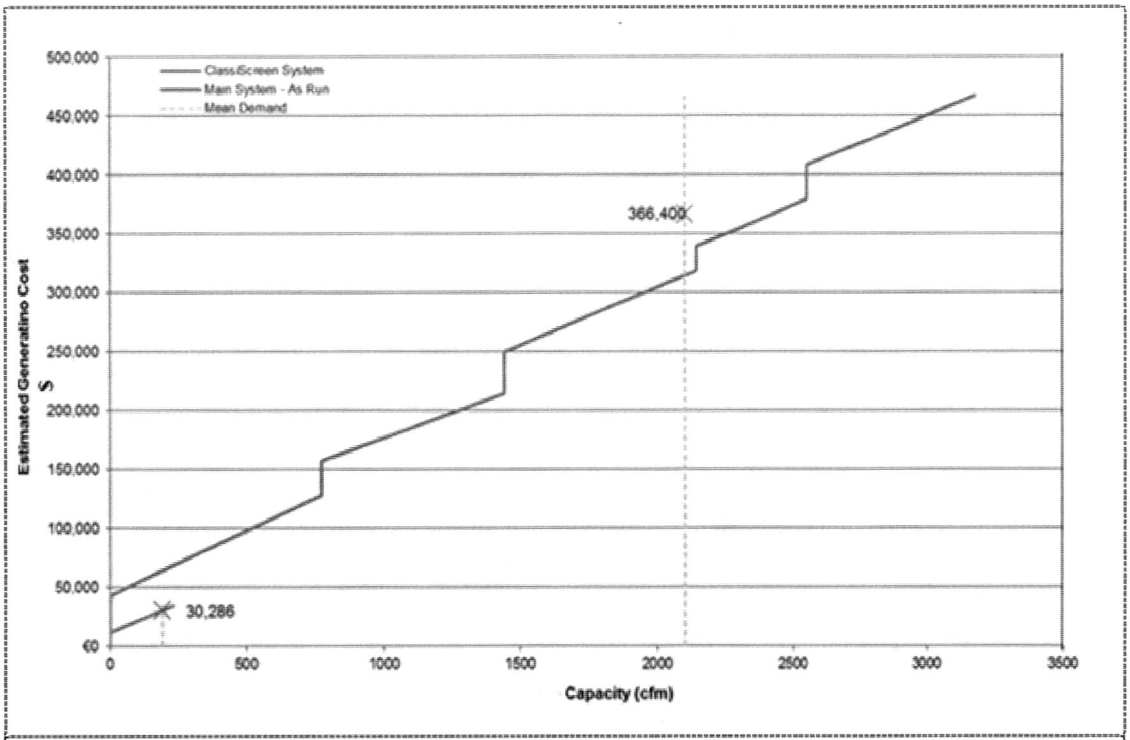

Figure 7. Annual Compressed Air Generation Cost

4.3. Determination of Energy Saving

In this study, in order to determine the potential for energy saving was conducted by optimization of compressed air system/compressed air system optimization, among others, by adjusting supply, eliminating or reconfiguring the use of operation inefficient, transform and equip equipment that fit the needs and improving efficiency operation. The approach was based on thus scenarios, among others:

- Optimization of existing compressors, among others, performed with switched off GA75 and GA110, running Bauer2 as last backup, running Bauer3 as penultimate backup, M160 as backup running at rated capacity.
- Installing new equipment, this was done partly by installing 2x132 kW fixed speed compressor and 1 x 132 kW VSD compressor.

Estimation of annual generation cost from scenarios illustrated in figure 8.

From figure 8 it could be seen that if GA110 and GA75 off, the obtained annual saving of USD 52,947 or around 455,714 kWh. If GA75 running light load or unloaded then obtained annual savings of USD 31,841 or around 270,685 kWh. If install new compressor 2x132 kW and 1x132 kW VSD obtained annual saving of USD 108,325 or around 928,500 kWh.

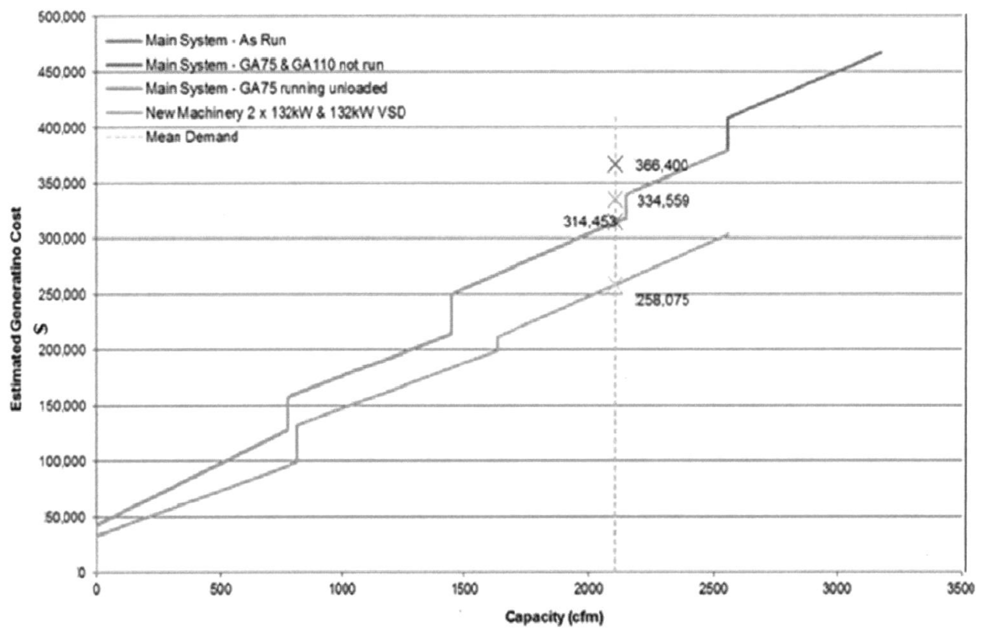

Figure 8. Annual Cost Scenario

5. Conclusion and Recommendation

5.1. Conclusion

The study of compressed air air system optimization was a step in the effort to save energy and cost savings on utility compressor and compressed air system. Compressed air system optimization approach included adjustment supply, removing or reconfiguring the use and operation of inefficient, transform and equip equipment that fit the needs and improve the efficiency of utility operations more effective because of this method that could be seen as asystem included; energy, cost and safety factors.

Based on the locus of study there were 3 (three) potential opportunities e.g. optimization of system pressure, reconfiguring equipment operation and adjustment of supply to demand. Estimation of cost and energy saving basd on the study was 52,947 around 455,714 kWh or USD 31,841 around 270,685 kWh or USD 108,325 around 928,500 kWh.

5.2. Recommendation

Some recommendations that could be delivered from the results of this study which aims to save energy or reduce the consumption of energy were:

a. Lowering system pressure of 7.5 barg to 6.8 barg will reduce energy consumption and running costs approximately 4.2%;
b. GA110 and GA75 off, the obtained energy and cost savings of electricity that was USD $ 52,947; 455 714 kWh;
c. GA75 running light load or unloaded then obtained electrical energy and cost savings was USD 31,841; 270,685 kWh;

d. Install new compressor 2x132 kW and 1x132 kW VSD acquired electrical energy and cost savings was USD 108,325; 928,500 kWh;
e. The results of this study needed further research such as carried out detailed audits and perform Cost Benefit Analysis or Investment Grade Audit (IGA).

6. Reference

[1] Abels and Kisscock (2011). Optimizing compressed Air Storage for Energy Efficiency: SAE International
[2] Arikunto, Suharsimi (2002). Prosedur Penelitian Suatu Pendekatan Praktis. Jakarta: Rineka Cipta.
[3] Beals C. E, (2009). Storage Volume for Air Compressor: Knowing Why, Where and How Much Hels to Minimize Operating Costs: Vectern Plan Services
[4] DESDM (2009). Hasil Studi Konservasi Energi di Indonesia: Jakarta DESDM-JICA.
[5] Harding, Paul and Moore, Ian (2012). Compressed Air System: UNIDO.
[6] Kane and Medaris (2003). Compressor and Compressed Air System: Fairmont Press, USA.
[7] Kazuhide Konitoku (2014). Energy Saving \in Food Industry: Best Practice: ECCJ, Japan.
[8] Thumman and Younger (2007). Handbook of Energy Audit: Fairmont Press, USA.
[9] Warent, Ernest; Galistky, Cristina; et.al (2010). Energy Efficiency Improvement and Cost Saving Opportunities: BNL, USA.

Performance Acceleration on Production Machines Using the Overall Equipment Effectiveness (OEE) Approach

A Mansur[1], R Rayendra[2] and MI Mastur[3]
[1,2,3] Industrial Engineering Department, Faculty of Industrial TechnologyUII, Yogyakarta, Indonesia

Email: agusmansur.am@gmail.com, rozar_rayendra@yahoo.com, 8352200102@uii.ac.id

Abstract. Mistakes during working can trigger a decrease in production level that may lead financial loss to the company. The factors that affect the mistakes are called losses, such as breakdown loss, set up/ adjustment loss, idling and minor stoppage loss, reduced speed loss, reduced yield loss, and rework loss. The objective of the research is to accelerate the performance of the JSW 330T machine in PT. YogyaPresisiTehnikatamaIndustri. JSW 330T is a machine that has the highest downtime numbers. The method for measuring the effectiveness is using the Overall Equipment Effectiveness (OEE). The results of the research show that the JWQ 330T has average rate of the effectiveness (OEE) of 52.66%, availability ratioof 73.43%, performance efficiency rate of 83.58% and quality rate of 84.6%. From the six big losses calculation, the factor that affects the most on the low score of OEE is the breakdown loss which is 58.85% with total time loss of 929.65 hours in a year.

1. Introduction

One of the improvement efforts on manufacturing industries is enhancing the equipments' utility as optimal as possible. Utilization of the existing equipments or machines in the manufacturing industries is good if at least it's about half of the machine's real capability [1]. Manufacturing companies must implement new technologies that aim to improve the plant performance. New technologies are often more complex to be maintained or repaired[2].Therefore, the machines need to be preserved to maintain their function. It will certainly contribute in total production costs, by having the maintenance expense for the equipments and machines of the manufacturing industries, either directly or indirectly [3]. To minimize and overcome the machines' impairment, the maintenance process should be done properly and regularly based on the predefined schedule. Maintenance is the conception of all the activities required to maintain the quality and performance of equipment and keep it functioning properly as in the previous condition to generate the desired products [4]. Therefore, we need a method that is capable to reveal maintenance problems clearly, in order to enhance the equipment performance optimally [5].

Overall Equipment Effectiveness (OEE) is a method used for implementing the Total Productive Maintenance (TPM) in order to keep the equipment in ideal condition by erasing its six big losses. Besides that, OEE is a quantitative metric that has been increasingly used in industry, not only for controlling and monitoring the productivity of production equipment, butalso as an indicator and driver of process and performance improvements[6].The OEE score is calculated by referring the existed concepts, that is determined the three main factors, i.e. the availability rate, performance rate, and quality rate [7]. It is used to identify losses related to equipment for the purpose of increasing the total asset performance and reliability[8].Failures identification from the most significant OEE score categories can affect the machines' effectiveness; therefore it needs the six big losses analysis to find any kind of losses that may occur.

Six big losses is a calculation engine that raises productivity losses for companies are often caused by the use of machine/ equipment that are not effective and efficient. The six big losses can be categorized into three namely downtime (time reduction), Speed losses (decrease speed), and defects (defects) [9]. The six big losses not related to the actual performance of maintenance, but also directly related to the speed and quality [10]. Therefore, to eliminate the six big losses needed improvement activities designed to increase the efficiency of the equipment [11].

The objective of this research is analyzing one of the machines in PT. YogyaPresisiTehnikatamaIndustri (PT. YPTI), that is JSW 330T, using TPM as a maintenance process developed for improving productivity by making the processes more reliable and less wasteful. TPM also facilitates the organizations to achieve higher levels of productivity as well as improve the customer service, morale, and profits [12].Therefore by using the OEE method, we can find the performance of the machine and the factors that contribute on the OEE score from the six big losses calculation.The purpose of OEE is to determine which of the six big losses that affect the decreasing effectiveness of the JSW 330T and the cost of losing for one year with the recommendation of using RCM [13].

2. Literature Review

2.1 Total Productive Maintenance (TPM)

Total Productive Maintenance was first defined in 1970s by Seiichi Nakajima from the Japan Institute of Plant Maintenance[1]. Maintenance is an activity to preserve and maintain the existing facilities as well as do the required reparation, adjustment, or replacement to get the ideal condition of production operation as planned. Total Productive Maintenance began from the idea of Preventive Maintenance and Production Maintenance, which came from America to Japan and developed into a typical Japanese new system and later it's called as Total Productive Maintenance (TPM). TPM is a maintenance concept which involves all the employees to reach effectiveness of the production system through participation as well as productive, proactive, and well-planned maintenance activities [15]. According to [14], the comprehensive definition of TPM covering five elements as follows:

- Its objective is creating a preventive maintenance (PM) system to extend the machines' or equipments' useful life.
- Its objective is optimizing the overall effectiveness of machines or equipments.
- It can be applied in several departments such as engineering, production, and maintenance department.
- It involves all of the internal stakeholders, from the highest management to the employees/ operator in production department.
- It is the development of maintenance system based on PM through motivational management.

2.2 Overall Equipment Effectiveness (OEE)

Overall Equipment Effectiveness is a method used as measuring instrument in TPM implementation program, to maintain the equipment in ideal condition by erasing the six big losses. The measurement of OEE score are determined by these three main ratios:

2.1.1. Availability Ratio
It is a ratio that illustrates the use of the available time for the machines' or equipments' operations. Availability ratio also measures the overall time in which the system is not in operation because of the occurrence damage to the equipment, production preparation and adjustment The formula to calculate this ratio is:

$$\text{Availability Ratio} = \frac{\text{Operation Time}}{\text{Loading Time}} \text{ X } 100\%$$

Loading Time = Total Planned Time – Planned Downtime

2.1.2. Performance Efficiency
It is a ratio that illustrates the machines' capability in producing products. Performance ratio is also a ratio of the actual operating speed of the equipment with ideal speed based on design capacity.The formula to calculate this ratio is:

$$\text{Performance Efficiency} = \frac{\text{Actual Product x Ideal cycle time}}{\text{Operation Time}} \text{ X } 100\%$$

Data required in the performance efficiency of the company in the form of percentage of working hours and production machine parts per month.

2.1.3. Rate of Quality
It is a ratio that illustrates the products' quality produced by the equipments in accordance with standards.Quality Ratio is intended on the loss quality of products such as of defects that occur related to equipment, for conversion into time with the intention of how much time equipment is consumed to produce the defective product.The formula to calculate this ratio is:

$$\text{Rate of Quality} = \frac{\text{Good Products} - \text{TotalDefect}}{\text{Gross Products}} \text{ X } 100\%$$

Data required in the rate of quality of the company in the form of a product that results are good and defective products.

2.3 Six Big Losses
The objective of the Six Big Losses calculation is to determine the OEE score. The low productivity of the machine causing harm to the company is often caused by the use of machinery / equipment that are not effective and efficient.

3. Methods
The research was conducted by using primary and secondary data. The focus of this research is on the JSW330T machine, which has the highest downtime numbers. The primary data was used for measuring the production process time, observing the amount of defected products and idle time. As explained above that to calculate the OEE we need to calculate the availability ratio, performance efficiency, an the rate of quality first. The formula of OEE score is as follows:

OEE (%) = Availability Ratio (%) x Performance Efficiency (%) x Rate of Quality (%)

After calculating the OEE score, the score can be compared with the standard of Japan Institute of Plant Management (JIPM) so the company's position can be seen towards the world class companies. This research will also calculate the main factors in the production plant that lead the company's financial loss, in order to conduct preventive actions for improvement in the future research.

4. Result and Discussion

4.1 Result

4.1.1 Production Data

Production and defect data from the Production Division of the PT. YPTI for one year starts on January to December 2014 is shown below:

Table 1. Production Data of the JSW 330T Machine

No	Months	Part Production (unit)	Good Product(unit)	Defect(unit)		
				Scrap	Rework	Total
1	January	2901	2474	427	0	427
2	February	7739	7092	657	0	657
3	March	17354	16139	1232	0	1232
4	April	4761	3983	790	0	790
5	May	18085	16509	1598	0	1598
6	June	23545	21879	1697	0	1697
7	July	8107	7112	1008	0	1008
8	August	3549	2999	569	0	569
9	September	9074	7459	1633	0	1633
10	October	4425	3448	990	0	990
11	November	10256	8773	1511	0	1511
12	December	2253	2004	252	0	252

4.1.2 OEE Score

The OEE of the JWS 330T machine in PT.YPTI during January to December 2014 are as follow:

Table 3. Overall Equipment Effectiveness Calculation

No	Months	Availability Ratio (%)	Performance Efficiency (%)	Rate of Quality (%)	OEE (%)	JIPM (%)
1	January	79.11	78.59	82.74	51.44	85
2	February	58.94	88.91	90.74	47.56	85
3	March	86.35	90.81	92.37	72.43	85
4	April	69.48	78.80	80.17	43.90	85
5	May	72.76	89.10	90.32	58.55	85
6	June	83.08	90.51	92.43	69.50	85
7	July	83.13	84.66	85.83	60.40	85
8	August	77.90	79.16	81.03	49.97	85
9	September	65.50	80.08	78.11	40.97	85
10	October	50.87	74.60	71.29	27.05	85
11	November	69.59	84.08	82.78	48.44	85
12	December	84.44	83.63	87.43	61.74	85

Average	**73.43**	**83.58**	**84.60**	**52.66**	**85**

4.1.3 Six Big Losses Identification on JSW 330T Machine

After calculating the six big losses percentage toward the effectiveness of the JSW 330T machine for one year, the average losses from January 2014 – December 2014 are as follow:

Table 4. Big Six Losses Percentage of the JSW 330T Machine

No.	Six Big Losses	Totals of Time Loss(hours)	Losses Percentage(%)
1	Breakdown Loss	929.65	58.85
2	Set up and Adjustment Loss	30.00	1.90
3	Idling Minor Stoppages	21.50	1.36
4	Reduced Speed Loss	343.39	21.74
5	Rework Loss	0.00	0.00
6	Scrap / Yield Loss	255.08	16.15
	Total	**1579.62**	

4.2 Discussions

From the result above, it can be seen that there were fluctuations in the availability time, it is because the down time machines tend to fluctuate due to frequent damage to the machine, which will directly affect the performance efficiency score.This company's rate of quality (84.6%) is lower than the JIPM standard (99.9%), while the performance ratio is 83.5% (JIPM standard is 95%) and the availability ratio is 73.43% which is too far from the JIPM standard.

The average OEE score of the company is 52.66%, it is also far from the JIPM standard which is 85%, although the score is considered proper if it is compared with the OEE score of some industries across the world that have been surveyed by the JIPM.

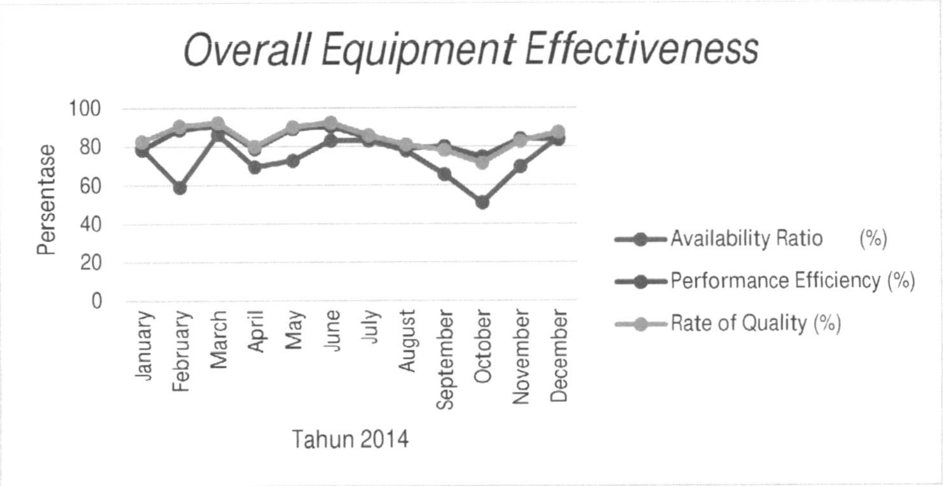

Figure 1. OEE Score of PT. YPTI

From the instable scores above, it can be concluded that the company's capability in conducting production process was not consistence. Therefore, it will be difficult for the company to get the guarantee assurance for its products as well as the production on time.

From the six big losses measurement, it shows that the machine experiences frequent breakdown, in which 929.65 hours or 116 days in a year. The reduced speed loss also often happens, in which 343 hours in 2014. It shows that the JSW 330T's reliability is considered low. Therefore, we recommend the company to conduct a more intensive maintenance or consider purchasing a new machine to replace it.

5. Conclusion And Recommendation

5.1 Conclusion

Based on the research on JSW 330T machine during January 2014 – December 2014, it can be concluded that:

- From the average rate of the OEE calculation is 52.66%, means the effectiveness of the JSW 330T is far from the required standard based on JIPM that is 85%.
- The most dominant factor of the six big losses which causes the decreasing on the production effectiveness or gives the biggest contribution on the OEE score is the breakdown losses (58.85%) with total time loss of 929.65 hours.

5.2 Recommendation

From the research results, we recommend the company to pay more attention to routine repairs to components of the machine so as to minimize the damage. As for further research can further optimize the increased effectiveness of the machine to determine the time interval determination of replacement parts for all machine components resulting in frequent downtime and determine total cost of losses to the company for one year on the machine JSW 330T so that the results obtained are more specific and more leverage in improving the effectiveness of the machine by using RCM.

6. Reference

[1] NakajimaS 1988*Introduction to Total Productive Maintenance* (Portland: Productivity Press Inc)

[2] Swanson L1997An Empirical Study of The Relationship Between Production Technology and Maintenance Management*International Journal of Production* Economics **53** 191-207

[3] Blanchard SB 1997 An Enhanced Approach for Implementing Total Productive Maintenance in The Manufacturing Environment *Journal of Quality in Maintenance Engineering* **3**

[4] Supandi 1990 *ManajemenPerawatanIndustri* (Bandung: Ganeca Exact)

[5] PatrikJ and Lesshammar M 1999 Evaluation and Improvement of Manufacturing Performance Measurement Systems The Role of OEE *International Journal of Operations & Production Management***19** 55 – 78

[6] Garza-Reyes JA, Eldridge S, Barber KD, and Meier H 2010Overall Equipment Effectiveness (OEE) and Process Capability (PC) Measures: A Relationship Analysis*International Journal of Quality & Reliability Management***27** 48 – 62

[7] Habib, Agil, Septiyan, andSupriyanto H 2012 PengukuranNilai Overall Equipment Effevctiveness (OEE) SebagaiPedomanPerbaikanEfektivitasMesinCNC Cutting, *JurnalTeknikPomitsJurusanTeknikIndustriFTI ITS***1** 1 – 6

[8] Muchiri P 2006*Performance Measurement Using Overall EquipmentEffectiveness (OEE): Literature Review and Practical Application Discussion*

[9] TriwardaniD,Rahman A, and Tantrika CFM 2012 *Analysis of Overall Equipment Effectiveness to Reduce Six Big Losses on Production of Dual Filter DD07 Machine*(Malang: UniversitasBrawijaya)
[10] Sherwin D 2000*A Review of Overall Models for Maintenance Management*(Sweden: Lund University Institute of Technology)
[11] Bamber CJ, Sharp JM, and Hides MT1999Factors affecting successful implementation of Total Productive Maintenance: A UK Manufacturing Case Study Perspective*Journal of Quality in Maintenance Engineering***5** 162 – 181
[12] Jain A, Bhatti R, Deep HS& Sharma SK 2012Implementation of TPM for Enchancing OEE of Small Scale Industry*International Research Journal Consortium***1**
[13] Wudhikarn R 2012Improving Overall Equipment Cost Loss Adding Cost of Quality*International Journalof Production Research***50**:12, 3434 – 3449
[14] Yoshikazu T and Osada T 2000*Total Productive Maintenance (TPM)*(Tokyo: Asian Productivity Organization)
[15] Suzaki K 2001 *PenerapanPerbaikanBerkesinambungantranslated by Ir. KristiantoJahja*(Jakarta:Productivity & Quality Management Consultant)

Analysis Clustering of Electricity Usage Profile Using K-Means Algorithm

[1] Yasirli Amri, [2]Amanda Lailatul Fadhilah, [3]Fatmawati, [4]Novi Setiani, [5]Septia Rani*

[1,2,3,4,5] Department of Informatics Engineering, Universitas Islam Indonesia, Yogyakarta, Indonesia

E-mail: septia.rani@uii.ac.id

Abstract. Electricity is one of the most important needs for human life in many sectors. Demand for electricity will increase in line with population and economic growth. Adjustment of the amount of electricity production in specified time is important because the cost of storing electricity is expensive. For handling this problem, we need knowledge about the electricity usage pattern of clients. This pattern can be obtained by using clustering techniques. In this paper, clustering is used to obtain the similarity of electricity usage patterns in a specified time. We use K-Means algorithm to employ clustering on the dataset of electricity consumption from 370 clients that collected in a year. Result of this study, we obtained an interesting pattern that there is a big group of clients consume the lowest electric load in spring season, but in another group, the lowest electricity consumption occurred in winter season. From this result, electricity provider can make production planning in specified season based on pattern of electricity usage profile.

Keywords—electricity usage; clustering; K-Means algorithm

1. Introduction

Electricity is one of the most important needs of the society. It serves as an economic resource which is the most needed in many activities. Demand for electricity will increase in line with population and investment growth. Electricity usage is an important factor in many sectors, such as in household sector, industrial sector, and also government sector.

Nowadays, the development of information technology has shown how big the role of data that we encounter every day. Existing data will have no mean if it is not treated with the correct methods. Treating the existing data with the correct methods will produce information that can support the work of human beings in all fields. It including with electricity data that became the topic of this paper. By using data mining, which is a study that can process a set of data into valuable information, we can determine the largest or smallest electricity consumption at certain seasons by calculating the used of power (kWh meters).

One of the tasks in data mining that can be used to determine the electricity usage profile by power calculation is clustering. We process the data of electricity usage consist of data power in kWh meter by grouping (clustering) client data using the *K*-Means algorithm. The intention of this grouping is to analyze the patterns of similarity about electricity consumption made by clients.

The purpose of analysis electricity consumption data is to help determining in which season the electricity at most and the least used. This information can help identifying patterns of clients. In the future, clients are expected to be aware of the amount of electricity power that they used. In addition, the analysis in this paper can give information to the provider of electricity in order to adjust the production of energy to the actual needs in the field. Adjustment of the amount of electricity production is important because storing electricity is expensive. Thus the amount of production must be adapted to the needs of consumption.

* To whom any correspondence should be addressed.

The rest of the paper is organized as follows. Section 2 describes the literature review. Section 3 describes the experiment. Section 4 describes the results and discussion. Finally, the conclusions of this works are described in section 5.

2. Literature Review

2.1. K-Means Algorithm

K-Means is one of the algorithm used for clustering which will split the data into several groups. *K*-Means algorithm is one of the method of data non-hierarchical clustering that can group the data into several clusters based on the similarity of the data. This mechanism enables data which have the same characteristics are grouped into one cluster and data that have different characteristics are grouped in other clusters.

To determine the cluster label of any data, calculated the distance between the data with each cluster centre. There are several ways that can be used to perform distance calculations, such as Euclidean distance, Manhattan distance, and Chebichey distance.

The *K*-Means method aims to minimize the sum of squared distances between all points and the cluster centre. This procedure consists of the following steps, as described in [1].

- Step 1: Select k out of the given n patterns as the initial cluster centres. Assign each of the remaining $n-k$ patterns to one of the k clusters; a pattern is assigned to its closest centre/cluster.
- Step 2: Compute the cluster centres based on the current assignment of patterns.
- Step 3: Assign each of the n patterns to its closest centre/cluster.
- Step 4: If there is no change in the assignment of patterns to clusters during two successive iterations, then stop; else, go to Step 2.

It is obvious in this description that the final clustering will depend on the initial cluster centres chosen and on the value of K. The latter is of the most concern since this requires some prior knowledge of the number of clusters present in the data, which, in practice, is highly unlikely.

K-Means algorithm was chosen because it is simple, easy to implement, widely used in many fields, and the most important is that it has the ability to cluster big dataset. In addition, *K*-Means algorithm is not affected to the order of objects. When using *K*-Means to cluster numeric data, it will be easier to find outlier and noise. The advantage of *K*-Means algorithm is in accordance with the characteristics of electricity data, in which the data is numerical and the size of dataset is quite big.

2.2. Outlier Detection

Outliers are data objects with characteristics that are considered to be very different from other data objects within the dataset. Outlier appeared in the form of extreme value to a single variable or combination of variables [2]. Data cleaning with removing outliers is very important. Including outliers in data mining algorithm is sometime will produce inaccurate result because the damaging characteristics of the data processed. Performing outlier detection and remove outliers will make the data more accurate, thus the data mining result will be more accurate.

2.3. Data Mining on Electricity Consumption

In the previous study, [3] addressing the segmentation of electricity clients according to social class, contracted power, family size and type of tariff for designing the energy efficiency solutions. [4]

employ support vector machine optimization to predict power load based on time series matrix. This prediction aims to be more easily regulate the use of electricity. In another study, it was mentioned that the genetic algorithm in the improvement of fuzzy clustering is used to predict changes in demand for electricity consumption for seasonal and monthly, especially in developing regions such as in China and Iran [5]. [6] proposed methodology using pattern recognition methodologies to recognize habitual electricity consumption behaviour given the intrinsic characteristics of the family. This approach could be useful to improve small scale forecast, and as a mechanism to enable the provision of tailor-made information to the families.

3. Experiment

3.1. Dataset

We use ElectricityLoadDiagrams2011-2014 dataset, taken from UCI Machine Learning Repository [7]. This dataset contains electricity consumption of 370 instances/clients, collected from 2011 through 2014. There were 140256 attributes and each attribute stating the value of electricity consumption in kW per 15 minutes from 2011-2014 (attribute type is numerical). The example of dataset can be seen in figure 1.

	X	MT_001	MT_002	MT_003	MT_004	MT_005	MT_006	MT_007	MT_008	MT_009	MT_010
36926	2012-01-20 15:30:00	17.7665	24.18208	2.606429	99.5935	48.78049	175.5952	2.826456	259.2593	104.8951	65.5914
36927	2012-01-20 15:45:00	16.49746	24.89331	2.606429	87.39837	42.68293	166.6667	2.826456	259.2593	104.8951	68.8172
36928	2012-01-20 16:00:00	17.7665	24.18208	2.606429	83.33333	43.90244	163.6905	2.261164	265.9933	99.65035	61.29032
36929	2012-01-20 16:15:00	16.49746	24.89331	2.606429	95.52846	42.68293	160.7143	2.826456	262.6263	76.92308	66.66667
36930	2012-01-20 16:30:00	15.22843	25.60455	2.606429	91.46341	39.02439	148.8095	2.826456	235.6902	85.66434	78.49462
36931	2012-01-20 16:45:00	16.49746	25.60455	2.606429	93.49593	39.02439	166.6667	2.261164	249.1582	92.65734	60.21505
36932	2012-01-20 17:00:00	35.53299	24.89331	2.606429	103.6585	39.02439	166.6667	2.261164	259.2593	62.93706	51.6129
36933	2012-01-20 17:15:00	26.64975	25.60455	2.606429	103.6585	39.02439	172.619	2.826456	276.0943	57.69231	50.53763
36934	2012-01-20 17:30:00	16.49746	26.31579	2.606429	115.8537	53.65854	208.3333	3.391747	276.0943	62.93706	74.19355
36935	2012-01-20 17:45:00	32.99492	26.31579	2.606429	115.8537	57.31707	187.5	4.522329	303.0303	59.44056	78.49462
36936	2012-01-20 18:00:00	35.53299	27.73826	2.606429	134.1463	69.5122	247.0238	5.08762	336.7003	73.42657	87.09677
36937	2012-01-20 18:15:00	17.7665	27.73826	2.606429	158.5366	80.4878	288.6905	5.652911	373.7374	90.90909	93.54839
36938	2012-01-20 18:30:00	19.03553	26.31579	2.606429	189.0244	85.36585	312.5	4.522329	390.5724	103.1469	104.3011
36939	2012-01-20 18:45:00	17.7665	26.31579	2.606429	180.8943	91.46341	333.3333	5.08762	427.6094	106.6434	100
36940	2012-01-20 19:00:00	19.03553	26.31579	2.606429	178.8618	92.68293	333.3333	6.218202	427.6094	111.8881	102.1505
36941	2012-01-20 19:15:00	17.7665	29.16074	2.606429	199.187	93.90244	324.4048	5.08762	434.3434	111.8881	106.4516
36942	2012-01-20 19:30:00	19.03553	29.87198	2.606429	225.6098	104.878	327.381	6.218202	491.5825	108.3916	110.7527
36943	2012-01-20 19:45:00	19.03553	31.29445	2.606429	209.3496	109.7561	345.2381	5.652911	474.7475	103.1469	113.9785
36944	2012-01-20 20:00:00	17.7665	27.02703	4.344049	184.9593	103.6585	330.3571	5.652911	430.9764	101.3986	112.9032
36945	2012-01-20 20:15:00	19.03553	32.00569	6.081668	189.0244	100	333.3333	6.218202	424.2424	103.1469	113.9785
36946	2012-01-20 20:30:00	20.30457	30.58321	4.344049	197.1545	104.878	348.2143	7.348785	400.6734	106.6434	93.54839
36947	2012-01-20 20:45:00	19.03553	29.16074	3.475239	195.122	95.12195	330.3571	7.348785	383.8384	94.40559	90.32258
36948	2012-01-20 21:00:00	17.7665	28.4495	3.475239	186.9919	87.80488	330.3571	7.914076	370.3704	83.91608	88.17204
36949	2012-01-20 21:15:00	19.03553	27.73826	3.475239	182.9268	91.46341	315.4762	6.218202	367.0034	82.16783	79.56989
36950	2012-01-20 21:30:00	17.7665	27.02703	3.475239	182.9268	86.58537	315.4762	8.479367	353.5354	80.41958	78.49462
36951	2012-01-20 21:45:00	11.42132	27.73826	5.212858	186.9919	91.46341	324.4048	6.783493	373.7374	87.41259	77.41935

Figure1. Example of dataset.

The original ElectricityLoadDiagrams20112014 dataset can be visualized using matrix D as follows:

$$D = \begin{bmatrix} x_{1,1} & \cdots & x_{1,370} \\ \vdots & \ddots & \vdots \\ x_{140256,1} & \cdots & x_{140256,370} \end{bmatrix}$$

The number of row is 140256 (stating the number of attributes) and the number of column is 370 (stating the number of instances). Data D is transposed so that rows declare instances and columns declare attributes.

$$D' = \begin{bmatrix} x_{1,1} & \cdots & x_{1,140256} \\ \vdots & \ddots & \vdots \\ x_{370,1} & \cdots & x_{370,140256} \end{bmatrix}$$

We perform dimension reduction to the dataset because the number of attribute is too much. Steps of dimension reduction are as follow:

- First, select only certain duration of time. We select the electricity consumption of 370 instances between March 21, 2013 and March 20, 2014.
- From the selected duration of time, we aggregated (summed) the electricity needs for each season. Dataset's information stated that the data is retrieved in Portugal time. Because in Portugal there are four seasons, we perform data aggregation as follows:
  - For each instance, all of the electricity consumption from March 21, 2013 until June 20, 2013 is added to obtain the value of total electricity consumption in spring.
  - For each instance, all of the electricity consumption from June 21, 2013 until September 22, 2013 is added to obtain the value of total electricity consumption in summer.
  - For each instance, all of the electricity consumption from September 23, 2013 until December 20, 2013 is added to obtain the value of total electricity consumption in autumn.
  - For each instance, all of the electricity consumption from December 20, 2013 until March 20, 2014 is added to obtain the value of total electricity consumption in winter.
- Thus the data dimension will be reduced. Each instance will have four attributes, as follows:

$$D_r \text{ (after dimension reduction)} = \begin{bmatrix} x_{1,1} & x_{1,2} & x_{1,3} & x_{1,4} \\ \vdots & \vdots & \vdots & \vdots \\ x_{370,1} & x_{370,2} & x_{370,3} & x_{370,4} \end{bmatrix}$$

Description: The first column is the amount of electricity consumption during spring, the second column is the amount of electricity consumption during summer, the third column is the amount of electricity consumption during autumn, while the fourth column is the amount of electricity consumption during winter.

Data D_r will be used for visualization and clustering.

3.2. Tools

In this study we used Rstudio version 3.2.2 [8] with several libraries such as xts package, ggplot, and outliers. We also used Weka version 3.7.4 [9] to perform clustering task.

4. Results and Discussion

4.1. Results

The example of result dataset after the process of dimension reduction can be seen in the figure2.

	row.names	V1	V2	V3	V4
1	MT_001	36017.77	64649.75	98902.28	19529.19
2	MT_002	251569	308256	240322.9	225683.5
3	MT_003	17935.71	16034.75	14913.12	14728.06
4	MT_004	839878	944004.1	981579.3	1212486
5	MT_005	352457.3	427929.3	460042.7	577119.5
6	MT_006	1503557	1546244	1738848	1951997
7	MT_007	31565.86	157960.4	33873.94	45366.87
8	MT_008	2151296	2434896	2236761	2277626
9	MT_009	381304.2	447903.8	516484.3	496896.9
10	MT_010	548790.3	438402.2	535288.2	508466.7
11	MT_011	293096.9	335868.9	350378.5	421040.2
12	MT_012	0	333385.1	1259196	1513834
13	MT_013	602344.9	714411.3	679792.7	652699.8
14	MT_014	356758.8	336899.2	407737.7	473116.8
15	MT_015	58722.62	176699.4	163892.6	448886.9
16	MT_016	254970.4	313702.2	281698.2	336478.9
17	MT_017	393219.2	454653.2	427071.9	449489.7

Figure 2. The data sample after dimension reduction.

Description:
MT_001, *MT_002*, etc. are the identification number for clients. Column *V1* states the total amount of electricity consumption during spring. Column *V2* states the total amount of electricity consumption during summer. Column *V3* states the total amount of electricity consumption during autumn. Column *V4* states the total amount of electricity consumption during winter.

4.2. Data Visualization

The visualization of electricity consumption from 370 clients in each season can be seen in the figure 3. The x-axis is the clients and y-axis is the electricity consumption in kW. Figure 3 shows that the majority of clients use electricity no more than 50 million kW in each season.

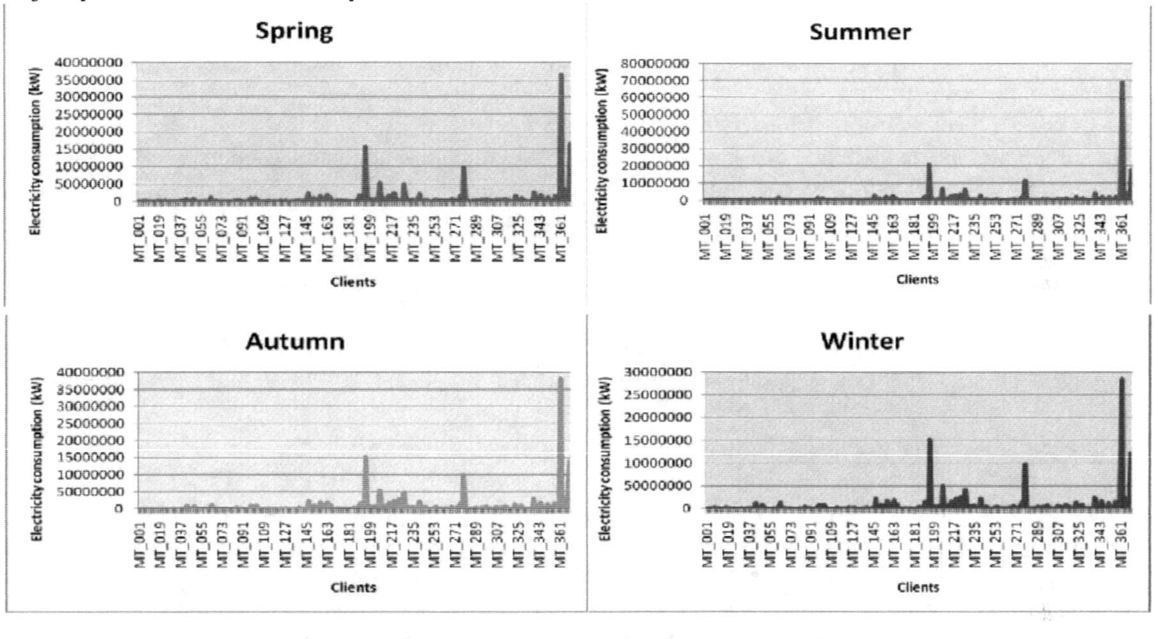

Figure 3. Visualization of electricity consumption in each season.

The comparison graph of electricity consumption between the four seasons can be seen in the figure 4. The x-axis is the clients and y-axis is the electricity consumption in kW. Figure 4 shows that for the majority of clients, the highest amount of electricity consumption is during the summer.

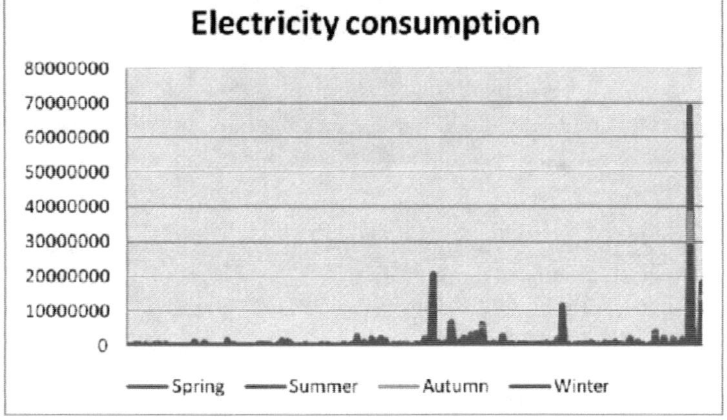

Figure 3. Comparison of electricity consumption between the four seasons.

4.3. K-Means Clustering

For clustering, we use data D_r (as explained in Section 2) as the input. We have four scenarios in the process of clustering, as follows:

- Scenario 1: Using the data D_r without outlier detection and then group the data into 4 clusters.
- Scenario 2: Using the data D_r with outlier detection (remove the outlier) and then group the data into 4 clusters.
- Scenario 3: Using the data D_r without outlier detection and then group the data into 5 clusters.
- Scenario 4: Using the data D_r with outlier detection (remove the outlier) and then group the data into 5 clusters.

For each scenario, we measure the Sum Square Error (SSE) and the number of iteration. SSE describes the value of standard deviation of each cluster to the data centre. The bigger SSE, it means that the degree of data similarity in one cluster is lower. Number of iteration describes the length of clusters formation process.

The clustering result from each scenario can be seen in table 1. From table 1, it can be seen that the largest SSE is obtained when we perform clustering with 4 clusters and outliers are not included (Scenario 2), in which, this scenario also has the most minimal number of iteration, that is 14.The best SSE is obtained when we use 5 clusters without eliminating outliers (Scenario 3).

Table 1. Clustering result.

No	Scenario	SSE	Number of iteration
1	4 clusters without eliminating outliers	0.174	20
2	4 clusters with eliminating outliers	0.752	14
3	5 clusters without eliminating outliers	0.134	21
4	5 clusters with eliminating outliers	0.509	22

The more detail result from Scenario 3 (5 clusters without eliminating outliers) can be seen in table 2. From table 2, it can be seen that the cluster 0 – 3 had the biggest average use of electricity in summer and the smallest average use of electricity in winter. There is only one distinct cluster, cluster number 4, which had the smallest average use of electricity in spring.

Table 2. Clustering result scenario 3.

Cluster Number	Number of cluster member	$\overline{v1}$	$\overline{v2}$	$\overline{v3}$	$\overline{v4}$
0	3	139.584.280,44	165.888.048,23	129.645.729,66	123.076.512,72
1	17	22.810.155,19	31.293.383,26	24.732.531,25	22.188.694,04
2	47	7.148.526,50	8.987.654,70	7.206.537,18	6.817.580,39
3	1	364.726.900,00	687.392.800,00	380.801.800,00	284.465.900,00
4	302	1.008.178,06	1.245.319,00	1.059.189,86	1.061.011,85

Table description:

$\overline{v1}$: the average use of electricity in spring (kW)
$\overline{v2}$: the average use of electricity in summer (kW)
$\overline{v3}$: the average use of electricity in autumn (kW)
$\overline{v4}$: the average use of electricity in winter (kW)
■ : the largest average use of electricity
□ : the smallest average use of electricity

4.4. Discussion

Based on the result from the third scenario, we got five clusters describing below:

1. Group of clients with very high electricity consumption (Cluster 3)
   There is 1 client with the highest consumption occurred on the summer season (687.392.800,00 kW) and the lowest consumption occured on the winter season (284.465.900,00 kW).
2. Group of clients with high electricity consumption (Cluster 0)
   There are 3 clients with the highest average of electricity consumption,165.888.048,23 kW, occurred on the summer season and the lowest average of electricity consumption, 123.076.512,72 kW, occurred on the winter season.
3. Group of clients with medium electricity consumption (Cluster 1)
   There are 7 clients with the highest average of electricity consumption, 31.293.383,26 kW, occurred on the summer season and the lowest average of electricity consumption, 22.188.694,04kW, occurred on the winter season.
4. Group of clients with low electricity consumption (Cluster 2)
   There are 47 clients with the highest average of electricity consumption, 8.987.654,70 kW, occurred on the summer season and the lowest average of electricity consumption, 6.817.580,39kW, occurred on the winter season.
5. Group of clients with very low electricity consumption (Cluster 4)
   There are 302 clients with the highest average of electricity consumption, 1.245.319,00 kW, occurred on the summer season and the lowest average of electricity consumption, 1.008.178,06 kW, occurred on the spring season.

This result shows that the most dominant group is the group of clients with very low electricity consumption (Cluster 4). The average electricity consumption from this group is very contrast to another client groups (Cluster 0, 1, 2, and 3). Another difference between cluster 4 and another cluster is they consume the lowest electricity on the spring season but another client groups consume the lowest electricity on the winter season.

5. Conclusions

Based on the analysis above, we obtained an interesting fact that is the biggest electricity consumption actually occurs in the summer. For some clients, the smallest electricity consumption is during winter. But for the majority of clients, the smallest electricity consumption occurs in the spring. This may be happened due to the differences in geographic location, the differences in economic level, and the differences in the use of electricity as the needs of industrial, household, office. The majority of clients are clustered in the group of the lowest electricity consumption. This group has very different electricity usage characteristic with other groups. There is an additional fact that exist a client who consume electricity almost half of the total electricity consumption.

References

[1] Murty MN and Devi VS 2011 Pattern recognition, An algorithmic approach (London: Springer)

[2] Hodge V H and Austin Jim 2004 A survey of outlier detection methodologies *J. Artificial Intelligence Review* **22** 85-126

[3] Pombeiro H Pina A and Silva C Analyzing residential electricity consumption patterns based on consumer's segmentation *Proc. Int. 1st. Workshop on Information Technology for Energy Applications (Lisbon, Portugal, 6-7 September 2012)* vol 923 ed P Carreira and V Amaral (Portugal: CEUR-WS) pp 29-38

[4] Dong-xiau N, Yong-li W and Xiao-yong M 2010 Optimization of support vector machine power load forecasting model based on data mining and Lyapunov exponents *J. Central South University of Technology* **17** 406-12

[5] Azadeh A, Saberi M, Ghaderi S F, Gitiforouz A and Ebrahimipour V 2008 Improved estimation of electricity demand function by integration of fuzzy system and data mining approach *J. Energy Conversion and Management* **49** 2156-77
[6] Abreu J M, Pereira F C and Ferrao P 2012 Using pattern recognition to identify habitual behavior in residential electricity consumption *J. Energy and Building* **49** 479–87
[7] https://archive.ics.uci.edu/ml/datasets/ElectricityLoadDiagrams20112014
[8] https://www.rstudio.com/products/RStudio/
[9] http://www.cs.waikato.ac.nz/~ml/weka/downloading.html

PERFORMANCE MEASUREMENT USING KPKU- BUMN IN X SCHOOL EDUCATION FOUNDATION

Sugih Arijanto[1], Ambar Harsono[2], Harsono Taroepratjeka[3]
Jurusan Teknik Industri, Institut Teknologi Nasional (Itenas) Bandung

Email: s_arijanto@yahoo.com

Abstract. The purpose of this research is to determine X School's Strengths and Opportunity of Improvement through performance measurement using KPKU-BUMN (Kriteria Penilaian Kinerja Unggul – Kementerian Badan Usaha Milik Negara). KPKU-BUMN is developed based on Malcolm Baldrige Criteria for Performance Excellent (MBCfPE). X school is an education foundation at Bandung that has provides education from kindergarten, elementary school, to junior and senior high school. The measurement is implemented by two aspects, Process and Result. The Process is measured by A-D-L-I approaches (Approach-Deployment-Learning- Integration), on the other hand The Result is measured by Le-T-C-I approach (Level-Trend- Comparison-Integration). There are six processes that will be measured: (1) Leadership, (2) Strategic Planning, (3) Customer Focus, (4) Measurement, Analysis and Knowledge Management, (5) Work Force Focus, and (6) Operation Focus. Meanwhile, the result are (a) product & process outcomes, (b) customer-focused outcomes, (c) workforce-focused outcomes, (d) leadership & governance outcomes, and (e) financial & market outcomes. The overall score for X School is 284/1000, which means X School is at "early result" level at "poor" global image.

Keywords: Performance Measurement, KPKU-BUMN, MBCfPE, Excellence

1. Introduction

Performance Measurement is important for an organization to identify its strength and opportunity for improvement *to increase its competitive advantages.* Malcolm Baldrige National Quality Award is a form of appreciation for companies in the United States was evaluated using the Malcolm Baldrige Criteria for Performance Excellence (MBCFPE). MBCFPE is a set of critical questions in the management and performance of the organization, which is compiled by the US Department of Commerce, National Institute of Standards and Technology, based on the best values and experience of US's superior world-class companies. At first the criteria used by companies in the United States to increase competitiveness to face global competition, especially with Japan.

[1,2,3] Industrial Engineering Department, Institut Teknologi Nasional (Itenas) Bandung Jalan PHH Mustofa 23 Bandung 40124 Indonesia; Indonesia Quality Award (IQA) Examiners; KPKU-BUMN Evaluator [(1)] ; Judge-Committee Indonesia Quality Award[(3)]

MBCfPE is built upon Eleven Concepts and Procedures Core Values are interrelated as follows: (1) leadership has the vision, (2) All-excellence-an-directed by the customer, (3) learning organization and employees, (4) Valuing employee and partners, (5) Agility, (6) Focus on the future, (7) The management of innovation, (8) Management based on facts, (9) The responsibility of society, (10) Focus on results and value creation, (11) System Perspective. Eleven Concepts and Administration Core Values is dissolved into the six criteria process and the criteria Results: (i) Leadership, (ii) Strategic Planning, (iii) Customer Focus, (iv) Measurement, Analysis and Knowledge Management, (v) Focus Power Work, (vi) Focus Surgery, (vii) results.

In Indonesia, MBCfPE adopted in Indonesia Quality Award which aims to raise awareness that performance excellence as an essential element for companies to compete and as a means to share information about successes in the implementation of the strategy performance and benefit from the use of this strategy. The Ministry of State-Owned Enterprises (SOEs) adapting MBCfPE into Excellent Performance Assessment Criteria –Kriteria Penilaian Kinerja Unggul, KPKU - to assessing the performance of state-owned companies.

X School is one of the private education foundation in Bandung that has provides education from kindergarten, elementary school, to junior and senior high school. To enhance its performance X School requires measuring overall performance. Recently X School evaluate it accreditation score to analyze its performance but one of the weaknesses of accreditation is its more focused to academic aspect not the health of organization as an entity. In the last several years X School develop key performance indicator to evaluate its performance, yet the measurement is only implemented on trial phase.

1.1. Problem Identification

X School needs to analyze its competitiveness level by performance measurement. The problems are the accreditation level isn't sophisticated enough to evaluate organization position compared to competitors, on the other hand the key performance indicator's that were built by X School is still on the beginning level of development. KPKU as one of the performance systems can be implemented to solve X School problem.

The research is held by team using collaborated researches. The prime researcher are Prof. Harsono Taroepratjeka, Ambar Rukmi Harsono and Sugih Arijanto, with several research assistant. Measuring the performance team are: Performance Measurement Leadership and Results of Leadership and Governance by Inayah, Measuring Strategic Planning and the result of Financial by Fitriani. Measuring Customer Focus and Results of Customer Focus by Triwahyuni, Measurement Analysis and Knowledge Management by Yuniarti, Performance Measurement The focus of workforce and Results of workforce by Fajarwati, Focus performance Measurement and Results of Operations and Education Services Program and Process by Mayani.

The objectives of the research is to determine X School score and to identify its Strengths and Opportunities for Improvement (OFI) based on the results of performance measurement using KPKU-BUMN based on Malcolm Baldrige Criteria for Performance Excellent (MBCFPE)

2. Literature

2.1 Performance Measurement

Performance measurement according to Lynch and Cross (1993) in Yuwono (2004), is a way to measure the direction and speed of change, or in other words the performance measurement is an action taken to measure the various activities that exist in the company or organization. Performance measurement serves to provide feedback to the company for the performance that has been done, the action taken in accordance with a predetermined plan or not, and can give hope of our future.

Measurements of performance by Christian (2010) is an assessment process work progress against objectives and targets have been defined previously, including the information above: efficiency of resource use in producing goods and services; quality of goods and services (how well the goods and services delivered to customers and the extent to which the customer is satisfied).

There are various methods for measuring the performance, for example: the Balanced Scorecard, performance Prism, Institutional Accreditation, and the Malcolm Baldrige Criteria for Performance Excellence. The method used in this study is the Malcolm Baldrige Criteria for Performance Excellence.

2.2. Malcolm Baldrige Criteria For performance Excellence (MBCFPE)

MBCfPE is a system to enhance overall organizational performance continuously that uses measurements with predetermined criteria and provides feedback on the performance of the organization as an entity in providing the excellent services.

The benefits of MBCFPE are: (1) To help improve the performance and effectiveness of the organization as a whole, (2) Facilitate good communication for all the elements that exist within the organization so that it can communicate well, (3) Fix the values that affect consumers and contribute to the success of the market share as well as improve performance and capability of the organization concerned, and (4) Can be used as a tool to understand and manage the performance of the organization, planning, and training of the workers.

MBCfPE criteria are the basis for assessment of self-organization, for the award, and for providing feedback to the applicant organization, in their efforts to improve overall performance. The criteria also have three important roles to strengthen the competitiveness of nations: (1) To help improve practice, the ability and the results of organizational performance, (2) Facilitating communication and sharing of information on best practices among all kinds of national organizations, and (3) serves as a work tool in the understanding and management of performance and as a guide for organizational planning and opportunities for learning.

Criteria MBCfPE were built on a foundation of 11 concepts and procedures for core values, namely: Leadership visionary, excellent leading to the student's interests and stakeholders, organizational learning and employee, valuing employees and partners, agility, focus on the future, information management, management by facts, social responsibility, focusing on results and value creation and systemic perspective.

Criteria MBCfPE has 7 based on literature category "Indonesian Quality Award Foundation" (Criteria for Performance excellence (Malcolm Baldrige Criteria for Performance Excellence) in the years 2013-2014 profit organization: (1) leadership, (2) strategic planning, (3) Customer Focus, (4) Measurement, Analysis and Knowledge Management, (5) The focus of Manpower, (6) Focus operation, and (7) results (results Products and Processes, Results Focus on customer, Results Focus Labor, Results Leadership and Governance and Financial Results and Market).

Malcolm Baldrige Criteria for Performance Excellence (MBCfPE) is a method for measuring the overall performance . MBCfPE has seven categories that indicate a systemic framework of interconnected and integrated as shown in Figure 1

IOP Conf. Series: Materials Science and Engineering **105** (2016) 012021 doi:10.1088/1757-899X/105/1/012021

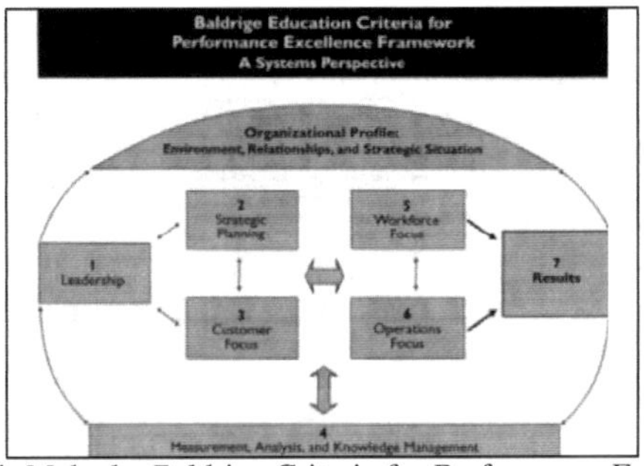

Figure 1 Malcolm Baldrige Criteria for Performance Excellence Framework (Resources: IQAF, 2011)

MBCfPE performance measurement method should be preceded by the preparation of application documents. The application documents are documents concerning matters to be measured by the organization with regard to its performance MBCfPE category. This document is based on the application questions, answers that exist in MBCfPE guidance. Sample questions for the category Work Force focus MBCfPE items can be seen in Figure 2, the format field can be seen in Figure 3.

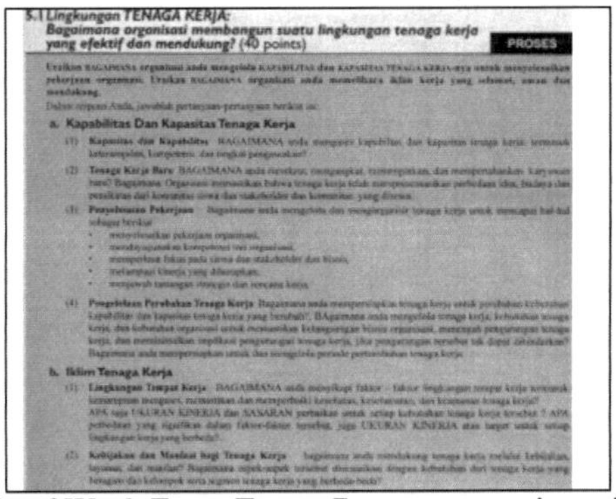

Figure 2 Example of Work Force Focus Category questionnaires List MBCfPE (Source: IQAF, 2013)

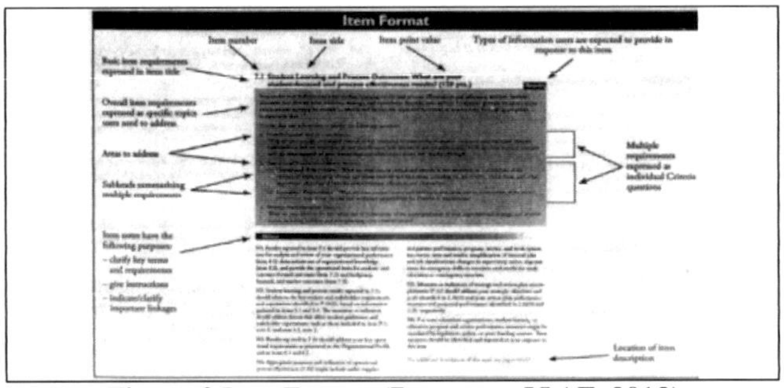

Figure 3 Item Format (Resource: IQAF, 2013)

2.3. Assessment System

The scoring system in MBCfPE based on two elements, the process and results. According to National Institute of Standard and Technology NIST approach, the process refers to the methods used and repaired by an organization in addressing the requirements of the field in the 1-6 category. Assessment process carried out by using a scoring guide of Kriteria Penilaian Kinerja Unggul – Badan Usaha Milik Negara, a Performance Excellent Measurement Criteria that developed by Ministry of State Enterprises. Four factors are used to evaluate Approach, Deployment, Learning, and Integration (ADLI). Four factors are used to evaluate the results of which level, tendencies (trends), comparisons, and integration (LeTCI).

3. Research Methodology

3.1. Problem Identification

The X School's problem is it doesn't know the overall performance level to implement some improvements in facing competition with other schools. Performance measurement can be done using the Malcolm Baldrige Criteria for Performance Excellence (MBCfPE).

3.2. Literature Study

The literature study is a review of the stages used in helping to resolve the problems that have been identified previously. Literature review of the methods used is the measurement of performance and methods of the Malcolm Baldrige Criteria for Performance Excellence (MBCfPE) including the categories and processes and outcomes assessment system.

3.3. Data Collection

Data mining in this research is be done by interviews and examination of documents. The data collection was based on the questionnaires about the profile of the organization and all categories.

3.4. Compilation of Organization Profile

The organization's profiles are performed based on guidelines in the field of education IQAF book version 2013-2014.

3.5. Compilation Application Document

Application Documents are compiled according to the guide book version IQAF education 2013-2014. In answering this question, the table support (Latham, 2013-2014). Incorporation documents and the application process is the process of combining the results of the entire application documents for each category.

3.6. Review Process

The application documents will be reviewed using Approach, Deployment, Learning and Integration (ADLI) . Example format ADLI assessment can be seen in Table1.

Table 1 Review ADLI

Key Factor	*Strengths / OFI*	*As Evidence By*	A	D	L	I	*Criteria Requireme nt*	*Link*

3.7. Result Review

Result Category is reviewed using Level, Trend, Comparison, Integration (LeTCI) Approach. Example of format review LeTCI can be seen in Table 2.

Table 2 Format Review LeTCI

EVIDENCE	ICON	Periode		Le			T				C					Requirements
		< 3 th	>= 3 th	Excellent	Good	Poor	Sustain	Fvrble	Flat	Advrs	Comparison (Competitor / Bencmark)	Leading	EQUAL	Lagging	I	
Total																
Jumlah Kinerja yang Dilaporkan																

3.8. Process and Result Categories Scoring

The scoring processes are conducted for each question according KPKU assessment guide. After the assessment is carried out for each question recapitulation of value to get the value of each item. After receiving the recapitulation value of each item then be converted to IQAF assessment guide which can be seen in Table 3 and determine the final value of each item. The Result is scored by reviewing conducted suitability of existing results to guide the assessment process in the field of results categories.

Table 3 KPKU-IQAF Scoring Conversion

KPKU	IQAF
1	0% or 5%
2	10%, 15%, 20%, or 25%
3	30%, 35%, 40%, or 45%
4	50%, 55%, 60%, or 65%
5	70%, 75%, 80%, or 85%
6	90%, 95%, or 100%

3.9. Malcolm Baldrige Overall Scoring
Malcolm Baldrige Scoring a whole is the sum of the values of all scoring categories.

3.10. Analysis
The analysis was performed on the results of a review ADLI and LeTCI to know the most strength and opportunities for improvement (most OFI) each category. Evaluation of most strength and OFI seen most of the scoring has been done before. At the time of the tagging process done scoring the greatest value to the candidate do most strength and marking the smallest value for the candidate most OFI . Than the selection fields and sub- fields which become most OFI and most strength .

3.11. Conclusions and Recommendations
Conclusions and suggestions are proposed for improvements suggested by the results of X School overall performance assessment organizations to make improvements

4. Data Mining and Processing

4.1 Data Mining
Data is obtained from interviews, surveys and inspection of documents. Data obtained from several sources such as foundations Supervisory X School, X School Board-trustee, Group Supervisors (KPS), Principal, Human Resources, , Finance, Research and Development and Drafting Team Strategic Plan. The data obtained from the answers to the questions from questions of each criteria MBCfPE, contained in the book "Indonesian Quality Award Foundation (IQAF); Performance criteria of excellence (Malcolm Baldrige Criteria for Performance Excellence); Year Edition 2013-2014. The next step create a document using the application Latham guide. After creating the application document that is carried out a review process based ADLI (Approach, Deployment, Learning, Integration) and review results based LeTCI (Level, Trend, Comparison, Integration). Furthermore, quantifies scores for each criterion so that the total score of X School performance can be obtained. Then determine the most strength or strength and most OFI (Opportunity For Improvement) or opportunities for improvement.

4.2 Data Processing
Data processing is the next stage after making application documents. Stages of data processing consists of a review process based ADLI, review result based LeTCI, scoring and scoring for the recapitulation of the whole category.

4.2.1 *Review and Scoring Proses*
The example of review process or assessment process for category Leadership by using ADLI can b

seen in Table 4.

Table 4 *Review Process ADLI*

Item Ref	Persyaratan	KE Relevan	Bukti dari Respon Aplikan	A		D		L			I	
				S		E	K	U	Ev	Ps	AL1	AL2
1.1a	**Visi, Tata Nilai dan Misi**											
1.1a(1)	**Visi dan Tata Nilai**											
	BAGAIMANA pimpinan senior bersama-sama menetapkan Visi dan Tata Nilai X SCHOOL?	2	Yayasan X SCHOOL menetapkan visi dan tata nilai dengan cara: • Para Pembina Yayasan X SCHOOL melakukan rapat pembina untuk menentukan visi yayasan X SCHOOL, kemudian dirumuskan bersama-sama dengan pengurus yayasan, pengawas sekolah, para ketua, dan team pengembang unit (team pengembang sekolah) Yayasan X SCHOOL melalui rapat pembuatan Rencana strategis. Selain membahas perumusan visi, dalam rapat tersebut membahas mengenai Tata Nilai yang akan digunakan X SCHOOL berdasarkan tata nilai yang ditanamkan oleh leluhur pendiri X SCHOOL. Hal tersebut dibuktikan dengan adanya catatan rapat pembina dan catatan rapat pembuatan rencana strategis. Sekolah X SCHOOL menetapkan visi dengan cara: • Kepala sekolah dan team pengembang sekolah membuat visi sekolah dengan rapat berdasarkan perumusan renstra X SCHOOL dan hasil EDS. Metode ini relevan dengan kategori 2 dan 4 yaitu pada 2.1a, 2.1b dan 4.1a.	√			√	√			√	

Analisa ADLI :

Pemimpin senior X SCHOOL sudah memiliki metoda yang jelas dan sistematis dalam pembuatan visi dan tata nilai X SCHOOL, metode tersebut sudah konsisten diterapkan di seluruh unit kerja, metode tersebut sudah ada tindakan perbaikan dari dari metode sebelumnya yaitu pembuatan visi dan misi sekolah yang berdasarkan perumusan renstra X SCHOOL, metode tersebut sudah selaras dengan kategori dan unit lain.

Draft Comment (sebagai kesimpulan dari analisa ADLI)

Strength :

Pemimpin senior X SCHOOL menetapkan visi dan tata nilai X SCHOOL dengan cara Para Pembina Yayasan X SCHOOL melakukan rapat pembina untuk menentukan visi yayasan X SCHOOL, kemudian dirumuskan bersama-sama dengan pengurus yayasan,

The evidence of process implementation as shown by column evidence is used to determine A-D-L-I. The checklist for column S is used when the company has Systematic Approach. (E: Effective, K: Consistence; U: implemented for all unit; Ev: evaluation; P: Improvement; Al: Align)

IOP Conf. Series: Materials Science and Engineering **105** (2016) 012021 doi:10.1088/1757-899X/105/1/012021

4.2.2 Review and Scoring Results

The example results for financial and market outcomes by using the Format Level-Trend-Comparison-Integration or LeTCI can be seen in Table 5.

Table 5 *Review Result* LeTCI

NO	Kode Indikator	NAMA INDIKATOR		Level				TREND			Comparison			INTEGRASI	
				Excellent	Good	Poor	Sust	Fav	Flat	Adv	Pembanding	Lead	Lagging	PROYEKSI	KEKURANGAN
7.1 HASIL KINERJA KEUANGAN DAN PASAR															
7.5.a(1) Kinerja Keuangan															
	1	Pendapatan DSP			1		1								
	2	Pendapatan PP			1		1								
	3	Pendapatan DU			1				1						
	4	Pendapatan SPP			1		1								
	5	Pendapatan lain-lain (Giro, tabungan deposito dan investasi)				1				1					
	6	Keteserapan anggaran dari rencana				1				1					
	7	Rencana anggaran terhadap penerimaan			1					1					
	8	Biaya operasional													1
	9	Biaya langsung													1
	10	Biaya tidak langsung													1
	11	Tabungan dan giro													1
11		Jumlah yang dilaporkan	7	0	5	2	3	0	1	3	0	0	0	0	4
		Presentase yang dilaporkan (%)		0.00%	71.43%	28.57%	42.86%	0.00%	14.29%	42.86%	0.00%	0.00%	0.00%	0.00%	57.14%
		Kekurangan	4	0.00%	45.45%	18.18%	27.27%	0.00%	9.09%	27.27%	0.00%	0.00%	0.00%	0.00%	63.64%
		Jumlah yang harus dilaporkan	11		45.45%				36.36%			0.00%			31.82%

The good level indicates the company achieves the target and has a better achievement compared by previous period. The sustain trend indicates the achievement of company performance is increase, on the other hand, adverse indicates the performance is decrease. Lead comparison indicates the company compares its result to benchmark company. The projection of integration indicates the indicator has been integrated with measurement and strategy to determine the next period target.

Review process begins with determining the key factors based on field research organization profile . After determining the key factors in accordance with the requirements criteria further define the strength or OFI requirements of the criteria that have been answered and show proof (As Evidence By) . Then proceed with the review process focused workforce are reviewed based Approach (A), Deployment (D), Learning (L), and Integration (I) .

Scoring of each category is performed by the ADLI reviews than has been done before. Scoring is done on each criteria requirements by providing a value between 1 and 6 in accordance with the guidelines KPKU which is then converted by the IQAF assessment (Table 3 KPKU-IQAF Scoring Conversion). Scoring is done to every question in every item, then do recapitulation for each multiple requirements.

Value multiple requirements recapitulation made reference to the area to address. Recapitulation value areas to address as a reference to recapitulation value items. In every part of the recapitulation, recapitulation value obtained by looking at the value of the most emerging. If there is a range of values in the results of the assessment needs to be reviewed range up to a maximum value minimum value, then reduce the range of values that is to be the difference between a value and select a point as the final value. The Example of recapitulation KPKU scoring for category 2 Strategic Planning can be seen in Table 6

Table 6. recapitulation KPKU scoring for category 2 Strategic Planning

	A	D	L	I
Sub Category 2.1 Strategy Development				
2.1 a. Strategy Development Process				
JUDGEMENT 2.1a(1)	4	4	3	3
JUDGEMENT 2.1a(2)	4	3	3	3
JUDGEMENT 2.1a(3)	3	3	2	2
JUDGEMENT 2.1a(4)	3	3	2	2
2.1 b. Strategic Objectives				
JUDGEMENT 2.1b(2)	4	4	3	3
HOLISTIC 2.1	4	3	3	2
		2-4		
JUDGEMENT 2.1 Band Score		3		
JUDGEMENT 2.1 Score (%)		30%-45%		
JUDGEMENT 2.1 Score point		35		
Sub Category 2.2 Strategy Implementation				
2.2 a. Action Program development and implementation				
JUDGEMENT 2.2a(1)	4	4	3	3
JUDGEMENT 2.2a(2)	4	3	3	3
JUDGEMENT 2.2a(3)	4	3	3	3
JUDGEMENT 2.2a(4)	3	3	2	3
JUDGEMENT 2.2a(5)	3	3	2	2
JUDGEMENT 2.2a(6)	3	3	2	2
2.2 b. Performance Projection				
JUDGEMENT 2.2b	4	4	3	3
HOLISTIC 2.2	4	3	3	3
		3-4		
JUDGEMENT 2.2 Band Score		3		
JUDGEMENT 2.2 Score (%)		30%-45%		
JUDGEMENT 2.2 Score point		40		

It can be seen from table 6, the score for 2.1. Strategy Development is 35%, while 2.2 Strategy Implementation is 40%

4.2.2 Scoring

The overall score can be seen at table 7 and the band score can be seen at table 8.

Table 7 The Overall *Score*

No.	Categories and Items	Point	*Value*	Score
1.	**Leadership**	**120**		
1.1	Senior Leadership	70	40	28
1.2	Governance and Societal Responsibilities	50	40	20
2.	**Strategic Planning**	**85**		
2.1	Strategy Development	45	35	15,75
2.2	Strategy Implementation	40	40	16
3.	**Student Focus**	**85**		
3.1	Voice of the customer	40	35	14
3.2	Customer engagement	45	30	13,5
4.	Measurement, Analysis, and Knowledge Management	**90**		
4.1	Measurement, Analysis, and Improvement of Organizational Performance	45	40	18

4.2	Management of Information, Knowledge, and Information Technology	45	30	13,5
5.	**Workforce Focus**	**85**		
5.1	Workforce Environment	45	30	13.5
5.2	Workforce Engagement	40	25	10
6.	**Operation Focus**	**85**		
6.1	Work Process	45	40	18
6.2	Operational effectiveness	40	40	16
7.	**Result**	**450**		
7.1	Product and process outcomes	120	15	18
7.2	Customer focused outcomes	85	5	4.25
7.3	Workforce focused outcomes	85	30	25.5
7.4	Leadership and governance outcomes	80	20	16
7.5	Financial and market outcomes	80	30	24
	TOTAL	1000		284

By table 7, The score of X School Education Foundation is 284 of 1000

Table 8 *Malcolm Baldrige Performance Level*

Level	*Band*	Score-Range	*Global Image*
8	*World Class Leader*	876-1000	
7	*Benchmark Leader*	776-875	*Excellent*
6	*Industry Leader*	676-775	
5	*Emerging Industry Leader*	576-675	
4	*Good Performance*	476-575	*Average*
3	*Early Improvement*	376-475	
2	*Early Result*	276-375	*Poor*
1	*Early Development*	0-275	

By table 8, the performance level of X School Education Foundation is 284 of 1000 which equal to Early Result Band

5. Analysis

5.1 The Most Strength

The most strength that can be seen in Table 9.

Table 9 *The Most Strength*

No.	**Categories and Items**	**item**	STRENGTH
1.	Leadership	1.2.a.1(4)	Governance System
2.	Strategic Planning	2.2a(3).1	Resource Allocation
3.	Student Focus	3.2.a.2	Customer Support
4.	Measurement, Analysis, and Knowledge Management	4.2.1	Knowledge Management
5.	Workforce Focus	5.2.c.1	Learning and development system
6.	Operation Focus	6.2a	Cost Control
7.	Result	**7.5**	Financial and market outcomes

5.2. The Most Opportunity for Improvement (OFI)

The most Opportunity for Improvement (OFI) which can be seen in Table 10.

Table 10 *The Most Opportunities for Improvement (OFI)*

No.	Categories and Items	item	*Opportunity for Improvement*
1.	Leadership	1.2c.(2)	Community Support
2.	Strategic Planning	2.1a(4).4	Work System and Core Competence
3.	Student Focus	3.1.b.1	Determination customer satisfaction and engagement
4.	Measurement, Analysis, and Knowledge Management	4.1.b	Performance analysis and review
5.	Workforce Focus	5.2.b.1	Assessment of workforce engagement
6.	Operation Focus	6.1b(3)	Process Improvement
7.	Result	7.2	Customer outcomes

6. Conclusion

6.1. Conclusion

The conclusion of the research is the total value of performance measurement X School Education Foundation amounted to 284 points from 1000 points so that the X School at the level of early results, which means Foundation X School is in the poor global image, which mean:

- **Approach**: An effective, systematic approach, responsive to the basic requirements of the Item, is evident.
- **Deploy**: The approach is deployed, although some areas or work units are in the early stages of deployment.
- **Learning**: The beginning of a systematic approach to evaluation and improvement of key processes is evident.
- **Integration**: The approach is in the early stages of alignment with the basic organizational needs identified in response to the Organizational Profile and other Process Items
- **Level**: Good organizational performance levels are reported, responsive to the basic requirements of the item.
- **Trend**: Some trend data are reported, and a majority of the trends presented are beneficial
- **Comparison**: Little or no comparative information is reported.
- **Integration**: Results are reported for a few areas of importance to the accomplishment of organization's mission.

The Contradictive result is The Accreditation level X School Education Foundation is A which mean in the best level for Indonesian School, but if it is compared to world class performance level, X School Education Foundation is on poor level.

6.2. Suggestions

The suggestions for X school to achieve better level are X school has to maintain and improve the strengths and the most important improvement is X school has fix the opportunities for improvement:

- X school should build systematic approaches for all process,
- X school should improve its deployment

- X school should evaluate and improve all of key processes
- X school should align all of key processes
- X school should report and evaluate all measurement
- X school should compare all of key measurement to benchmark competitors

7. References

[1] Apandi, Tany Triwahyuni. 2015. "Pengukuran Performansi Berdasarkan MBCfPE Pada Kategori

[2] Proses Fokus Pelanggan dan Kategori Hasil Item Fokus Pelanggan Di Yayasan x". Laporan Tugas Akhir. Institut Teknologi Nasional Bandung.

[3] Blazey, Mark L.,2011, *Insights to Perfomance Excellence 2011-2012,* ASQ Quality Press, United States of America.

[4] Christian, Engelbert., (2010). *Pengukuran Kinerja Perusahaan Jasa Penerbangan di Indonesia dengan Metode Performance Prism*, Tesis Fakultas Teknik Universitas Indonesia, Depok

[5] Fajarwati, Andini Indah. 2015. "Pengukuran Performansi Berdasarkan MBCfPE Pada Kategori Proses Fokus Tenaga Kerja dan Kategori Hasil Item Fokus Tenaga Kerja Di Yayasan x". Laporan Tugas Akhir. Institut Teknologi Nasional Bandung.

[6] Gaspersz, Vincent, 2011, *Malcolm Baldrige Criteria for Perfomance Excellence*, Penerbit Vinchristo Publication, Bogor.

[7] Inayah, Nurfitriani. 2015. "Pengukuran Performansi Berdasarkan MBCfPE Pada Kategori Proses Kepemimpinan dan Kategori Hasil Item Kepemimpinan dan Tata Kelola Di Yayasan x". Laporan Tugas Akhir. Institut Teknologi Nasional Bandung.

[8] *Indonesian Quality Award Foundation* (IQAF), 2013-1014, "*Kriteria Kinerja Ekselen (Malcolm Baldrige Criteria for Performance Excellence) Bidang Organisasi Profit*", Edisi Kedelapan,Yayasan Indonesia *Quality Award*, Jakarta.

[9] Latham, John., John Vinyard, 2011, *Organization Diagnosis, Design, and Trasformation*, Monfort Institute, Northern Colorado.

[10] Mayani, Regi. 2015. "Pengukuran Performansi Berdasarkan MBCfPE Pada Kategori Fokus Operasi dan Kategori Hasil Item Produk dan Proses Di Yayasan x". Laporan Tugas Akhir. Institut Teknologi Nasional Bandung.

[11] Neely dan Adam. 2000. "Mengukur Kinerja Dengan Metode Performance Prism". <http://download.portalgaruda.org/article.php?article=21632&val=1263>. Diakses Kamis, 30 Juli 2015. 09.00.

[12] Lynch dan Cross, 1993, *Pengertian Pengukuran Kinerja,* <http://www.sarjanaku.com/2013/04/pengertian-pengukuran-kerja-tujuan.html>. Diakses Senin, 1 Juni 2015.

[13] Yuniarti, Dameria. 2015. "Pengukuran Performansi Berdasarkan MBCfPE Pada Kategori Proses Pengukuran, Analisis, dan Manajemen Pengetahuan". Laporan Tugas Akhir. Institut Teknologi Nasional Bandung.

AUTHOR BIOGRAPHIES

Sugih Arijanto is a lecturer in Industrial Engineering Dept., Faculty of Industrial Tecnology, Itenas, Bandung. He Received his Magister Management from Magister Management Business and Administration ITB in 1998. His research interests are in Performance Management, Corporate Strategy, and Organization Development. He is an evaluator in Indonesia Quality Award and an examiner of KPKU-BUMN (Ministry of State Enterprises, Republik Indonesia). His email address is s_arijanto@yahoo.com or sugih@itenas.ac.id . His office address is Jurusan Teknik Industri Itenas Bandung Jalan PHH Mustofa 23 Bandung 40124 Indonesia

Classification of Texture Using Multi Texton Histogram and Probabilistic Neural Network

Agus Eko Minarno[1], Yuda Munarko[1], Arrie Kurniawardhani[2] and Fitri Bimantoro[3]

[1]Department of Informatics, Universitas Muhammadiyah Malang, Malang, Indonesia
[2]Department of Informatics, Faculty of Industrial Technology, Universitas Islam Indonesia, Jl. Kaliurang Km. 14,5 Yogyakarta, 55584, Indonesia
[3]Universitas Mataram, Mataram, Indonesia

email: agoes.minarno@gmail.com

Abstract. Image classification plays an important rule for other image domains such as image retrieval, object recognition, image annotation and relevance feedback. In this paper, we describe our work in image classification using Multi Texton Histogram (MTH) and Probabilistic Neural Network (PNN). The result shows that the proposed method reaches 92% accuracy for Batik dataset and 98% for Brodatz dataset. This indicates that the use of MTH and PNN for image classification for Batik dataset and Brodatz dataset are effective.

1. Introduction

A number of research have been conducted for pattern recognition and classification of images. One of those research is the texture classification of Songket images by Cheong and Loke. They used six multispectral co-occurrence matrix in order to extracting colour features. After that, they incorporated Tchebichef orthogonal polynomial to compute coefficient moment and then extracting features based on co-occurrence. They showed that the used method is powerful for differentiate texture colour of Songket [1]. Continue their research, Cheong and Loke tried to reducing features using Principal Component Analysis (PCA). It may reduce 2% of features without any issue in classification accuracy [2]. A similar research is conducted by Nurhaida et al. [3], which also focus on Batik features extraction. They compared three methods, those are Gray Level Co-occurrence Matrices (GLCM), Canny Edge Detection, and Gabor filters. They found that the performance of GLCM is best to other methods for image classification.

In general, image classification is built based on colour, texture and shape features which are used independently or in combinatorial. From those features, colour feature becomes the most dominant and is broadly used for discriminating images. Almeida et al. has been proposed a method that used colour feature and distance between pixels by creating image index on three dimensional table. The table illustrates the spatial relationships in the image colour change. The objective of colour indexing is compare an image query to images in the database [4].

Research on texture features has also been widely proposed in the field of pattern recognition clustering, classification and object detection. Julesz, in his research on the analysis of texton interaction to discriminate textures, found that by using texton combined with simple statistical methods, visual perception to distinguish the texture of the image can be obtained significantly

[5][6][7][8][9]. Research on texture has also been done by Haralick using Gray Co-Occurrence Matrix (GLCM). GLCM is using statistical methods of order one and two to produce 14 features. Including in these features are mean, variance, correlation, energy and homogeneity [10][12].
From some research above, we see that the accuracy of the image classification using either statistical or texton methods could be improved. Therefore, this study was proposed to develop a method for feature extraction using Multi Texton Histogram (MTH) and Probabilistic Neural Network (PNN) as a classifier. PNN is chosen as the classifier because it has a relatively lower complexity compared to Neural Network [11].

2. Dataset

The datasets in this study are the 950 images of Brodatz colour dataset and the 300 images of Batik dataset. Brodatz dataset consists of 50 classes where each class consists of 19 images. While Batik dataset consists of 50 classes where each class consists of 6 images. Figure. 1 is an example of the Brodatz's images (a) and Batik's images (b). The dimensions of each image is 128x128 pixels.

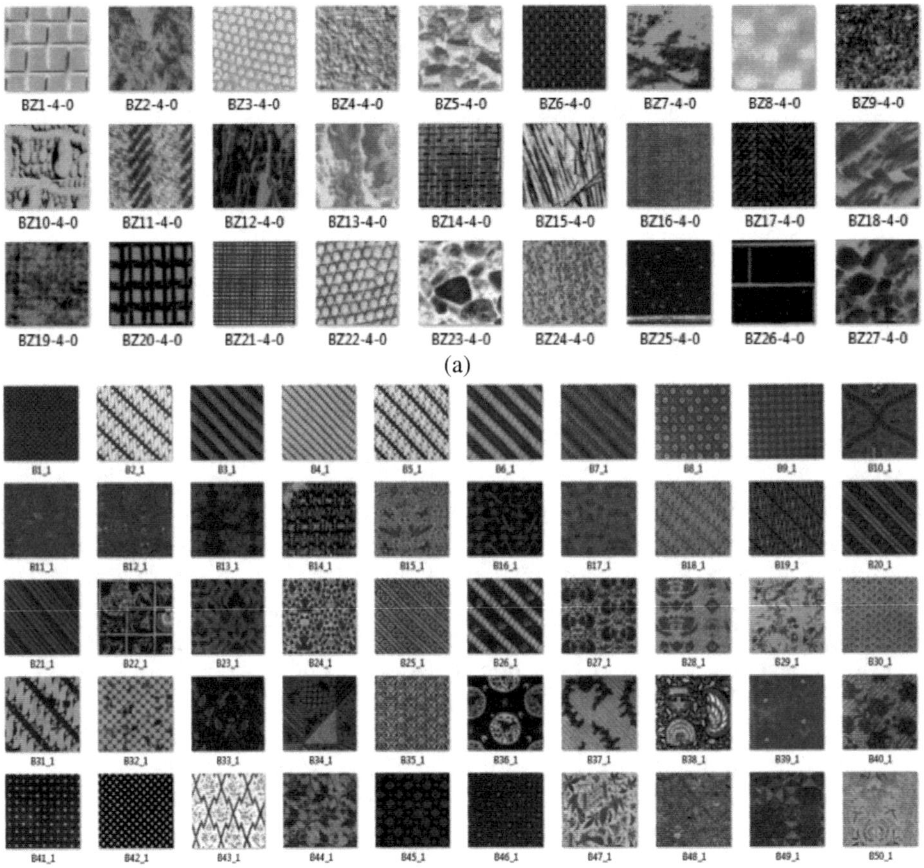

(b)

FIGURE 1. An example of images (a) Brodatz and (b) Batik

3. Multi Texton Histogram

MTH basic idea is texton theory proposed by Julesz. MTH is using four types of texton to detect the micro structure of an image. Those four types of texton are shown in Figure. 3. MTH approach does not involve segmentation process and training data. MTH extracts image features by utilizing a colour histogram in RGB colour space and detects the edge orientation of an image using Sobel operator. MTH stages can be described into four steps. First, perform edge orientation detection using Sobel operator. Edge orientation detection results are then quantized into 18 bins. Since MTH is using 1-180 degrees in representing edge orientation value, so if the value is quantized into 18 bins, then the quantized value is a multiplication of 10. Second, do a quantization to the image's colour in RGB colour space. For each RGB component, the value is quantized into 4 bins; R = 4 bin, G = 4, bins, and B = 4 bins. Third, detect texton on the results of quantized edge orientation and quantized colour using four different textons. Detection process is conducted from left to right and from top to bottom, by two pixels of each shift. The result of texton detection is colour histogram and edge orientation histogram. Finally, colour histogram and edge orientation histogram are combined becoming a single histogram. The combined histogram is consisted of 82 features which composed by 4x4x4 = 64 colour features and 18 edge orientation features.

4. Probabilistic Neural Network

Probabilistic Neural Network (PNN) was chosen because of its speed and accuracy in patterns recognition and classification. PNN is developed from Neural Network method which based on Bayes Classification Parzen. PNN architecture as can be seen in Figure. 2. The PNN consists of four layers, namely input layer, pattern layer, the summation layer, and the layer decision.

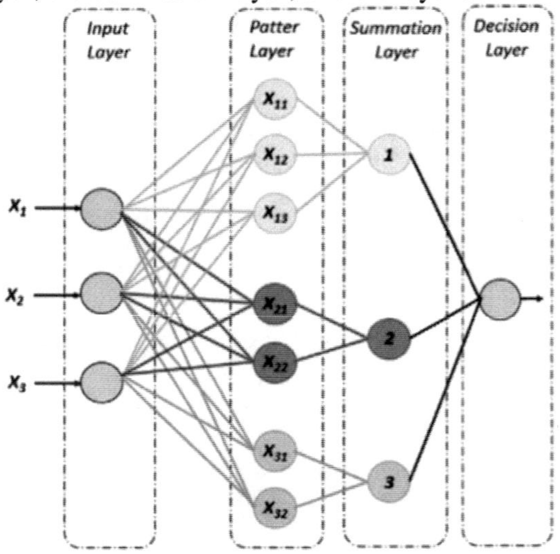

FIGURE 2. PNN Architecture

The input layer is consisted of several node which are vectors. The vector is a hallmark of input image. The pattern layer is consisted of the sum of the total number of neuron of training data. In the pattern layer, the input features are calculated using Gaussian function and probability density function (PDF) based on Parzen window. The summation layer is a layer that calculates total of the highest probability from pattern layer. Lastly, the decision layer, which responsible for determining an image to the most similar target and decide to a specific class.

5. Feature Extraction and Classification

5.1. Extraction of Edge Orientation Feature

The first stage is performing an extraction of edge orientation feature using Sobel operator. Sobel operator produces vector and magnitude which are then quantized into 18 bins. The selection of bin size is based on previous studies by [5][7][8][9][10][12]. This stage is resulting 18 edge orientation features.

5.2. Extraction of Colour Feature

The second stage is the extraction of RGB colour feature. In this stage, colour is separated into RGB colour component. Then, each value of RGB colour component is quantized into 4 bins, so there are 4x4x4 = 64 bins. These features will be used in the next stage along with edge orientation features.

5.3. Texton Detection

The next stage is the detection of texton on the result of colour quantization and edge orientation quantization. The 2x2 element size is used in this paper to increase texture discrimination, because the texture gradients contain only at texture boundaries [6]. The types of texton that detected can be seen in Figure. 3., named T1, T2, T3 and T4. Those texton types are choosen because the co-occurring probability of two same-valued pixels is bigger than that of three or four same-valued pixels in a 2x2 grid element size.

Features are extracted by convoluting each texton on pixels that contains quantized colour and quantized edge orientation information. The convolution is performed from the top left of image to the right side of image and from the top downwards by two pixels per detection. MTH uses grids with dimension 2x2 pixels which are marked as V1, V2, V3 and V4. When a grid contains two similar pixels which have similar pattern with a particular texton, then, that grid is calculated as a texton. Two pixels are considered similar when the value of quantized colour or quantized edge orientation are equal. For example, a grid is considered to texton T1 when the 2 top pixels have quantized colour value 10. Consequently, the number of texton T1 with quantized colour value 10 is increased by 1. Total number of texton such as T1 with quantized colour value 10 is then stored in a histogram. An illustration of texton detection is shown in Figure. 4.

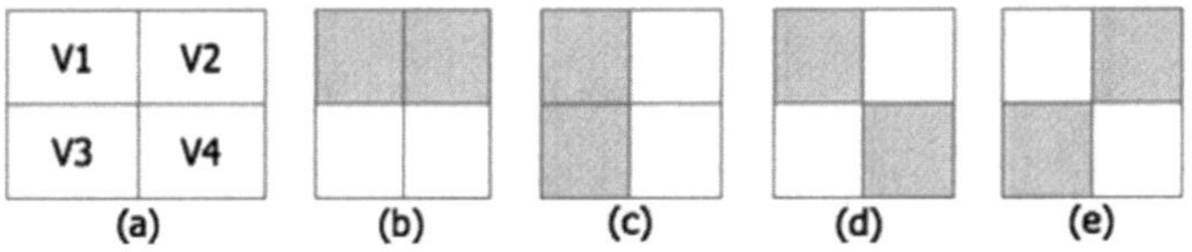

FIGURE 3. Four types of MTH's texton. (a) 2x2 Grid; (b) T1 ; (c)T2 ; (d) T3; (e) T4

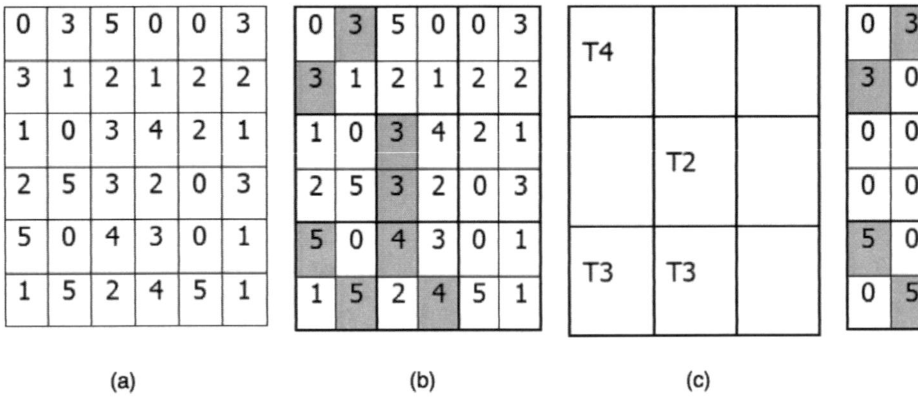

FIGURE 4. An illustration of texton detection in MTH (a) original image; (b) 4 texton detection; (c) texton type; (d) result of detection.

6. PNN Classification

The classification experiment is conducted using 950 images of Brodatz dataset, which 900 images is set for training data and 50 image as testing data. PNN is selected as classification method because PNN is relatively faster and may perform good accuracy compared to Neural Network. PNN is a combination of Neural Networks and Probability Density Function (PDF). This method is fast since it uses probability approach for activation. PDF is calculated using Equation 1 and Equation 2.

$$\phi_{ki}(x) = \frac{1}{\sqrt{(2\Pi)^d \sigma^d}} \exp(-\frac{\|x_i - x_{ki}\|^2}{2\sigma^2}) \quad (1)$$

Where x_i is a vector of features, x_{ik} is a set of training data, k indicates the class and σ is smoothing parameter.

$$g_k(x) = \frac{1}{\sqrt{(2\Pi)^d}} \frac{1}{n\sigma^2} \sum_{i=1}^{n} \exp(-\frac{\|x_i - x_{ki}\|^2}{2\sigma^2}) \quad (2)$$

Where n is the number of training data, then likelihood can be calculated from features which classified to class k. The last calculation is to compute the maximum probability at the output layer of the PNN using Equation 3.

$$C(x) = \arg\max\{g_k(x)\}, k = 1,2,...,m \quad (3)$$

Where k is a class and m is the number of classes and x is a feature.

7. Performance Measure

The accuracy of features extraction and classification is calculated using Equation 4.

$$Accuracy = {}^{m}/_{n} \cdot 100\% \quad (4)$$

Where M is the number of data test which are classified correctly, whereas N is the total number of data test.

8. Result and Discussion

To measure the performance of the proposed method, an experiment was conducted using Brodatz dataset and Batik dataset. The total number of Brodatz dataset is 950 images, consists of 50 classes, each class composed of 19 images. 900 images are used for training data and 50 images are used for testing data. While the total number of Batik data is 300 images, consists of 50 classes, each class composed of 6 images. For each class, 4 images are used for training data and two images are used for testing data.

Features that feed on PNN are detected using texton. The number of features that related to colour is 64 while the number of features that related to edge orientation is 18. Those features are represented as a histogram. An example of the result of edge orientation detection using Sobel operator can be observed in Figure. 5. Figure. 5(a) is the original image while Figure. 5(b) is the result of edge detection using Sobel operator. An example of MTH histogram that are detected using texton is shown in Figure. 6. MTH histogram represents the number of co-occurrence pixel successful detected using 4 types of texton. The Y axis is representing the number of co-occurrence of a particular feature with a particular value in a pair of pixels.

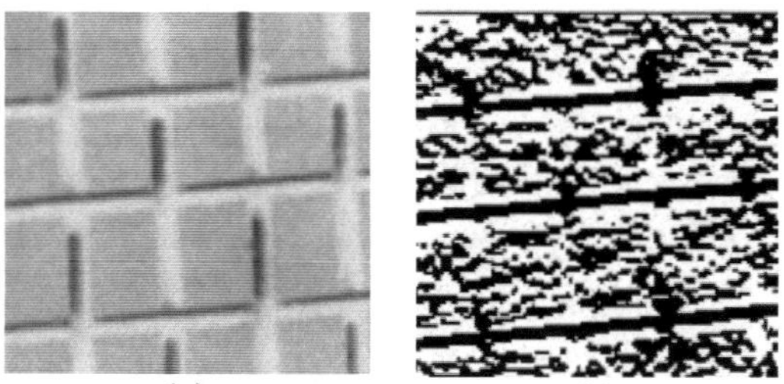

(a) (b)

FIGURE 5. An example of the result of edge orientation extraction using Sobel operator (a) original image; (b) the result of Sobel operator.

Table 1 is the result of classification using MTH and PNN. From the experimental results in Table 1, we can see that features extraction method using MTH and PNN classifier is able to classify the image properly. For Batik dataset, the accuracy reaches 92% for 250 training data and 50 testing data. While, for Brodatz dataset, the accuracy is 6% higher by Batik dataset, reaches 98% for 900 training data and 50 testing data. The nature of Brodatz dataset and Batik dataset are a reach variety of feature values, so these datasets are appropriate to use for examining the reliability of the proposed method. Still based on Table 1, the number of training data that used does not contribute a significant influence. Therefore, we can see that MTH and PNN are able to present a good performance.

TABLE 1. The result of PNN classification

Testing Scenario	*Dataset*	*Accuracy*
Batik 250 training data, 50 testing data	Batik	92%
Brodatz 900 training data,50 testing data	Brodatz	98%

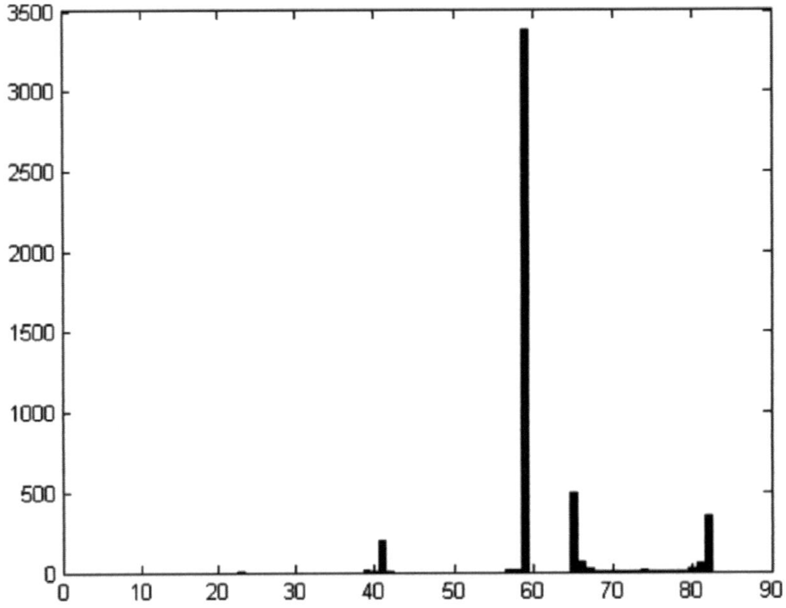

FIGURE 6. An example of MTH histogram that extracted from Figure. 5(a)

Moreover, Table 1. indicate that the 2x2 texture element can represent the Batik features well. The 2x2 texture elements can discriminate the texture patterns within an image based on texton gradients. Texton gradients are used to represent texture pattern because those represent the texture bundaries.

9. Conclusion

This study was elaborated a features extraction method using Multi Texton Histogram and an image classification method using Probabilistic Neural Netwrok on Brodatz dataset and Batik dataset. The accuracy of classification is relatively high which reaches 92% for Batik dataset and 98% for Brodatz dataset. Based on the experiment results, we may conclude that Multi Texton Histogram and Probabilistic Neural Network are reliable for extracting features and classifying images, especially images that contains texture character. Future studies can be developed to improve performance of the detection of co-occurrence of pixels and edge orientation.

References

[1] Cheong M. and Loke K.S., 2008, An approach to texture-based image recognition by deconstructing multispectral co-occurrence matrices using tchebichef orthogonal polynomials,

ICPR 2008. 19th International Conference on. IEEE. 1–4.

[2] Loke K.S. and Cheong M., 2009, Efficient textile recognition via decomposition of co-occurrence matrices, *ICSIPA 2009 IEEE International Conference on. IEEE.* 257–261.

[3] Nurhaida I., Manurung R., and Arymurthy A. M., 2012, Performance comparison analysis features extraction methods for batik recognition, *ICACSIS 2012 International Conference on. IEEE.* 207–212.

[4] de Almeida C. W. D., de Souza R. M., and . Candeias A. L. B, 2010, Texture classification based on co-occurrence matrix and self-organizing map, *SMC 2010 IEEE International Conference on. IEEE.* 2487–2491.

[5] Huang J ., . Kumar S. R, Mitra M., Zhu W.J., and Zabih R., 1997, Image indexing using color correlograms, *in Computer Vision and Pattern Recognition, 1997. Proceedings., 1997 IEEE Computer Society Conference on. IEEE.* 762–768.

[6] Liu G.-H., Zhang L., Hou Y.K., Li Z.Y., and Yan J.Y. g, 2010, Image retrieval based on multi-texton histogram, *Pattern Recognition.* **43** 2380–2389.

[7] Haralick R. M., 1979, Statistical and structural approaches to texture, *Proceedings of the IEEE.* **67** 786–804.

[8] Minarno A. E., Munarko Y., Bimantoro F., Kurniawardhani A., and Suciati N., 2014, Batik image retrieval based on enhanced micro-structure descriptor, *APCASE 2014 Asia-Pacific Conference on. IEEE.* 65–70.

[9] Minarno A. E., Kurniawardhani A., Munarko Y., Bimantoro F., and Suciati N., 2014, Texture feature extraction using co-occurrence matrices of sub- band image for batik image classification, *ICoICT.* 244–249.

[10] Minarno A.E. and Suciati N., 2014, Batik image retrieval based on color difference histogram and gray level co-occurrence matrix, *TELKOMNIKA.* **12**, no. 3.

[11] Specht D. F., 1990, Probabilistic neural networks, *Neural networks.* **3** 109–118.

[12] Minarno A.E. and Suciati N., 2014, Image Retrieval using Multi Texton Co-Occurence Descriptor, *Journal of Theoretical and Applied Information Technology.* **67** 103-110.

Analysis of Muscle Contraction on Pottery Manufacturing Process Using Electromyography (EMG)

Hartomo Soewardi[1] and Amalia Azka Rahmayani[2]
[1]Senior lecturer of Industrial Engineering Department, Islamic University of Indonesia, Yogyakarta
[2]Graduated Student in Industrial Engineering Department, Islamic University of Indonesia, Yogyakarta

Email : hartomo@uii.ac.id, azkamalia@gmail.com

Abstract. One of the most common problems in pottery manufacturing process is musculoskeletal disorders on workers. This disorder was caused by uncomfortable posture where the workers sit on the floor with one leg was folded and another was twisted for long duration. Back, waist, buttock, and right knee frequently experience the disorders. The objective of this research is to investigate the muscle contraction at such body part of workers in manufacturing process of pottery. Electromyography is used to investigate the muscle contraction based on the median frequency signal. Focus measurements is conducted on four muscles types. They are lower interscapular muscle on the right and left side, dorsal lumbar muscle, and lateral hamstring muscle. Statistical analysis is conducted to test differences of muscle contraction between female and male. The result of this research showed that the muscle which reached the highest contraction is dorsal lumbar muscle with the average of median frequency is 51,84 Hz. Then followed by lower interscapular muscle on the left side with the average of median frequency is 31,30 hz, lower interscapular muscle on the right side average of median frequency is 31,24 Hz, and lateral hamstring muscle average of median frequency is 21,77 Hz. Based on the statistic analysis result, there were no differences between male and female on left and right lower interscapular muscle and dorsal lumbar muscle but there were differences on lateral hamstring muscle with the significance level is 5%. Besides that, there were differences for all combination muscle types with the level of significance is 5%.

1. Introduction

Pottery manufacturing industry in Indonesia is one of strategic industries that have contributed to government income and also provide the big opportunity of employment [8]. In 2009, there were 441 pottery shop and 2.367 workers that always increase in every year [2]. This enhancement was also followed by an increasing in production and marketing [6]. In pottery manufacturing, the workers must work effectively in 7 hours for a day which they did it repetitively and used a static posture for long duration. This condition will cause musculoskeletal disorders that are impairments of body structures especially in muscles, joints, tendons, ligaments, nerves, bones or localized blood circulation system [3].

According to a preliminary study that was conducted to workers in pottery manufacturing industry showed that 28.3% of workers experienced a complaint at back, 53.5% at waist, 51.7% at buttock, and 48.3% at right knee. This fact describes a problem on musculoskeletal system. But, how deep of this

IOP Conf. Series: Materials Science and Engineering **105** (2016) 012023 doi:10.1088/1757-899X/105/1/012023

problem occur is still questionable. Thus, it is important to analyze this state in detail based on muscle contraction study.On the preview research, the researcher already used electromyography to analyze the muscle contraction but not for pottery workers. So, the objective of this research is to analyze muscle fatigue by using electromyography.

2. Research method

Experimental study was conducted in the ergonomics laboratory to record signal of muscle contraction.

2.1. Subjects

Fifteenuniversity students, which consist of 8 males and 7 females who had an average age of 20-22 years old, were participated. This subjectwas selected for experiment based on twocriteria that are healthy and no physical disability.

2.2. Apparatus

The tools that used in this research were electromyography (EMG ME3000P8) where its function was to record the signal of muscle contraction.A set of personal computer was used to display the raw signal recorded and Megawind 700046 software version 2.4 (Mega Electronics, Ltd.)was also used to transform the raw signal into median frequency signal.The signal was identified by electrodes as sensor that it was adhered on the skin.

2.3. Design of Experiment

Each participant in a sitting position on the floor as comfortable as possible. Participant was sit in front of the pottery maker with 30 cm in distance withposition of two handsand right leg parallel to the pottery maker (Fig 1.(a)). And Fig 1.(b) show the placement of electrodes on skin of the body part.

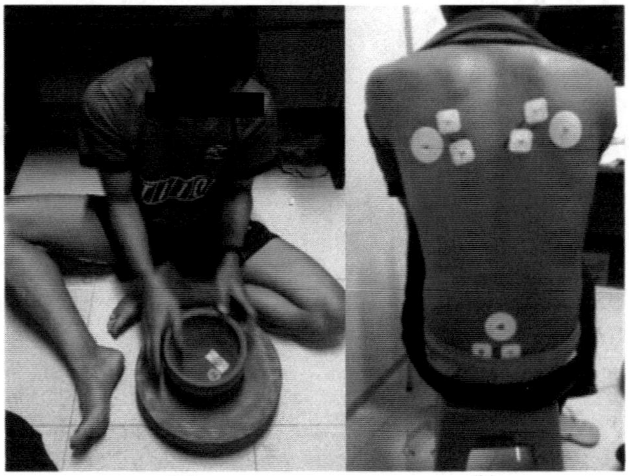

(a) (b)

Fig 1. (a) Design of experiment and (b) placement of electrodes

The placement of electrodes were based on Reference [1], where the location of lower *interscapular* on the left and right side (Fig 2.(a)) was at inferior medial border of the scapula on an oblique angle, approximately 5 cm down from the scapular spine. For the *dorsal lumbar* muscle (Fig 2.(b)), the two electrodes are placed at the T-10 level and L-3 level of the spine, approximately 2 cm lateral from spine over the muscle belly. And the lateral hamstringmuscle (Fig 2.(c)) may be placed parallel to the muscle in the center of the back of the thigh, approximately half the distance from gluteal fold to the back of the knee.

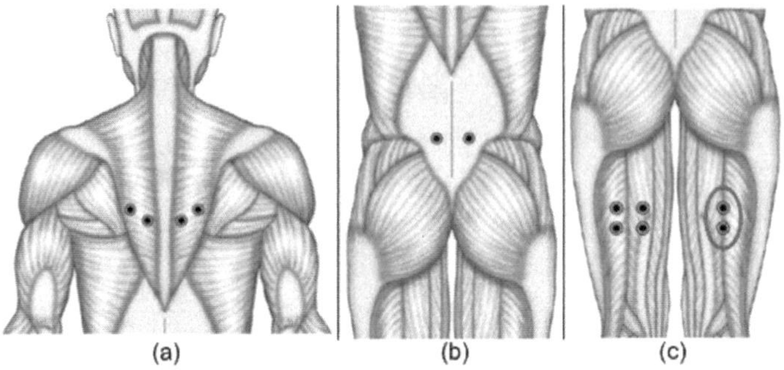

Fig 2. (a) Lower *interscapular* muscle on the left and right side, (b) *dorsal lumbar* muscle, and (c) lateral hamstring muscle.

2.4. Task
The participants were instructed to make product from clay using pottery maker where the left hand take the clay and then directed to the pottery maker. While the right hand manufactures the clay into a pot and right leg twisted the pottery maker. At the same time, the muscle contraction was recorded.

2.5. Experimental procedure
The experiment was conducted in the ergonomics laboratory. Before doing the experiment, pilot study was conducted to determine locations of electrodes at the contracted muscles. Each participant was given 5 minutes to do exercise to make familiar with the task.The real experiment was conducted to identify the muscle contraction data by using electromyography for 15 minutes each participant. So, the total time of experiment for each participant is 20 minutes.
After completing the experiment, the recorded signals as araw signal weretransferred into median frequency signal using Megawin software based on the formulation as follow:

$$\int_0^{MF} P(f)df = \int_{MF}^{\infty} P(f)df = \frac{1}{2}\int_0^{\infty} P(f)df \qquad (1)$$

Where p(f) is the power spectrum density. This transformation is done to change in the time domain into the frequency domain.

2.6. Data analysis
Statistical analysis was used to determine the differences for each muscle contraction between female and male by using Wilcoxon and independent sample t-test based on a value of median frequency. Normality test was used to determine which one the appropriate test at 5% of significance level.

3. Result and discussion

3.1. Analysis of electromyographysignal
Table 1 showed the result of average of median frequency signal for lower *interscapular* muscle on the left side (LIL), lower *interscapular* muscle on the right side (LIR), *dorsal lumbar* muscle (DL), and lateral hamstring muscle (LH). One period of time was 20 seconds for experiment for each female and male respectively.

Table 1. Result of Average of Median frequency

	Average of Median Frequency (Hz)			
	LIL	LIR	DL	LH
Female	30.98	21,74	55.65	15.47
Male	31.62	30.75	48.02	28.06

	Average of Median Frequency (Hz)			
	LIL	LIR	DL	LH
Female	30.98	21,74	55.65	15.47
Male	31.62	30.75	48.02	28.06
Average	**31.30**	**31.24**	**51.84**	**21.77**
St. Dev	3.72	2.20	2.98	3.30

Median frequency ishalf of the total spectrum energy of the muscle [5]. Thus, the median frequency signal can be used as a valid indicator to determine muscle fatigue [7]. The decrease in the median frequency indicates a fatigue in the muscle [4].

Fig 3 presents a graph of average of the median frequency signal for contraction of each muscle type during 49 periods for both males and females.

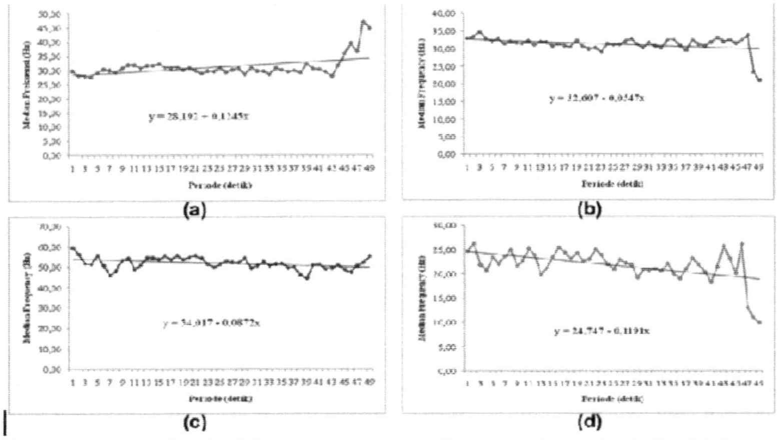

Fig 3. Median frequency graph of (a) lower *interscapular* muscleon the left, (b) lower *interscapular* muscle on the right, (c) *dorsal lumbar* muscle, and (d) lateral hamstring muscle.

The change in the graph describes contraction of the muscle. Decrease in the median frequency shows that the muscle starts to not be able to contracted. It means the fatigue will be occurred. Otherwise, the muscle will increase in contraction. This state will also cause the fatigue in long period.

Fig 3 (a) shows that contraction of the lower *interscapular* (left) muscle in the first four periods was decreased.It because no activity was done by the muscle, but the contraction is still took place for holding the position of left hand. At fifth period until 49th period, the contraction was increase. It means that the muscle works harder than before. So, the fatigue will be happen immediately.

While Fig 3 (b) shows thatthe contraction of lower i*nterscapular* (right) muscleinfourth period until the end of period was decrease. It means that the muscle start to be active for works. So, this decrease indicates there was a symptom of fatigue. And by the same token for *dorsal lumbar* muscle (Fig 3 (c)) and lateral hamstring muscle (Fig 3 (d)). However, the *dorsal lumbar* muscle has the highest median frequency of 51.84 Hz (see Table 1). Meanwhile the lower *interscapular* (left) muscle is 31.30 Hz, lower *interscapular* (right) muscle is 31.24 Hz, and lateral hamstring muscle is 21.77 Hz.

This condition can occured because dorsal lumbar muscle has function for holding the sit and movement of waist for long period. And the lower *interscapular* (left) muscle has function for take the clay into the pottery maker.While the lower *interscapular* (right) muscle has function to form a pot, and lateral hamstring muscle to twisted the pottery maker.

3.2. Statistical analysis

Table 2 show the result of normality test between male and female. While table 3 show the result of normality test both male and female. The significance which less than 0.05 means that the data was not normal and if the data more than 0.05 then means normal.

Table 2. Result of Normality Test between Male and Female

Muscle (M)	Sig	Sig	Muscle (F)
Lower *Interscapular* (left)	0.000	0.000	Lower *Interscapular* (left)
Lower *Interscapular* (right)	0.062	0.200	Lower *Interscapular* (right)
Dorsal Lumbar	0.200	0.200	*Dorsal Lumbar*
Lateral Hamstring	0.200	0.200	Lateral Hamstring

* P (0.000)< 0.05
P ≥0.05

Table 3. Result of Normality Test both Male and Female

Muscle	Sig
Lower *Interscapular* (left)	0.000
Lower *Interscapular* (right)	0.000
Dorsal Lumbar	0.200
Lateral Hamstring	0.002

* P (0.000)< 0.05
P ≥0.05

Table 4 shows the result of independent t-test for all the muscles between male and female. For lower *interscapular* (left) muscle, lower *interscapular* (right) muscle, and *dorsal lumbar* muscleshow that there were no differences on muscle contraction between male and female.It means that three muscles work with the same load for both workers.
While the lateral hamstring muscle shows that there was differenceon muscle contraction between male and female. It means that there was different load when muscle works where the contraction for female was bigger than male.

Table 4. Result of Independent Sample T-test

Muscle	Sig
Lower *Interscapular* (left)	0.566
Lower *Interscapular* (right)	0.531
Dorsal Lumbar	0.027
Lateral Hamstring	0.000

* P (0.000) < 0.05
P ≥0.05

Table 5 shows the result of Wilcoxon test for testing differences in muscle contraction on some combination of muscle types.From the result, all of the combination of muscles show that there were differences in muscle contraction. This can occured because each muscle has different function for making pottery.

Table 5. Result of Wilcoxon Test

Muscle	Sig
Lower *Interscapular* (left) Vs Lower *Interscapular* (right)	0.013
Lower *Interscapular* (left) Vs Lateral Hamstring	0.000
Lower *Interscapular* (right) Vs	0.000

Muscle	Sig
Lateral Hamstring	

* P (0.000)< 0.05

4. Conclusion

Based on the analysis, it can be concluded that:

1. The muscle which reached the highest contraction is *dorsal lumbar* muscle with the average of median frequency is 51,84 Hz. Then followed by lower *interscapular* muscle on the left side with the average of median frequency is 31,30 hz, lower *interscapular* muscle on the right side average of median frequency is 31,24 Hz, and lateral hamstring muscle average of median frequency is 21,77 Hz.
2. There were no differences in muscle contraction for lower *interscapular* on the right and left side and *dorsal lumbar* muscle between male and female, but there was a difference significant muscle contractionon lateral hamstring muscle at 5% of significance level.
3. There were differences of muscle contraction for among the combination muscle types at 5% significance level.

References

[1] Criswell, E. (2011). *Surface Electromyography, Second Edition*. Canada : Jones and Bartlett.
[2] Dinas Perindustrian Perdagangan dan Koperasi Bantul. (2009). Pendataan *Recovery* Gempa. Yogyakarta.
[3] European Agency for Safety and Health at Work. (2008). *Work-related musculoskeletal disorders : Prevention report*. Luxembourg : Office for Official Publications of the European Communities.
[4] Oliveira, R.S.; Pedro, R.E.; Bortolotti, H.; da Silva, R.A.; Abrão, T.; Altimari, J.M.; Camata, T.V.; Moraes, A.C.; Altimari, L.R. (2012). *Spectral Analysis of Electromyographic Signal in Supramaximal Effort in Cycle Ergometer using Fourier and Wavelet Transforms : a Comparative Study*. Rev Andal Med Deporte. 2012;5(2):48-52.
[5] Phinyomark, A.; Limsakul, C. & Phukpattaranont, P. (2009). *A Novel Feature Extraction for Robust EMG Pattern Recognition*. Journal of Computing, Vol.1, No.1, pp. 71-80, ISSN 2151-9617.
[6] Prasudi, M. Fajar. (2007). Perkembangan Sentra Gerabah Kasongan (Tinjauan Historis Sosiologis). Yogyakarta : Widyaiswara PPPPTK Seni dan Budaya.
[7] Su-Young, Lee., Hwa-Kyung, Shin., Sang-Hyun, Cho. (2003). *Characteristics of Median Frequency According to the Load During Fatiguing Isometric Exercise*. KAUTPT Vol. 10 No. 3 2003.
[8] UPNYK. (2007). Abstrak. (*online*) : http://repository.upnyk.ac.id/1119/1/ABSTRAK.pdf (14 Oktober 2014).

Author biographies

Hartomo Soewardi is a senior lecturer of Industrial Engineering Department, Faculty of Industrial Technology, Islamic University of Indonesia, Yogyakarta-Indonesia. Currently he is Ph.D in Engineering Design and Manufacture. His teaching and research interest are industrial ergonomic design, product design, management and quality design. His email address is hartomo@uii.ac.id

Amalia Azka Rahmayani is a graduated student in Department of Industrial Engineering, Faculty of Industrial Technology, Islamic University of Indonesia. She was born in Jakarta, December 5th 1992. Her email address is azkamalia@gmail.com

Prediction of Elderly Anthropometric Dimension Based On Age, Gender, Origin, and Body Mass Index

P Indah[1], A D Sari[2], M R Suryoputro[3] and H Purnomo[4]

[1,2,3,4]Industrial Engineering, Faculty of Industrial Technology, Islamic University of Indonesia, Yogyakarta, Indonesia

E-mail: indahpriliaty@yahoo.com, amarria.dila@gmail.com,mragils@gmail.com, hari_pnm@yahoo.com

Abstract. Introduction: Studies have indicated that elderly anthropometric dimensions will different for each person. To determine whether there are differences in the anthropometric data of Javanese elderly, this study will analyze whether the variables of age, gender, origin, and body mass index (BMI) have been associated with elderly anthropometric dimensions. Age will be divided into elderly and old categories, gender will divide into male and female, origins were divided into Yogyakarta and Central Java, and for BMI only use the normal category. Method: Anthropometric studies were carried out on 45 elderly subjects in Sleman,Yogyakarta. Results and Discussion: The results showed that some elderly anthropometric dimensions were influenced by age, origin, and body mass index but gender doesn't significantly affect the elderly anthropometric dimensions that exist in the area of Sleman. The analysis has provided important aid when designing products that intended to the Javanese elderly Population.

KEYWORDS
Elderly, anthropometric dimension; Body Mass Index, Javanese

1. Introduction

Worldwide, elderly population (aged over 60) is growing very fast, even the fastest among other age groups. In 2025, there were an estimated 1.2 billion in 2050 and the elderly will be 2 billion (21% of total population). Around 80% of elderly people living in developing countries and the Asia-Pacific region represent the fastest growing parts of the world [1].

Indonesia is a country that entered the era of structured aging population because Indonesia is one of the Asian countries with relative rapid growth of elderly population. In 2010, the number of elderly is 14.439.967 people (7,18%), the number increased to 23.992.553 people (9,77 %) in 2010, while in 2011, the number of elderly people decreased to 20 million people (9,51 %) with life expectancy of 67,4 years. Meanwhile, in 2020, the number was estimated at 28,8 million (11,34 %) with life expectancy of 71,1 years [2]. Since 2000, the population of Indonesia has been classified as structurally old (elderly > 7 % of total population) (WHO). There are 11 provinces where the population has entered the old structure, namely: Yogyakarta (14,04 %) Central Java (11,16%), East Java (11,14%), Bali (11,02%), South Sulawesi (9,05%), West Sumatera (8,74%), North Sulawesi

(8,62%), West Nusa Tenggara Barat (8,21%), West Java (8,08%), Lampung (7,78%), and East Nusa Tenggara (7,68%) [3]. Elderly, according to the WHO classification, includes: middle age (45-49 years), Elderly (60-74 years old), Old (75-90 years), Very Old (above 90 years).

Today, with the increasing elderly population, a lot of products are invented to help the elderly, such as mobility aids for elderly users move in a public place [4]. There is also a design of walking aids and seating for the elderly [5]. With these products around, the elderly anthropometric data are needed for the needs of product design to enable the conformity between the product and the user. There are more products are designed by putting the needs of anthropometric people who will use the product in mind, including products that intended to the elderly.

There are some researches on the measurement of elderly anthropometric data to see the effect of variable such as age, gender, ethnicity, and BMI as did [6] on anthropometric measurements in 5462 in the Italian elderly around the age of 65-84 years with reference to how they differ according to age and gender variables. The purpose of this study is to investigate the anthropometric characteristics and their relation to gender and age in elderly Italian population [7],[8],[9].

Elderly anthropometric data will be grouped by the category of age, gender, origin and body mass index (BMI). The results of these groupings can describe the state of the elderly in the area of Sleman, and it can be seen from the relationship between each of the variables of age, gender, origin and BMI of the elderly anthropometric dimensions. The results of grouping the elderly body dimensions can be used for designing tools intended to the elderly and can be tailored to the target of designing tools.

2. Method

In this study, 35 body dimensions that consist of 16 hand dimensions, 9 sitting dimensions, and 10 standing dimensions. Detail of elderly anthropometric dimension that used in this study listed in the Table 1.

Table 1. Elderly Anthropometric Dimension

Number	Dimension
1	Thumb length
2	Forefinger length
3	Middle finger length
4	Ring finger length
5	Little finger length
6	Metacarpal thickness
7	Palm thickness
8	Thumb thickness
9	Forefinger thickness
10	Grip hand length
11	Hand length
12	Palm length
13	Distance of thumb to little finger
14	Maximal grip diameter
15	Minimal grip diameter
16	Fist height
17	Sitting elbow height
18	Sitting shoulder height
19	Sitting eye height
20	Sitting height

21	Vertical grip reach (sitting)
22	Popliteal height
23	Buttock-knee length
24	Buttock-popliteal length
25	Sitting shoulder-fingertip length
26	Standing shoulder-fingertip length
27	Body thickness
28	Hip height
29	Knee height
30	Standing knuckle height
31	Standing elbow height
32	Standing shoulder height
33	Standing eye height
34	Body height
35	Vertical grip reach (standing)

Data of anthropometric dimensions are taken from 45 elderly people in Sleman, aged 62-88 years, divided into 26 men and 19 women and 34 people from Yogyakarta, and 11 people from Central Java and from 45 respondents all had normal BMI category. Body Mass Index (BMI) is calculated using the World Health Organization (WHO) standard method. There are 5 types of statistical tests that are performed on the data that are normality test, Pearson product moment correlation, correlation of coefficient contingency c, linear regression and independent 2 samples t-test.

Normality test aims to determine whether the sample comes from a population with normal distribution or not. Normality test uses Kolmogorov-Smirnov. Normality test is performed 7 times against a group of elderly, old, male, and female, Yogyakarta, Central Java and normal BMI.

Pearson Product Moment Correlation is performed to find the relationship or influence between the independent variable (X) and the dependent variable(Y). In this research, the independent variable (X) is the age that divides into Elderly and Old and BMI while the dependent variable (Y) is the elderly anthropometric dimensions of the elderly. In the Pearson Product Moment correlation, there are interpretations of correlation coefficient r values that indicate the level of relationship between independent variables and the dependent variable.

To see the correlation between gender variables and elderly anthropometric dimensions, as well as the origin variables and elderly anthropometric dimensions, the analysis is performed using contingency coefficient c analysis. Contingency coefficient c is conducted to determine the level of association between variables X that are gender and origin of the elderly anthropometric dimensions as variable Y.

Linear regression analysis was used to determine how the dependent variable (Y) that is the value of each elderly anthropometric dimensions can predictable through the independent variable (X), there are age and BMI partially or jointly (simultaneously).

Independent 2 samples t-test is performed to compare whether the elderly anthropometric dimensions for each sub-variables are the same or different. Independent 2 samples t-test is performed three times to compare between Elderly and Old, men and women, and the elderly from Yogyakarta and Central Java.

3. Results and Discussion

Results of normality tests indicated that all the significant value of each elderly anthropometric dimensions showed values greater than 0.05(p>0,05), which means all dimensions are normally distributed.

Test results of Pearson product moment correlation between categories elderly age and elderly anthropometric dimensions are 13 hand dimension, those are thumb length, middle finger length, ring finger length, Little finger length, Metacarpal thickness, Palm thickness, Grip hand length, Hand length, Palm length, Distance of thumb to little finger, Maximal grip diameter, Minimal grip diameter and Fist height; 7 sitting dimensions those are Sitting elbow height, Sitting shoulder height, Sitting eye height, Sitting height, Popliteal height, Buttock-knee length and Sitting shoulder-fingertip length; and 9 standing dimensions are Standing shoulder-fingertip length, Hip height, Knee height, Standing knuckle height, Standing elbow height, Standing shoulder height, Standing eye height, Body height, and Vertical grip reach (standing) show that there is no relationship between variables X and Y. Meanwhile the 2 hand dimensions are Forefinger length and Forefinger thickness; 2 sitting dimensions are Vertical grip reach (sitting) and Buttock-popliteal length; and 1 standing dimension is Body thickness. There is a very low correlation between variables X and Y. For Thumb thickness dimension has a low correlation between the variables X and Y. For significance value of all elderly anthropometric dimensions except dimension Thumb length, Little finger length, and Fist height have significance value ≥0,05 which means that elderly age has no significant relationship with elderly anthropometric dimensions, or in other words Ho is accepted and Ha is rejected. For old age category of 7 hand dimensions are Forefinger length, Middle finger length, Ring finger length, Little finger length, Grip hand length, Maximal grip diameter, and Fist height; 4 sitting dimensions are Sitting elbow height, Sitting shoulder height, Popliteal height, and Buttock-popliteal length; and 1 standing dimension is Standing shoulder-fingertip length shows that there is no relationship between variables X and Y. Meanwhile the 6 hand dimensions are Thumb length, Metacarpal thickness, Thumb thickness, Hand length, Distance of thumb to little finger, and Minimal grip diameter; 5 sitting dimensions are Sitting eye height, Sitting height, Vertical grip reach (sitting), Buttock-knee length, and Sitting shoulder-fingertip length; and 7 standing dimensions are Body thickness, Standing elbow height, Hip height, Standing knuckle height, Standing shoulder height, Standing eye height, and Body height have very low relationship between the variables X and Y. For Palm thickness, Forefinger thickness, Palm length, Knee length, and Vertical grip reach (standing) dimension has low correlation between variables X and Y. All of the significant values of elderly anthropometric dimensions have a significance value ≥0,05, which means the Old age category doesn't have a significant relationship with elderly anthropometric dimensions, or in other words Ho is accepted and Ha is rejected. As for the results of the correlation between BMI variable and elderly anthropometric dimensions showed that as many as 5 hand dimensions are Thumb length, Thumb thickness, Forefinger thick, Distance of thumb to little finger, and Minimal grip diameter; 9 sitting dimensions are Sitting elbow height, Sitting shoulder height, Sitting eye height, Sitting height, Vertical grip reach (sitting), Popliteal height, Buttock-knee length, Buttock-popliteal length and Sitting shoulder-fingertip length; and 8 standing dimensions are Standing shoulder-fingertip length, Hip height, Knee height, Standing knuckle height, Standing elbow height, Standing shoulder height, Standing eye height, and Body height indicate that there is no relationship between variables X and Y. While the 9 hand dimensions are Forefinger length, Middle finger length, Ring finger length, Little finger length, Grip hand length, Hand length, Palm length, Maximal grip diameter, and Fist height and 1 standing dimensions is Vertical grip reach (standing), there is a very low correlation between variables X and Y. For Metacarpal thickness, Palm thickness, and Body thickness dimension have low correlation between the variables X and Y. For a significance value of all elderly anthropometric dimensions except Sitting shoulder-fingertip length, Standing shoulder-fingertip length, and Body thickness have a significance value of ≥ 0,05, which means BMI has no significant relationship with elderly anthropometric dimensions or in other words Ho accepted and Ha is rejected. Test results of coefficient contingency c correlation between gender and origin of the elderly anthropometric dimensions shows that all significance values in gender variableare ≥0.05, which means the variable gender has a significant relationship with the dimensions

or in other words Ha is received and Ho is rejected, while the origin variable only Thumb thickness dimension which has a significance value ≤ 0.05, which means the original variable has significant relationships with the elderly of Thumb thickness dimension while the other dimension is not.
For linear regression results are presented in Table 2.

Table 2. Table of Linear Regression

Dimension	Equation (Age)	Equation (BMI)
Thumb length	8,227 - 0,318X	6,873 - 0,110X
Forefinger length	8,700 - 0,256X	6,838 + 0,018X
Middle finger length	9,484 - 0,298X	6,383+ 0,147X
Ring finger length	8,454 - 0,242X	6,939+ 0,005X
Little finger length	6,919 - 0,239X	4,599 + 0,147X
Metacarpal thickness	1,667 + 0,093X	0,500 + 0,234X
Palm thickness	4,163- 0,074X	1,203 + 0,214X
Thumb thickness	0,847 + 0,176X	1,731- 0,010X
Forefinger thickness	0,891 + 0,135X	1,444- 0,007X
Grip hand length	15,213 - 0,237X	8,687+ 0,107X
Hand length	20,767 - 0,300X	16,875+ 0,020X
Palm length	11,736 - 0,253X	9,334+ 0,060X
Distance of thumb to little finger	25,361 - 0,275X	25,303 - 0,236X
Maximal grip diameter	5,011 - 0,120X	2,676 + 0,078X
Minimal grip diameter	0,618 + 0,144X	3,312- 0,260X
Fist height	5,983 - 0,025X	3,141+ 0,156X
Sitting elbow height	20,090 - 0,028X	19,685 - 0,019X
Sitting shoulder height	59,016 - 0,153X	49,165 - 0,034X
Sitting eye height	78,581 - 0,232X	66,867 - 0,089X
Sitting height	91,844 - 0,232X	74,662 - 0,048X
Vertical grip reach (sitting)	109,649 - 0,130X	90,810 - 0,027X
Popliteal height	35,973 + 0,035X	46,849- 0,205X
Buttock-knee length	61,524 - 0,198X	57,774 - 0,093X
Buttock-popliteal length	47,628 - 0,066X	46,296 - 0,029X
Sitting shoulder-fingertip length	62,915- 0,047X	85,160 + 0,347X
Standing shoulder-fingertip length	76,356 - 0,180X	86,436 - 0,309X
Body thickness	26,386 - 0,122X	0,929+ 0,334X
Hip height	106,239 - 0,249X	97,117 - 0,104X
Knee height	44,654 - 0,037X	62,212 - 0,288X
Standing knuckle height	63,964 - 0,005X	89,458 - 0,209X

Standing elbow height	107,049 - 0,123X	118,158 - 0,188X
Standing shoulder height	149,493 - 0,271X	140,112 - 0,159X
Standing eye height	170,797 - 0,295X	146,497 - 0,086X
Body height	186,790 - 0,311X	168,556 - 0,150X
Vertical grip reach (standing)	211,615 – 0,289X	156,333 – 0,024X

Those are the equations that can be used to predict a variable Y or elderly anthropometric dimensions if known variable X those are age or BMI variable. That linear regression equation consists of constants and beta. For example, as listed in Table 2, the linear regression equation of Thumb length dimension for age variable is 8.227 to 0.318X, it means constant at 8.227 states that if there is no addition of age, then the value of Thumb length dimension in the elderly is 8.227 cm. Regression coefficient of 0.318 states that each additional 1 year of age will reduce the value of 0.318 cm Thumb length dimension. Conversely, if the age of 1 year reduced, the Thumb length dimension value increases by 0.318 cm. So, the sign - stating the opposite direction of the relationship, where the addition of the independent variable (X) will result in a reduction in the dependent variable (Y), while the reduction of the independent variable (X) will result in the addition of the dependent variable (Y), if the + then declare the direction of the relationship unidirectional, where the addition or subtraction of the independent variable (X) will result in the addition or reduction of the dependent variable (Y).
The results of independent 2 samples t-test between elderly and old age categories are only Grip hand length and Knee height dimension which has a significance value <0.05, which means that there are significant differences between Grip hand length and Knee height dimensions between elderly and old people or variance not homogeneous. For the gender categories, male and female, the result is Thumb thickness, Forefinger thickness, and Buttock-popliteal length dimension which has a significance value <0.05, which means there are significant differences in the dimensions of Thumb thickness, Forefinger thickness, and Buttock-popliteal length between groups of elderly men and women or the group variance was not homogeneous. Lastly, for the origin category of Yogyakarta and Central Java, shows that only the Palm length dimension which has a significance value <0.05, which means there are significant differences between the groups on the dimensions Palm length group of elderly men and elderly women or variance not homogeneous.

4. Conclusion

This study has shown that only age variables significantly influence dimension of Thumb length, Middle finger length, Hand length, standing eye height, and Body height of elderly while the other dimensions are not affected by the age variable. As for the Body Mass Index (BMI) variable concluded that a significant effect on the Sitting shoulder-fingertip length, Standing shoulder-fingertip length, and Body thickness of the elderly, while the other dimensions are not affected by BMI variable. While the results of the correlation age variable age Elderly (60-74 years) and Old (75-90 years) and BMI variables on the elderly anthropometric dimensions for ages ' Elderly ' has no correlation to the 29 elderly anthropometric dimensions, 5 dimensions correlate very low , and only 2 dimensions that have a low correlation. As for the age of ' Old ' has no correlation to the 12 elderly anthropometric dimensions , 18 dimensions correlate very low, and only 5 dimensions that have a low correlation. BMI variable has no correlation to 22 elderly anthropometric dimensions, 10 dimensions.

5. References:

[1] National Commission for the Elderly2012Kondisi dan permasalahan penduduklansia.Retrieved fromhttp://www.komnaslansia.go.id/modules.php?name=News&file=article&sid=24.

[2] Health Department. 2012.

[3] National Commission on the Elderly 2012

[4]Simanjuntak, P,A.& Syarief, A 2010 DesainAlat Bantu MobilitasPenggunaLanjutUsiaUntukBeraktivitas Di TempatUmum1: 1.

[5]Rahmasari, F,A.& Sriwarno, A,B 2012 DesainAlat Bantu BerjalanDanDudukUntukLansiaDenganAplikasiElektromiografi 1: 1.

[6] Perissinotto, E,. Pisent, C,. Sergi, G,. Grigoletto, F & Enzi, G 2002 Anthropometric Measurements In The Elderly: Age And Gender Differences.*British Journal of Nutrition87:*177-186.

[7] Rosnah, M,Y,. Rizal, M,H.& Norazizan, S,S,A,R.2009 Anthropometry Dimensions Of Older Malaysians: *Comparison Of Age, Gender And Ethnicity.Asian Social Science 6: 133-140*

[8]Bisai, S., Bose, K., Khatun, A & Bauri, H 2009Age-Related Anthropometric Changes And Undernutrition Among Middle Aged And Older Savar Tribal Females Of Keonjhar District, Orissa, India.*J Life Science1*: 21-26.

[9] Lopez, P,M,. Ballesteros, R,F,. Zamarron, M,D. & Lopez, S,R 2011 Anthropometric, Body Composition And Health Determinants Of Active Ageing: A Gender Approach.*J, Biosoc, Science* 43:597-610.

Optimization of Physical Working Environment Setting to Improve Productivity and Minimize Error by Taguchi and VIKOR Methods

Fety Ilma Rahmillah

Department of Industrial Engineering
Universitas Islam Indonesia
Yogyakarta, Indonesia

E-mail: fety_rahmillah@yahoo.com

Abstract. The working environment is one factor that has contribution to the worker's performance, especially for continuous and monotonous works. L_9 Taguchi design experiment for inner array is used to design the experiment which was carried out in laboratory whereas L_4 is for outer array. Four control variables with three levels of each are used to get the optimal combination of working environment setting. Four responses are also measured to know the effect of four control factors. Results shown that by using ANOVA, the effect of illumination, temperature, and instrumental music to the number of ouput, number of error, and rating perceived discomfort is significant with the total variance explained of 54,67%, 60,67%, and 75,22% respectively. By using VIKOR method, it yields the optimal combination of experiment 66 with the setting condition of A_3-B_2-C_1-D_3. The illumination is 325-350 lux, temperature is 24^0-26^0C, fast category of instrumental music, and 70-80 dB for intensity of the music being played.

Keywords: Working environment, illumination, temperature, music, ANOVA, Taguchi, VIKOR

1. Introduction

Physical working environment is one factor affecting productivity [1][2]. Physical working environment factors such as coloring in clean and clear working environment, good lighting and ventilation, music played, noise, and temperature are necessary to be well designed suits with the workers characteristics so that humans can perform work activities effectively and efficiently. But, if the workers are not comfort with their working environment, their performance will not be maximum. Even, it can make them faster to get bored and fatigue, moreover for continuous and monotonous works [3]. Activities which are categorized as monotonous job are sewing, typewriting, proofreading, office task, and visual inspection tasks in manufacturing settings. A task is monotonous when its stimulants do not change or the changes are predictable or there is a high type of repetition [4]. Headache and occurrence of accidents like needle-piercing because of the visual strain are caused by insufficient light at the working area [5].

Reference [5] stated that 42.5% of respondents in textiles and clothing sector felt of eye strain because of poor working condition. This data actually has similarities with the condition of workers in garment industry [6]. Scholar [2] had been done a research related with physical working environment and its influence to the civil servants in Malaysia while scholar [1] did the research to the workers in selected oil and gas industry in Nigeria. Questionnaire is used to collect the data and the result showed that there was significant relationship between physical environments (comfort level, temperature) with civil servants productivity [2]. Reference [7] found that human error is significantly affected by four major factors explored, which are stress, repetition, fatigue, and work environment by approximately 48.8%. A friendly physical working environment can be created through application of background music. Reference [8] has proved it by doing this research to the workers in garment manufacturing company in Sri Lanka. According to [9], the music should be only instrument; because if there was a lyrics, people will concentrate to the lyrics, not to the music; and it could be annoying. Even, the study [9] contributes to the field of utilizing music to influence human performance in the workplace.

The Taguchi method is a powerful method of solving quality problems in various fields of engineering.The Taguchi method can be utilized to find the sequence of dominant factors that contributed to the productivity of the operator at the specified production work station. Even, it can significantly reduce the time required for experimental investigation, as it is effective in investigating the effects of multiple factors on performance as well as studying the influence of individual factors to determine which factor has more influence, and which less. Research [10] was done in automotive components assembly factory, and examined environmental factors such as illuminance, humidity and WBGT while the response factor is production rate. The study reveals that the dominant factor that contributed to the productivity was humidity, followed by illuminance and WBGT.

The recent study aims at knowing the relationship among some factors such as temperature, illumination, beat of instrument, and intensity of music by using Analysis of Variance (ANOVA). The second objective is finding the optimum comfort level of sewing working environment setting by using VIKOR to improve productivity and minimize error by considering four responses which are number of output, average of eye blink, number of error, and rating perceived discomfort (RPD).

2. Research Method

The experiments were carried out in Climate Room of Work Design Analysis and Ergonomic Laboratory Universitas Islam Indonesia. The subject was asked to stitch beads in the veil during one hour experiment. They have to pass the training until reach certain amount of production so that the skilled and unskilled factor can be minimized to prevent bias. Eye blink and rating perceived discomfort are used to know when the fatigue happens. Those two kinds of measurements are used to mutually complete each other whereas the number of error is to indicate the accuracy level of the worker's job. Before doing experiment, subject was asked about their healthy condition and RPD. That's because the unhealthy condition can affect the experimental result. RPD is used to know whether the subject was in fresh eye condition or sleepless. The maximum score is 0.5, if the subject has more than 0.5, so that they asked to come at another day. Each of experiments was done for one hour. During the experiment, subject are not allowed to do other activities or talking with other people because it could affect the experiment result.

2.1. Data Requirement

The data required in this research are:

1. Number of output. It can be calculated from how many unit of beads that can be stitched in the veil during one hour experiment. This relates to the productivity of workers.
2. Average of eye blink. It is measured by direct observation and then make an average of the amount of eye blink for every 5 minutes of 0–5th minute, 5th-10th minute, 10th-15th minute, 45th-50th minute, 50th-55th minute, and 55th-60th minute. Blinking is a common facial motion and

reflecting a person's emotional or cognitive state [11]. There are many factors that can affect blink rate such as loud noises, flashing lights, tasks, and environment (room temperature and humidity) [12].

3. Number of error. It is calculated from the amount of error in stitching beads such as sewing result is not straight, the thread is not locked, and the beads sewed is unsymmetric.
4. Rating Perceived Discomfort (RPD) is based on Borg's General Scale (Table 3). The subject is asked to mention the eye fatigue scale that is experienced after one hour experiment.
5. Working area temperature. First type (20^0-22^0C) is a cold comfortable working area where some jobs are good when it is done in this temperature; second type (24^0-26^0C)] is the comfort zone for Indonesian climate is 24^0-26^0C [13][14], and third (28^0-30^0C) is the real condition.
6. Illumination. First type is 175-200 lux, second type is 250-275 lux, and third is 325-350 lux. Those all of intensity actually already fulfilled the requirement of illumination for sewing activity [14]. It is measured by using Luxmeter.
7. Beat of instrumental. First type is fast, second type is slow, and the third is middle.
8. Intensity of instrumental. It variates from 40-50 dB, 55-65 dB, and 70-80 dB. The good effect of background music can reduce wrong finger touch in typing task [9].

Table 1. Control factors and types

No.	Control Factors	1	2	3
1.	Illumination	175-200 lux	250-275 lux	325-350 lux
2.	Temperature	20^0-22^0C	24^0-26^0C	28^0-30^0C
3.	Beat of instrument	Fast	Slow	Middle
4.	Intensity of music	40-50 dB	55-65 dB	70-80B

Table 2. Noise factor and type

No.	Noise factor	1	2
1	Age	18-25 years old	26-33 years old

Table 3. Borg's General Scale [14]

Score	Type of Fatigue
0	Nothing at all
0.5	Extremely weak (just noticeable)
1	Very weak
2	Moderate
3	Somewhat strong
4	Strong
5	
6	
7	Very strong
8	Strong
9	
10	Extremely strong (almost maximal)

Table 4. Responses

No	Response
1	Number of Output (Larger the Better)
2	Average of Eye Blink (Smaller the Better)
3	Number of Error (Smaller the Better)
4	RPD (Smaller the Better)

The data is collected by doing some experiments based on L9 design experiment from Taguchi as shown in Table 5 with twice replication. Then, SNR values for all responses are calculated. If the combination experiment for all responses is different, then VIKOR is used to find the result. The next step is doing confirmation experiment to validate the result obtained.

Tabel 5. Experimental Design using L9 Orthogonal Array

					L4 OA (Outer Array)	
				E	1	2
L9 IA (Inner Array)					Experiment Data	
	A	B	C	D	Y1	Y2
	Column Number					
Run	1	2	3	4		
1	1	1	1	1		
2	1	2	2	2		
3	1	3	3	3		
4	2	1	2	3		
5	2	2	3	1		
6	2	3	1	2		
7	3	1	3	2		
8	3	2	1	3		
9	3	3	2	1		

2.2. Technique for Analizing Data

2.2.1. Signal Noise Ratio (SNR)

The number of output will use Larger the Better (LTB) whereas average of eye blink, number of error, and RPD use Smaller the Better (STB). The equation to calculate SNR is below.

$$SNR_{LTB} = -Log\left[\frac{1}{n} \sum_{i=n}^{n} \frac{1}{yi^2} \right] \quad (1)$$

$$SNR_{STB} = -10\, Log\left[\frac{1}{n} \sum_{i=n}^{n} yi^2 \right] \quad (2)$$

where: n = number of tests in the experiment (trial)

y_l = response value for each replication

2.2.2. Multiple Linear Regression

The common model is:

$$y_0 = b_0 + b_1x_{1i} + b_2x_{2i} + \ldots.. + b_kx_{ki} + e_i \quad (3)$$

2.2.3. VIKOR

VIKOR requires performance rating of each alternative on each criterion A_iC_j wrote normalized calculated by:

$$r_{ij} = \frac{x_{ij}}{\sqrt{\sum_{i=1}^{m} x_{ij}^2}} \quad (4)$$

where :$i = 1, 2, \ldots, m$ and $j = 1, 2, \ldots.., n$

Positive ideal solution A^+ and the negative ideal solution A^- rating can be determined based on normalized weights (y_{ij}) as follows:

$$y_{ij} = w_{ij}\, r_{ij} \qquad (5)$$

where: $i = 1, 2, \ldots\ldots, m$ and $j = 1, 2, \ldots\ldots, n$.

$A^+ = (y_1^+, y_2^+, \ldots\ldots, y_n^+)$

$A^- = (y_1^-, y_2^-, \ldots\ldots, y_n^-)$

where

$$y_j^+ = \begin{cases} \max_i y_{ij}; & \text{if } j \text{ is profit attribute} \\ \min_i y_{ij}; & \text{if } j \text{ is cost attribute} \end{cases} \qquad (6)$$

$$y_j^- = \begin{cases} \min_i y_{ij}; & \text{if } j \text{ is profit attribute} \\ \max_i y_{ij}; & \text{if } j \text{ is cost attribute} \end{cases} \qquad (7)$$

The next step is to determine the utility of measurement (S_i) and the measurement of regret (R_i).

$$S_i = \sum_{j=1}^{n} \frac{(y_j^+ - y_{ij})}{(y_j^+ - y_j^-)} \qquad (8)$$

$$R_i = \max \frac{(y_j^+ - y_{ij})}{(y_j^+ - y_j^-)} \qquad (9)$$

Then calculate the index VIKOR, which is:

$$Q_i = v\left[\frac{(S_i - S_i^-)}{(S_i^+ - S_i^-)}\right] + (1-v)\left[\frac{(R_i - R_i^-)}{(R_i^+ - R_i^-)}\right] \qquad (10)$$

The smallest the value of Q_i (VIKOR index) shows that best alternative A_i then be selected.

2.2.4. Confirmation Experiment
This step aims to validate the conclusion resulted by using Wilcoxon Signed Test. It is used to test the difference of nonparametric repeated measured data. SPSS software is used to do this test.

2.2.5. Analysis of Variance
It is used to identify the contribution of each factor to all responses. The table is as follow:

Table 6. Analysis of Variance

Source	Sq	V	Mq	F_{ratio}	Sq'	ρ%
Factor *A*	S_A	v_A	$M_A = S_A / v_A$		$S_A - v_A . v_E$	$\frac{SA'}{ST} x100\%$
....	....	....	....		....	....
Factor *n*	S_n	v_n	$M_n = S_n / v_n$		$S_n - v_n . v_E$	$\frac{Sn'}{ST} x100\%$
Error	S_E	v_E	$M_E = S_E / v_E$	*1*		
Total	S_T	*N*	-			

3. Results And Discussion
The objective of the experiment is to optimize the environmental parameters (temperature, illumination, beat of instrument, and intensity of music) in order to obtain a better productivity and minimum error and therefore the optimum characteristics should be quantified. The second one is to

know the relationship among four responses and control factors. All the experimental data are already tested their normality and categorized as normal.

4. Effect of Factors to Responses

Table 7 and 8 show the optimal combination for each of responses. A_3-B_2-C_4-D_1 is the optimal combination for number of output whereas the optimal combination for average of eye blink consists of A_3-B_1-C_4-D_2. Based on those table, it can be seen that the rank is different each other.

Table 7. SNR effect to number of output and average of eye blink.

	Output				Eye blink			
	A	B	C	D	A	B	C	D
Type 1	87.00	82.17	88.67	87.42	65.67	84.63	69.35	70.78
Type 2	82.00	89.75	83.25	80.50	73.76	59.57	74.89	64.92
Type 3	87.50	84.58	84.58	88.58	73.82	68.72	68.68	77.22
Delta	5.50	7.58	5.42	8.08	8.15	25.06	6.21	12.31
Rank	3	2	4	1	3	1	4	2

Table 8. SNR effect to number of error and RPD

	Output				Eye blink			
	A	B	C	D	A	B	C	D
Type 1	8.33	8.08	6.25	9.25	4.13	3.96	3.88	3.67
Type 2	6.50	4.50	7.75	4.33	4.13	3.50	4.17	3.96
Type 3	6.33	8.58	7.17	7.58	3.50	4.29	3.71	4.13
Delta	2.00	4.08	1.50	4.92	0.63	0.79	0.46	0.46
Rank	3	2	4	1	1	2	4	3

2.2. Multiple Linear Regression

This step is used to predict the correlation and it follows equation (3). The results are:

$$Y_{11} = 103.89 - 6.5X_1 + 2.5X_2 - 2.67X_3 - 4.33X_4 \quad (R^2=0.892)$$

$$Y_{12} = 68.11 + 2.67X_1 + 5.67X_2 - 3X_3 + 0.67X_4 \quad (R^2=0.895)$$

$$Y_{13} = 57.22 - 10.67X_1 + 2.83X_2 - 2.17X_3 + 4.83X_4 \quad (R^2=0.611)$$

$$Y_{14} = 112.78 - 5.83X_1 - 6.17X_2 - 0.33X_3 + 2.5X_4 \quad (R^2=0.805)$$

Meanwhile the multiple linear regression for number of error are:

$$Y_{31} = 117.78 - 3X_1 - 1.17X_2 + 0.00X_3 + 0.67X_4 \quad (R^2=0.851)$$

$$Y_{32} = 13.44 + 0.67X_1 + 0.17X_2 + 0.5X_3 - 4.17X_4 \quad (R^2=0.639)$$

$$Y_{33} = 4.22 + 1.33X_1 + 2X_2 + 2.17X_3 - 3.67X_4 \quad (R^2=0.705)$$

$$Y_{34} = 7.78 - 3X_1 + 0.00X_2 - 0.83X_3 + 3.83X_4 \quad (R^2=0.805)$$

By using the same way in SPSS, the multiple linear regression for RPD are:

$$Y_{41} = 7.67 - 0.17X_1 - 0.83X_2 - 0.5X_3 - 0.67X_4 \quad (R^2=0.810)$$

$$Y_{42} = 1.72 + 0.08X_1 + 0.5X_2 - 0.08X_3 + 0.58X_4 \quad (R^2=0.820)$$

$$Y_{43} = 2 - 0.5X_1 + 0.67X_2 + 0.17X_3 + 0.67X_4 \qquad (R^2=0.837)$$

$$Y_{44} = 4.28 - 0.67X_1 + 0.33X_2 + 0.08X_3 + 0.33X_4 \qquad (R^2=0.840)$$

The R^2 value of eye blink response is not reach 0.8, so that eye blink response is not used.

2.3. Analysis of Variance (ANOVA)

The purpose of ANOVA is to investigate which of the factors significantly affect the workers' productivity by using F-test, statistically. The larger the value of F, the greater the effect on the performance characteristics. When F>4, it means that the change of operating factors has a significant effect on the quality characteristics.

Table 9. Analysis of Variance for Number of Output

Factors	Sq	V	Mq	F	Sq'	P (%)
Illumination	10.10	2	5.05	0.66	3.22	0.53
Temperature	91.27	2	45.64	6.9	84.39	13.9
Beat of Instrumental	229.26	2	114.63	23.65	222.38	36.62
Intensity of Music	28.90	2	14.45	1.95	22.02	3.63
Error	247.78	72	3.44	1		45.33
ST	607.32	80				100

Table 10. Analysis of Variance for Number of Error

Factors	Sq	V	Mq	F	Sq'	P (%)
Illumination	54.26	2	27.13	17.41	52.53	29.88
Temperature	6.4	2	3.2	1.47	4.67	2.66
Beat of Instrumental	11.88	2	5.94	2.83	10.15	5.78
Intensity of Music	41.04	2	20.52	11.88	39.31	22.36
Error	62.23	72	0.86	1		39.33
ST	175.8	80				100

Table 11. Analysis of Variance for Rating Perceived Discomfort (RPD)

Factors	Sq	V	Mq	F	Sq'	P (%)
Illumination	1.29	2	0.64	4.63	1.21	9.99
Temperature	1.64	2	0.82	6.07	1.56	12.85
Beat of Instrumental	3.5	2	1.75	15.79	3.42	28.2
Intensity of Music	3.01	2	1.51	12.87	2.94	24.19
Error	2.71	72	0.04	1		24.77
ST	12.15	80				100

It can be seen on Table 9 that only temperature and beat of instrumental music which statistically have significant influence toward number of output (F>4). From this result, it is known that there is other factor that also give influence since the error is 45.33%. While illumination and intensity of music being played influence the number of error significantly as shown in Table 10 (p value is 29.88% and 22.36%). Table 11 presents that all factors have significant effect toward rating perceived discomfort of the subjects. The accumulation of contribution of those four factors reach 75.23%, it can be concluded that those four factors have to be really considered in order to make a good working environment. Eventhough Rating Perceived Discomfort is a subjective measurement, but sometime it can really represent the condition of someone because they felt it by theirself. But of course, the other measurement should also be used to complete in order to get the real effect. Based on ANOVA test, the effect of total factors which are illumination, temperature, beat of instrument and intensity of music to the number of output, number of error, and RPD are 54.67%, 60.67%, and 75.22%, respectively. It means that those factors are necessary to be considered to find the best working environment setting.

2.4. VIKOR

Tabel 7 and 8 show the different optimal condition for all responses so that it is necessary to decide the optimal combination experiment by using VIKOR method. To calculate the score of SNR of each response, equation (1) is used to calculate response number of output whereas equation (2) is for response number of error and Rating Perceived Discomfort. The example is shown below.

SNR_{11}=-10 Log [1/4*(0.0005)] = 3.88 (Number of output)

SNR_{11}=-10 Log [1/4*(507)] = -20.69 (Number of error)

SNR_{11}=-10 Log [1/4*(74.25)] = -11.99 (RPD)

Equation (4) is to determine the normalized decision score.

$r_{ij} = \frac{x_{11}}{|x_1|} = 0.115$ (Number of output)

$r_{ij} = \frac{x_{11}}{|x_1|} = -0.143$ (Number of error)

$r_{ij} = \frac{x_{11}}{|x_1|} = -0.164$ (RPD)

While equation (5) is to define the positive and negative ideal solution by considering the weight for each response which is equal to 0.33.

y_{11} = 0.115 x 0.33 = 0.038 (Number of output)

y_{11} = -0.143 x 0.33 = -0.048 (Number of error)

y_{11} = -0.164 x 0.33 = -0.054 (RPD)

Then, the results will be summed at the similar position:

y_{11} = y_{11} (number of output) + y_{11} (number of error) + y_{11} (RPD) = 0.038 + (-0.048) + (-0.054) = -0.064

The positive ideal solution (A^+) is calculated based on equation (6) whereas equation (7) is to count the negative ideal solution.

y_1^+ = max (-0.064; -0.044; -0.029; -0.046) = -0.029

y_1^- = max (-0.064; -0.044; -0.029; -0.046) = -0.064

The next step is determining utility measurement (S_i) by using equation (8) and regret measurement (R_i) by using equation (9).

$$S_1 = \frac{(-0.029 - (-0.064)) + \cdots + (-0.029 - (-0.046))}{(-0.029) - (-0.064))}$$

$$S_{imax} = 2.941 \; and \; S_{imin} = 1.204$$

$$R_1 = Max\left(\frac{(-0.029-(-0.064))+\cdots+(-0.029-(-0.046))}{(-0.029-(-0.064))+\cdots+(-0.029-(-0.046))}\right)=1$$

$$R_{imax} = 1 \; and \; R_{imin} = -0.031$$

VIKOR score is the closest value of each responses to the ideal solution and it can be quantified based on equation (10).

$$Q_i = v\left[\frac{(S_i - S_i^-)}{(S_i^+ - S_i^-)}\right] + (1-v)\left[\frac{(R_i - R_i^-)}{(R_i^+ - R_i^-)}\right] = 0.5\left[\frac{(1.903-1.204)}{(2.941-1.204)}\right] + 1.5\left[\frac{(1-(-0.031))}{(1-(-0.031))}\right] = 0.215$$

Based on VIKOR calculation, it can be concluded that the optimal combination is experiment 66 (A_3-B_2-C_1-D_3) which is illumination of 325-350 lux, temperature of 24-26^{0}C, fast instrumental music, and 70-80 dB for intensity of the instrumental music being played. This experiment produced the number of output, number of error, and RPD score of 101.75, 3.75, and 3 respectively. Table 12 shows the comparison of the prediction result and confirmation experiment. Wilcoxon test which is used to test the experiment confirmation shown value of 0.321 means that there is no significant difference since it is more than 0.05.

Table 12. Comparison of result and confirmation experiment

	Output				Error				RPD			
	Y_1	Y_2	Y_3	Y_4	Y_1	Y_2	Y_3	Y_4	Y_1	Y_2	Y_3	Y_4
Prediction	87	92	119	105	2	0	6	6	2	4	3	4
Confirmation	90	95	114	108	2	1	7	5	2	3	3	4

For the illumination, it is already relevant since the higher the intensity is better for hand sewing which is need high accuracy. A good illumination will make the worker easily to see the object or the tools used in doing the job and it is in such a way that the visibility of the needlepoints will be optimum [14]. Based on the observation while experiment conducted, the amount of illumination intensity is not appearing glare. For the temperature, it suits with some researches [13][14], in which 24^0-26^{0}C is categorized as comfortable working area, but the duration should be considered. If it is for a long duration of work, some negative effect will start to appear such as psychical or physiological problems [14]. The result of this research find that the best is fast beat of instrumental music for sewing activity in which it is not a complex task. If the beat is low, it can make workers to easily get bored or sleepy. However, it was not inline with reference [15] since it said that in the workplace, faster tempos of background music may cause declines in worker output, especially if the employee is engaged in a complex task. Reference [15] also stated that playing background music in the classroom is beneficial and teachers need to be cognizant of the tempo of the background music. For the intensity of instrumental music, that amount is permitted. It is still safe, not becoming a noise factor since the louder the sound, it can disturb the worker. Listening to music can make a positive mood and enhanced perception on design while working [16]. Basically, the influence of music on this performance is difficult to be measured because many factors are involved such as preferences what kind of music which they like.

CONCLUSION

This study was done to empirically prove the perception of the effect of work environment factors towards productivity and error. Based on ANOVA test, the effect of total factors which are illumination, temperature, and instrumental music to the number of output, number of error, and RPD is 54.67%, 60.67%, and 75.22%, respectively. From all of processes that have been done in this research, those can be concluded that the optimal combination of working environment setting for sewing activity as one of monotonous work is experiment 66, with the setting condition of A_3-B_2-C_1-D_3 or illumination of 325-350 lux, temperature of 24^0-26^{0}C, beat of instrumental music is fast, and 70-80 dB for intensity of the instrumental music being played. The findings will also be useful to engineers in the design of working environment system for other monotonous work to improve the comfort of the work station area and control productivity of workers.

REFERENCES

[1] Taiwo A 2010 The influence or work environment on workers productivity: A case of selected oil and gas industry in Lagos, Nigeria *African Journal of Business Management* ***4*** 299-307

[2] Ismail J, Ladisma M, Amin S H, and Arapa A 2010 The Influence of Physical Workplace Environment on the Productivity of Civil Servants: The case of the Ministry of Youth and Sports, Putrajaya, Malaysia *Voice of Academia* ***5*** 71-78

[3] Mahasusiawanto, Frihartati M, Saliana R 2006 Study on Factors Affecting Bussiness Development of SMEs in North Sumatera *Jurnal Pengkajian Koperasi dan UKM Nomor 1*

[4] Sullivan J M 2008 *Visual Fatigue and The Driver* Research Report, The University of Michigan Transportastion Research Institute

[5] McAtamney L, Oxborrow L, and Hague J 2001 A Study on Working Conditions in theTextiles and Clothing Sector Undertaken by FIA-UGT, Spain", *Report of Musculoskeletal Disorders and Work Organisation in the European Clothing Industry, European Trade Union Technical Bureau for Health and Safety*

[6] Jahan M 2012 Woman Workers in Bangladesh Garment Industry: A Study of the Work Environment *International Journal of Social Science Tomorrow* **1**

[7] Yeow J A, Ng, P K, Tan K S, Chin T S, and Lim W Y 2014 Effects of Stress, Repetition, Fatigue and Work Environment on Human Error in Manufacturing Industries *Journal Applied Science* ***14*** 3464-3471

[8] Padmasiri M K, and Dhammika K A 2014 The Effect of Music Listening on Work Performance: A Case Study of Sri Lanka *International Journal of Scientific & Technology Research* ***3*** 118-122

[9] Jiang X, and Sengupta A K 2011 Effect of Music and Induced Mental Load in Word Processing Task. *IEEE* 3261-3266

[10] Haniff M, Ismail A, Deros B, Rahman M, and Kardigama K 2011 The Taguchi Approach In Optimizing Environmental Factors Affecting Productivity In The Automotive Industry. *International Journal of Automotive and Mechanical Engineering (IJAME)* ***3*** 306-317

[11] Bacher L, and Smotherman W 2004 Spontaneous Eye Blinking in Human Infants: A Review. *Developmental Psychobiology* **44** 95-102

[12] Crnovrsanin T, Wang Y, and Ma K-L 2014 Stimulating a Blink: Reduction of Eye Fatigue with Visual Stimulus *CHI.* Toronto ACM

[13] Wijaya I 2012 Word Effect of Temperature, the Lighting, Workload, Noise against Eye Fatigue, General Fatigue and Stress Affect Learning Outcomes the Student Computer Users *International Journal of Computer Applications* ***58***

[14] Grandjean E 1986 *Fitting the Task to the Man* London: Taylor and Francis Inc

[15] Miller Jeffrey M and Peden Blaine F 2003 Complexity and Degree of Tempo Modulation as Factors in Productivity *Psi Chi Journal of Undergraduate Research* **8** 21-27

[16] Lesiuk Teresa 2005 The effect of music listening on work performance *Psychology of Music* **33** 173-191

IOP Conf. Series: Materials Science and Engineering **105** (2016) 012026 doi:10.1088/1757-899X/105/1/012026

Removal of free fatty acid in Palm Fatty Acid Distillate using sulfonated carbon catalyst derived from biomass wastefor biodiesel production

Arif Hidayat[1,2], Rochmadi[2], Karna Wijaya[3] and Arief Budiman[2,4]

[1] Chemical Engineering Department, Indonesia Islamic University, JalanKaliurang km 14,5 Yogyakarta 55584, Indonesia
[2] Chemical Engineering Department, GadjahMada University JalanGrafika 2 Yogyakarta 55281, Indonesia
[3] Chemistry Department, GadjahMada University Sekip Utara Bulaksumur Yogyakarta 55281, Indonesia
[4] Center for Energy Studies, GadjahMada University, Sekip K1A, Yogyakarta 55281, Indonesia

E-mail: abudiman@ugm.ac.id

Abstract. In this research, the esterification of PFAD using the sulfonatedcoconut shell biochar catalyst was studied. Carbon solid catalysts were prepared by a sulfonation of carbonized coconut shells. The performances of the catalysts were evaluated in terms of the reaction temperatures, the molar ratios of methanol to PFAD, the catalyst loading and the reaction times. The reusability of the solid acid carbon catalysts was also studied in this work. The results indicated that the FFA conversion was significantly increased with increasing catalyst loading and reaction times. It can be concluded that the optimal conditions were an PFAD to methanol molar ratio of 1:12, the amount of catalyst of 10%w, and reaction temperature of 60°C.At this optimum condition, the conversion to biodieselreached 88%.

1. Introduction

Increased demand for energy, global warming due to emission of greenhouse gases, environmental pollution, and fast diminishing supply of fossil fuels are the major key factors leading to search for alternative sources of energy [1]. Biodiesel is an-ecofriendly fuel for diesel enginee that is produced from vegetable oil (edible or non-edible oil) and animal fats. It has been gaining worldwide popularity as alternative energy source because it is a renewable, biodegradable, nontoxic fuel and non-flammable [2]. Conventionally, biodiesel fuel is made through transesterificating oils where the process involves a catalyst and an alcohol. The biodiesel production can also be obtained through the esterification of free fatty acids (FFA).Crude biodiesel can be further purified by washing and evaporating to remove any remaining catalyst and methanol.Base catalyst such as NaOH, KOH etc. usually used for transesterification reaction, while acid catalyst such as H_2SO_4, HCl, etc. used for esterification [3, 4,5,6].

The homogeneous catalytic systems, however, suffer from many problems such as difficulty in removing the basic catalysts after the reaction, production of large amount of waste water and emulsification. Other disadvantage is associated with the catalysts because they are hazardous, caustic and hygrosocopic [6]. Then in order to eliminate the problems, several heterogeneous basa catalysts for biodiesel have been studied, include $CaZrO_3$ [7], Al_2O_3–SnO [8], Li/MgO [9], CaO/Al_2O_3 [10], KOH/Al_2O_3 and KOH/NaY [11], KF/Ca-Al hydrotalcite [12], alumina/silica supported K_2CO_3 [13]. Several reports in the literature have discussed the use of heterogeneous acid catalysts, including zirconia supported mixed metal oxides [14], sulphated lanthanum oxide (SO_4^{2-}/La_2O_3) [15], organophosphonic acid-functionalised silica [16], Amberlyst-15 [17] and ZnO/SBA-15 and MgO/SBA-15 [18]. The esterification of FFAs by solid acid catalysts has also been explored. Sulphated zirconia (SO_4^{2-}/ZrO_2), sulphated tin oxide (SO_4^{2-}/SnO_2) and sulphated titanium oxide (SO_4^{2-}/TiO_2) are representative catalysts with good catalytic activities [19, 20]. Several of these solid acid catalysts showed high catalytic activity and good stability during the simultaneous catalysis of both the esterification and trans-esterification reactions [20].

However, inorganic-oxide solid acids such as zeolite and alumina catalysts were prepared complexly and expensive synthesis routes demonstrated poor reusability and were non-biodegradable. In particular, these catalysts have small pore and thus are not suitable for biodiesel production because of the diffusion limitation of the large fatty acid molecules. In the other hand the inorganic-oxide solid acids commonly is hydrophilic, and its activity will be decreased by the water produced from the esterification of FFA.

To overcome these issues, catalyst research for the production of cost effective biodiesel has been focused towards low cost renewable "green catalyst". Recently, a novel carbon-based solid catalyst was reported as promising catalysts for the production of biodiesel. Carbon-based solid catalysts have several distinct advantages over inorganic-oxide solid acid catalysts; they are stable under acidic basic conditions, can have very high surface area (200–1500 m^2/g), renewable biomass sources can be used to generate the carbon, and the non-polar nature of the support matrix may reduce adsorption of polar molecules (e.g., water or glycerol) that can deactivate the catalyst. Carbon-based solid catalysts also are considered as catalysts for biodiesel production due to their chemical inertness and superior mechanical and thermal stability [21].

Carbon material has been widely used as a catalytic support that exhibits multiple desirable properties, including the following characteristics: (1) the catalyst can be used in both acid and base conditions; (2) the catalyst can be used in high temperature conditions (650°C); (3) the pore structure can be modified for different purposes; (4) the surface properties can be modified to become hydrophobic; (5) the spent catalysts can be recycled for metal by burning; and (6) the materials are inexpensive [22, 23, 24, 25]. Commercial activated carbon can be obtained from various carbonaceous materials, such as coal, lignite, nutshells, wood, and peat. Considered to be agricultural by-products, the abundance and availability of these products represent good sources of raw materials for activated carbon materials.

Exploring new methods to produce biodiesel from low-cost raw materials is the main interest in recent biodiesel research. Many studies have proposed waste and non-edible oils for biodiesel production such as sludge palm oil [26], Jatrophacurcas oil [27], rubber seed oil [28], mahua (Madhucaindica) [29], Karanja (Pongammiapinnata) [30] and tall oil [31]. Palm Fatty Acid Distillate (PFAD) is the byproduct during the refining of palm oil and is mainly produced in the fatty acid stripping and deodorization stage. PFAD is low-cost and non-food source of feedstock. PFAD is a by product from production of consumable palm oil, with high FFA content (75-93 % wt.) and can be used as a feedstock for the production of biodiesel [25]. Compared with the relatively expensive Crude Palm Oil (1 USD per litre), the Palm Fatty Acid Distillate (PFAD) costs only 0.4 USD per litre, suggesting that using PFAD for FAMEs production with the esterification reaction would reduce the production cost of biodiesel to compete economically with petroleum-based fuels.

In this research, the esterification of PFAD on solid acid carbon catalysts prepared by a sulfonation of the carbonized char of coconut shell was studied. The performances of the solid acid carbon catalysts were evaluated in terms ofthe molar ratios of PFAD to methanol, the catalyst loading and the reaction temperatures.

2. Experimental

2.1. Experimental

Coconut shells were used as the raw materials to produce the char in this study (Sleman Regency, D.I. Yogyakarta province, Indonesia). The starting materials were manually chosen, cleaned with deionised water, dried at 110°C for 24 hours and ground with a mill to obtain samples with particle sizes of 3-5 mm. PFAD was obtained from the Palm Oil Industry Public Company Limited, Indonesia. A light yellow solid at room temperature, PFAD consists of 93% of FFA (56.74% palmitic, 29.51% oleic, 1.89% hexadecanoic, 0.75% linoleic, 0.51% myristic, 0.38% stearic, 0.27% octadecanoic acids) with triglycerides, diglycerides (DG), monoglycerides (MG) and traces of impurities. All chemicals, including methanol (99.5%) and sulphuric acid (98%), were analytical grade.

2.2. Synthesis of the carbon solid catalysts

Prior to use, coconut shells were crushed and sieved to obtain particles with sizes less than 3 mm. The crushed coconut shells were then placed inside a stainless steel tubular reactor, heated to the carbonization temperature of 400°C and held for 2 hours at the carbonization temperatures under N_2 gas flow. After carbonization, the samples were cooled to room temperature under N_2 flow. The carbonized samples were then crushed and sieved to obtain particles with sizes less than 1-1.5 mm.

The sulfonation of the coconut shell carbonized char was performed using the following procedure. Carbonized coconut shell (5.0 grams) and a concentrated H_2SO_4 (98%) solution (100 mL) were placed into a 100 mL Teflon-tube autoclave. In this procedure, 5 grams of the coconut shell carbonized char was immersed in 100 ml concentrated H_2SO_4 (98%) at 423 K for 12 hours. The product was then washed with hot distilled water until no sulphate ions were detected in the wash water and then dried.

The surface areas (based on the BET equation) and pore volumes for the carbon solid acid catalyst were determined using the nitrogen gas sorption technique on a Quantachrome Nova Win2. Fourier transform infrared (FT-IR) spectroscopy was used to characterise the surface functional groups using a Shimadzu FTIR-8400. A resolution and aperture setting of 4 cm^{-1} with a scan speed of 2.5 kHz over a wave number range of 4000–650 cm^{-1} were used to collect the spectra. The acid density of the catalysts was estimated by a titration analysis.

2.3. Performance Evaluation

The batch esterification was performed at a laboratory scale. A hot plate with a magnetic stirrer was used for heating the mixture in a three-neck 250 mL round bottom-flask fitted with a water-cooled reflux condenser. The round bottom-flask was held in a water bath to maintain the temperature. The mixture was stirred at the same speed for all test runs. The experimental setup for the esterification reaction of the biodiesel production using carbon sulfonated catalyst was shown in Figure 1. After the esterification reaction, the product mixture was poured into a separating funnel and then allowed to settle into two phases. The bottom FAME layer was separated and purified by a water washing process prior to the analyses.

The FFA conversion wasanalyzed by titration procedure for the evaluation of free residual acidity. Once the sample was taken from the reactor, it was washed with water in order to stop the reaction and separates the catalyst and the alcohol from the oil phase. A weighted amount of the sample was dissolved in ethanol and diethyl ether in order to be able to make the titration analyses, and some drops of phenolphthalein as indicator were added to the system to be able to measure the conversion. The titration was done with a 0.05 mol/L KOH solution. The acid value was calculated using the following equation:

$$AV = \frac{V.c.56.1}{m} \quad \dots (1)$$

where AV is the acid value of the sample, V is the volume of the KOH solution of the titration (mL), c is the concentration of the KOH solution (mol/L), m is the weight of the sample (g), and 56.1 is the molecular weight of KOH. The conversion of free fatty acid was calculated using the following equation:

$$X_F = \left(1 - \frac{AV_2}{AV_1}\right) \times 100 \qquad \text{... (2)}$$

where X_F is the FFA conversion, AV_1 is the initial acid value of the mixture and AV_2 is the acid value of mixture after reaction.

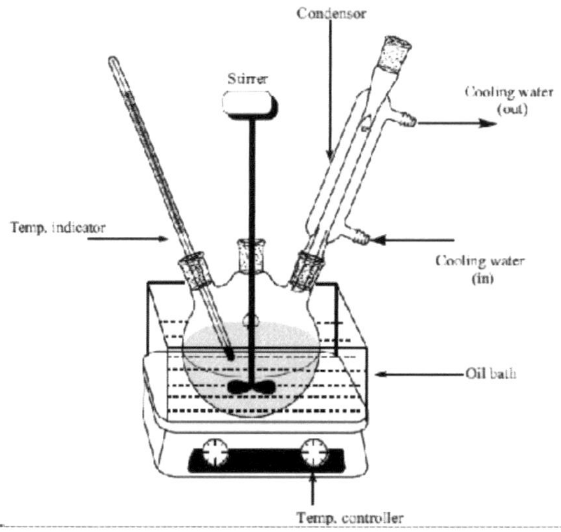

Figure 1. The experimental setup for the esterification reaction of the biodiesel production using carbon sulfonated catalyst.

3. Results and Discussion

3.1. Catalyst characterization

The catalyst characteristics, including the BET surface area, the average pore size, and the total pore volume of the carbon solid acid catalystsboth before and after sulfonation, are presented in Table 1. The coconut shell carbonization charhad low surface areas. The acid treatment by sulfonation improved the surface area characteristics by opening the pores of the material. The acid treatment of the partially carbonized material improved the surface area from19.86 m^2/g to 21.47 m^2/g. Meanwhile, the average pore size diameter increase from 3.53 to 2.56 nm. The larger pore size diameter would make the reactant molecules easily to access to the internal catalyst. The total acid density of the carbon solid acid catalyst was measured using a back titration method. The SO_3H density of the catalyst was significantly increasing after sulfonation. The catalyst exhibits high catalytic acid density.

Table 1. Characteristics of carbon solid acid catalysts

Sample	BET surface area ($m^2\ g^{-1}$)	pore size (nm)	total pore volume ($cm^3\ g^{-1}$)	Acid density ($mmol\ g^{-1}$)
Coconut shell carbonization char	19.86	3.53	0.103	-
catalyst	21.47	2.56	0.109	0.95

The FT-IR spectra of the coconut shell carbonization char and the carbon acid catalystsare shown in Figure2. The FT-IR spectra of the coconut shell carbonization char exhibiteda peak at 744 cm^{-1}that can beattributed to the out-of-plane bending of the ring C–H bonds of both aromatic and heteroaromatic compounds. In addition, two peaks at approximately 1457 and 1660 cm^{-1}can be assigned to the aromatic ring modes and aromatic acidic groups (C–O in carbonyl group). A broad peak at approximately 3400 cm^{-1}can also be assigned to the –OH stretching vibration mode of the –COOH group and phenolic –OH.The spectrum of the carbon acid catalyst (after sulfonation) was similar to the results obtained for the coconut shell carbonization char. In this latter spectrum, the strong band at 1701 cm^{-1} and the weak bands at 1052 cm^{-1}can be attributed to the stretching modes of the sulphate groups. These bands may be assigned to the SO_2 asymmetric and symmetric stretching vibration mode from the $-SO_3H$ groups.The bandsat 1660 and 1361 cm^{-1} also confirm the presence of poly aromatic hydrocarbons in the carbon acid catalyst. The overall FT-IR analysis suggests that the carbon acid catalystconsisted of polycyclic aromatic carbon sheets containing $-SO_3H$, –COOH, and –OH functionalities.

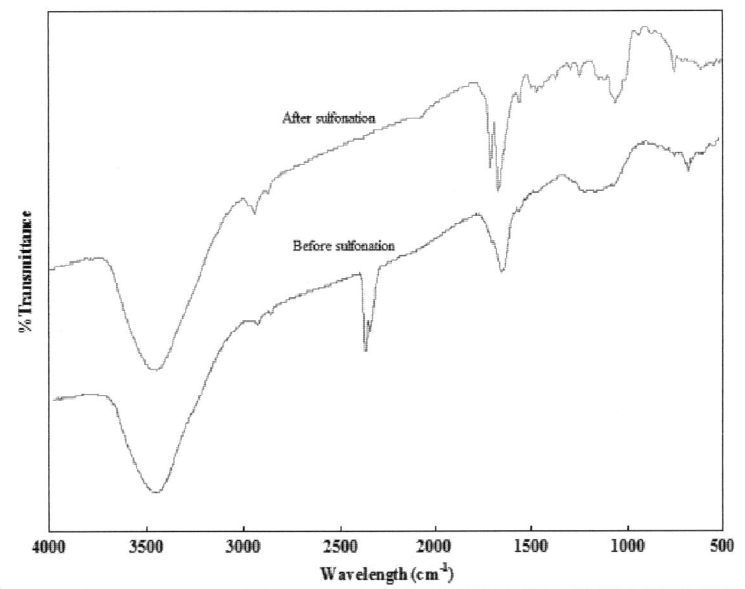

Figure 2. FTIR Spectra of samples before and after sulfonation.

3.2. Performance Evaluation

3.2.1. Effect of PFAD to methanol molar ratio.

Generally, the esterification reactions were performed in excess of methanol or ethanol to favour the forward reaction, since the esterification reactions of fatty acids with alcohols is reversible. Molar ratio is one of the important factors affecting the conversion of FFA. Molar ratio is the ratio of number of moles of alcohol to number of moles of oils in the oil. In order to study the effect of molar ratio PFAD to methanol on esterification reaction, different experiments were carried out using 1:6, 1:8, 1:10, and 1:12, at 60°C.Figure 3shows the effect of the molar ratio on the conversion of the FFA content of PFAD.It was observed that the FFA conversion increased upon increasing the oil to methanol molar ratio from 1:6 to 1:12, possibly due to a greater availability of alcohol molecules for further attack at the carbonyl group of FFA molecules at the acidic sites on the catalyst's surface. Further increases in methanol content had the effect on the FFA conversion.

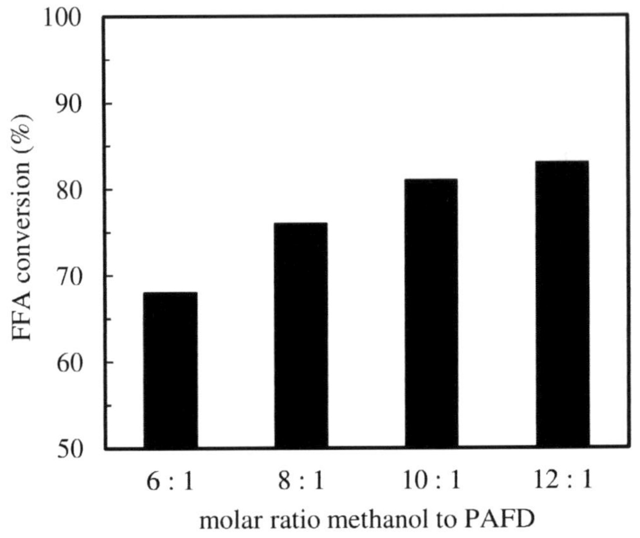

Figure 3. Effect of molar ratio methanol to PAFDon FFA conversion (Reaction conditions: reaction temperature 60°C, stirring speed 800 rpmand catalyst loading 10% wt. of PFAD).

3.3. Effect of catalyst loading

Catalyst loading is an important parameter that needs to be optimized to increase the FFA conversion. Figure4 shows the effect of catalyst loading on FFA conversion. The value of catalyst loaded was varied from 1 to 10%wt. of PFAD. From the Figure4, it can be observed that there is a significant increase in FFA conversion as the catalyst loading increased from 1 to 10%. This is explained that the total number of available active sites increased resulted in faster reaction rate to reach reaction equilibrium, with more catalyst addition. This might be due to the immiscibility of PFAD and methanol, which causes the reaction to be the rate limiting step at the beginning of the reaction.

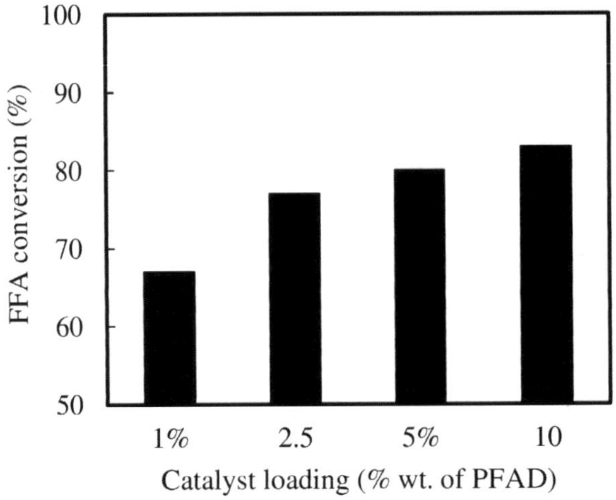

Figure 4. Effect of catalyst loading on FFA conversion (Reaction conditions: reaction temperature 60°C, stirring speed 800 rpm, and molar ratio of PFAD to methanol 1:12).

3.4. Effect of reaction temperature

The rate of reaction is strongly affected by the reaction temperature. A higher reaction temperature can decrease the viscosities of oils and result in an increase in reaction rate as more energy is being supplied for the reaction to occur. Thus the yield of the biodiesel product is improved. However, the reaction temperature must be less than the boiling point of alcohol (boiling point of methanol is at 60–70°C at atmospheric pressure) to ensure the alcohol will not be lost through vaporization.Figure 5 shows that the FFA conversion increases from 64to 84% as the temperature increases from 40 to 60°C. The increased of FFA conversion might be not only due to the effect of increase of the reaction rate by increasing temperatures but also some improvement of the mass transfer limitation between reactant and catalyst. On the other hand, the increase in the FFA conversion when reaction temperature increased from 40 to 60°C could be justified as follows. PFAD and methanol are immiscible. Therefore reaction with heterogeneous catalyst would create a-phase system at which the mass transfer rate of reactant molecules between the three phases is very limited. As the reaction temperature is increased, all reactant molecules will gain more kinetic energy that will eventually accelerate the mass transfer rate among the oil–methanol–catalyst phases that resulted in the formation of more FAME in a shorter time.

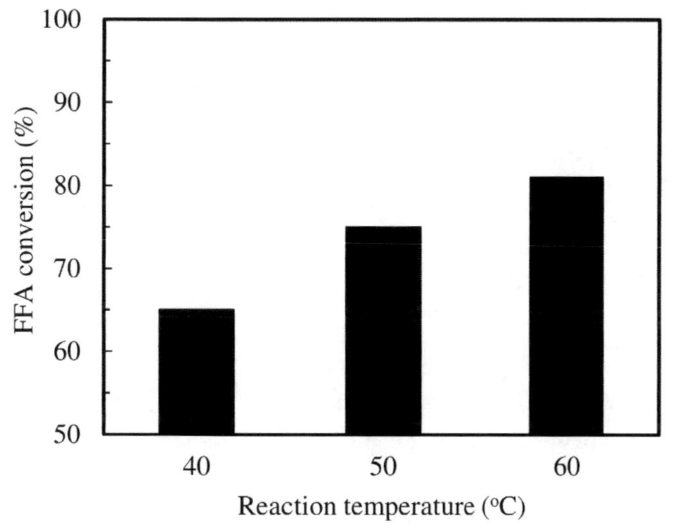

Figure 5. Effect of reaction temperature on FFA conversion (Reaction conditions: stirring speed 800 rpm, molar ratio of PFAD to methanol 1:12,catalyst loading 10%wt. of PFAD).

4. Conclusions

The esterification of the free fatty acid on PFAD catalyzed by sulfonatedcarbon catalyst derived from biomass waste was investigated. The performances of the catalysts were evaluated in terms of the reaction temperatures, the molar ratios of methanol to PFAD, the catalyst loading and the reaction times. The reusability of the solid acid carbon catalysts was also studied in this work. The results indicated that the FFA conversion was significantly increased with increasing catalyst loading and reaction times. It can be concluded that the optimal conditions were anmolar ratio of PFAD to methanol of 1:12, the amount of catalyst of 10%w, and reaction temperature of 60°C. At this optimum condition, the conversion to biodiesel reached 83%.

IOP Conf. Series: Materials Science and Engineering **105** (2016) 012026 doi:10.1088/1757-899X/105/1/012026

5. References

[1]. Fatimah I, Hidayat A, Setiawan K H 2010*Asian J. Chem.* 22 (5) 3793.
[2]. Demirbas A 2009 *Energ. Convers. Manage.*50(1)14.
[3]. Freedman B, Pryde EH and Mounts TL 1984 *J. Am. Oil Chem. Soc.*61(10)1638.
[4]. Di Serio M, Tesser R, Dimiccoli M, Cammarota F, Natasi Mand Santacesaria E 2005 *J. Mol. Cat. A Chem.*239(1-2)1111.
[5]. Dmytryshyn SL, Dalai AKand Chaudhari ST*Bioresour. Technol.* 200492(1)55.
[6]. HelwaniZ, OthmanMR, AzizN, KimJ, and FernandoWJN2009*Appl.Catal. A: Gen.* 3631.
[7]. Kawashima A, Matsubara K, Honda K. 2008 *Bioresour. Technol.* 99(9)3439.
[8]. Macedo CCS, Abreu FR, Tavares AP, Alves MB, Zara LFandRubim JC2006*J.Braz. Chem. Soc.*17 (7)1291.
[9]. Wen ZZ, Yu XH, Tu ST, Yan JYandDahlquist E 2010 *Appl.Energ.*87(3)743.
[10]. Umdu ES andSekerE2012*Bioresour. Technol.*106178.
[11]. Noiroj K, Intarapong P, Luengnaruemitchai Aand Jai-In S 2009 *Renew.Energ.*34(4)1145.
[12]. Takase M, Zhang M, Feng W, Chen Y, Zhao T, Cobbina SJ, Yang L, Wu X2014*Energ. Convers. Manage.*80117.
[13]. Lukic I, Krstic J, Jovanovic D and Skala D. 2009*Bioresour. Technol.*100(20)4690.
[14]. Kim M, DiMaggio C, Salley SO, Ng KYS2012*Bioresour. Technol.*11837.
[15]. Vieira SS, Magriotis ZM, Santos NAV, Saczk AA, Hori CEandArroyo PA 2013*Bioresour.Technol.*133248.
[16]. Yin P, Chen L, Wang Z, Qu R, Liu X, Ren S2012*Bioresour.Technol.* 110258.
[17]. Park JY, Kim DK and Lee JS. 2010*Bioresour. Technol.*101(1)S62.
[18]. Barros SDT, Coelho AV, Lachter ER, Gil RASS, Dahmouche K, da Silva MIPandSouza ALF2013*Renew.Energ.* 50585.
[19]. Kiss AK, Dimian AC andRothenberg G2006*Adv. Synth.Catal.*348(1-2)75.
[20]. Furuta S, Matsuhashi H andArata K 2004 *Catal.Commun.*5(12)721.
[21]. Hidayat A,Rochmadi,Wijaya K,HinodeHandBudimanA 2013 *Asian J. Chem.* 25 (3)1569.
[22]. Illan-Gomez MJ, Linares-Solano A, Radovic LR, Salinas-Martinez de Lecea C. 1995*Energy Fuels*9 (1) 97.
[23]. de la Puente G and Menendez JA 1998*Solid State Ionics* 112103.
[24]. Goncalves F andFigueiredo JL2004*Appl.Catal.B Environ.* 50(4)271.
[25]. HidayatA, Rochmadi, WijayaK, NurdiawatiA, KurniawanW, HinodeH, YoshikawaKandBudimanA 2015*Energy Procedia*75 969.
[26]. Hayyan A, Alam MZ, Mirghani MES, Kabbashi NA, Hakimi NINM, Siran YM and Tahiruddin S 2010*Bioresour. Technol.*1017804.
[27]. Patil PD and Deng S. 2009 *Fuel*881302.
[28]. Ramadhas AS, Jayaraj S andMuraleedharan C 2004*Fuel*84335.
[29]. Ghadge SV and Raheman H2005*Biomass Bioenergy*28601.
[30]. Naik M, Meher LC, Naik SN Das LM. 2008*Biomass Bioenergy* 32354.
[31]. Demirbas A. 2008*Energy Sour. A* 301896.

Preparation and characterization of activated carbon from sugarcane bagasse by physical activation with CO_2gas

BachrunSutrisno[1], Noni AyuRizka[2], AnnisaSolichaHidayat[3] and ArifHidayat[4]*

[1,2,3,4] Chemical Engineering Department, University of Islam Indonesia, JalanKaliurang km 14,5 Yogyakarta 55584, Indonesia

*E-mail:arif.hidayat@uii.ac.id

Abstract. A series of experiments have been conducted to study the effects of different carbonization temperatures (400, 600, and 800°C) on characteristics of porosity in activated carbon derived from carbonized sugarcane bagassechar at activation temperature of 800°C. The results showed that the activated carbon derived from high carbonized temperature of sugarcane bagassechars had higher BET surface area, total volume, micropore volume and yield as compared to the activated carbon derived from low carbonized temperature. The BET surface area, total volume and micropore volume of activated carbon prepared from sugarcane bagassechars obtained at 800°C of carbonized temperature and activation time of 120 min were 661.46m^2/g, 0.2455cm^3/g and 0.1989cm^3/g, respectively. The high carbonization temperature (800°C) generated a highly microporous carbonwith a Type-I nitrogen adsorption isotherm, while the low carbonization temperature (400 and 600°C) generated a mesoporous one with an intermediate between types I and IInitrogen adsorption isotherm.

1. Introduction

Activated carbons with highly developed surface area are usually used worldwide for water treatment, wastewater reclamation, gas purification and also as catalyst support [1, 2]. Activated carbonsare produced from a variety of carbonaceous source materials. The choice of precursor is largely dependent on its availability, cost, and purity, but the manufacturing process and intended application of the product are also important considerations [3].In the recent years, many other agricultural by-products have been used as sources for activated carbons. Agricultural biomass wastes have proved to be promising raw materials for the production of activated carbons because of their availability at a low price, abundantly available, and renewable resources [4]. They were used for the production of activated carbon due to high carbon content, a high adsorption capacity, high density and considerable good mechanical strength. They also have low ash content which is ideal for creating highly porous structures within the activated carbon matrix. Many agricultural by-products such as coconut shell [5,6,7], grain sorghum [8], coffee bean husks [9], rubber wood sawdust [10], chestnut wood [11], and fruit stones [12], have been discovered to be suitable precursors for activated carbon due to their high carbon and low ash contents.

In general, there are two main steps for the preparation of activated carbon: (1) the carbonization of raw materials below 800°C, in the absence of oxygen, and (2) the activation of carbonized product (char), either using physical or chemical activation methods [13]. Generally physical activation is a two-step process which involves carbonization of raw materials followed by activation of the resulting char at elevated temperature in the presence of suitable oxidizing gases such as carbon dioxide, steam,

air or their mixtures. In the chemical activation process, both steps are carried out simultaneously, with the precursor being mixed with chemical activating agents. In chemical activation, dehydrating reagentssuch as phosphoric acid, sulfuric acid, zinc chloride, potassium hydroxide are used to impregnate the raw materials [14, 15, 16]. Chemical activation offers several advantages which include single step activation, low activation temperatures, low activation time, higher yields and better porous structure. However, the process involves a complex recovery and recycle of the activating agent, which generates liquid discharge that demands effluent treatment.
Sugarcane bagasse (SCB) is suitable for preparing activated carbons due to their excellent natural structure and low ash content. SCB is a byproduct of sugarcane industries obtained after the extraction of juice for production of sugar. Conversion of sugarcane bagasse into activated carbons which can be used as adsorbents, ion exchange, carbon molecular sieve, catalyst would add value to these agricultural commodities, help reduce the cost of waste disposal, and provide a potentially cheap alternative to existing commercial carbons. Studies on CO_2 activation of sugarcane bagasse have not been reported in literature and hence the present study attempts to prepare activated carbon with well-developed porosity. The aims of this work were to utilize SCB, biomass waste, for the preparation of activated carbon by physical activation using CO_2 gas as an activating agent. There are several critical parameters in the preparation of activated carbon that would affect its structure, one of which is carbonization temperature. The effects of carbonization temperature on the physico-chemical characteristics of the activated carbon were studied in order to obtain high adsorption capacity and surface area of the product. The one-step physical activation process using CO_2 gas is desirable due to lower energy consumption, capital expenditure, processing time that can significantly improve the process economics.

2. Methodology

2.1. Materials
Sugarcane bagasse used as raw materials to produce activated carbon in this study was obtained from sugarcane industries in Bantul, D.I. Yogyakarta province. The starting materials were manually chosen, cleaned with deionized water, dried at 110°C for 48 h and ground with a roller mill to obtain samples of 1–2 mm particle size. The schematic diagram of the experimental setup for the preparation of activated carbons using physical activation using CO_2 was shown in Figure 1.

2.1.1. Carbonization of sugarcane bagasse.
Carbonization was conducted in a vertical tube furnace. Subsequently, sugarcane bagasse was loaded on a stainless steel tubular reactor which is placed inside vertical tube furnace. The sugarcane bagasse were heated up to a carbonization temperature (400, 600, and 800°C) at a heating rate of 10°C/min and were held for about 2 hours at the carbonization temperatures under N_2 gas flow (20 cm^3/min). After this period, the products were cooled down to room temperature with nitrogen flowing into the reactor.

2.1.2. Activationofsugarcane bagasse char
10 g sugarcane bagasse chars were loaded inside the reactor situated in the furnace, and then heated up at a heating rate of 20°C/min to a carbonization temperature of 800°C under the nitrogen (N_2) gas flow (100 cm^3/min) and then the CO_2 gas (100 cm^3/min) was introduced to activate the samples for 2 hours. After activation, solid samples were cooled to room temperature under the N_2 gas flow. The products were kept in a desiccator for further characterization experiments.

2.1.3. Characterization of activated carbon
The determinationoftheporosityof activated carbon samples was carried out using N_2adsorption at −196°C using surface area analyzer to assess the pore morphology of the produced activated carbon. The Brunauer–Emmett–Teller (BET) surface area, micropore and total pore volume, and pore size distribution were obtained by analyzingN_2 adsorption. The specific surface areas were determined

IOP Conf. Series: Materials Science and Engineering **105** (2016) 012027 doi:10.1088/1757-899X/105/1/012027

according to the BET method at the relative pressure in the range of 0.05–0.30. Pore volume was directly calculated from the volume of nitrogen held at the highest relative pressure (P/Po = 0.99). The pore size distribution (PSD) was obtained by the Barrett Joyner-Halenda (BJH) method, which was applied to the desorption branch of the isotherms. Fourier transform infrared (FT-IR) spectroscopy were conducted to characterize the surface functional groups using a Varian 3100 FT-IR Excalibur Series spectrometer. Resolution and aperture setting of 4 cm^{-1}, with a scan speed of 2.5 kHz over a wavenumber range of 4000–650 cm^{-1} were applied to collect the spectra. An elemental analysis was carried out using by X-Ray Fluorescence.The surfaces of ACs were chemically characterized by Boehm titration and the pH zero point charge (pH_{ZPC}).

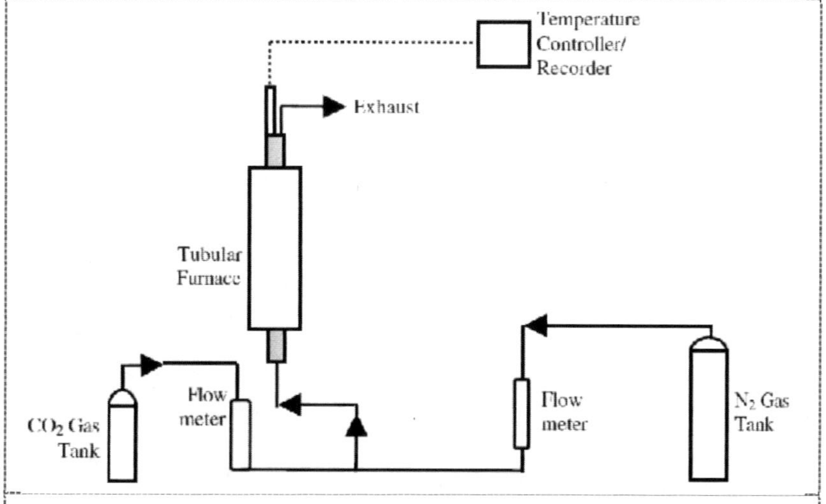

Figure1. Schematic diagram of the experimental setup for the preparation of activated carbon using physical activation.

3. Results and Discussion

3.1. Yield

The activated carbon yield was defined as the final weight of product after carbonization, washing, and drying. The percent yield was determined from the relation:

$$\text{Yield (\%)} = \frac{Wc}{Wo} \times 100 \qquad \text{... (1)}$$

whereWc and Wc are the final activated carbon dry weight (g) and the precursor dry weight (g), respectively.The yield values of carbonized SCB chars were 32.15%, 26.12% and 20.46% for the carbonization temperature 400, 600 and 800°C, respectively. According to the results, the increase in the carbonization temperature caused a decrease in the yield values. With increasing temperature, more volatiles in samples would be released, causing a lower yield of chars. These findings are consistent with the general concept that increasing the carbonization temperature decreases the amount of the unstable volatiles on the carbon samples. The decrease in the yield for activated carbonsis also justified by action of the dehydrating reagent, which provided elimination and dehydration reactions, breaking the bonds C–O–C and C–C of the raw material.

3.2. Textural characterization

Figure2 shows the Nitrogen adsorption isotherms of activated carbonsfrom SCBproduced at different carbonization temperatures. The different shape of the adsorption isotherm is an indication of the different pore size distribution of the samples. It can be seen that the volume of nitrogen adsorbed increases with increase in carbonization temperature.At carbonization temperature of 800°C, the Nitrogenadsorption isotherm showsan intermediate between types I and II of the referred IUPAC classification.Type I isotherms are given by predominantlymicroporous adsorbents having a relatively small externalsurface area.The adsorption capacity at relatively low relative pressures (P/Po) is very high, which indicates the presence of a well-developed microporous structure.Hysteresis is detected between adsorption and desorption isotherms, indicating the presence of a little proportion of mesopores and macropores.The isotherms of chars prepared at temperatures 400 and 600°C belong to type II IUPAC classification. As shown in Figure2, the initial part of the type II isotherm for the chars represents micropore filling, and the slope of the plateau at high relative pressure is due to multilayer adsorption on the nonmicroporous surface, i.e., in mesopores, in macroporesand on the external surface.The desorption isotherms presents a hysteresis loop at high relative pressure. The desorption isotherms is useful for determination of the mesoporous nature of activated carbons, and the presence of a hysteresis between the adsorption and desorption curves implies the existence of mesopore pores.

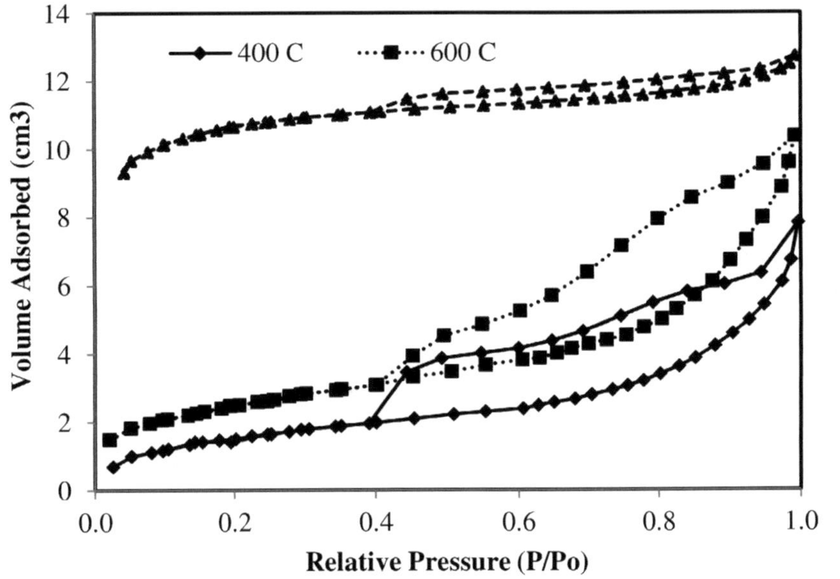

Figure 2. Adsorption/desorption isotherms of N_2 at 77 K for SCB activated carbons prepared different carbonization temperatures.

In accordance to the classification adopted by the InternationalUnion of Pure and Applied Chemistry (IUPAC), adsorbent poresare classified into three groups: micropores (size <2 nm),mesopores (2–50 nm), and macropores (>50 nm). Microporescan be divided into ultramicropores (width less than 0.7 nm) andsupermicropores (width from 0.7 to 2 nm). The size of pores thatare formed would have an effect on the porosity, the total surfacearea that is available for adsorption and most importantly, the sizeof molecules that can diffuse into the solid. Thus the developmentof an appropriate pore structure is necessary if the activated carbonproduced is to be used for a particular application.

The effect of carbonization temperatures on porosity development of activated carbon is shown in Table1. With increase in carbonization temperatures the continual devolatilization and carbon–carbon dioxide reaction, results in the formation of fresh micropores, which increases the BET surface area and micropore volume of activated carbons. The higher BETsurface area is an indication of the existence of a significantamount of microporosity.With increasing carbonization temperatures, theadsorption capacity of the activated carbon increasedenhancing the amount of microporosity. At highcarbonization temperatures, amore ordered structure is likelyto be developed in the char that leads to a slower rate of gasificationin the interior of the particle. From Table 1, it was observed that the BET surface area, total volume and micropore volume of activated carbon prepared increased, while average diameter decreased.The highest the BET surface area, total volume and micropore volume were obtained at carbonization temperature of 800°C using activation temperature of 800°C and holding time 120 min.

Table 1. Characteristics of porosity in SCBactivated carbons prepared different carbonization temperatures

Carbonization temperatures (°C)	S_{BET} (m^2/g)	V_{tot} (cm^3/g)	V_{micro} (cm^3/g)	D_{avg} (nm)
400	225.27	0.1026	0.0636	4.24
600	405.81	0.1590	0.1177	3.63
800	661.46	0.2455	0.1989	2.48

Figure3 illustrated the pore size distributions of activated carbons prepared at different carbonization temperatures. Pore-size distribution is the principal factor determining theadsorption characteristics of activated carbons.The structural heterogeneity of porous materials is generallycharacterized in terms of the pore size distribution. The pore sizedistribution represents a model of solid internal structure andregularly shaped model pores which represent the complex voidspaces within the real solid. Thisfigure confirmed that the activated carbons prepared at higher carbonization temperature had more microporous structures compared to activated carbons prepared at lower carbonization temperature.

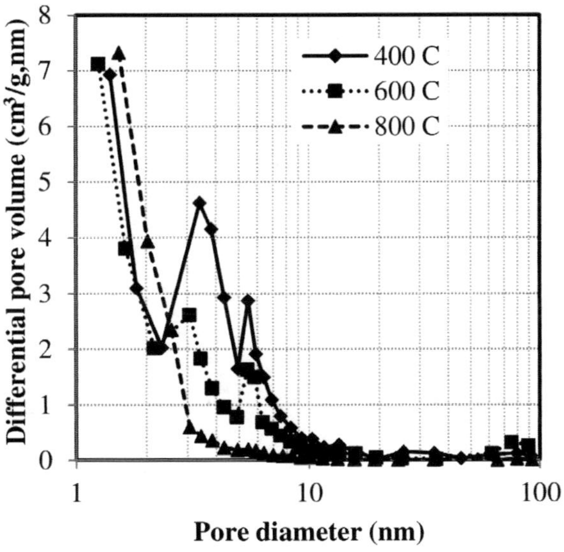

Figure 3. Pore size distributions of activated carbons prepared at carbonization temperatures.

3.3. Chemical surface characterization

Table 2 shows the elemental analyses of SCB activated carbonproduced at different carbonization temperatures. Results indicate the presenceof many elements with predominance of carbon in the threesamples. The percentageof carbon increased after carbonization and activation. These increases in carbon percentage areessentially due to the loss of volatile matter and the decompositionof cellulose and hemicellulose during pyrolysis, and, on the otherhand, to the decomposition of lignin during activation.Table 2 also shows a decrease in the H/C and O/C ratios with increasingactivation temperature. The increasing temperatures cause more heteroatoms toevaporate from the carbon rings and an increasingnumber of C-H bonds are ruptured as the activation ofthe char support progresses.

Table 2. Elemental analysis of the theSCB activated carbonsproduced at different carbonization temperatures.

Carbonization temperature (°C)	elemental (wt %)				
	C	H	O	H/C	O/C
400	63.42	0.84	23.47	0.0132	0.3701
600	69.19	0.41	15.96	0.0059	0.2307
800	81.38	0.39	6.04	0.0048	0.0742

The oxygen functional groups are very importantcharacteristics of the activated carbons because they determinethe surface properties of the carbons and hence their quality asion exchangers, adsorbents, catalysts, and catalyst supports. The contents of oxygen-containing functional groups with variousacidity strength (total acidic groups: carboxyl, lactonic, hydroxyland carbonyl) as well as, the total amount of the basic groups arepresented in Table 3. It shown in Table 3, for activated carbons prepared at carbonization temperature of 400°C, the total oxygen functional groups is 0.196 mmol.g^{-1} with a breakdown of 0.196 mmol.g^{-1}, 0.196 mmol.g^{-1}, and 0.196 mmol.g^{-1}, individually, for carboxylic, lactone and phenolic groups.When thecarbonization temperature increases, the total oxygen containing functional group decreases to 0.240 (600°C) and 0.196 mmol.g^{-1}(800°C). Decrease in oxygen containing functional groups, takes mostly in the carboxylic, lactone and phenolic groups. The phenolic group decreases from 0.117 to 0.240 (600°C) and 0.196 mmol.g^{-1}(800°C). Meanwhile the lactone group decreases from 0.117 to 0.240 (600°C) and 0.196 mmol.g^{-1}(800°C). While the concentration of the carboxylic groups decrease from 0.117 to 0.240 (600°C) and 0.196 mmol.g^{-1}(800°C).

Table 3. Results of the Boehm and pH_{ZPC} methods for the SCB activated carbons produced at different carbonization temperatures.

Carbonization temperature (°C)	Acid (mmol.g^{-1})			Basic (mmol.g^{-1})	pH_{pzc}
	Carboxylic (mmol.g^{-1})	Lactonic (mmol.g^{-1})	Phenolic (mmol.g^{-1})		
400	0.798	0.017	1.196	1.963	5.21
600	0.664	0.012	1.143	1.635	5.36
800	0.427	0.011	0.745	0.867	5.86

The FTIR spectroscopy provides information on the chemical structure and functional group of SCB and its chars prepared at different carbonization temperatures, shown in Figure 4. It can be seen that the intensity of all bands present in the raw SCB declined more or less after the carbonization process. Typical infrared band assignments for SCB indicate that, SCB contains a number of atomic groups and structures. The functional groups O–H, aromatic C=C, aliphatic C–H, olefinic C=C, C–O, C–O–C, C=O, CH_3, C–O–H and Si–O–Si bands are contained in the SCB. The band at 3415 cm^{-1} O–H is corresponding to (O–H) vibrations in hydroxyl groups. The position and shape of this band suggest that the hydroxyl groups are involved in hydrogen bonding. Water residual in SCB could take part in the formation of hydrogen bonds. According to the four types of hydrogen-bonded structures, the pre dominant one in SCB is self-associated OH groups. The fact that the band at 3415 cm^{-1} is somewhat broader towards lower wavenumbers suggests that SCB also contains OH-ether hydrogen bonds. The band at 2919 cm^{-1} representing the presence of(C–H) vibrations. The appearance of(C–H) adsorption bands indicating the presence of methyl and methylene groups in SCB. However, the peak at 1459 cm^{-1} was attributed to (CH_3) stretching vibration. The characteristic absorption of methyl group at 1375 cm^{-1} is valuable for methyl group estimation. Hence the band at 1459 cm^{-1} is slightly weaker than the band at 1375 cm^{-1}, proving the presence of acetoxy groups (–O–$COCH_3$) in SCB. The band at 1732 cm^{-1} indicating the presence of (C=O) vibrations in carbonyl groups, shows the presence of acetyl derivative groups, aldehydes groups, and others. The olefinic (C=C) and aromatic (C=C) vibrations were detected at 1635 cm^{-1} and (1606–1459 cm^{-1}) region, respectively. While, alkyl groups were detected from the C–CH_3 blending vibration at 1326 cm^{-1}. Moreover, the bands between 1100 and 1300 cm^{-1}, the (C–O–H) vibrations in phenolic and (C–O–C) vibrations can be related to asymmetric vibrations in single graphitic sheet and between two such sheets; oxygen can act as a crosslinking agent between aromatic sheets. The absorbance band at 1000–1100 cm^{-1} are attributed to (C–O) vibrations in primary C–OH, secondary O–H and the presence of silica, Si–O–Si. Furthermore, some bands are observed in the region 470–900 cm^{-1}, which are band positions compatible with C–H vibrations in olefinic or aromatic structures.

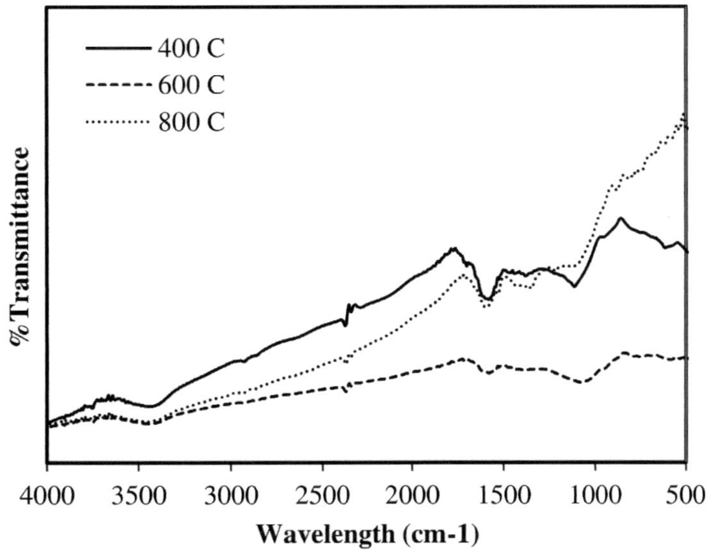

Figure 4. The FT-IR spectra of the SCB activated carbons produced at different carbonization temperatures.

4. Conclusions

The carbonization temperature was determined to significantly influence the porosity of the activated carbon from SCB chars. The high carbonization temperature (800°C) generated a highly microporous carbonwith a Type-I isotherm, while the low carbonization temperature (400 and 600°C) generated a mesoporous one with an intermediate between types I and IIisotherm. At a higher temperature, theBET specific surface and total pore volume were higher due to enhancement of the poredevelopment.

5. References

[1] AbdullahM O, TanI A W and LimL S2011*Renew. Sust. Energ. Rev.* 15, 2061.
[2] Hidayat A, Rochmadi, Wijaya K, Nurdiawati A, Kurniawan W, Hinode H, Yoshikawa K andBudiman A 2015 *Energy Procedia* 75 969.
[3] StavropoulosGG and ZabaniotouAA 2005 *Micropor. Mesopor. Mater.* 83 79.
[4] HidayatA, Rochmadi, WijayaK, HinodeH and BudimanA, 2013*Asian J. Chem.* 25 (3), 1569.
[5] HuZ and SrinivasanMP, 1999*Micropor. Mesopor. Mater.*27 11.
[6] MozammelHM, MasahiroO and BhattacharyaSC2002*Biomass Bioenergy* 22 397.
[7] SekarM,SakthiV and RengarajS2004*J. Colloid Interface Sci.* 279 307.
[8] DiaoY,WalawenderWP and FanLT2002*Bioresour. Technol.* 81 45.
[9] BaqueroMC, GiraldoL, MorenoJC, Suarez-GarciaF, Martinez-AlonsoA and TasconJMD2003*J. Anal. Appl. Pyrolysis* 70 779.
[10] SrinivasakannanC and BakarMZA2004Biomass Bioenergy 27 89.
[11] Macias-GarciaV2005*Mater. Lett.* 59 846.
[12] PuziyAM, PoddubnayaOI, Martinez-AlonsoA, Suarez-GarciaF, TasconJMD2005*Carbon* 43 2857.

[13] Ioannidou O andZabaniotou A2007 *Renew. Sustain.Energ. Rev.* 11 1966.
[14] Kim J, Sohn M, Kim D, Sohn S and Kwon Y2001 *J. Hazard. Mater.* 85 301.
[15] Ucar S, Erdem M, Tay T and Karagoz S2009 *Appl. Surf. Sci.* 255 8890.
[16] Molina-Sabio M and Rodriguez-Reinoso F2004 *Colloids Surf. A* 241 15.

IOP Conf. Series: Materials Science and Engineering **105** (2016) 012028 doi:10.1088/1757-899X/105/1/012028

Prototype of annotation tools for microscopic digital images on Android devices

Muhimmah I and Nugraha D DC

Department of Informatics, Faculty of Industrial Technology, Universitas Islam Indonesia, JL.Kaliurang KM 14.5, Sleman, D.I. Yogyakarta

E-mail: izzati@uii.ac.id

Abstract. Reading a slide under a microscope manually is very complicated. An expert may spend 3-4 hours to read a single slide. Moreover, the intra- and inter-observer variability is known to be high. This prototype was developed to simplify the slide-reading process on Android devices in order to accelerate the reading process and generate more accurate information.The prototype allows users to annotate the boundaries of an object. Moreover, the proposed prototype has successfully reconstructed multiple object boundaries into simple closed curves from a limited amount of user input.Thecoordinates of the annotated objects are stored in a text file (*.txt) that can be usedfor further analysis.The prototype's performance with respect to time and memory usage are included.

1. Introduction

Knowledge of object recognition on microscopic images is very useful for future study or to diagnose various kinds of diseaseIn a clinical setting, such knowledge is important to differentiate between normal and abnormal cells.Thus, it is mandatory to gain the correct knowledge.

An expert in object recognition usually uses a slide that is observed through a microscope. It takes 3-4 hours to read a single slide. The intra- or inter-observer variability is known to be high,so it is possible that errors will occur that couldlead to serious consequences.

Microscopic image recognition requires precision and skill. Precision usually takes a long time. So, doing research or diagnosing diseasesthatrequiremicroscopic image recognition takes a long time. A kind of tacit knowledge is obtained by reading objects through a microscope, and it is difficult to pass this knowledge on to young doctors. Hence, a tool that allows this reading process shown in big screen, rather than under microscope only, is needed. To help the experts perform their duty, a tool that can annotate amicroscopic image is required. In aprevious study, a similar tool wasdeveloped [1]. The tool does not have ability to perform object reconstruction with limited object boundary information. Also,the tool was designed as a PC-based application, so it does not support doctor mobility. Aportable tool with similar specificationscapable of object reconstruction is needed.

The remainder of this paper is outlined as follows: the mathematical model is described in Section II. Section III presents an overview of the proposed system. The data and evaluation scenario are presented in Section IV. Thediscussionsin Section V. The conclusion and further works are highlighted in Section VI.

IOP Conf. Series: Materials Science and Engineering **105** (2016) 012028 doi:10.1088/1757-899X/105/1/012028

2. Mathematical Model

2.1. Annotation

An annotation is an explanatory note or comment added to a text or diagram. The purpose of an explanation or comment is to facilitate a better understanding of the representation of a text or a diagram. In other words, annotations are made to improve a text or diagram. According toanother definition, annotation is the process of enhancing selected images with information to convey amedical object[2], and includesthe use of arrowsthat point toanobject, visualization of anobject's boundaries, and so on.

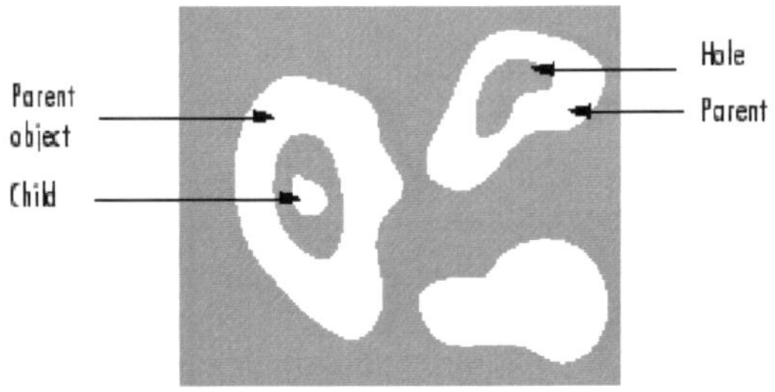

Figure 1. Image annotated with arrows and text

2.2. Digital Images

A digital image is a digital version of a document such as a photograph, literature, art, and so on. The image is represented by a matrix of N rows and M columns, where the elements of the matrix contain gray levels or the colors of each pixel inan image [3].

0	10	0	10	0	10	0
20	0	20	0	20	0	20
0	30	0	30	0	30	0
40	0	40	0	40	0	40
0	50	0	50	0	50	0
60	0	60	0	60	0	60
0	70	0	70	0	70	0
80	0	80	0	80	0	80
0	90	0	90	0	90	0
100	0	100	0	100	0	100

Figure 2. 10 x 10 matrix of an image, the value of each element is in the range of 0 - 2^8-1

Figure 3. The image generated by the matrix in Figure 1

2.3. BézierCurve

A Béziercurveis acurve that wasintroducedbyPierreBézierin 1962. At that time, Bézier curveswere usedto design automobile bodies atRenault. Thesefamouscurvesare now widely used inthe field ofcomputer graphics. The curveis madeby adjustingcertaincontrol points [4], as shownin Figure4.

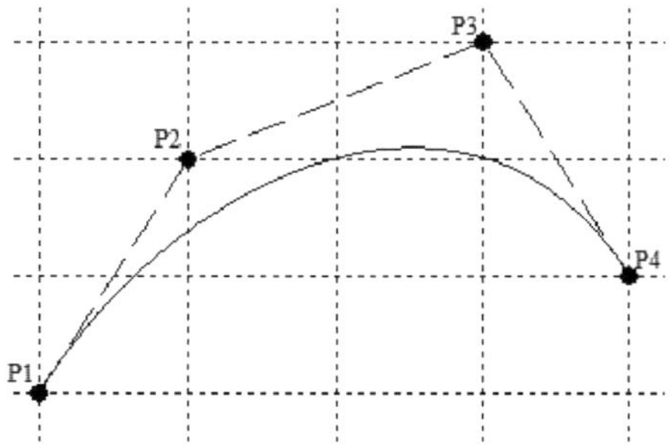

Figure 4.Bézier curve with four points

Given n + 1 control points P_0, P_1, …,P_n, the Bézier curve of degree n is defined as

$$C(t) = \sum_{i=0}^{n} P_i B_{in}(t) \tag{1}$$

where $B_{i\,n}\ (t)$ istheBernsteinpolynomial. The equation is as follows:

$$B_{in}(t) = \binom{n}{i} t^i (1-t)^{n-i} \tag{2}$$

$$t \in [0,1] \tag{3}$$

$$\binom{n}{i} = \frac{n!}{i!(n-i)!} \tag{4}$$

Thecoefficients matrix of Bézier curves is as follows:

$$\begin{bmatrix} 1 & -3 & 3 & -1 \\ 0 & 3 & -6 & 3 \\ 0 & 0 & 3 & -3 \\ 0 & 0 & 0 & 1 \end{bmatrix} \tag{5}$$

So, to get the curve, the equation is as follows:

$$\begin{bmatrix} x1' & x2' & x3' & x4' \\ y1' & y2' & y3' & y4' \end{bmatrix} = \begin{bmatrix} x1 & x2 & x3 & x4 \\ y1 & y2 & y3 & y4 \end{bmatrix} \begin{bmatrix} 1 & -3 & 3 & -1 \\ 0 & 3 & -6 & 3 \\ 0 & 0 & 3 & -3 \\ 0 & 0 & 0 & 1 \end{bmatrix} \begin{bmatrix} t^0 \\ t^1 \\ t^2 \\ t^3 \end{bmatrix} \tag{6}$$

For the smooth curves, t is as follows:

$$t = \{0, 0.001, 0.002, ..., 1\} \tag{7}$$

2.4. Interpolation by Relaxed Cubic Splines

Interpolation is a process of estimating intermediate values between precise data points [5]. A cubic spline is a spline constructed of piecewise third-order polynomials with a continuous second derivative [6]. A cubic spline is relaxed if its second derivative is zero at each endpoint. Data points $S_0, \ldots, S_n$ are given and the cubic spline curve is $P(t)$ for $0 \leq t \leq n$,such that $P(i) = S_i$; that is, the curve goes through the data points. One approach is to use B-splines as an intermediate step. The appropriate control points $B_0, \ldots, B_n$must be computedprior to the computation of the Bézier control pointsforthe individual pieces.

To compute the B-spline control points, the matrix form for linear equations is as follows (for $n = 5$):

$$\begin{bmatrix} 4 & 1 & 0 & 0 \\ 1 & 4 & 1 & 0 \\ 0 & 1 & 4 & 1 \\ 0 & 0 & 1 & 4 \end{bmatrix} \begin{bmatrix} B_1 \\ B_2 \\ B_3 \\ B_4 \end{bmatrix} = \begin{bmatrix} 6S_1 - S_0 \\ 6S_2 \\ 6S_3 \\ 6S_4 - S_5 \end{bmatrix} \tag{8}$$

The coefficients matrix of the cubic spline (which is usually called the "1 4 1 matrix") for this relaxed cubic splinewill have an$M$ square matrix, where $M = n - 1$. To compute the Bézier control points, the equation is as follows:

$$\begin{bmatrix} S_i - 1 \\ \frac{2}{3}B_{i-1} + \frac{1}{3}B_i \\ \frac{1}{3}B_{i-1} + \frac{2}{3}B_i \\ S_i \end{bmatrix} \tag{9}$$

A relaxed cubic spline is shown in Figure 5.

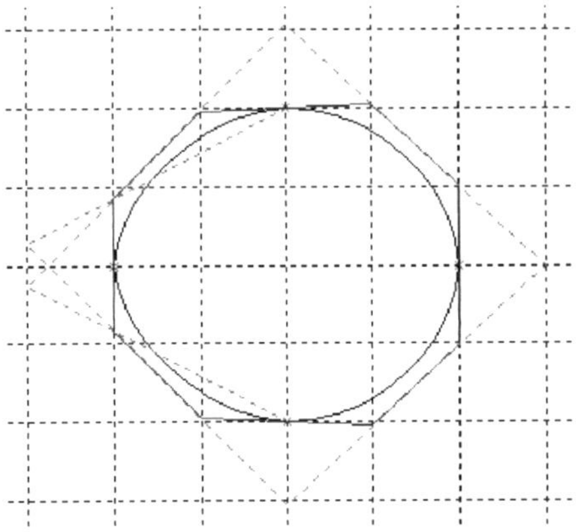

Figure 5. Relaxed cubic spline

3. Proposed System Overview

This prototype will be developed for Android devices.The use ofAndroiddevices is intended to support doctor mobility.
This prototype uses a common image format like .jpg, .bmp, and so on. Users can run the prototype in the following order.

- Choose an image that will be annotated
- Mark the boundaries of the object by dragging on the desired location
- Users can use the zooming facilities by pinching the desired location
- The system will automatically reconstruct the object
- After users complete the annotation and save it, the system automatically stores all the object coordinates in a .txt file.

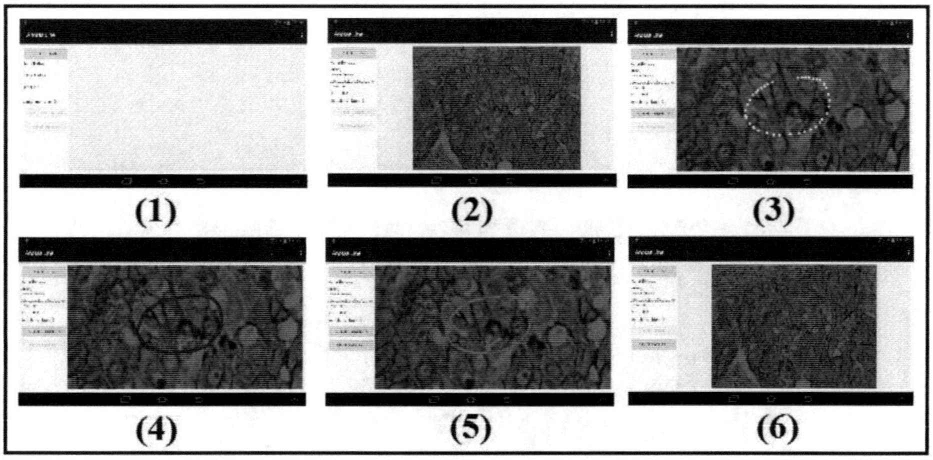

Figure 6. System flow: (1) the initial view of the prototype, (2) displaying the image that will be annotated, (3) user's given object boundaries (with zooming facility), (4) system's generated closed curve, (5) curve that has been saved, (6) final appearance of the object that has been annotated (zooming out)

4. Data and Evaluation

The data used in this research were HER2 microscopic images obtained from the UniversitasIslam Indonesia Pathology Laboratory. These data will be used for testing the performance of this prototype. The Android device used in this research is a Samsung Galaxy Tab GT-P3110, and the operating system is Android Jelly Bean version 4.1.2. The IDE used to develop this prototype is Android Studio 1.3.2.
The performance to be tested is the ability of the system to reconstruct an object using the limited information of object boundaries.Figure 7 is the image that was used in this test.

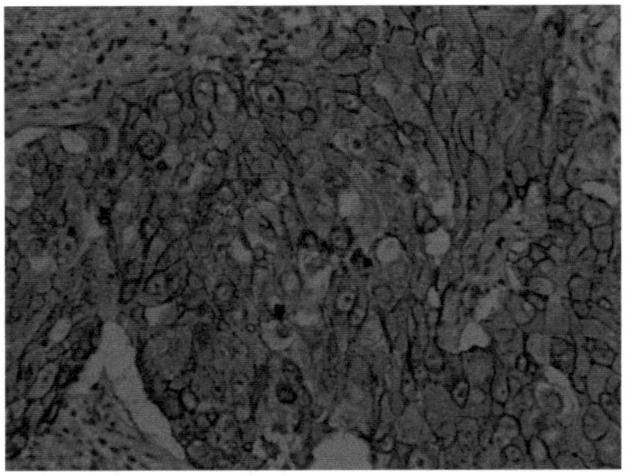

Figure 7. Microscopic image that was used has a resolution of 1175 x 881 pixels and is 299 KB.

Two objectsinthe image in Figure 7were annotatedtotestthe performance ofthe prototype, as shown in Figure 8.

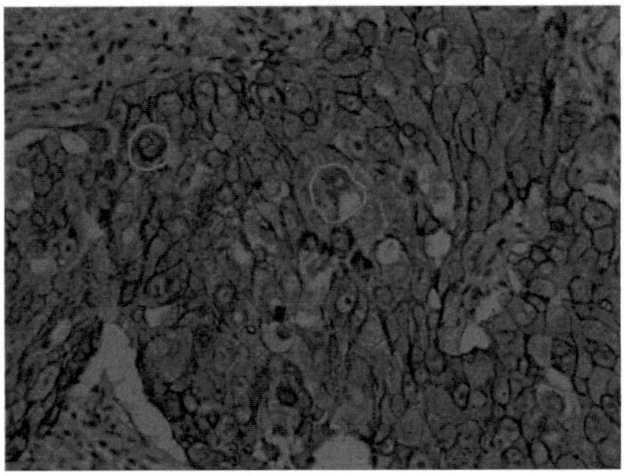

Figure 8. Two objects are annotated in the imageinFigure 7: object one (left, Figure 9), object two (right, Figure 10).

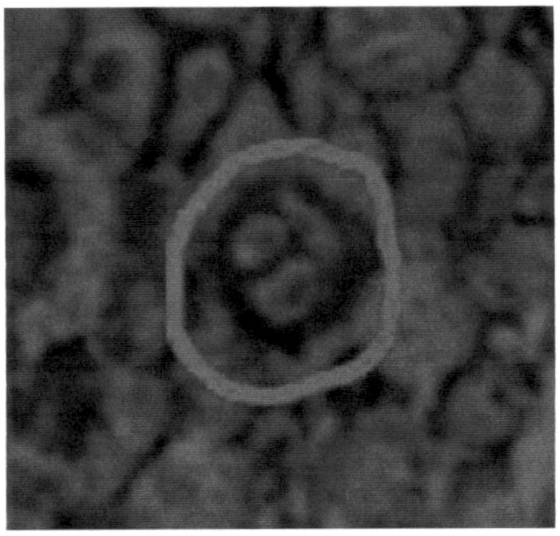

Figure 9. Object one

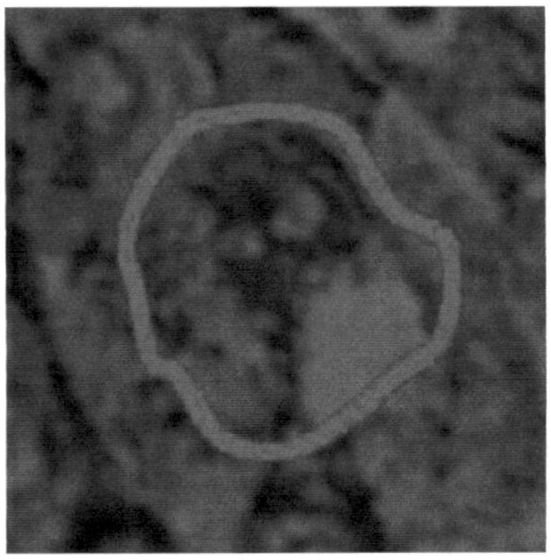

Figure 10. Object two

Table 1 shows the performance details of theproposed prototype whenperforming object reconstruction.

Tables 1. Performance detailsforobject reconstruction

Process	Object 1	Object 2
CPU Usage	51.39 %	51.79 %
Memory Usage	17.04 MB	19.82 MB
User Input Point	45 points	33 points
B-Spline Control Point	47 points	35 points
Bézier Control Point	176 points	128 points
Reconstructed Point	316 points	414 points
Time to reconstruct	2428 milliseconds	1835 milliseconds
Time to save object	45 milliseconds	56 milliseconds

5. Discussion

The proposed system has been runningaccording to the required functions. The systemhas beenable todisplaythe selectedimage. The systemhas also successfully responded tothe userdraggingon the screenandhas translatedthatintoobjectboundaryinformation, as shown in Figure 11. From that information, the system has been able to reconstruct the object into a simple closed curve, as shown in Figure 12.

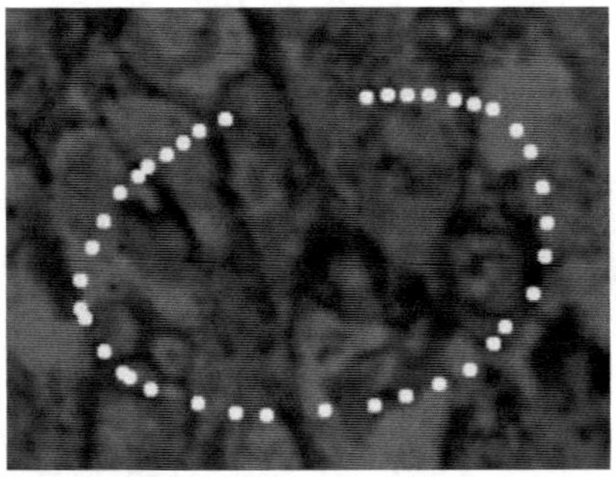

Figure 11. System responses to a user's continuous touch(dragging) on the image in Figure 7.

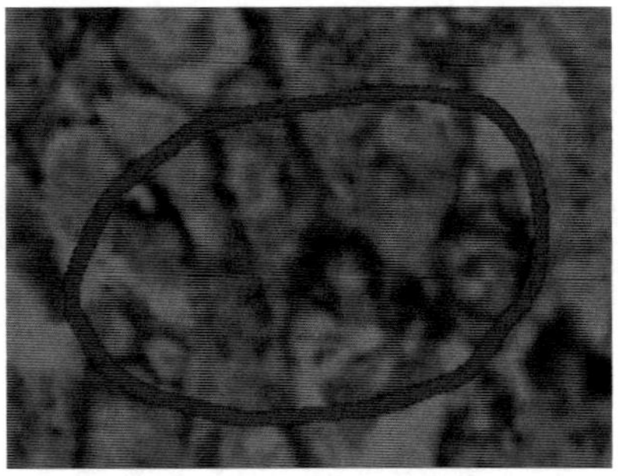

Figure 12. Simple closed curve generated by thesystemonthe image in Figure 7.

The system has also been able to store the results of the object reconstruction in a .txt file. The stored resultshave demonstratedthat the systemcan perform automatic incremental numbering of objects.In other words, this prototype can be used to annotate multiple objects in a session, as can be seen in Figure 8.As the stored coordinateswere provided by experts, theycan be used further as "truth"whencomparison between any automatic segmentation results were needed. Moreover, this information can also be used as boundary in any feature extraction steps.

6. Conclusions and Further Works

The proposed prototype annotation system has been able to annotate an object and reconstruct it. The reconstructed curve was in the form of a smooth simple curve. The object can be reconstructed even if the boundaries of the object given by the user are limited. Thedevelopment ofthis systemhas also

demonstratedthatAndroid devices arecapable of performingcomputing processesadequately. It canbe seen thatAndroid devicescando workin the field ofcomputer graphics.
While the curves are programmed to be closed, the resulting shape should also accurately represent the original shape of the object. The expert opinions on the reconstructed objects will be evaluated. The proposed prototype can also be used by doctors to collect images of individual cells or objects. These collections can serve as a knowledge base of the selected objects.

Acknowledgement
A research grant from HibahPUPT-DIKTI, contract no. 001/HB-LIT/III/2015, is gratefully acknowledged.

References

[1] Muhimmah I and Pranajaya 2012 Alat anotasi untuk sel epitel pada citra pap smear *Teknoin* **18** pp 26–31
[2] Preim B and Bartz D 2007 *Visualization in Medicine, 1st Edition. Theory, Algorithms, and Applications* (San Fransisco, CA: Morgan Kaufmann)
[3] Low A 1991 *Introductory Computer Vision and Image Processing* (London: McGraw-Hill)
[4] Marsh D 2005 *Applied Geometry for Computer Graphics and CAD 2nd Edition* (London: Springer)
[5] Chapra S C and Canale R P 1990 *Numerical Methods for Engineers 2nd Edition* (London: McGraw-Hill)
[6] Bartels R H, Beatty J C and Barsky B A 1998 *An Introduction to the Use of Splines in Computer Graphics* (San Fransisco, CA: Morgan Kaufmann)

The Effect of Considering Environmental Aspect to Distribution Planning: A Case in Logistics SME

Yudha Prambudia[1], **Andri Andrian Nur**
Center for Sustainable Systems – Faculty of Engineering
Widyatama University –Cikutra 204A, Bandung, Indonesia
[1]Email: yudha.prambudia@widyatama.ac.id

Abstract. Environmental aspect is often neglected in traditional distribution planning process of a product. Especially in small-medium enterprises (SME) of developing countries where cost efficiency is the predominant factor. Bearing in mind that there is a large number of SME's performing logistics activities, the consideration of environmental aspect in their distribution planning process would be beneficial to climate change mitigation efforts. The purpose of this paper is to show the impact of environmental aspect should it be considered as a contributing factor in distribution planning. In this research, an adoption of CO2-emission factor in an SME's distribution planning in Indonesia was simulated. The outputs of distribution planning with and without the factor consideration are then compared. The result shows that adoption of CO2-emission factor would change the priority of delivery route.

Keywords: CO2 emission, Distribution Planning, SME.

1. Introduction

Economic development has led mankind to face global warming and climate change challenges. It is generally agreed that the impacts of using resources (for economic goals) to the environment and social life need to be balanced [1]. That has been the concept of Sustainable development [2] which is a common term used to describe the attempt to realize a profitable economic activities without making any damage to the environment for the future generation.

In the context of logistics business, cost aspect is basically the main driver of any attempt to improve efficiency. It is more often than not that environmental impact of logistics activities is neglected or put into the bottom of priority. However, attempts to promote environmental aspect as part of factors in improving logistics performance can be found under the topic of green logistics and has been progressing increasingly in developed countries [3]. However, in many developing countries, the progress has not been as much.

In Indonesia, logistics service providers are mostly in the category of small and medium enterprise (SME) and many are operated informally. They tend to neglect their contribution to environmental problems in favor of lower operational cost. This is partly because of high competitive market which determined that cost is their main key success factor [4]. In addition, there is not yet any government regulation that regulates how their operation should be conducted in order to lessen environmental impact.

In this research, distribution planning of a logistic SME in Indonesia is studied by simulating vehicle routing prioritization. The simulation was run with and without environmental aspect in several conditions. This research aims to show that small logistics providers will be facing a dilemma should the government enforce a strict regulation related to environmental aspect of logistics operation. The government should take notice on this and think of something that will make environmental aspect more interesting to be considered in distribution planning process of every logistics SME.

2. Logistic in Indonesia

Logistics has strong relationship with Indonesia's economic development. Along with economic growth, logistics activities in Indonesia are growing as well [5]. In 2009, the volume of national goods movement is at 44 million items. The volume of freight movement on land alone is around 39 millions trucks [6].

The number of companies providing logistical service is growing rapidly in the last few years. A report in 2014 mentions that there are 60 3PL companies and about 4200 transporter and expedition companies operating in Indonesia. In addition, many of the companies are SME and operated informally [7]. Looking at the numbers above, it is easy to say that a small change of consideration from cost oriented aspects towards environmental aspect would contribute greatly to climate change and global warming mitigation target.

3. Research Methodology

This research is conducted in two parts; the first part is model development, second is model simulation. In the developing the model part, there are three steps required; (1) Define the basic routing model that fit to the problem, (2) Define the measurement factors, (3) Define data requirement and collect the data. In the second part, two steps were conducted; (1) Develop scenarios and (2) Run the model. Afterwards the simulation result then analyzed. The methodology is presented in figure 1 below.

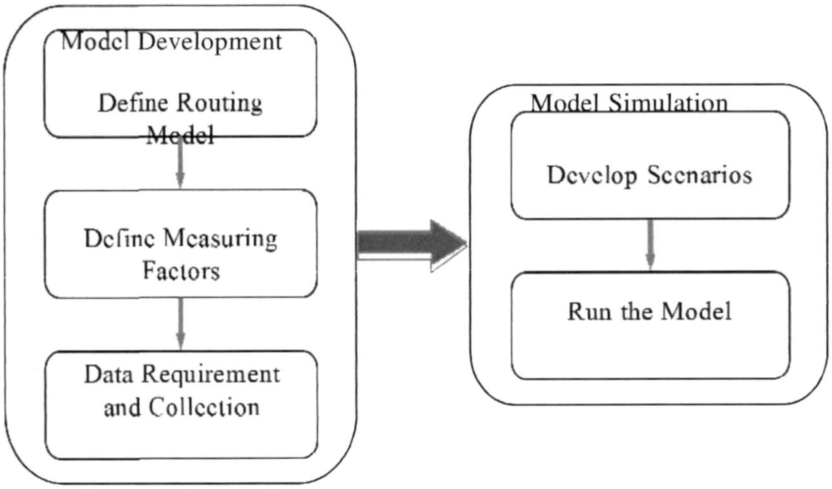

Figure 1. Research Methodology

3.1 Routing Model

The routing model used in this research is based from the combination of vehicle routing problem with time windows or VRPTW [8] and vehicle routing problem with multiple trip (VRPMT). The decision on this is related to the fact that VRPTW cannot address a multiple trip problem which is actually faced in real situation. On the other hand, the problem can be approached by the VRPMT model. Figure 2 below illustrates the basic setup of a vehicle routing problem.

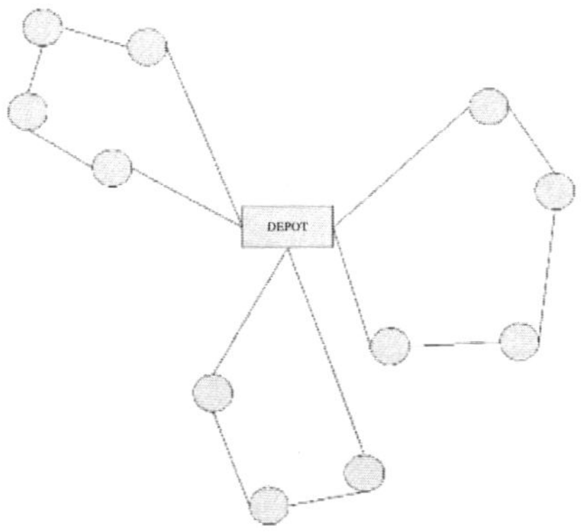

Figure 2. Basic Vehicle Routing Problem Solution [9]

The algorithm being used to find the best route is saving heuristics algorithm [10]. The basic concept of this method is to find the highest "cost saving" route by joining two or more destination as a route The figure 3 below illustrates the saving heuristics mechanism.

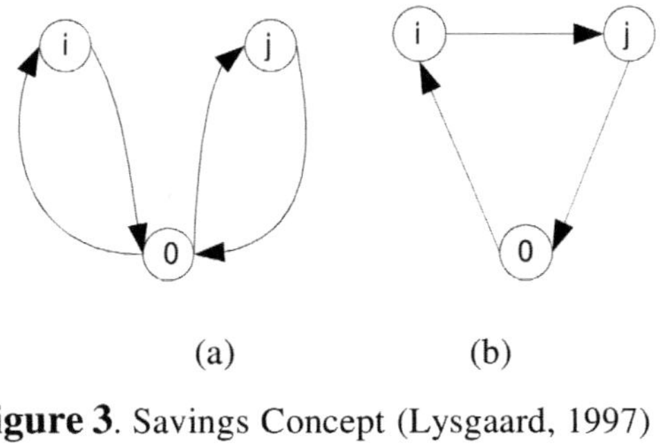

Figure 3. Savings Concept (Lysgaard, 1997)

The algorithm proceeds as follow:

- *Allocation step:* Start with allocating all possible vehicles to destinations.
- *Calculation step:* Calculate the saving made by joining two or more destination iteratively.The calculation is Sij=Ci0+C0j-Cij where *S* is savings, *C* is cost, *i* is first destination point, *j* is subsequent destination and 0 is the starting point. Sort the results in increasing direction.
- *Decision step:* In each iteration, find the highest saving where *i* and *j* does not use the same path and cross the capacity and time constraints.
- *Connection step:*Connect *i* and *j* to create a new route

3.2 Factors

The factors used in this research can be categorized into cost and environmental factors. The traditional factors related to cost considered in this research are operational cost and lost sale. Operational cost is defined as a function of driver-cost, fuel and maintenance cost and overtime cost. Lost sale is defined as the amount of product not sold to the customer when there is a demand. On the other hand, CO2-emission is used as a factor representing the environmental aspect. CO2 emission in this case is expressed as the amount of CO2 emitted from the use of different type of fuel used by the delivery vehicle.

The factors are weighted in proportion to the importance of the factors. Since, the interest of this research is the environmental aspect; CO2-emission factor is given higher weight than cost and lost sale factors. The weight was assigned base on a series of interviews with the distribution manager of the company and experts from academic and government.

3.3 Data

The SME in this research is a transport company. They specialized in distributing drinking water product in gallon container. The company profitability is depending on client satisfaction and the amount of cost they can save.

The operational characteristic of the company is as follow:

- 5 vehicles to operate their distribution plan.
- The vehicles includes 1 driver and 1 assistant
- Distribution area is only in West Bandung.
- 30 customers to serve their needs
- Delivery can only be done in opening hour of clients.

Data are collected by observations and interviews in the company. Detailed operation data such as number of vehicles available, demand, loading and unloading time, customers address, working hours, opening hours, vehicles costs and driver costs are recorded.

Data related to distance was obtained through the use of Google Earth to increase its accuracy. In order to do this, customer address is used as the basis for latitude and longitude coordinate. Figure 4 below shows a representation of distances between the depot as starting point and customers, and customers to customers.

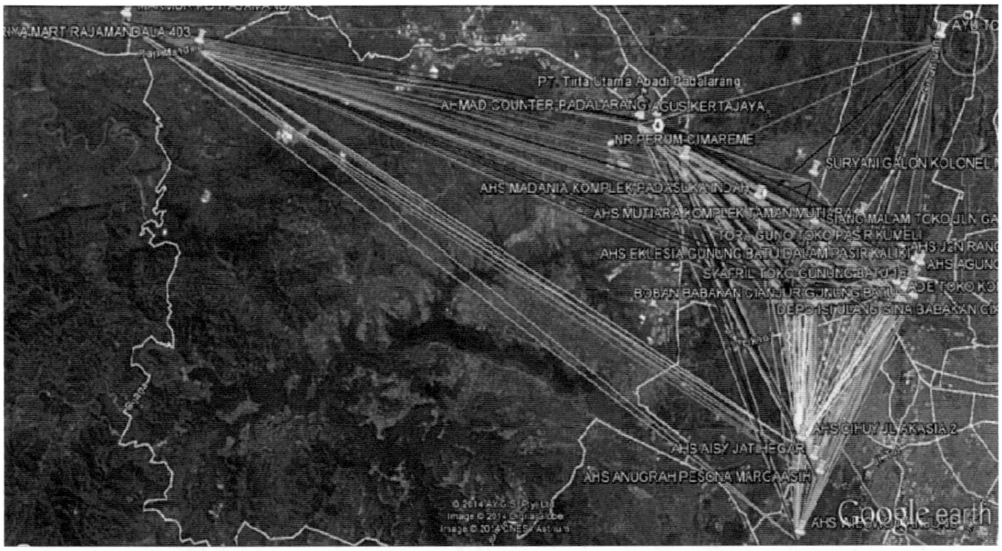

Figure 4. Distribution Network in The Study

3.4 Simulation

The simulation was implemented in optimization software. Several scenarios were developed and ran using two parameters; number of vehicles and whether or not overtime is permitted. Scenario 0 is the base scenario. The following table 1 shows the scenarios.

Tabel 1. The scenarios

Scenarios	Number of Vehicles	*Overtime*
0	5	No
1	5	No
2	4	No
3	3	No
4	2	No
5	1	No
6	5	Yes
7	4	Yes
8	3	Yes
9	2	Yes
10	1	Yes

4. Simulation Results and Analysis

The results of the simulation are presented in table 2 below. Total cost and lost sale are express in monetary term while CO2 emission in liter per day. Looking at the factors respectively, scenario 0 or the base scenario is the worst scenario in term of both total cost and emission. The best scenario is selected based on the highest score scenarios.

Tabel 2. Raw Result from simulation

Scenarios	Total Cost	Lost Sale	Emission CO^2 (litre/day)
0	Rp 1.071.654	Rp -	108.2933
1	Rp 934.706	Rp -	36.9782
2	Rp 765.489	Rp -	36.9782
3	Rp 597.041	Rp -	36.9782
4	Rp 416.208	Rp 2.773.000	34.3369
5	Rp 207.734	Rp 5.062.200	15.8478
6	Rp 934.706	Rp -	36.9782
7	Rp 765.489	Rp -	36.9782
8	Rp 597.042	Rp -	36.9782
9	Rp 479.576	Rp -	36.9782
10	Rp 346.208	Rp 3.894.000	31.6956

Scoring the scenarios was done by normalizing the results by min-max method. The results of Total Cost, Lost Sale and CO2 Emission were converted to points. So when scenarios have a total point, scenarios with highest point will be selected as chosen scenario. The score of scenarios are presented in the table 3 below.

Table 3. Scenarios Scores

Scenarios	Total Cost (30%)	Lost Sale (10%)	Total point	Emission CO^2 (60%)	Total point with CO2
0	0.00	1.00	1.00	0.00	0.10
1	0.16	1.00	1.16	0.77	0.53
2	0.35	1.00	1.35	0.77	0.59
3	0.55	1.00	1.55	0.77	0.65
4	0.76	0.45	1.21	0.80	0.67
5	1.00	0.00	1.00	1.00	**0.80**
6	0.16	1.00	1.16	0.77	0.53
7	0.35	1.00	1.35	0.77	0.59
8	0.55	1.00	1.55	0.77	0.65
9	0.69	1.00	**1.69**	0.77	0.69
10	0.84	0.23	1.07	0.83	0.69

The score in Table 2 shows that if CO2 emission is not considered than the best scenario is scenario 9 (score = 1.69). It should be noted that the weight of Total Cost and Lost Sale in this case is the same. On the other hand, when CO2 emission is considered, the weight is changed as follow; Total Cost is 30%, Lost Sale is 10% and CO2 emission is 60%. Therefore, the best result is changed to scenario 5 (score = 0.80).

The weights were changed due to the fact that there was a change of view from the SME viewpoint to the government viewpoint. The SME does not consider CO2 emission as an important factor and regards Total Cost and Lost Sale as equally important. Meanwhile, the government view CO2 emission as an important factor in mitigating GHG, hence give more weight in it.

5. Conclusion

The government has been promoting integration of environmental aspect in logistics activities. By considering environmental aspect in distribution planning, the contribution of logistics activities to CO2 emission, thus global warming and climate change, can be reduced. Bearing in mind that there are great number of logistic SME operating in Indonesia, this small change would multiply greatly and deliver a significant impact to GHG reduction.

However, from the SME perspective, this can lead to higher cost. It is almost certain that the SME will consider the cost perspective more important than the environmental aspect. This may render efforts to improve global warming and climate change situation become less effective. Therefore, in order to improve the situation, the government may opt to put forward policies that give a kind of appreciation (e.g. incentives) to SME for the saved CO2 not emitted.

6. References

[1]. Sutamihardja, 2004, *PerubahanLingkungan Global*; Program Studi Pengelolaan Sumber Daya Alam dan Lingkungan Sekolah Pascasarjana; IPB

[2]. World Commission on Environment and Development-WECD, 1987, *Our Common Future*. Oxford: Oxford University Press.

[3]. International Energy Agency, 2009, *Transport, Energy and CO2, Moving Towards Sustainability*. IEA.

[4]. Gunasekaran, A., Ngai, E.W.T., 2003, *The Successful Management Of A Small Logistics Company*, International Journal of Physical Distribution & Logistics Management, Vol. 33 Iss: 9, pp.825 - 842

[5]. Reza, M., *The Relationship between Logistics and Economic Development in Indonesia: Analysis of Time Series Data*. Jurnal Teknik Industri, Vol. 15, No. 2, Desember 2013, 119-124.

[6]. Perpress 26, 2012, *Cetak Biru Pengebangan Sistem Logistik Nasional,* Sekretars Kabinet Indonesia.

[7]. SWA, 2014, *Berebut Kue Logistik yang Menggiurkan*, SWAOnline, http://swa.co.id/business-strategy/management/, Accessed; 1 August 2015.

[8]. Solomon, M.M., 1987, *Algorithms for Vehicle Routing and Scheduling Problems with Time Window Constraints*. Operation Research, Vol. 35, no. 2.

[9]. Ballou, R.H., 2004, *Business Logistik/Supply Chain Management: Fifth Edition,* Prentice Hall, New Jersey

[10]. Lysgaard, J., 1997, *Clarke & Wright's Savings Algorithm*. Department of Management Science and logistics. Aarhus.

7. Acknowledgement

The authors would like to give much appreciation to all the experts in Center for Sustainable Systems(CSS) – Faculty of Engineering-UTama for fruitful discussions and inputs to the paper.

IOP Conf. Series: Materials Science and Engineering **105** (2016) 012030 doi:10.1088/1757-899X/105/1/012030

Low Cost RF Amplifier for Community TV

Syafaruddin Ch[1], Sudi Mariyanto Al Sasongko[1], I Made Budi Suksmadana[1], Cahyo Mustiko Okta Muvianto[1], Suthami Ariessaputra[1]
[1]Electrical Engineering Faculty of Engineering Mataram University
Majapahit Street 62, Mataram, Indonesia

E-mail: mariyantosas@unram.ac.id

Abstract. The capability of television to deliver audio video makes this media become the most effective method to spread information. This paper presents an experiment of RF amplifier design having low-cost design and providing sufficient RF power particularly for community television. The RF amplifier consists of two stages of amplifier. The first stage amplifier was used to leverage output of TV modulator from 11dBm to enable to drive next stage amplifier. CAD simulation and fabrication were run to reach optimum RF amplifier design circuit. The associated circuit was made by determining stability circle, stability gain, and matching impedance. Hence, the average power of first stage RF amplifier was 24.68dBm achieved. The second stage used RF modules which was ready match to 50 ohm for both input and output port. The experiment results show that the RF amplifier may operate at frequency ranging from 174 to 230MHz. The average output power of the 2nd stage amplifier was 33.38 Watt with the overall gain of 20.54dB. The proposed RF amplifier is a cheap way to have a stable RF amplifier for community TV. The total budget for the designed RF amplifier is only a 1/5 compared to local design of final TV amplifier.

1. Introduction

TV programs take a role in spreading information in form of voice and picture to society. TV programs are advantageous. Hence, they can be functioned to entertain, to advertise some products of local home industries as well as to improve the human resources quality. The capability of television to deliver voices and pictures makes this media become an effective tool to educate people.
TV broadcasts are grouped into commercial TV and community TV. The programs in the community TV are independent, no advertisement, and have special target of viewers. Community TV is built by a particular community such as campus or some group with same interest [1].

Solid state of RF amplifier at UHF and VHF analog TV Transmitter has been introduced. It uses FET transistor for RF amplifier [2]. Optimizing performance of RF buffer by using current-reusing with active inductor technique has been investigated. It can reduce power dissipation and increase the power generated [3]. A reliable linearization method is used to reduce noise in power amplifier transmitter. It has been investigated for frequency UHF (225 MHz-400 MHz) and VHF (118 MHz-144 MHz) [4].

The design and implementation of DVB-T technology with high power amplifier for television broadcast community have developed. It has worked in UHF frequency with omnidirectional radiation pattern [5][6]. Community TV Broadcast is designed to serve the campus community.

In this paper, a series of experiments is presented related to design RF amplifier. The RF amplifier consists of two stages of amplifier, and uses MOSFET transistor. The first stage amplifier is used to leverage output of TV modulator to drive next stage amplifier. RF amplifier works in band III TV VHF at frequency of 174 – 230 MHz. It is easy to be used and operated by a particular community, and low cost in manufacturing.

2. Method

The RF amplifier especially for community television should be at low cost, and exhibits sufficient RF output power. Originally, TV modulator has relatively covered short distance due to weak power of the modulator output. RF power amplifier is functioned to boost the signal in order to extend the coverage area with the sufficient power.

In this work, the RF amplifier consisted of two stage of amplifier. The first stage amplifier was used to gain output of TV modulator from less than 11dBm to enable to drive next stage amplifier. Such CAD simulation and fabrication were run to reach optimum RF amplifier design circuit. CAD simulation was provided by using free evaluation software of ADS (Analog Design System). The associated circuit was made by determining stability circle, stability gain, and matching impedance.

The diagram block of experiment is shown in figure 1.

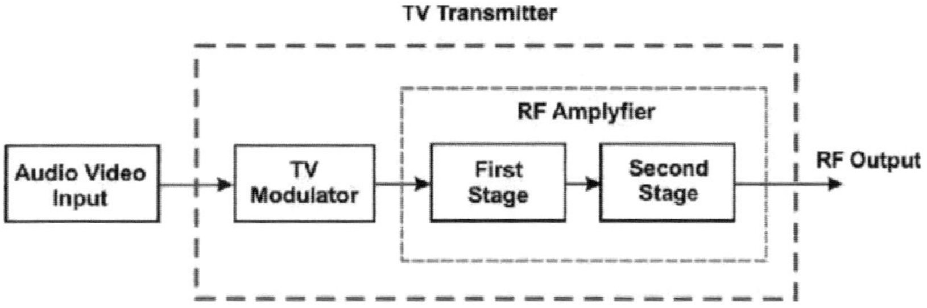

Figure 1. Diagram block of TV transmitter.

Before designing the matching network for the amplifier to reduce the input signal reflection at the input interface of the amplifier, an appropriate biasing circuit was needed to make the transistor operate in proper condition. The detailed procedures to design the biasing circuit and input matching network are shown below step by step. The first stage of RF amplifier circuit used RD15HVF1 type transistor. The transistor was specifically designed for VHF/UHF High power amplifiers applications with High Efficiency: 60% type and 55% type respectively on VHF and UHF Band. The biasing circuit, V_{gs} (The gate-source voltage) and V_{dd} (drain-drain Voltage), was chosen to be 5 V and 12.5V respectively as datasheet stated. This biasing circuit was to guarantee the sufficient current I_{dd} (drain- drain current) to be less than 1 Ampere.

After designing the suitable biasing circuit, the procedures were: designing impedance matching network for both input and output. Firstly, we needed to extract S-parameter of the RD15HVF1 as input parameter of CAD simulation, then checked the stable region, S11 (return loss) position, and gain circles for the transistor as shown in figure 2. Hence, the first stage met the specified goals of minimizing the input reflection coefficient and constant gain while the stability requirement fulfilled.

In figure 2, the region below the blue lines is the stable region. The region within the pink circles is gain circle at 20 dB from 50 to 550 MHz. The red line is the location of S11 plot. Notice that at center frequency operation of 200 MHz, the amplifier will be stable (m2 marker) and the gain stays relatively flat (m3 marker), but the impedance input (m4) Zin = Z0*(0.14-j0.035) = 7 – j1.7 ohm is not near 50 ohm.

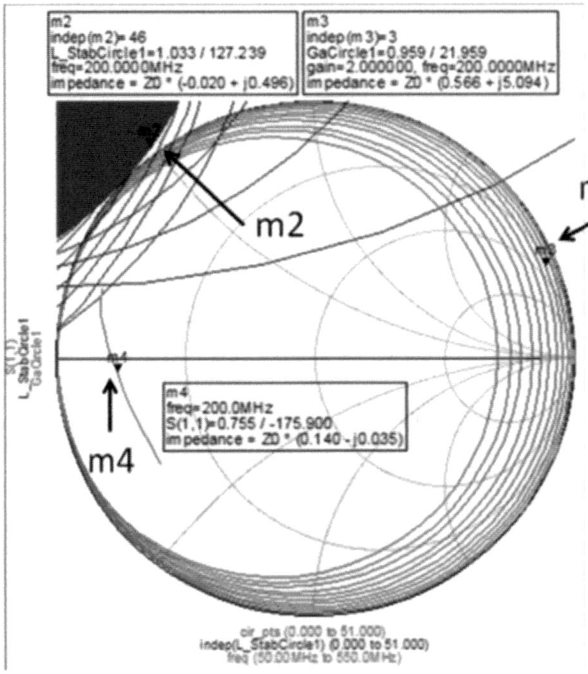

Figure 2. Simulation for transistor stability, gain and impedance.

The design of impedance matching network could be started from input part or output part with frequency center at 200 MHz. Passive inductor (L) and capasitor (C) networks are usually used to match impedances between the source (generator) and load. In this work, input matching networks were designed by using combinations of C (capacitors) and TL (transmission line) instead of inductors as depicted in figure 3.

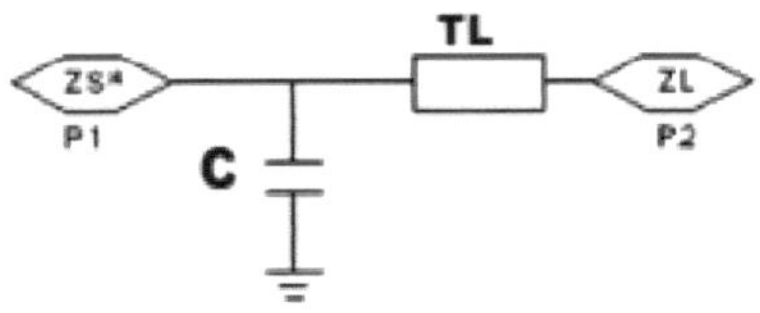

Figure 3. Matching network C and TL topology.

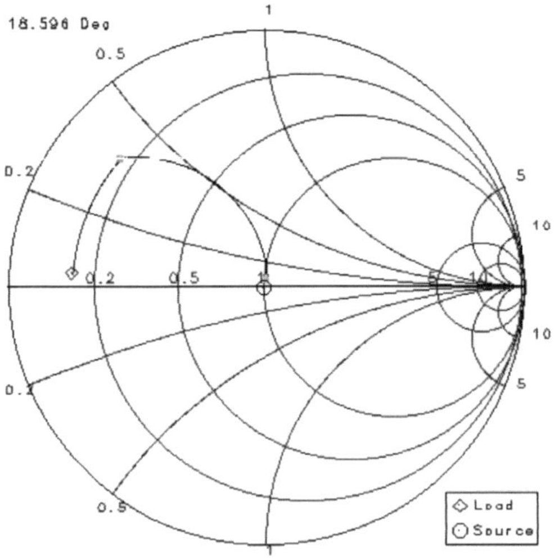

Figure 4. Input matching network using transmission line TL and capacitor C.

As illustrated in figure 4, a transmission line (TL) Z0= 50 ohm moved toward Z source 50 ohm starting from Zin about 18.595 degree long. The 50 Ohm match was achieved by placing C capacitor 32

IOP Conf. Series: Materials Science and Engineering **105** (2016) 012030 doi:10.1088/1757-899X/105/1/012030

pf at the end of the transmission line. Similar procedure was applied to reach better impedance matching for the transistor output part. The appropriate capacitor and transmission line values of the matching networks are important. After several times of the optimization, the complete input and output matching network for amplifier is shown in figure 5.

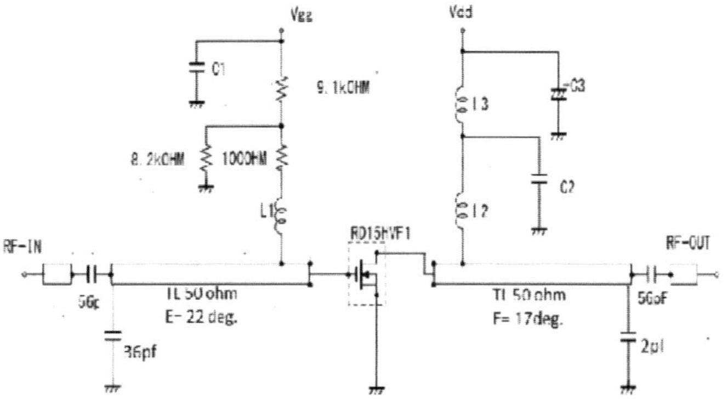

Figure 5. Complete circuit of matching network and biasing for both input output part.

To clearly see the response of this network, the following graph, shown in figure 6, provides the simulated gain and input reflection coefficient of the designed amplifier after optimization. It shows that S11 was significantly matched from -2.44dB to -25dB. This means all RF power from source was mostly delivered to the transistor. First stage amplification S21 had improved from 15 dB to 19 dB gain at 200 MHz. Therefore it can be summarized that both of the gain (S21) and input return loss (S11) fulfilled the previous set goals.

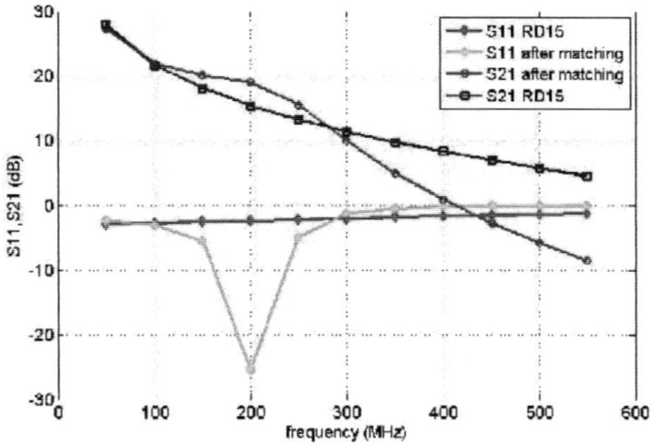

Figure 6. Simulation result of S21-S11 for first stage circuit were applied matching network.

The output of the first stage amplifier circuit was further amplified by second stage circuit. This circuit used RF module type RA 30H1721M as shown in figure 7. This transistor is classified as MOSFET and may operate at frequency range of 175 – 215 MHz when it is supplied with 12.5 Volt [8]. The maximum power that can be generated is more than 30 Watt. The RF module is ready match to 50 ohm for both input and output port. There is no need matching impedance for both ports.

IOP Conf. Series: Materials Science and Engineering **105** (2016) 012030 doi:10.1088/1757-899X/105/1/012030

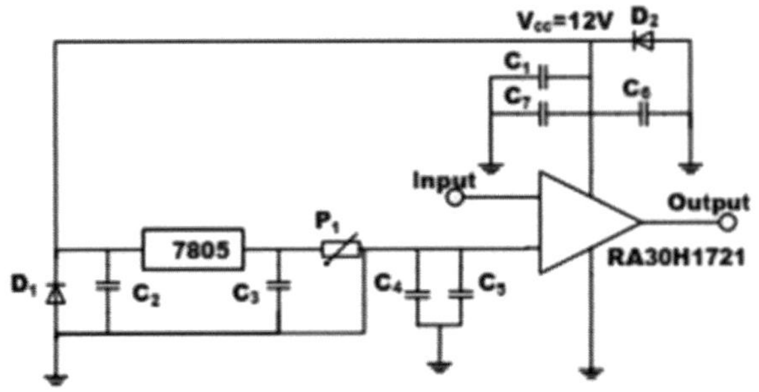

Figure 7. RF Module

Fabrication for the circuit were made on double layer PCB FR4 and placed on top of proper heatsink to desipate heat. The measurement set up is shown in figure 8.

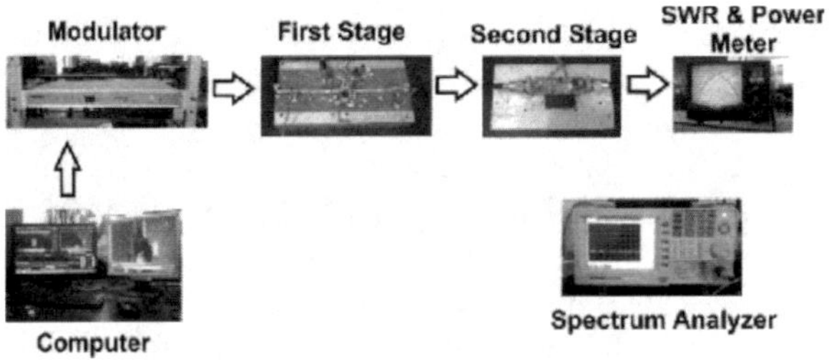

Figure 8. Experimental setup

Firstly, output power level of modulator was observed before and after modulation by using spectrum analyzer GW INSTEK GSP-830. Spectrum analysis was carried out by comparing the spectrum frequency before and after modulation. Secondly, output power of first stage amplifier was measured and analyzed as RF output of modulator was applied. Finally, stage 2 was examined to see the stability and clearness of RF signal by using power meter and spectrum analyzer. Meanwhile, the quality of broadcasted program was examined by TV monitor.

3. Result and Discussion

The output of modulator was observed before and after modulation applied. Figure 9 displays spectrum of modulator output before modulation. The level of video carrier was 11 dBm at 189.25 MHz. The level of audio carrier was -15 dBm at 194.5 MHz. Meanwhile, after modulation level of video carrier remained constant at 11 dBm but the level of audio increased to -10 dBm as presented in figure 10.

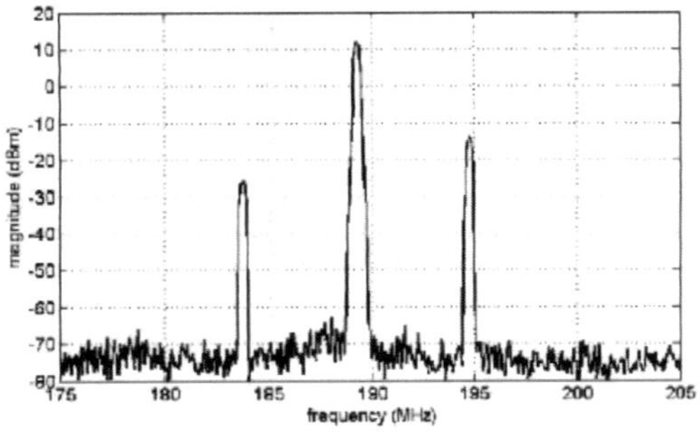

Figure 9. Spectrum of output modulator with carrier only

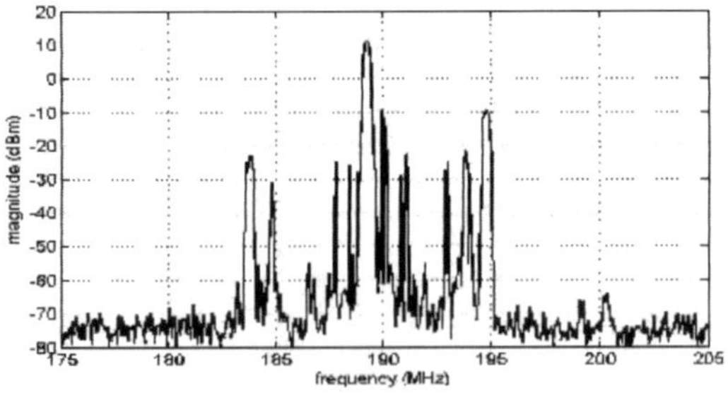

Figure 10. Spectrum of output modulator with carrier plus information.

IOP Conf. Series: Materials Science and Engineering **105** (2016) 012030 doi:10.1088/1757-899X/105/1/012030

The measurement result for RF amplifier stage 1 is displayed in figure 11. It was measured at the frequency range of 175.25 to 224.25 MHz. The lowest power at frequency 175.25 MHz was 53.07 mwatt (17.28 dBm), while the highest power reached 522.38 mwatt (27.18 dBm) at frequency of 196.25 MHz. Hence, the average power that could be achieved was 24.68 dBm. Therefore, stage 1 had amplification gain around 13.5dB.

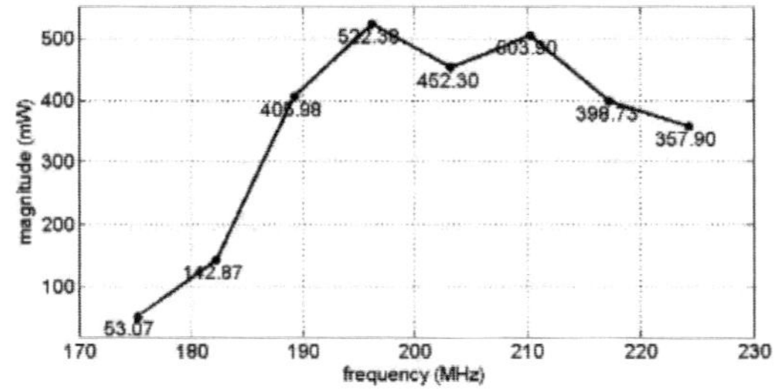

Figure 11. Level power output buffer.

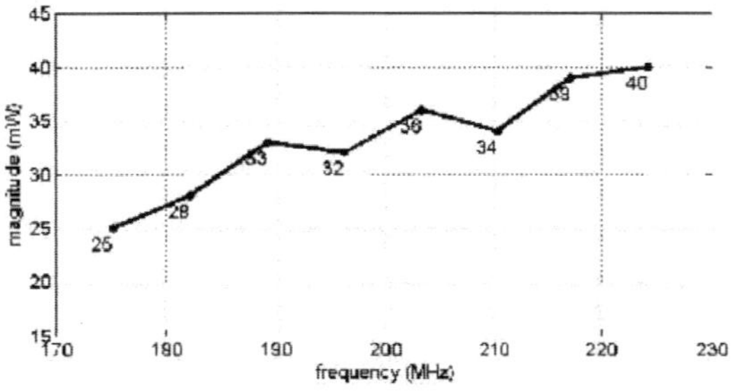

Figure 12. Level output power Amplifier.

Next procedure was applying RF output of stage 1 to stage 2 to verify RF amplifier performance. The measurement result is shown in figure 12. The average power that could be achieved was 33.38 Watt (45.23 dBm). The highest power that could be achieved at frequency 224.25 MHz was 40 Watt (46.02 dBm), while the lowest power was 25 Watt (43.98 dBm) at frequency 175.25 MHz. Table 1 shows gain calculation of stage 2 for eight channels at band III VHF TV. Parameter gain of RF amplifier could be calculated using (1).

$$G = 10^{10} log \frac{P_{out}}{P_{in}} \quad (1)$$

G is gain in logarithmic scale (dB). P_{out} is level output power RF amplifier stage 2 and P_{in} is level input power Amplifier stage 2.

Table 1. Gain for stage 2 RF amplifier.

Frequency (MHz)	Gain (dB)
175.25	26.73
182.25	22.92
189.25	19.10
196.25	17.87
203.25	19.01
210.25	18.29
217.25	19.90
224.25	20.48
average	20.54

Based on the resulting power, it can be inferred that the RF amplifier stage 2 has met the minimum required power for the community TV, i.e less than 50 Watt. When comparing to local brand design of RF amplifier for community television, the total cost for a proposed designed RF amplifier is about only 1/5.

4. Conclusions

The designed RF amplifier especially for community television is presented. Capability of television to deliver voices and pictures makes this media become the most effective method to spread information. Television is beneficial to improve education. Programs with educative content are useful to improve the quality of human resources. The proposed RF amplifier consists of two

IOP Conf. Series: Materials Science and Engineering **105** (2016) 012030 doi:10.1088/1757-899X/105/1/012030

stages of amplifier. The first stage amplifier was used for amplified output of TV modulator less then 11dBm so that the first stage output able to drive next stage amplifier. Such CAD simulation and fabrication were run to reach optimum RF amplifier design circuit. The associated circuit was made by determining stability circle, stability gain, and matching impedance. Hence, the average power of first stage RF amplifier was 24.68 dBm achieved.

The second stage used RF modules which was ready match to 50 ohm for both input and output port. There was no need matching impedance for both ports. The experiment result shows that the RF amplifier may operate at Band III TV VHF ranging from 174 to 230 MHz. The resulting average power of the 2nd stage amplifier was 33.38 Watt with the average gain of 20.54 dB. The designed RF amplifier is cheap way to have very stable and clean signal of the RF amplifier for community TV. The total budget for a designed RF amplifier is about only 1/5 compared to local design RF final amplifier.

References

[1] Ministry of Communication and Information 2012 *Procedures for the Establishment and Implementation of Broadcasting Community broadcasting agency* No. 39. (Jakarta)

[2] Miki N 1988 *Technologies for all solid state TV transmitter* (London: IET).

[3] Tai-Cheng Lee and Yen-Chuan Huang 2005 *An optimization technique for RF buffers with active inductors* (Taiwan: IEEE)

[4] Badarou A, Reed S, Ndong D, Pastore M and Prasad R 2013 *Power amplifier design challenges in UHF and VHF transmitters* (Paris: IEEE)

[5] Suraperwata A V, Hendrawan, Suryana J and Haryadi S 2014 *Design and implementation of DVB-T digital TV broadcast antenna for community television in Bandung Institute of Technology campus* (Kuta: IEEE)

[6] R Sopian, B Prasetya and Y Wahyu 2014 *e-Proc of Engineering* Vol 1 (Bandung: Telkom University)

[7] Mitsubishi Electric Corp 2011 *datasheet Silicon RF Power MOSFET RD15HVF1* (Mitsubishi Electric Corp)

[8] Mitsubishi Electric Corp 2011 *datasheet Silicon RF Power Modules RA30H1721M* (Mitsubishi Electric Corp)

IOP Conf. Series: Materials Science and Engineering **105** (2016) 012031 doi:10.1088/1757-899X/105/1/012031

Developing Baby Bag Design by Using *Kansei* Engineering Method

D Janari[1]A Rakhmawati[2]
[1,2]Industrial Engineering Department, Faculty of Industrial Technology, Universitas Islam Indonesia.Yogyakarta

Email: kerjadian@gmail.com, rakhmawati.annisa@gmail.com

Abstract. Consumer's preferences and market demand are essential factors for product's success. Thus, in achieving its success, a product should have design that could fulfill consumer's expectation. Purpose of this research is accomplishing baby bag product as stipulated by *Kansei*. The results that represent *Kansei*words are; neat, unique, comfortable, safe, modern, gentle, elegant, antique, attractive, simple, spacious, creative, colorful, durable, stylish, smooth and strong. Identification value on significance of correlation for durable attribute is0,000 < 0,005, which means significant to baby's bag. While the value of coefficient regression is 0,812 < 0,005, which means that durable attribute insignificant to baby's bag.The result of the baby's bag final design selectionbased on the questionnaire 3 is resulting the combination of all design. Space for clothes, diaper's space, shoulder grip, side grip, bottle's heater pocket and bottle's pocket are derived from design 1. Top grip, space for clothes, shoulder grip, and side grip are derived from design 2.Others design that were taken are, spaces for clothes from design 3, diaper's space and clothes' space from design 4.

1. Introduction

Consumer's preferences and market demand are essential factors for product's success. Thus, in achieving its success, a product should have design that could fulfill consumer's expectation. A good product needs to fulfill two basic characteristics. The first isfulfilling basic performance, perceivedquality, capacity, and appearance. The second isaestethics design that willattract customers. First characteristic is described as physical factor, while the second is mental factor[6].

Appearance of the product could be identified from the design itself. Design, in general, is the form, model, shape or pattern that can figure a construction, planning and contains purposes [12]. Things that should be considered in designing is the usability, appearance, maintenance, low cost and communication [1]. While, product design is a scheme where all functional elements and product are arranged into groups of physical components [14]. One's perception on design product of an object consists of single or multi parts [9]. In literature of engineering development, Schutte 2002 stated, factors that involve psychological part are called as human *kansei*.Since consumers always find attractive and comfortable products, the company could interpret this information into appropriate design to develop new products [10].

Fashion industriesare known as a time-changing creative industry that willnever stops to innovate by creating various styles and designs in every single season. It is always create excitement and vibrancy in the market. Same thing goes for baby bag. Baby bag is also categorized as a fashion items.Designs have become a key point to meet customer's expectation of the baby bag. Currently, manufacturing company starts to applyconcept of market strategyto marketits new products. The principle of production that based on customers' preference is the core concept of market strategy. Once the customers feels satisfaction, automatically the company will gain a good brand image of the products. There is a high chance that this brand image could affect the sales of this product.

Yet, every costumer has different preference.Althought there are various preference, customers have the same thing in common, they tends to select the efficient and effective kind of products [11].One of the methods to consider human feeling or emotion towards product is *Kansei* Engineering. Firstly introduced by Mitsuo [5] as a consumer-oriented technology in developing products. *Kansei* Engineering is a method of product development that interprets customer feeling to be the element of product design. *Kansei* in Japanese is translated as entire feeling or emotion that experienced by a person through his senses on certain thing, environment or situation [10]. *Kansei* Engineering often applies for both product design and product development. It is employed on automotive industry, for example Miata from Mazda [4], steering wheel [4], car interior [3,13] and other products such as bag, office chair [8], and washing machine [2].

A customer-oriented approach to product development with particular consideration of emotions is represented by *Kansei* Engineering, characterized by the possibility to capture customer perceptions and to translate them into product features. As one of the new methodology which aimed at making simpler the product/market fit process, *Kansei* become one of user-centered approach that advocate for customization of products guided by users' expectations[16].In order to uncover what users think but also what they feel and desire when facing new products. That is elicitation of the so-called *Kansei*s or "psychological feelings". Then, a new generated prototype based on this method could be truly matter to people because they fit the deepest users' demands[17]. Thus, improving innovation and marketing success rate. In this paper, authors illustrated the details of the proposal in a case study related to baby bag. In order to assess these perceptions, a semantic space is generated. For this purpose it is necessary to identify *Kansei* words that are gathered up through research in marketing concepts, and consumer's thought; represented by surveys and interviews.

Designs that include mental satisfaction as well as emotional are increasing. The company applies it on product differentiation and positioning [15]. Hence, to design a baby bag that oriented on costumer's affective side, *kansei* words are required. One of the methods that could be used is *Kansei* Engineering, to find out the consumer's preferences from psychological perspectives. *Kansei* Engineering was firstly introduced in Japan to combine emotional factor (aesthetics or "*kansei*") on product design.*Kansei* Engineering is defined as technology that translates consumer's emotion in design specification [7].

Research and development team in a company captures consumer's feelings under the name of *Kansei*; analyzing the data by using method of psychology, ergonomics, medical and engineering; designing new products based on analysis of information. *Kansei* process/effective engineering should contain several questions, as follows: who are the consumers, what are their needs, how to evaluate *Kansei*'s consumers. After evaluation, engineer should analyze the Kansei data by using statistical analysis or psycho physiological measurement and transfer the data to be analyzed in design field. It takes a developed model such as *Kansei* engineering model to interpret human psychological feelings as parameters to a product.

*Kansei*designated for product design could be used as example in automotive product, wrist watch product as well as development in the field of information technology by using psychology software with the approach of artificial intelligence. The main problem of this research is formulatingbaby bag design in accordance to*Kansei* words. The research is purposed to provide the design for baby bag as *Kansei* words by using *Kansei* engineering.

2. Method

Research is conducted by questionnaire and interview that will be explained later on this section. First, researchers determined two selected baby bags namely A and B that was obtained randomly from market. Picking two random baby bags and showing it as a sample to be shown for answering questionnaire is meant to create a product that fit multi-function design baby bags. Preliminary stage is initiated by identifying the function of each part of baby bag that could be separated as bottle's pocket, space for clothes, bottle's heater pocket, space for diapers, top grip, side grip and shoulder grip. Bottle's pocketis designed to store main and alternative bottles. Space for clothes is used to keep the alternative clothes. Top grip is used to pull and grab the bag from upper side. Side grip is used to pull the bag horizontally. Meanwhile, the shoulder grip is designed to place the bag on the back and the grip was placed on the shoulder.

There are two types of models implemented for each product. *Kansei* words are collected from survey's result on users. The *Kansei* words are selected by assumption of 90% reliability level. *Kansei* words that will be applied are chosen by > 90 % of total respondent. The questionnaire distribution is performed in 3 (three) stages. Questionnaire 1 is conducted to gain data on respondents' opinion of product's importance towards *Kansei* words. Purpose of questionnaire 2 is finding out respondents' opinion on relationship between *Kansei* words to its elements of design category. Furthermore, questionnaire 3 is applied to recognize the respondents' opinion on relationship between *Kansei* words toward the selected type of design elements categories.Determining which design will be used is dependedon the result of data processing on *Kansei* engineering by using double binary logistic regression. Hence, the researchers collect which elements that affecting respondents' opinion and used the data collection to determine design and type of model.Questionnaires were distributed to 40 women as respondents, common users for baby's bag A and baby's bag B.There were only 39 questionnaires out of 40 that completely filled. Then only 39 questionnaires will be forwarded to next processes. The data obtained was processsed using IBM® SPSS® version 22.

The second method used in this research is interviews that was carried out with women interviewee. Women are considered as the main target market and common users for baby's bag. It was performed to identify basic need and detail specifications for users. Interviews was condusted to 40 women as respondents as well.

3. Results and Discussions

3.1 Baby Bag Design Process

Bag's design is determined by surveys and interviews. Survey is considered as locating baby's bag in market to be further analyzed for its initial function, material and price. Interviews were carried out toward the women, common users for baby's bag. It was performed to identify basic need and detail specifications for users.

3.2 Stages to Identify Kansei Words

Kansei words is illustrated as human's psychological feelings, the words reveal the consumer's preferences on baby's bag. The feelings are represented by certain words. *Kansei* words are derived from direct interviews. Initial surveys expose 22 words, which are: neat, safe, elegant, attractive, creative, stylish, unique, modern, antique, casual, colorful, smooth, comfortable, soft, simple, spacious, durable, strong, special, regular, small and complicated. The selected *Kansei* words are determined by assumption of 90% reliability level. *Kansei* words that will be applied are chosen by > 90 % of total respondent. While the words that selected by < 90% respondent will be ignored.

Therefore, the selected words are calculated as 18 words and other 4 words are omitted, as shown by table below:

Table 1. Selected *Kansei*words

No	Kansei	No	Kansei	No	Kansei
1	Neat	7	Unique	13	Comfortable
2	Safe	8	Modern	14	Soft
3	Elegant	9	Antique	15	Simple
4	Attractive	10	Casual	16	Spacious
5	Creative	11	Colorful	17	Durable
6	Stylish	12	Smooth	18	Strong

Table 2. Unselected *Kansei*words

No	Kansei	Number of respondents that select
1	Special	21 < 90% total respondent
2	Regular	18 < 90% total respondent
3	Small	22 < 90% total respondent
4	Complicated	23 < 90% total respondent

3.3 Data Collection for Questionnaire 1

Questionnaires were distributed to 40 women as respondents, common users for baby's bag A and baby's bag B. There were only 39 questionnaires out of 40 that completely filled. Then only 39 questionnaires will be forwarded to next processes. It was composed to identify the initial consumers' preferences on design, function and material of baby's bag. Each respondent was ordered to assess the statement by scaling, there are 5 criteria of importance level, which are: score 1 (not important), 2 (less important), 3 (important enough), 4 (important) and 5 (very important). The result of data collecting will be later processed to identify level of self-sufficiency, validity and reliability. Once all data are processed and considered as sufficient, valid and reliable, hence those *Kansei* words are eligible for standards in composing questionnaire 2 and 3.

3.4 Data Collection for Questionnaire 2

Questionnaire 2 is created to recognize relationship between *Kansei* words with categorization of design's elements. The following is the category of design's elements of baby's bag based on selected *Kansei* words that passed the validity test. Questionnaire 2 is employed to identify the category of design's elements that effected by *Kansei* words. The Questionnaire will be filled by binary numbers, which are 1 and 0. Value of 1 will be attached if the design's elements effected by *Kansei* words and value of 0 if it is not effected by one. Below is the result of Questionnaire 2.

Table 3. Result of Processing Data on *Kansei* words

No	V1	V2	V3	V4	V5	V6	V7	V8	V9	V10	V11	V12
1	1	1	1	1	1	1	1	1	1	1	1	1
2	1	1	1	1	1	1	1	1	1	1	1	1
3	1	1	1	1	1	1	1	1	1	1	1	1
4	1	1	1	1	1	1	1	1	1	1	1	1

Table 3. Result of Processing Data on *Kansei* words

No	V1	V2	V3	V4	V5	V6	V7	V8	V9	V10	V11	V12
5	0	1	0	1	0	1	1	1	1	1	1	1
6	1	1	1	1	1	0	1	1	0	1	1	0
7	1	0	1	0	1	1	0	0	1	0	0	1
8	1	1	1	1	1	1	1	1	1	1	1	1
9	1	1	1	0	1	0	1	1	1	1	1	1
10	0	1	1	1	1	1	0	1	1	1	0	1
11	1	1	0	1	1	1	1	1	1	1	1	1
12	1	0	1	1	0	1	1	1	1	0	1	0
13	1	1	1	1	1	1	1	0	1	1	1	1
14	1	1	1	1	1	1	1	1	0	1	1	1
15	1	1	1	1	1	1	1	1	1	1	1	1
16	1	1	1	1	1	1	1	1	1	1	1	1
17	1	1	1	1	1	1	1	1	1	1	1	1
18	1	1	1	1	1	1	1	1	1	1	1	1
23	1	1	1	1	1	0	1	1	0	1	1	0
19	1	1	1	1	1	1	1	1	1	1	1	1
20	1	1	1	1	1	1	1	1	1	1	1	1
21	1	1	1	1	1	1	1	1	1	1	1	1
22	0	1	0	1	0	1	1	1	1	1	1	1
23	1	1	1	1	1	0	1	1	0	1	1	0
24	1	0	1	0	1	1	0	0	1	0	0	1
25	1	1	1	1	1	1	1	1	1	1	1	1
26	1	1	0	1	1	0	1	1	1	1	1	0
27	1	1	1	1	1	1	0	1	1	1	0	1
28	1	1	1	0	1	1	1	0	1	0	1	1
29	1	1	1	1	1	1	1	1	1	1	1	1
30	1	1	1	1	1	1	1	1	1	1	1	1
31	1	1	1	1	1	1	1	1	1	1	1	1
32	1	1	1	1	1	1	1	1	1	1	1	1
33	1	1	1	1	1	1	1	1	1	1	1	1
34	1	1	1	1	1	1	1	1	1	1	1	1
35	1	1	1	1	1	1	1	1	1	1	1	1
36	1	1	1	1	1	1	1	1	1	1	1	1
37	1	1	1	1	1	1	1	1	1	1	1	1
38	1	1	1	1	1	1	1	1	1	1	1	1
39	1	1	1	1	1	1	1	1	1	1	1	1

Description:

V1 = Neat	V5 = Casual	V9 = Simple
V2 = Safe	V6 = Smooth	V10= Spacious
V3 = Unique	V7 = Comfortable	V11= Durable
V4 = Antique	V8 = Soft	V12= Strong

Score: 1 = Yes ; 0 = Not

3.5 Relation Identification between Design's Elements with Kansei Words

By using statistical test, double binary logistics is used to identify relationship between categories of product design's elements that effected by *Kansei* words. Data from questionnaire 1 will be used as input for regression processing by employing Likert scale with *Kansei* words as Y (dependent) and data from questionnaire 2, categories of design's element as X (independent). There are two steps are conducted for regression processing, which are:

- *Result of Correlation Significance*
  By considering the significance value on table of *Model if Term Removed*, which turns to be the output of regression analysis on binary logistics, that could determine which independent variables effected to *Kansei* words. Hence, the decisions could be determined by following hypotheses:
  H0 : independent variable (category of product's design elements) is not significant to dependent variable (*kansei* words).
  H1 : independent variable (category of product's design elements) is significant to dependent variable (*kansei* words). The rejection criteria is set as follows; if $P > 0{,}05$ hence H0 is acceptedand if $P < 0{,}05$,then H0 is rejected. *Kansei* word that is taken is durable towards function of baby's bag, by showing the value of $0{,}000 < 0{,}005$, means that the word durable has significant value towards baby's bag.
- *Result of Regression Coefficient*
  To determine whether regression coefficient is appropriate or not, a test is needed to determine whether there is linier relation between dependent variable (*Kansei* words) towards independent variable (category of product's design elements) by using probability value (Sig) on table of *variabel inthe equation*. Therefore, the decisions could be determined by following hypotheses:
  H0 : Regression Coefficient is not significant. H1: Regression coefficient is significant. The rejection criteria is set as follows; if $P > 0{,}05$ hence H0 is acceptedand if $P < 0{,}05$,then H0 is rejected. *Kansei* word that is taken is durable. The result of regression coefficient is 0,812. It suggests that durable is not coefficient significant.

3.6 Data Collection on Questionnaire 3

Data collection on questionnaire 3 is designated to identify the relationship between *Kansei* words with category of design elements, bottle's pocket, space for clothes, bottle's heater pocket, space for diapers, top grip, side grip and shoulder grip, whether they meet the consumer's preferences. Questionnaire 3 is used to recognize the selected model of design's elements categorization that fits the *Kansei* words. The questionnaire will be filled by Binary numbers, which are 1 and 0. Value of 1 will be attached if the design's elements effected by *Kansei* words and value of 0 if it is not effected by one. Dependent variable is resulted from Likert scale of questionnaire 1 and independent variable (model of product's design elements categorization) from questionnaire 3. The selection of influenced product design's elements is based on category of design's elements and model of baby's bag design elements categorization, as shown by table 4 below.

Table 4. Parts of Influenced Baby's Bag Design

No	Kansei	Parts of Influenced Design	
		Category of Design	**Selected Design**
1	Neat	Space for Clothes, Space for Diapers, Baby's bottle pocket	Design 1
2	Safe	Space for Clothes	Design 3
3	Unique	Top Grip, Side Grip	Design 2

4	Modern	Shoulder Grip	Design 1
5	Smooth	Space for Clothes	Design 3
6	Comfortable	Bottle's Heater	Design 1
7	Soft	Space for diapers, space for Clothes	Design 4
8	Simple	Shoulder Grip	Design 2
9	Spacious	Space for Clothes	Design 3
10	Durable	Shoulder Grip, Side Grip	Design 1
11	Strong	Top Grip	Design 2
12	Casual	Baby's bottle Pocket, Bottle's heater	Design 1

4. Conclusion

From the results, it can be concluded that *Kansei* words that included are: Neat, unique, comfortable, safe, modern, smooth, elegant, antique, simple, attractive, casual, spacious, creative, colorful, durable, stylish, soft and strong. *Kansei* words that excluded from the data are: special, regular, small and complicated. Returned questionnaires are 39, with scale assessment from 1-5. Identification on correlation significance for durable towards baby's bag design is valued as 0,000 < 0,005, which means significant. While, the result of regression coefficient is valued as 0,812 < 0,005,which is not coefficient significant. The result of selection on parts of baby's bag design for questionnaire 3 is resumed as follows; space for clothes, space for baby's diaper, baby's bottle are suggested to design 1, space for clothes, design for top grip and side grip are suggested to design 2, shoulder grip is suggested to design 1, space of clothes is suggested to design 3, bottle's heater is suggested to design 1, space for baby's diaper, space for clothes are suggested to design 4, shoulder grip is suggested to design 2, space for clothes is suggested to design 3, shoulder grip, side grip are suggested to design 1, top grip is suggested to design 2, bottle's pocket, bottle's heater pocket are suggested to design 1.

Acknowledgement

Authors would like to express gratitudeto assistance and encouragement in accomplishing this researchfrom Muhamad Mawan, Ricky Ano and Industrial Engineering Department, Universitas Islam Indonesia. Honorable mention to authors' colleaguesand families for infinite support.

References

[1] Dreyfuss, H. (1967). Designing for People. New York: Paragraphic Books.
[2] Ishihara S, Ishihara K, Nakagawa R, Nagamachi M, Sako H, Fujiwara Y, Naito M. Development and improvement of a washer-dryer with *kansei* ergonomics. *Proceedings of The International MultiConference of Engineers and Computer Scientists. Conference*. 2010
[3] Jindo T, Hirasago K. "Application studies to car interior of kansei engineering". *International Journal of Industrial Ergonomics 19*. 105-114. 1997
[4] Nagamachi M. "Kansei engineering as a powerful consumer-oriented technology for product development". *Applied Ergonomics 33*. 289-294. 2002
[5] Nagamachi M. Kansei engineering: "A new ergonomic consumer-oriented technology for product development". *International Journal of Industrial Ergonomics 15*. 3-11. 1995
[6] Nagamachi, M. (1999). Kansei Engineering; the Implication and Applications to Product Development", Proceeding of System, Man, and Cybernetic.
[7] Nagamachi, M., & Lokman, A. M. (2011). Innovation of Kansei Engineering. Boca Raton: CRC Press.
[8] Park J, Han SH. "A fuzzy rule-based approach to modeling affective user satisfaction towards

office chair design". *International Journal of Industrial Ergonomics 34*. 31-47. 2004 Product Development. Sweden: Linkoping University.
[9] Santoso, S. (2003). Buku Latihan SPSS Statistik Multivariat. Jakarta: Elex Media Komputindo.
[10] Schütte S. "Designing feelings into products: Integrating kansei engineering methodology in product development". Linköping, Sweden. Linköping Universitet. 2002
[11] Sudianto, A. (2010). Pengembangan Produk Handphone Berdasarkan Konsep Kansei Engineering. Yogyakarta: Skripsi S1 Teknik Industri Universitas Islam Indonesia.
[12] Tahid, S., & Nurcahyanie. (2007). Konsep Teknologi dalam Pengembangan Produk Industri. Jakarta: Kencana Prenada Media Group.
[13] Tanoue C, Ishizaka K, Nagamachi M. "Kansei engineering: A study on perception of vehicle interior image". *International Journal of Industrial Ergonomics 19*. 115-128. 1997
[14] Ulrich, K. T., & Eppinger, S. D. (2001). Perancangan dan Pengembangan Produk. Jakarta: Salemba Teknika.
[15] Yoshimura, M., & Papalambros, P. Y. (2004). Kansei Engineering in Concurrent Product Design: a Progress Review. Proceedings of the TMCE 2004 (pp. 177-186). Lausanne, Switzerland: Millpress
[16] Alonso, J. M., Pancho, D. P., & Magdalena, L. (2014, January). Customization of Products Assisted by Kansei Engineering, Sensory Analysis and Soft Computing. In *Information Processing and Management of Uncertainty in Knowledge-Based Systems* (pp. 616-625). Springer International Publishing.
[17] Steinberg, F., Tursch, P., & Woll, R. (2015). The Improvement of Kansei Engineering by Using the Repertory Grid Technique. *Management and Production Engineering Review*, *6*(3), 77-82.

Designing Performance Measurement For Supply Chain's Actors And Regulator Using Scale Balanced Scorecard And Data Envelopment Analysis

Elisa Kusrini[1], Subagyo[2] and Nur Aini Masruroh[3]

[1] Department of Industrial Engineering, Indonesia Islamic University, Jl Kaliurang Km 14,5 Yogjakarta 55584 Telp (+62) 274 898444 , Indonesia,
Mechanical and Industrial Engineering, Gadjahmada University, Jl. Grafika No. 2, Yogyakarta, 55281, Indonesia, (+62) 274 521673, Email : elisa_kusrini@yahoo.com

[2] Mechanical and Industrial Engineering, Gadjahmada University, Jl. Grafika No. 2, Yogyakarta, 55281, Indonesia, (+62) 274 521673, Email : subagyo@ugm.ac.id

[3] Mechanical and Industrial Engineering, Gadjahmada University, Jl. Grafika No. 2, Yogyakarta, 55281, Indonesia, (+62) 274 521673, Email : aini@ugm.ac.id

Abstract. This research is a sequel of the author's earlier conducted researches in the fields of designing of integrated performance measurement between supply chain's actors and regulator. In the previous paper, the design of performance measurement is done by combining Balanced Scorecard - Supply Chain Operation Reference - Regulator Contribution model and Data Envelopment Analysis. This model referred as B-S-Rc-DEA model. The combination has the disadvantage that all the performance variables have the same weight. This paper investigates whether by giving weight to performance variables will produce more sensitive performance measurement in detecting performance improvement. Therefore, this paper discusses the development of the model B-S-Rc-DEA by giving weight to its performance'variables. This model referred as Scale B-S-Rc-DEA model. To illustrate the model of development, some samples from small medium enterprises of leather craft industry supply chain in province of Yogyakarta, Indonesia are used in this research. It is found that Scale B-S-Rc-DEA model is more sensitive to detecting performance improvement than B-S-Rc-DEA model.

Keywords: Performance Measurement, Supply Chain's Actors and Regulator, Balanced Scorecard, Data Envelopment Analysis, Scale BSC-DEA

1. Introduction

Performance measurement is defined as the process of quantifying the level of efficiency and effectiveness of an activity into a measured value [1]. Performance measurement is a process of assessing the progress of work towards the goals and objectives that have been defined previously. The measurement of supply chain performance becomes the main concern of not only the supply chain

---

1 elisa_kusrini@yahoo.com

actors but also the regulator/government. For supply chain actors, performance measurement can be used as a basis to improve its performance. For regulators, performance measurement can be used to identify the effectiveness and the impact of its program in facilitating the improvement of supply chain performance. All this time, performance measurements by supply chain actors and regulator/government are conducted separately. It causes inefficiency, as well as obstructs both parties in understanding the effectiveness of each respective performance. Therefore, an efficient and effective integrated model of performance measurement by supply chain actors and a regulator is required.

Currently, the measurement model is developed separately. Supply chain performance measurement takes more attention from both practitioners and academics. The performance measurement model of supply chain actors is improved by using various approaches and different focuses. Several models have been proposed and employed such as *Balanced Scorecard* [2,3,4,5,6,7,8,9] , *Supply Chain Operation Reference* (SCOR) [10,11,12,13,14,15], *Component of measurement*, seperti *quality, cost, delivery time* dan sebagainya [16, 17, 18,19,20, 21], *Data Envelopment Analysis* (DEA) [22, 23,24,25,26]. Until recently, there has been no agreement of researchers which model is most flattering to be used and which key indicator that important to be maintained. Recent models demonstrate its complex performance measurement due to wide range of supply chain that becomes cause of weaknesses. The complexity will put management in difficulties for selecting the most important key indicator to be maintained. The existing models could not yet be employed for best performance management since it have not based on optimization model.
While, the models of performance measurement on regulator/government contribution towards development of supply chain actors are commonly performed but only on certain aspects, i.e. regulator's role in technology, research and development [27,28,29], regulator's role in improving financial performance [30,31,32], regulator's role in improving human resources capability [33,34], regulator's role in improving networking and marketing [35,36]. The measurement model for regulator/government's contribution on the whole process of chain operation still limited, especially those which specifically effect on performance of financial, customers, internal business process and learning and growth of supply chain.

Kusrini et al. (2014) [37]. proposed a model of integrated measurement between actors and regulators by combining Balanced Scorecard - Supply Chain Operation Reference - Regulator Contribution model and Data Envelopment Analysis. This model referred as B-S-Rc-DEA model. This B-S-Rc-DEA model has been tested in an innovative industry of leather craft in Indonesia and has managed to measure the performance of the regulator and supply chain's actor. However, B-S-Rc-DEA model has the disadvantage that all variables have the same weight. This paper investigates whether by giving weight to performance variables will produce more sensitive performance measurement in detecting performance improvement.
This paper is divided into 4 sections . The first part is introduction. The second part contains discussion of performance measurement using B-S-Rc-DEA model. The third part shows the development of scale B-S-Rc-DEA model. The final section contains conclusions.

2. Performance Measurement For Supply Chain Actor And Regulator Using B-S- Rc-Dea Model

The following is a brief explanation of the model B-S-Rc-DEA. Development of B-S-Rc-DEA model is performed in three stages, as follows [37]:

1. Developing BSC model to obtain valid indicators
   The basic model is Balance Scorecard (BSC) model. It is selected based on its frequent use in various industries and classified as the model with the most comprehensive approach by involving all aspects in balance. Widely, it applied also to assess regulator/government's performance. In this stage, basic model is improved, from BSC (corporate) model to BSC's

supply chain model (BSC-SC). Later, process activity measurement that conducted based on SCOR (Plan-Source-Make-Deliver-Return) is integrated into the perspective of internal process business originated from BSC-SC. Furthermore, integration is continued by involving performance measurement on regulator contribution towards supply chain actors on each BSC-SC perspective. Then, this model is known as BSC-SCOR-Regulator Contribution integration model. Later on, it is abbreviated as B-S-Rc (Balanced Scorecard- SCOR- Regulator Contribution) as illustrated by Figure 1. The indicators in this proposed model are consisted of 31 KPIs,i.e 19 indicators to measure performance of supply chain's actors and 12 indicators to measure the regulator's performance.

2. Integrating developed B-S-Rc model with Data Envelopment Analysis (DEA).
   This model integration is required to cover weaknesses on a model based on BSC, which are the absence of standard methodology and benchmarking facility. These could be accomplished by combining B-S-Rc model with DEA, whereas DEA is a mathematic optimization model that calculates actors' level of efficiency and might be used for benchmarking with clear and objective methodology [38,39]. This concept integration of B-S-Rc with DEA is carried out by indentifying variables in B-S-Rc into input and output variables. Those variables will be inputted into mathematical optimization calculation of DEA. Model of B-S-Rc-DEA is illustrated by Figure 2.

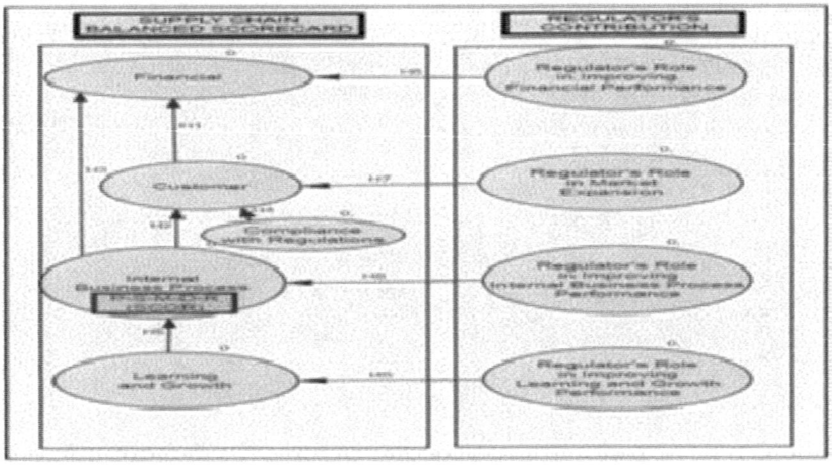

Figure 1. B-S-Rc model

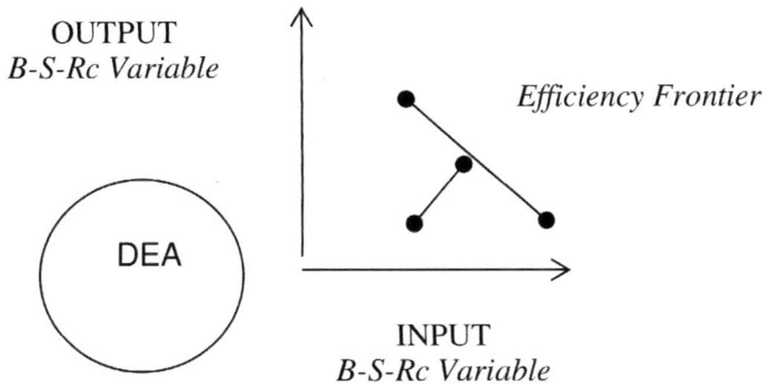

Figure 2. Integrated b-S-Rc-DEA model

Performance measurement using B-S-Rc-DEA model is initiated with identification of input and output variables from B-S-Rc model. Those variables will be inputted into mathematical optimization calculation of DEA model. DEA using primal and dual model with CRS (Constant Return to Scale) approach. Input model is consisted of 7 variables, which are regulator's role in improving financial performance (X1), regulator's role in market expansion (X2), regulator's role in improving internal business process performance (X3), regulator's role in improving learning and growth performance (X4), customer's performance (X5), internal business process performance (X6), learning and growth performance (X7). Model output is consisted only one variable, which is financial performance (Y1) [40].

Performance measurement towards Small Medium Enterprises (SME) SC is performed on 40 innovative SME's product of leather's craft in Yogyakarta, Indonesia by measuring the indicators include in B-S-Rc-DEA model during year 2013. Regulator's measurement is held based on 'LAKIP' data (Report of Performance Accountability for Government Institution) and interview with Department of industry and trade, province of Yogyakarta, Indonesia. Each indicator measurement is normalized with SNORM and generates value ranged from 0-100.

B-S-Rc-DEA uses linear programming model of primal and dual forms. The model solved by Linear Programming with the help of Software Win QS. Score (value) reflects the performance efficiency values are given in Table 1. Score of actors and regulators' performance with B-S-Rc-DEA model is ranged between 40,81 – 100 with average score of 60,61. Base on the measurement result using B-S-Rc - DEA model, it is obtained only two SMEs that have a value of performance with 100% efficiency is Decision Making Unit (DMU/SME) 19 and 44. This shows that the two businesses can manage their input to get the optimum output [41]. The role of the regulator in these DMU is optimum with the assistance and facilitation that can be managed effectively by both DMU 's. Other DMUs can refer to both the DMU's in managing performance.

Table 1. Perfomance score using model B-S-Rc–DEA (in %) [40]

DMU	A2	A3	A4	A5	A6	A7	A8	A9	A10	A11
Performance score using B-S-Rc - DEA	40.81	40.81	40.81	72.30	77.72	43.29	60.02	40.81	40.81	68.02
DMU	A12	A13	A14	A15	A16	A17	A18	A19	A20	A21
Performance score using B-S-Rc - DEA	40.81	72.79	72.51	55.18	64.14	54.05	48.17	100.00*	80.53	74.15
DMU	A22	A24	A27	A28	A29	A30	A31	A32	A33	A35
Performance score using B-S-Rc - DEA	40.81	45.99	49.62	96.31	61.45	63.61	42.58	85.90	83.28	45.01
DMU	A36	A37	A38	A40	A41	A42	A43	A44	A45	A46
Performance score using B-S-Rc - DEA	61.50	50.02	48.91	40.81	57.38	60.55	68.70	100.00*	74.15	60.27

*= efficient

From the above explanation , it can be concluded that the B-S-Rc-DEA model has been tested in an innovative supply chain industry and has managed to measure the performance of the regulator and supply chain's actor. However, B-S-Rc-DEA model has the disadvantage that all variables have the same weight. The following section will discuss the development of a model B-S-Rc-DEA by giving weights to the variables. This model is called the Scale B-S-Rc-DEA.

3. Performance Measurement For Supply Chain Actor And Regulator Using Scale B-S-Rc-Dea Model

Model combination of B-S-Rc-DEA excludes the consideration on indicator's weighting. In fact, each indicator possibly provides difference contribution on performance. Hence, to acquire more sensitive result on performance improvement, then B-S-Rc-DEA will be more developed by weighting its each variable (later called as Scale B-S-Rc-DEA model). This is a model with weighting input and output variables. The concept of Scale B-S-Rc-DEA model is presented in Figure 3.

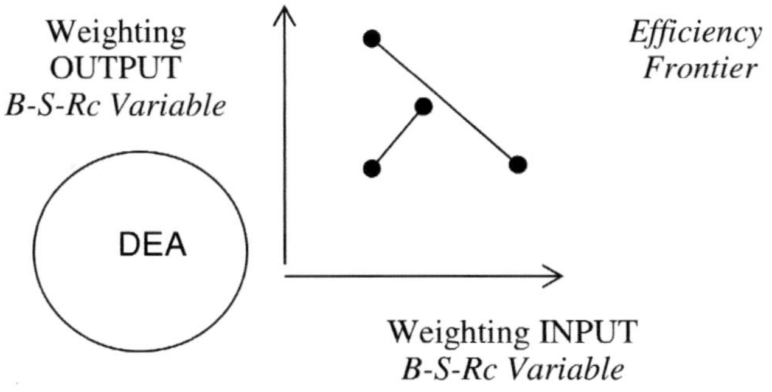

Figure3. Integrated Scale B-S-Rc-DEA

This measurement is performed by weighting the input and output variables. The performance variable weighting is carried out by conducting a survey and tested by using Analytical Hierarchy Process (AHP) technique. AHP is a tool in decision-making by many criteria. A total of 74 questionnaires distributed and filled out by the respondents with the composition of the respondents are given in Table 2. The questionnaire contains pair wise comparisons between perspective and an indicator of the scale of 1-9. Designed questionnaire was designed to determine the relative importance of each indicator and perspectives. The hierarchical structure is given in Appendix 1.

The result of AHP analysis shows that the actors' performance weighted as 67,3% while the regulators weighted as 32,7%. The actors' performance that provide highest contribution is customer's performance (27,9%), followed by, consecutively financial performance (26,8%), internal business process performance (24,3%) and learning and growth performance (21%). While, for regulator, the performance that provides highest contribution is the role in improving new market expansion (29%), followed by, consecutively, the role in improving financial performance (28,8%), the role in improving internal business process (23%) and the role in improving learning and growth (19%).

Calculation of performance using B-S-Rc-DEA model is done by first multiplying the value of each indicator with weights obtained from AHP. The performance of the weighted value is then analyzed using a mathematical model of DEA. Results of the calculation of the performance of the Scale B-S-Rcs –DEA model is given in Table 3. Score of actors and regulators' performance with Scale B-S-Rc-DEA model is ranged between 39,69 – 100 with average score of 59,98.

To find out whether there is a significant differences in the value of the performance between B-S-Rc-DEA and Scale B-S-Rc- DEA, then the independent t test is conducted. Results of statistical test t test showed that both methods confirmed no significant difference in performance values. However, The average value (mean) of Scale B-S-Rc - DEA (59.98) is lower than B-S_Rc-DEA model (60.61). This means that Scale B-S-Rc-DEA model more sensitive in assessing the performance due to low efficiency values indicate greater gap value (towards 100 % efficiency). It will provide information about opportunities for improvement [42].

Table 2. Respondent Composition

Number	Responden	Total
1.	The Cooperative, Industrial and Trade service of Yogyakarta and Center for Leather, Rubber and Plastics Yogyakarta, Indonesia	23
2.	Leather's craftsmen, trader and administrators of cooperative or association of leather craftsmen	47
3.	SCM's Consultant	4
Total		74

Table3. Performance score using Scale B-S-Rc-DEA model

DMU	A2	A3	A4	A5	A6	A7	A8	A9	A10	A11
Performance score using Scale B-S-Rc - DEA	40.01	39.91	39.89	73.10	83.44	42.80	59.68	39.69	39.69	67.74
DMU	A12	A13	A14	A15	A16	A17	A18	A19	A20	A21
Performance score using Scale B-S-Rc - DEA	39.69	73.54	69.00	52.88	60.96	55.18	47.92	100.00*	80.06	72.99
DMU	A22	A24	A27	A28	A29	A30	A31	A32	A33	A35
Performance score using Scale B-S-Rc - DEA	39.69	44.67	51.24	89.74	61.39	62.47	43.67	87.87	83.39	44.12
DMU	A36	A37	A38	A40	A41	A42	A43	A44	A45	A46
Performance score using Scale B-S-Rc - DEA	57.71	50.60	46.27	40.39	59.31	60.20	67.66	100.00*	72.99	57.48

4. Conclusion

The B-S-Rc-DEA model excludes the consideration on indicator's weighting. In fact, each indicator possibly provides difference contribution on performance. Hence, B-S-Rc-DEA model is developed by weighting its input and output variable using AHP technique. This model referred as Scale B-S-Rc-DEA model. Base on case study in small medium enterprises of leather craft industry supply chain in province of Yogyakarta, Indonesia, it can be conclude there is no significant difference in performance values between B-S-Rc-DEA and scale B-S-Rc-DEA model. However Scale B-S- Rc-DEA model is more sensitive in assessing the performance than B-S-Rc-DEA model. The limitation of this study is that the sample is limited to Supply chain of innovativeproducts, thus it is required consideration and prudent to generalize the results to apply to supply chain of functional product.

Reference

[1] Neely, Agregory, M., and Platts, K. 1995 Performance Measurement System Design: A Literature Review and Research Agenda, *International Journal of Operation and Production Management,* Vol 15, pp 80-116.

[2] De Sousa, T.B., Camparotti, C.E.S., Esposto, K.F., and Guerrini, F.M. 2014 Alignment of Balanced Scorecard Perspectives with Supply Chain Management Objectives: a Literature Review, *Independent Journal of Management & Production (IJM&P)* ISSN: 2236-269X, Vol. 5, No. 4, pp. 1050-1070.

[3] Barber, E. 2008 How to Measure the "Value" in Value Chains, *International Journal of Physical Distribution & Logistics Management,*Vol. 38 No. 9, pp. 685-698.

[4] Bhagwat, R. and Sharma, M.K. 2007 Performance Measurement of Supply Chain Management: a Balanced Scorecard Approach, *Computers & Industrial Engineering* ,Vol. 53 pp.43–62.

[5] Brewer, P.C. and Speh, T.W. 2000 Using Balanced Scorecard to Measure Supply Chain Performance, *Journal of Business Logistics,*Vol 21, No.1.

[6] Chia, A., Goh, M., and Hum, S.H. 2009 Performance Measurement in Supply Chain Entities: Balanced Scorecard Perspective, *Benchmarking: An International Journal,* Vol. 16 No. 5, pp. 605-620.

[7] Chang, H. H. 2009 An Empirical Study of Evaluating Supply Chain Management Integration using Balanced Score Card in Taiwan', *The Service industries Journal,*Vol.29,No.2, pp.185-201.

[8] Gunasekaran, A., Patel, C., and Mc Gaughey, R.E. 2004 A Framework for Supply Chain Performance Measurement', *International Journal Production Economy,* Vol 87, 2004, pp.333–347.

[9] Park, J.H., Lee, J. S., and Yoo, J. S. 2005 A Framework for Designing the Balanced Supply Chain Scorecard, *European Journal of Information Systems*, Vol. 14, pp. 335–346.

[10] Georgise ,F.,B., Thoben, KD., and Seifert, M. 2012 Adapting the SCOR Model to Suit the Different Scenarios: A Literature Review & Research Agenda, *International Journal of Business and Management,*Vol. 7, No. 6; pp. 1-17.

[11] Chae, K. 2009 Developing Key Performance Indikators for Supply Chain: an Industry Perspective', *Supply Chain Management: An International Journal,* 14/6, pp. 422–428.

[12] Huan, S., Sheoran, S.,Unil K., and Wang, G. 2004 A Review and Analysis of Supply Chain Operations Reference (SCOR) Model, *Supply Chain Management,*Vol 9, No.1. pp 23-29.

[13] Lockamy, A. and McCormack, K. 2004 Linking SCOR Planning Practices to Supply Chain Performance: An Exploratory Study', *International Journal of Operations & Production Managemen, Vol 24, No.* 11/12, pp 1192- 1217.

[14] Supply Chain Council 2001 Supply Chain Operation Reference Model, http://www.supply-chain.org [online] (accessed 14 December 2010).

[15] Theeranuphattana, A. and John C.S. Tang. 2008 A Conceptual Model of Performance Measurement for Supply Chains Alternative Considerations, *Thailand Journal of Manufacturing Technology Management*, Vol. 19 No. 1, pp.125-148.

[16] Beamon, B,M. 1999 Measuring Supply Chain Performance, *International Journal of Operation & Production Management,* Vol 19, pp. 275-292.

[17] Espinoza, O., Bond, B.H., and Kline, E. 2010 Supply Chain Measures of Performance for wood products manufacturing', *Forest Production Journal,* 60(7/8):700–708.

[18] Farris, M.T. and Hutchison, P.D. 2002 Cash-To-Cash: The New Supply Chain Management Metric, *International Journal of Physical Distribution & Logistics Management*, Vol 32/4, Pp 288-298.

[19] Hum, S.H. and Parlar,M. 2014 Measurement and Optimization of Supply Chain Responsiveness, *IIE Transactions,* Vol 46 Number 1, Taylor & Francis Group, Philadephia.

[20] Marthin, R. and Patterson, J, W. 2009 On Measuring Company within A Supply Chain, *International Journal of Production Research*, Vol.47, No.9, pp.2449 - 2460.

[21] Pohlen, T., L. and Coleman, B., J. 2005 Evaluating Internal Operations and Supply Chain Performance Using EVA and ABC, *Advanced Management Journal;* spring, 70, 2; ABI/INFORM Global 45-58.
[22] Chen,Y.S. and Chen, B.,Y. 2009 Using Data Envelopment Analysis (DEA) to Evaluate the Operational Performance of the Wafer Fabrication Industry in Taiwan, *Journal of Manufacturing Technology Management*, Vol. 20, No. 4, 475-488.
[23] Wong,W.P. and Wong,K.Y. 2007 Supply Chain Performance Measurement System using DEA Modeling, *Industrial Management & Data Systems,* Vol. 107, No. 3, pp. 361-381.
[24] Yang, Y., Feng, D., Wu, L., Liang, G., and Wu, D. D. 2011 Supply Chain DEA: Production Possibility Set and Performance Evaluation Model, *Annals of Operations Research*, Vol. 185, no. 1, pp. 195–211.
[25] Liang, L., Yang, F., Cook, W.D., and Zhu, J. 2006 DEA Models for Supply Chain Efficiency Evaluation, *Published Online: Annual Operation Research,* 145:35–49.
[26] Deng, Z., Wu, C., Zhang, J., and Wang, J. 2014 Study on the Supply Chain Efficiency of Rural Public service in China Based on Three Stage DEA Model, *Asian Agricultural Research*, 691, 6-13.
[27] Park, J.Y. and Kim, S.W. 2010 An Empirical Model To Assess The Influence Of The Government's Research And Development Program On Korean Small And Medium Enterprise (SME) Performance, *The Asian Journal on Quality,* Vol. 11 No. 3, pp. 288-302.
[28] Nam, M. 2010 The Impact Of Government R&D Subsidies On Smes In Korea Do Government R&D Subsidies Make Smes More Competitive In The Market?, *A Thesis submitted to the Faculty of the Graduate School of Arts and Sciences of Georgetown University, UMI publishing, ProQuest LLC.*
[29] Cin, B.,C., Kim, Y.J., and Vonortas, N.S. 2013 The Impact of Government R&D Subsidy on Firm Performance: Evidence from Korean SMEs, *The Asian Research Policy Symposium*, Seoul, Korea.
[30] Saad, M.N. and Idris, A. 2014 A Profile of Malaysian International Small and Medium Enterprises: Mapping Current Performance on National Policies and Strategic Objectives, *The IUP Journal of Business Strategy*, Vol. XI, No. 3, Pp.60-83.
[31] Xiang, D. and Worthington, A.C. 2013 The Impact of Government Financial Assistance on SMEs in Australia During the GFC, *Disscussion Paper Financial, Griffith Business School*, No. 2013-07.
[32] Tesfayohannes, M. 2010 Streamlining Government Financing Programs for SME's in the Sub-Saharan Africa: The Case of Botswana, **ICSB World Conference Proceedings:** 1-37. Washington: International Council for Small Business (ICSB).
[33] Čučković, N. and Vučković, V. 2013 Education-Research-Innovation Policy Triangle And Competitiveness Of The Croatian SME Sector, *The Tenth International Conference Challenges of Europe: The Quest for New Competitiveness.*
[34] Shariff, M.N.M., Peou, C., and Ali, J. 2010 Moderating Effect of Government Policy on Entrepreneurship and Growth Performance of Small-Medium Enterprises in Cambodia, *International Journal of Business and Management Science*, 3(1),Pp. 57-72.
[35] Hussain, I., Farooq, Z., and Akhtar, W. 2012 Smes Development And Failure Avoidance In Developing Countries Through Public Private Partnership, *African Journal of Business Management*, Vol. 6 (4), pp. 1581-1589.
[36] Hadiyati, E. 2015 Marketing and Government Policy on MSMEs in Indonesian: A Theoretical Framework and Empirical Study, *International Journal of Business and Management;* Vol. 10, No. 2,Pp. 128-141.
[37] Kusrini, E., Subagyo., and Masruroh, N.A. 2014 A New Approach to Design Supply Chain Key Performance Indicator for Actors and Regulator : A Case Study in Innovative Product In Indonesia, *International Journal of Business Performance Management,* (in press).

[38] Coelli, T. 1996 A Guide to DEAP version 2.1. A Data Envelopment Analysis (Computer) Program, *Departement of Economic University of New England.*

[39] Rickards, R.C. 2003 Setting Benchmarks and Evaluating Balanced Scorecards with Data Envelopment Analysis, *International Benchmarking Journal*, Vol. 10, No. 5, pp. 62-86.

[40] Kusrini, E., Subagyo., and Masruroh, N.A. 2015 Combining Balanced Scorecard and Data Envelopment Analysis to Design Performance Measurement for Supply Chain Actor and Regulator: A Case Study in Innovative Product in Indonesia, *Lectures Note in Electrical Engineering, Industrial Engineering, management Science and Appliccation 2015*, *Springer,* Vol 349, pp 551-562.

[41] Banker, R.D., Charness, A., and Cooper, W.W. 1984 Some Models for Estimating Technical and Scale Inefficiencies in Data Envelopment Analysis, *Management Science*, Vol. 30 No. 6,1078-92.

[42] Lee, J.Y. 2012 Combining Balanced Scorecard and Data Envelopment Analysis in Kitchen Employees Performance Measurement - An Exploratory Study - A Dissertation Hospitality Management, Iowa.

Appendix 1. The Hierarchical Structure of Supply Chain's Actors and Regulator's Performance

Identification of Behavior Based Safety by Using Traffic Light Analysis to Reduce Accidents

A Mansur[1] and M I Nasution[2]

[1,2]Industrial Engineering Department, Universitas Islam Indonesia, Jl. Kaliurang km 14, 5 Yogyakarta 55584, Indonesia
E-mail: agusmansur.am@gmail.com, m.ihsan17@yahoo.com

Abstract. This work present the safety assessment of a case study and describes an important area within the field production in oil and gas industry, namely behavior based safety (BBS). The company set a rigorous BBS and its intervention program that implemented and deployed continually. In this case, observers requested to have discussion and spread a number of determined questions related with work behavior to the workers during observation. Appraisal of Traffic Light Analysis (TLA) as one tools of risk assessment used to determine the estimated score of BBS questionnaire. Standardization of TLA appraisal in this study are based on Regulation of Minister of Labor and Occupational Safety and Health No:PER.05/MEN/1996. The result shown that there are some points under 84%, which categorized in yellow category and should corrected immediately by company to prevent existing bad behavior of workers. The application of BBS expected to increase the safety performance at work time-by-time and effective in reducing accidents.
Keywords: *Behavior-Based Safety; Traffic Light Analysis; Oil and Gas Industry; Safety Assessment, Fishbone Diagram.*

1. Introduction

Safety at work is widely accepted as an accurate action to increase worker performance and reduce accidents [1]. Accidents at work occur either due to lack of knowledge or training, a lack of supervision, or a lack of means to carry out the task safely, or alternatively, due to an error of judgment, carelessness, apathy or downright reckless. Safety at work suggested as complex phenomenon that related with the subject of safety performance and safety behavior.In previous study [2], unsafe behavior is the most significant factor in the cause of site accidents and therefore provides evidence of a poor safety culture and may cause dangerous accidents at industrial companies.Among leading industry, oil and gas industry is one of the most risky industry in worldwide. Oil and gas industry jobs has highest number of accidents and facility rate such as explosion, struck by object, ambient heat condition, highway crash and other accidents [3]. Safety of workers in oil and gas industry is critically important. Figure 1 shows that the national average for the rate of fatal job injuries owned by oil and gas industry currently in high number year-by-year [4].

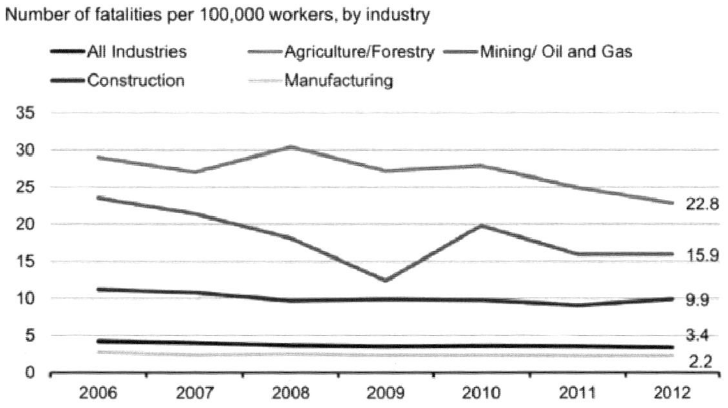

Figure 1. The national average for the rate of fatal job injuries

The figure implies a special challenge for safety improvement in oil and gas industry. The features of safety assessment should be determined when implementing any type of safety initiative to reduce accident rate. Therefore, the specific objectives of this study are first, to measure the behavior performance through worker that that exist in the company (in this study in oil and gas industry). Second, to determine the group of factors that has the most effect on site safety, utilizing traffic light analysis and root-cause diagram.

2. Literature Review

2.1 Occupational Safety in Oil and Gas Industry

Workers in Oil and Gas industry are generally susceptible to following agents, which lead to various health and illnesses hazards including; chemical, physical hazards, biological hazards, ergonomic hazards and psychosocial hazards [5]. The aim of occupational safety and health risk management is to identify and assess safety and health hazard existing at the workplace and to define appropriate control/reaction and retrieval steps [6].

2.2 The Behavior Based-Safety Approach

Al-Hemoud and Al-Asfoor suggests that Behavior-Based safety (BBS) initiative are the current trend toward improving safety performance [7]. The definition of BBS itself is a proactive accident prevention effort that focus at riskbehavior/dangerous behavior that is likely to cause accidents. The quality of BBS implementation is not only a program, but also an integrated management process that

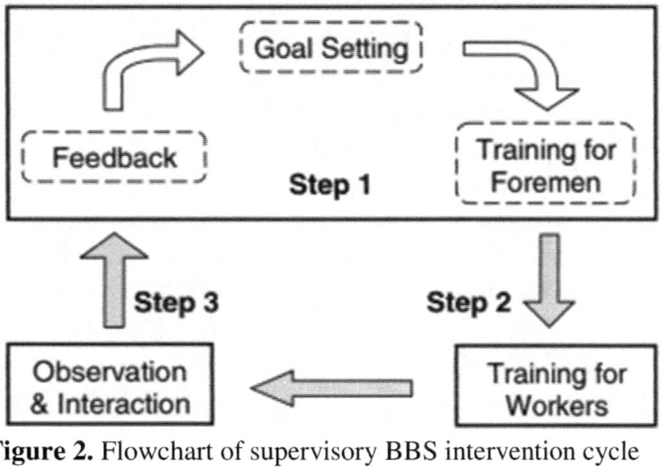

Figure 2. Flowchart of supervisory BBS intervention cycle

IOP Conf. Series: Materials Science and Engineering **105** (2016) 012033 doi:10.1088/1757-899X/105/1/012033

undertaken continuously by company/organization [8].

According to J Agnew and Ashworth [9],to be successfulBBS process must include all employees, from the CEO to the front line workers. Including but not limited to hourly, salary, union employees, contractors and sub-contractors. To achieve changes in behavior, a change in policy, procedures and/or systems most assuredly will also need some change. Those changes cannot accomplished without buy-in and support from all involved in making those decisions[10]. Zhang and Fang also stated that in this recent year, many researchers in various setting such as clinical medicine, manufacturing, petro-chemistry, mining/oil and gas and institution office have applied BBS successfully [11]. Implementation of BBS within field area of oil and gas industry may affect significant organizational benefit include:

- Significant reduction in number of accidents caused by inappropriate behavior
- Good opportunity for management and workforce participation
- Improving visibility of manager
- Behaviors and actions influence culture through attitudes and perceptions.

2.3 Ishikawa "Fishbone" Diagram

Many companies repeatedly experience the same type of incidents. Investigation indicate what seem to be right solution to reducing accidents. A cause and effect diagram, often called a "fishbone" diagram can help in brainstorming to identify possible cause of a problem and in sorting ideas into useful categories[12]. Fishbone analysis begins with a problem and the fishbone provides a template to separate and categorize the causes. Usually there are six categories, but the number can be change depending on the problem (Figure 3). This method allows problems to be analyze and, if it this used with employees, it gives everybody an insight into the problem so solutions can be develop collaboratively [13].

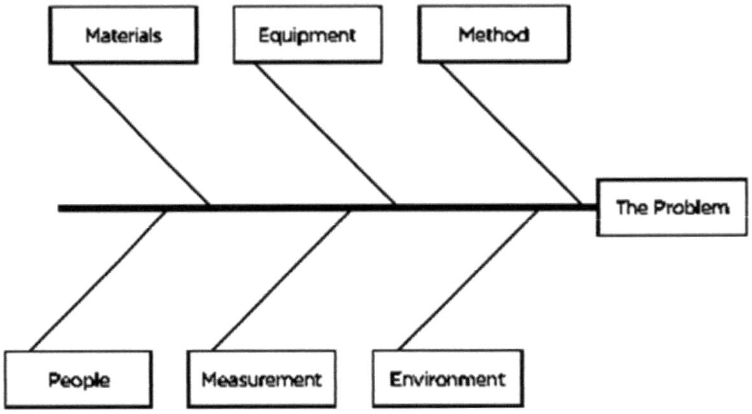

Figure 3. Typical fishbone analysis

Company in which employees are encouraged to evaluate practice, risk and mistakes when they occur tend to have a culture where root cause analysis or fishbone analysis is used. This helps to really understanding the cause of problems and clarify the issues.To support in identifying the root causes from all the ideas generated, observers consider a multi-voting technique such as having each team member identify the top three root causes. Behavior and feedback comment from employees are generated to establish the root –cause of problem in BBS appraisal.

2.4 Traffic Light Analysis

The Traffic Light Assessment (TLA) is a high-level indicator of the potential risks that can occur on government/division projects. According to Rochmoeljati [14], the achievement of TLA implementation expressed in three categories; green, yellow, and red refers to National Regulation of Minister of Labor Occupational Safety and Health No: PER.05/MEN/1996. The description for each category mentioned as follows:

I. Green
This indicatorstates that theimplementationwell done achieved. The range ofvaluesforthis indicatoris85% -100%.
II. Yellow
Thisindicatorstates that the value isclose to the target. The range ofvaluesfor theperformance indicatorsthis indicatoris 60% -84%.
III. Red
The red colorof thisindicatorstates that theimplementationis doneis below the targetthatshouldbe fixedas soon as possible. The range ofvaluesforthis indicatoris0% -59%.

To measure TLA's performance indicator, the raw data that obtained will converted into performance score and score of specific behavior calculated as follows:

Score = (Number if safe behavior/Total number of behaviors observed) x 100% (1)

3. Method

3.1 Data collection
As per the plan, the company implemented the BBS management system with the company's management team and writer was involved in development of the BBS technique analysis, which using TLA indicator. One division of the company that focuses on operational and maintenance work selected as participant in this study. A number of questionnaires survey were distributed to seek the current implementation of BBS and the barriers towards its implementation from safety officers, safety supervisors, project manager, engineers, contractor, sub-contractor and other managerial and technical staff. The entire participant from these division were asked to fill out the question but were not required to complete all questions and return these questionnaires as they were for their information and recording the explanation of how they need to act safely during conducting their tasks. Through questionnaires, encountered problem of work behavior will detected.

3.2 BBS Questionnaire
The company has sets the list of questionnaire previously, it is depend on what kind of work behavior that will be ask to the workers. In this study, the questionnaire are divided into 2 categories, first related with BBS checklist items and second is an essay to clarifying the behavior of the worker at work. Checklist of questionnaire distributed to the entire participant of the division. These steps, company's observer required to submit completed checklists and writer responsible for entering these results from the checklist into a computer data for compiling the result. A sample checklist items for BBS questionnaire that used in research shown in Table 1.

Table 1 .Checklist of behavior-based safety observation form

No	Point of Behavior	Behavior Description	At-risk/safe?
1	Body Position	Avoiding Line of Fire	
		Walking/Moving with Eyes on Path	
		Keeping Eyes on Task	
		Ascending/Descending	

No	Point of Behavior	Behavior Description	At-risk/safe?
2	Body Use	Lifting/Lowering/Pushing/Pulling Avoid Twisting Responding to Industrial Ergonomic Risks	
3	Tools & Equipment	Selecting and Using Tools/Equipment Using Guards/Barricades / Warning Devices	
4	Work Procedure	Job Preparation and JHA Doing-Hot Work Entering a Confined Space Communication to Co-Workers	
5	Work Area	Office Ergonomics Cleaning Up/Storing Tools and Equipment (Housekeeping) Working in a Well-Lighted Environment	
6	Office Ergonomics	Taking Rest Breaks Phone Posture, Back Support Shoulder Posture, Wrist and Arm Position Holding/Moving Mouse Recognizing and Reporting Discomfort	
7	Environmental Stewardship	Preventing Spills Preparing for Spill Clean-up Managing Waste	
8	Personal Protective Equipment	Protecting Head, Hearing, Respiration, Hands Protecting Eyes and Face Fall Protection Protective Clothing Protecting Feet	
9	Driving	Journey Planning Pre-Trip Inspection and Seat Belt Driving at Proper Speed Following Distance Changing Lanes	
10	Marine Operations	Preparing for a Vessel Trip Moving/Breaking Vessels Entering Inspections	

The second BBS questionnaire emphasize the question in detail why those bad behaviors occurred during work. This question focused to seek-out the reasons of the worker by explaining the case of bad behavior by giving feedback, suggestion etc. A sample of questionnaire that used shown in Table 2.

Table 2.BBS clarifier form/observation comments

Behavior Comment	Condition	Date	Location	Company (observed)	Observer
Feedback					
While					
Was					
Because					
Suggestion/try					

Agree behavior at
occurred?
At Risk behavior?

The Checklist of BBS observation form (Table 1) will converted to TLA indicator to determine what point of behavior categories that experienced in highest levels of risk. The highest-level point of identified work behavior will used to encounter the problem of improper work behavior by using fishbone diagram. Meanwhile BBS clarifier form (Table 2) are used to supporting and identifying the root-cause of a problem.

4. Result and Analysis

This section describes result of the BBS research that carried out at operation and maintenance division in oil and gas industry that located on Sumatra. An experiment conducted to measure the safety score of behavior of based safety at the company. Data collected over a period of 12 months for experimental observation. By used equation formula (1), the result of performance indicator for BBS questionnaire shown in Table 3.

Table 3. Result of performance indicator of BBS observation form

No	*Point of* Behavior	Percentage (%)
1	Body Position	75.42
2	Body Use	56.84
3	Tools & Equipment	81.94
4	Work Procedure	71.11
5	Work Area	85.55
6	Office Ergonomics	85.59
7	Environmental Stewardship	81.48
8	Personal Protective Equipment (PPE)	94.07
9	Driving	95.83
10	Marine Operations	95.83
	% Average	**82.37%**

Table 3 recapitulated that there are several points that classified in yellow indicator based on Traffic Analysis Assessment (TLA) and indicated as bad behavior at work. To ease the identification point, graphic of radar indicator presented as follows:

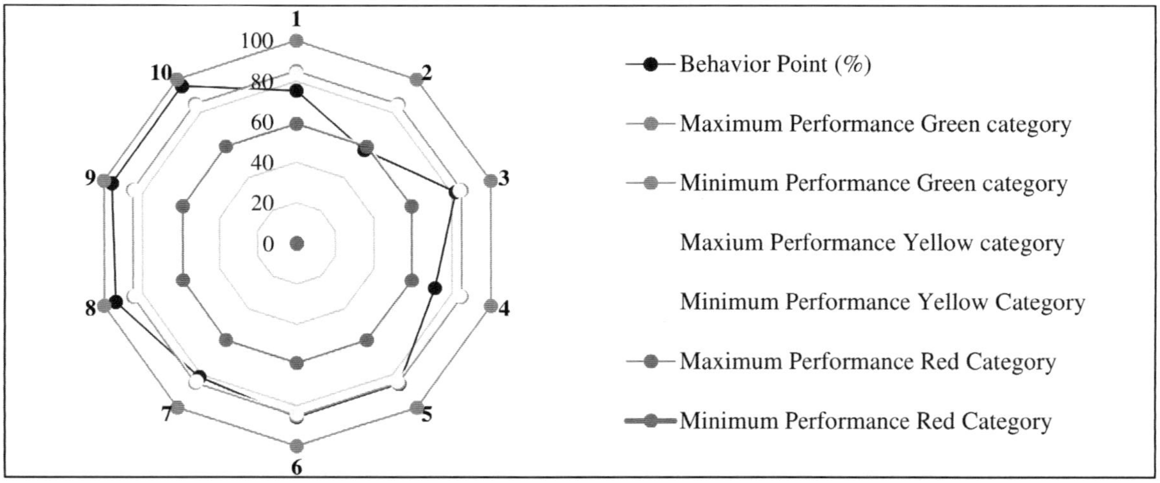

Figure 4. BBS result in radar indicator. Appointment of green line (85%-100%), yellow line (60%-84%) and red line (0%-59%) refers to National Regulation of Minister of Labor Occupational Safety and Health No: PER.05/MEN/1996

In behavior point of body position, body use, tools & equipment, work procedure, and environmental stewardship categories, measured scores of safety performance at this division were below 84%, which stationed in yellow and red boundary line (Table 3).In body position, ascending and descending are the most complained position by the worker. In tools and equipment, selection and its utilization are not being consideration at work. Bad behavior point also occur at work procedure, the worker did not fully give their attention to job preparation, job hazard analysis (JHA) and isolating energy (LOTO).

In environmental stewardship, the worker also ignoring to managing waste. However, the most critical point showed on point 2, which "body use" was 56.84%. This point categorized in dangerous behavior and stationed in red boundary line. Concerning with this behavior point, common workers in this division did not give full attention with the proper of body utilization like ignoring the movement of body twisting, lifting/lowering/pushing/pulling load etc. To identify what is the root of problem why those bad behavior happen, the use of fishbone diagram are providing as follows:

Five problem categories depicted on Figure 5. Policy, material, personnel, environment and method are figure-out to encounter the root of bad behavior that occur in the division. In sum, overall root cases that appear in the fishbone diagram concluded that the most problem occurs in human resources issues. This happens due to lack of awareness and commitment of the workers themselves. By overcoming the problems that described in fishbone diagram, improper/bad behavior at work can be reduce. Therefore, engagement from both employees and management system for continuous improvement in this division being critical point that must considered by the company.

5. Discussion

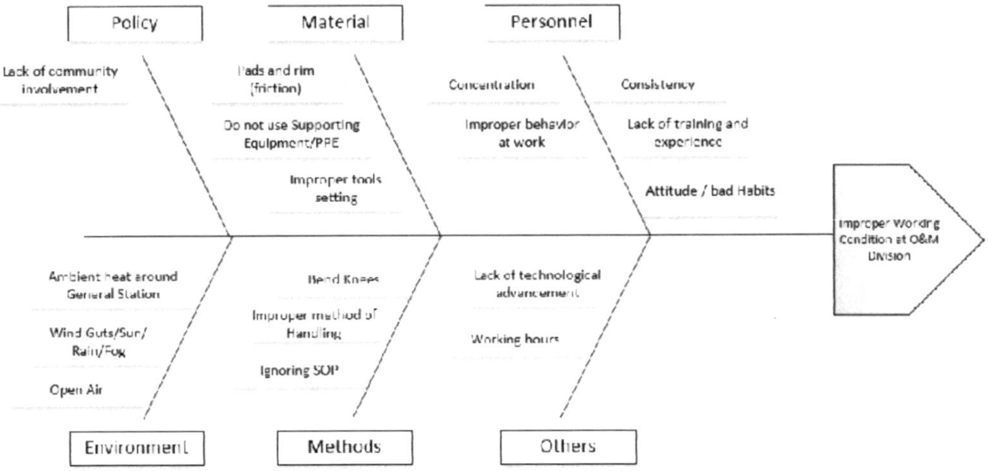

Figure 5. Fishbone diagram of improper work behavior

The concept of Behavior Based Safety (BBS) approach revealed as to change bad/unsafe behavior to safe behavior and to eliminate the occurring accident at workplace. The result of questionnaire based on 10 points of behavior related with BBS appraisal was 82.37%. Referring to the concept of Traffic Light Assessment (TLA), are know that overall work behavior in operation & maintenance division at the company is in yellow category. Unsafe work behavior that encountered in this study discovered in body position, body use, tool & equipment, work procedure and environmental stewardship. 5/10 point of work behavior inserted in *"quite safe"* category, which mean the company must conduct continuous controlling and regular evaluation for this division.

To deal with those five core issues, result that obtained from fishbone diagram also giving the same direction to encountered improver work condition/unsafe behavior. The problems encountered has been identified such as; less involvement of management, staff and workers, workers worried on blaming and penalties, lack of community involvement, improper method of handling, ignoring SOP and lack of commitment. As dynamic case problems of accident at work, the use of BBS study is not only for the company, but also should improve the wellbeing of the workers as they worked together, provided feedbacks and social praises, and motivated each other toward achieving the goal of reducing accidents[16].

The result clearly indicated that the BBS intervention useful to encountered bad behavior within the oil and gas industry. Intervention within BBS and evaluation based on fishbone diagram can used effectively to reduce work accidents that derived from bad behavior at work. Thus, behavior based approach can be implemented towards safety performance improvements in all industries including the oil and gas industry with full commitment and involvement at all levels. Although this paper fully focused on BBS appraisal, psychology study can include for further research.

Acknowledgment

Appreciation given to Mr. Kazuerman as supervisor for assisting in the data collection exercise. The authors thank the Islamic University of Indonesia for giving the opportunity to participate in this conference. The invaluable comments by the reviewers on this paper owned special thanks by the authors.

References

[1] Ridley, J., & Channing, J. (2008). *Safety at Work:Seventh Edition.* Jordan Hill, Oxford, UK, pp 48-49.

[2] Sawacha, E., Naoum, S., and Fong, Daniel. (1999). *Factors Affecting Safety Performance On Construction Sites.*International Journal of Project Management Vol. 17, No. 5, pp. 309-315

[3] Richard, P. (2015). *A Time of Risk and Opportunity – Challenges Facing the Oil and Gas Industry in 2015.* Retrieved on 25 September 2015 from https://www.blplaw.com/expert-legal-insights/articles/time-risk-opportunity-challenges-facing-oil-gas-industry-2015

[4] Trottman, M & Berzon, A. (2012). *With North Dakota's Job Boom, A Sharp Uptick in Work-Related Deaths* Retrieved on 25 September 2015 from http://blogs.wsj.com

[5] Chauhan, N. (2013). Safety and Health Management System in Oil and Gas Industry. Wipro Technologies, Sarjapur Road, Bangalore, India.

[6] Angkat S. 2008. *Analisa Upaya Pencegahan Kecelakaan Kerja pada Pekerja Bangunan Perusahaan X*, Tesis, Program Pasca Sarjana,Universitas Sumatera Utara, Medan.

[7] Al-Hemoud, A. M., & Al-Asfoor, M. M. (2006). *A Behavior Based Safety Approach at a Kuwait Research Institution. Journal of safety research, 37*(2), 201-206.

[8] Phillip, W. & William, L. (2003). *Selecting an Effective BBS Process.* Professional safety. pp 39-41

[9] J Agnew, J., & Ashworth, C. (2012). *Behavior-based safety: Setting the record straight.* Aubrey Daniels International Inc. http://aubreydaniels.com/pmezine/behavior-based-safety-setting-record-straigh

[10] Locke, E, A (1968). *Toward A Theory of Task Motivation and Incentives, Organizational Behavior and Human Performance.* (3) 157–189

[11] Zhang, M. & Fang, D. (2012), *Continuous Behavior-Based Safety Strategy for Persistent Safety Improvement in Construction Industry.* Tsinghua University, Beijing, 100084, China.

IOP Conf. Series: Materials Science and Engineering **105** (2016) 012033 doi:10.1088/1757-899X/105/1/012033

[12] Ishikawa, Kaoru (1986). *Guide of Quality Control.* Tokyo, Japan: Asian Productivity Organization

[13] Phillips, J. & Simmonds, L. (2013). *Change Management Tools Part 1: Using Fishbone Analysis to Investigate Problems.* Nursing Times; 109: 15: 18-20, www.nursingtimes.net

[14] Rochmoeljati. (2009). *Implementasi Program Keselamatan dan Kesehatan Kerja (K3) dan Perangkingan Hazard dengan Pendekatan Manajemen Resiko,* Jurnal of Universitas Pembangunan Nasional Veteran JawaTimur, Surabaya.

[15] Anonymous. 1996. *Peraturan Menteri Tenaga Kerja dan Transmigrasi Nomor KEP.05/MEN/1996.* Sistem Manajemen Keselamatan dan Kesehatan Kerja,Jakarta.

[16] Choudhry, R. (2014). *Behavior-Based Safety on construction sites: A Case Study.* Department of Civil and Environmental Engineering, College of Engineering, King Faisal University, Al-Hofuf, Al-Ahsa 31982, Saudi Arabia.

Author Biographies

Agus Mansur, S.T., M.Sc. is a senior lecturer of Industrial Engineering Department, Faculty of Industrial Technology, Islamic university of Indonesia. He finished his master degree at University of New South Wales, Australia. His teaching and interest are economics and finance, risk analysis and project management.His email address is agusmansur.am@gmail.com

Muhammad Ihsan Nasution is a final student of Industrial Engineering Department, Faculty of Industrial Technology, Islamic university of Indonesia. He is an assistant of ergonomic laboratory expert in product developmentand occupational safety analysis. His email address is m.ihsan17@yahoo.com

Work Physiology Evaluation of Laundry Workers

A D Sari[1], M R Suryoputro[2], M D Pramaningtyas[3], P S Putra[4] and S B Maulidyawati[5]
[1,2,4,5] Work System Design & Ergonomics Laboratory, Industrial Engineering Department, Faculty of Industrial Technology, Universitas Islam Indonesia, Yogyakarta, Indonesia
[3]Medical Department, Universitas Islam Indonesia

E-mail: amarria.dila@gmail.com[1], mragils@gmail.com[2], mirantidewip@gmail.com[3], perdanasuteja@gmail.com[4], sitibarirohm@gmail.com[5]

Abstract. This study aimed to assess cardiovascular strain during laundry operations in terms of physical workload, based on heart rate changes and level of pain complaints. Researchers measured resting and working heart rates and calculated cardiovascular load (%CVL), cardiovascular strain (%CVS), reserve heart rate (%RHR), energy expenditure, oxygen consumption, and measure level of pain complain in 6 laundry workers using Nordic Body Map questionnaire (NBM). Based on the result of %CVL and %CVS, the work in laundry was classified as acceptable level. Similarly, a high-level category was recorded for %RHR in moderate of energy expenditure. However, there are very pain level complain for hand using NBM questionnaire. Thus, there is a need to redesign the work content of equipment used and keep the physical workload in acceptable level, as this will increase their productivity and reduce their health risk.

1. Introduction

Ergonomics field of study includes Ergonomics of physical science, Cognitive Ergonomics, Ergonomics of Organization [1]. Physical ergonomics include human anatomy, anthropometry, work physiology and biomechanics. Work physiology approach is an aspect in industrial engineering that focuses on the metabolic requirements, the performance of the functions of the body and its components in the design of work and workplaces. Measurement of human physical activity aiming to do a job that can be measured by two criteria: operational criteria and physiology criteria [2]. Based on the physiology criteria, energy that is needed for human to work in the industry with 8 hours work should not exceed 30% - 40% of maximal aerobic capacity (VO2 max) of workers concerned. Physical workload that can exceed those limit will result in a number of adverse effects, for example: decrease of work performance, excessive fatigue, including the potential for injuries and accidents[2].

The laundry business developed in many towns where there are many students, workers and households using the laundry services to facilitate everyday activities [3]. Especially in Jogjakarta, the industry development is also due to tourism and education. The high human need for laundry services gives good impact to economic development and employment equity. Some work elements in the laundry job includes weighing and sorting dirty clothes, washing, drying, ironing and packing the clean clothes. Based on preliminary survey of Tarwaka et al [4], the work process of ironing is the most exhausting job among other work processes. This was due to the static work position of the worker at work, the work station and facilities did not correspond with the user's anthropometry, and the ambient temperatures are quite hot.This study aimed to assess cardiovascular strain during laundry operations in terms of physical workload, based on level of pain complaints and heart rate changes.

2. THEORITECAL REVIEW

2.1 Physical Workload

The workload is the number of activities that must be completed by a person or group of people during a specific time period under normal circumstances [5]. The workload arising from the interaction between the needs of the task / job, working conditions, skills, behavior, and perception of the operator. While capacity is the ability of man to do a job. The workload is the difference between workload demands of an assignment with a maximum capacity of a person's mental load in motivated condition [6]. The workload includes physical workload and mental workload[7]. Physical workload can be found in workers who are more physically utilize to complete their work such as lifting, carrying, and holding.

2.2 Measurement of Physical Workload

Physical work is something that can be done if a worker has energy, because it is useful to support muscle contraction. The result of extensive research on work physiology have shown that energy expenditure rate of a work is linearly related to the amount of oxygen consumed by the body and to heart rate [8].

2.2.1 Oxygen Consumption

The human body needs energy to sustain/maintain the basic functions of life eventhough no similar activities carried. The lowest energy expenditure that is needed to maintain the basic functions of life are called basal metabolism (basal metabolism) [8]. Each individual has different basal metabolic rate, some of the factors that influence these differences are gender, age, and weight.

Based on the explanation, physical workload can be grouped into several levels based on expended energy (energy expenditure). Several studies of work physiology showed that the energy expended to work directly proportional to the amount of oxygen consumption and heart rate [8]. Dr. Lucien Brouha hasclassification tables of workload in physiological reactions, to determineseverity of a job, as shown in Table 1.

Table1. Classification of Workload and Physiological Reaction

Work Level	Energy expenditure		Heart Rate	VO_2
	Kcal / min	Kcal / 8h	Rate / min	Liter / min
Unduly Heavy	>12.5	>6000	>175	**>2.5**
Very Heavy	10.0 – 12.5	4800 – 6000	150 – 175	**2.0 – 2.5**
Heavy	7.5 – 10.0	3600 – 4800	125 – 150	**1.5 –2.0**
Moderate	5.0 – 7.5	2400 – 3600	100 – 125	**1.0 – 1.5**
Light	2.5 – 5.0	1200 – 2400	60 – 100	**0.5 – 1.0**
Very Light	**< 2.5**	**< 1200**	**< 60**	**< 0.5**

2.2.2 Heart Rate

Heart rate, the number of heart beats per minute, is another commonly used physiological measure of physical workload. Heart rate is commonly used to estimate energy expenditure or physical strain during sports, work or daily activities [9]. Extensive research has shown that for moderate work, heart rate is linearly related to oxygen consumption [10]. Based on Rodahl (1989), an increase in the potential of the pulse of rest until maximum employment measured by HR reserve.

$$\% \text{ HR Reserve} = \frac{\text{HR}_\text{W} - \text{HR}_\text{R}}{\text{HR}_\text{Max} - \text{HR}_\text{R}} \text{ x } 100 \quad (1)$$

2.2.3 Physical Strain

The classification of workload based on the comparison between working heart rate and maximum heart rate that caused by cardiovascular load (%CVL) [11]. It can be stated as:

$$\%CVL = \frac{100\, x\, (HR_W - HR_R)}{(H_{max} - H_R)} \quad (2)$$

Furthermore, cardiovascular strain (CVS) was determined as the ratio of change in heart rates between heart rate of work and heart rate of rest to the initial (resting heart rate) as stated to the formula below [9].

$$CVS = 100 \left[\frac{HR_W - HR_R}{HR_R}\right] \quad (3)$$

Where HR_W= working heart rate and HR_R= resting heart rate, Astrand and Rodahl'scategories of work intensity were used to classify %CVS: light ($HR_W < 90$); moderate ($90 \leq HR_W < 110$); heavy ($110 \leq HR_W < 130$), very heavy ($130 \leq HR_W < 150$) and extremely heavy ($150 \leq HR_W < 170$). Consequently, %CVS was classified as follows: 0%–50% = acceptable, no action required; 51%–80% = moderate, action required within a few months; 81%–120% = high, action required within a few weeks; 121%–150% = very high, action required within a few days and 151%–180% = intolerable, action required immediately [10]

2.2.4 Nordic Body Map

Nordic Body Map is a subjective measurement tools like questionnaire used to determine the parts of muscle experiencing symptoms ranging from discomfort (mildly ill) to very pain [12], here are four indicators in the level of complaints that indicate the pain level of body; A indicates no pain felt; B indicates moderate pain; C indicates painful; D indicates very painful. This questionnaire using the image of the human body which is divided into 9 main body part that is the neck, shoulders, upper back, elbows, lower back, hips, knees and ankles. From 9 body parts are then broken down into 28 parts of the body.

3. METHOD

3.1. Subject and Object

The research was conducted among 6 healthy laundry workers in range of age between 25 to 40 years old who volunteering on research. Participants are female worker who work in average period of 8 hours a day. They worked in Cuisine Laundry, Sleman, Yogyakarta. All participants were engaged in operations that involved at work such as washing, material handling, ironing, cleaning and so on. In addition, the object of research is the heart rate of work and rest.

3.2 Research Procedure

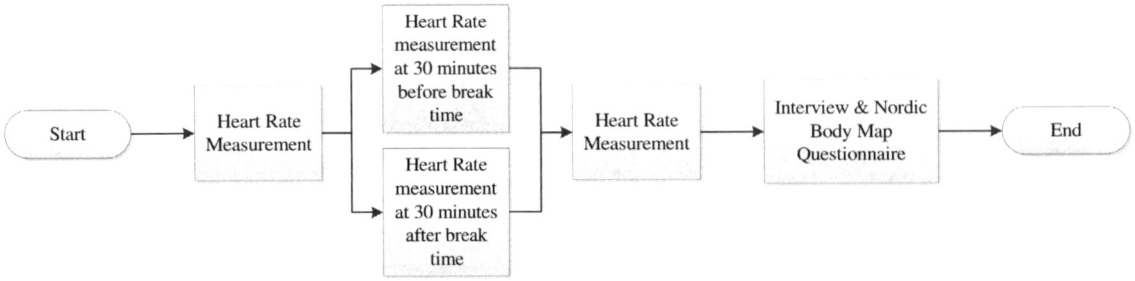

Figure 1. Research Procedure

Based on Figure 1, there are three times measurements to determine the value of working heart rate. The pulse rate measured by placed the second, third, and fourth fingers upper the radials artery, the examiner then count the pulse in one minute [13]. The value of working heart rate was measured at 30 minutes after the worker starts the job, 30 minutes before break time, and 30 minutes before the worker finish the job. Each measurement was conducted twice measurement for 1 minute for each and calculate the average value in order to get the final value. In addition, resting heart rate was measured at 30 minutes after break time while the worker does nothing. On the research process, the worker does the job naturally. Based on the result of working heart rate and resting heart rate, it can be calculated as cardiovascular load and strain in order to know the emergence level of heart, heart rate condition, energy expenditure and oxygen consumption needed by workers to do the job. In addition, the interview directly and fulfilling Nordic body map questionnaire was conducted at the end of session in order to identify the level of complaints for each part of body and its cause.

4. RESULT AND DISSCUSSION

Based on the data collected, it can be shown the result as following Table 2 that shows the basic statistics of range, mean, median and standard deviation value. The participants' mean age was 31.67 years (in range 29-38),mean of body height is 1.575 m (in range 1.53-1.62), mean of body mass is 53.33 kg (in range 47-56), while mean of body mass index (BMI) is 21.50 (in range 20.08-22.77). Normal BMI is 18.5–24.9, while 20.08–22.77 is regarded as normal [14], so none of the workers were underweight or more than slightly obese. In addition, mean of heart rate of rest is 90.33 bpm (in range 88.5-91.5 bpm), heart rate of work is 102.81 bpm (in range 99.17-105.84 bpm), and maximal heart rate is 168.33 bpm (in range 162-173 bpm).

The median of age is 30.5 years (standard deviation: 4.18), median of body height 1.58m (standard deviation: 0.034), median of body mass is 54.5 kg (standard deviation: 3.44) and body mass index' median is 21.96 bpm (standard deviation: 1.28). In addition, median for heart rate of rest is 90.25 bpm (standard deviation: 1.125), heart rate of work is 103.17 bpm (standard deviation: 2.57), and maximal heart rate is 169.5 bpm (standard deviation: 4.18).

Table 2. Data collected of laundry workers

Statistics	Age (years)	Body Height (m)	Body Massa (Kg)	BMI (Kg/m2)	HRrest (bpm)	HRwork (bpm)	HRmax (bpm)
Range	29 - 38	1.53 - 1.62	47 – 56	20 - 22.7	88.5 - 92	99.2 - 106	162 - 173
Mean	31.67	1.575	53.33	21.50	90.33	102.81	168.33
Median	30.5	1.58	54.5	21.96	90.25	103.17	169.5
StDev	4.18	0.034	3.44	1.28	1.125	2.57	4.18

The table 3, represents value of %CVL, %CVS, %RHR, Energy expenditure and Oxygen consumption.The mean value of %CVL is 15.99% (in range 11.32-23.6), it means the %CVL data was classified on the acceptable level or in range <30%. Then, value of %CVS is 13.69% (in range 9.8-19.6%) that can be classified on acceptable level or in range 0-50%(Astrond & Rodahl, 1986). On the other side, %RHR is 15.99% (in range 11.3-23.6%), energy expenditure is 5.292 Kcal/min (in range 4.99-5.58 Kcal/min), and oxygen consumption is 1.057 liter/min (in range 0.99-1.116 liter/min). In addition, the median value of %CVL is 15.61% (standard deviation: 4.55), %CVS is 13.69% (standard deviation: 3.76), %RHR is 15.61% (standard deviation: 4.55), energy expenditure is 5.32 Kcal/min (standard deviation: 0.24), and oxygen consumption is 1.063 liter/min(standard deviation: 0.0494).

Table 3. Cardiovascular load, strain, reserve heart rate, energy expenditure and oxygen consumption

Statistics	%CVL	%CVS	%RHR	Energy Expenditure (Kcal/min)	Oxygen Consumption (liter/min)
Range	11.32-23.6	9.8-19.6	11.3-23.6	4.99-5.58	0.99-1.116
Mean	15.99	13.81	15.99	5.292	1.057
Median	15.61	13.69	15.61	5.32	1.063
StDev	4.55	3.76	4.55	0.24	0.0494

The table4 represents the results of Nordic body map questionnaire in order to identify the level of complaint for each part of body that has complaint during completing the work. The research was collected data for 18 parts of body that has complained by respondents. The result was following the table below that shows the percentage of participants to fulfill the questionnaire.

Table 4. Results of Nordic body map questionnaire

No	Location	Level of Complaints (%)			
		A	B	C	D
1	Left shoulder	33.33	16.67	50.00	0
2	Right shoulder	33.33	16.67	50.00	0
3	Back	50.00	0	50.00	0
4	waist	50.00	16.67	33.33	0
5	left elbow	0	33.33	66.67	0
6	right elbow	0	33.33	66.67	0
7	Left lower arm	0	50.00	50.00	0
8	Right Lower arm	0	50.00	50.00	0
9	left wrist	33.33	33.33	33.33	0
10	right wrist	33.33	33.33	33.33	0
11	left hand	33.33	16.67	33.33	16.67
12	right hand	33.33	16.67	33.33	16.67
13	left knee	0	50.00	50.00	0
14	right knee	0	50.00	50.00	0
15	left ankle	83.33	0	16.67	0
16	right ankle	83.33	0	16.67	0
17	left foot	16.67	50.00	33.33	0
18	right foot	16.67	50.00	33.33	0

Based on the table 4 above, A indicates no pain felt; B indicates moderate pain; C indicates painful; D indicates very painful. It can be seen that the highest level of complaints in the very painful level is left hand (16.67%) and right elbow (16.67%). The complaint of moderate pain are highest rate on left lower arm, right lower arm, left knee, right knee, left foot and right foot for each part of body is in percentage 50% of respondent's complaints. In addition, part of body that listed in no pain complaint as shown on the table above.

5. CONCLUSION

The work physiology content of laundry workers usually was normal and not pose high risk to the cardiovascular strain of the workers, that were classified as acceptable level in range < 50%. On the other hand, there are two part of body which need attention that were right hand and left elbow. Therefore, it is necessary to redesign the working steps of laundry worker to reduce the level of painful and increase productivity of the workers. Two main improvements suggested based on this research are to minimize lifting activity and provision equipment which could make laundry workers activity easier. Thus, it would reduce pain of the workers during laundry activity.

REFERENCES

[1] Scott, P., Kogi, K., & McPhee, B 2010 Ergonomics guidelines for occupational health practice in industrially developing countries. *Darmstadt: International Ergonomics Association.*

[2] Sitohang, D. R., Winaningthias, M., &Iridiastadi, H 2010EvaluasiBebanFisiologisPadaIndustriManufaktur (IndustriPembuatanKomponenPesawatTerbang Dan Industri Sepatu). *J@ TI (TEKNIK INDUSTRI)*, ***5*(2)**, 119-126.

[3] Husain, S., Lisi, I. Z., &Susanti, E 2013TanggungJawabPelaku Usaha Kecil JasaCuciPakaian (Laundry)(Studi di KecamatanSamarindaUlu). *JURNAL BERAJA NITI*, *2*(10).

[4] Tarwaka, S. H., &Sudiajeng, L 2004Ergonomiuntukkeselamatan, kesehatankerjadanproduktivitas*UNIBA, Surakarta.*

[5] Herrianto2010 inGinting, S. *Pengaruh Beban Kerja terhadap Status Gizi Pekerja Peternakan Ayam di Desa Silebo-lebo Kabupaten Deli Serdang.* Medan: Politeknik Kemenkes. 2013.

[6] Henry R. Jex1988 in Akriyanto, L. A. *PengukurandanAnalisisBebanKerjaFisikdan Mental Pengemudi Bus AKDP Rute Solo-Semarang.* Surakarta: UniversitasMuhammadiyah Yogyakarta. 2014.

[7] Wignjosoebroto & Wiratno 2000 *Evaluasi Ergonomis Dalam Proses Perancangan Produk. Proceeding Seminar Nasional Ergonomi*, 1-4..

[8] Wicken, C. D. 2004An Introduction to Human Factors Engineering. New Jersey: Pretince Hall

[9] Ismaila S O, Oriolowo K T, and Akanbi O G 2013 "Cardiovascular Strain of Sawmill Workers in South-Western Nigeria."*International Journal of Occupational Safety and Ergonomics* **19**, no. 4 (2013): 607-611.

[10] Rodahl&Astrand 1986 Text Book of Work Physiology New York: McGraw-Hill Book Company

[11] Tarwaka, Solichul HA, and Sudiajeng L 2004 "Ergonomiuntukkeselamatan, kesehatankerjadanproduktivitas" *UNIBA, Surakarta*

[12] Corlett, E.N 1992 Static Muscle Loading and the Evaluation of Posture Edited by Wilson. J.R. &Corlett, E.N Evaluation of Human Work a Practical Ergonomics Methodology London :Tailor & Francis.

[13] Swartz, Mark H 1995 "Bukuajardiagnostikfisik." EGC

[14] Singh, D., Park, W., & Levy, M. (1-7). 2009. Obesity does not reduce maximum acceptable weights of life. ApplErgon

Red Dyeing Silk in Room Temperature Using Fermented Rice (*Oryza Sativa*) and Yam Tuber (*Pachyrhizus erosus*) by *Monascus purpureus* as an Alternatives of an Eco-friendly Textile Dyes

I N Mauliza[1,2], Mardiyati[1], and B Sunendar[3]

[1]Material and Science Engineering, ITB, Bandung, Indonesia
[2]Textile Chemistry, Politeknik STTT, Bandung, Indonesia
[3]Physics Engineering, ITB, Bandung, Indonesia

E-mail: mardiyati@material.itb.ac.id

Abstract. Potential dyes to be developed derived from fermentation by *Monascus purpureus*. As a staple food, rice can be replace to the yam tuber lees as a substrates. The purpose of this study was to compare the dyeability between fermented rice and yam tuber by *Monascus purpureus* on silk fabrics at the room temperature in any different pH of dyebath. *Monascus purpureus* first cultured on Potato Dextrose Agar for 7 days. Yam tuber peeled, grated and squeezed. The material is taken from yam tuber lees, then inoculated with *Monascus purpureus* for 14 days until an uniform red color obtained. The substrate is dried and then characterized by UV-Vis spectrophotometry and Fourier Transform Infrared Spectroscopy (FTIR). Dyes obtained from fermented rice and yam tuber then extracted and used for dyeing silk at room temperature with various pH of the dyebath. Results showed that dyeing silk with fermented yam tuber having the same color characteristics as fermented rice. The optimum color absorption at a wavelength of 520 nm for both, except on the results of dyeing using fermented yam tuber extract with pH 6 and pH 7. The maximum absorption is achieved at pH 3 with values dyeing K/S 5.840. Color fastness to rubbing are excellent (5 point) in dry rub, while the wet rub still good at the range of 4 to 4/5.
Keywords—*Monascus*; rice; yam tuber; fermentation; dyeing; silk

1. Introduction

Red is one of the primary colors that are rarely obtained from the dye derived from natural materials. There are 32 natural red dyes listed in Colour Index, they are extracted from the roots or bark of plants, or stands in the bodies of insects such as cochineal. However, the sources of these dyes are limited. Natural red dyes have been of significant and importance [1]. This resources are difficult to cultivate in large quantities. The difficulties arise mainly in terms of area planted, harvesting time, and degree of uniformity both of these natural ingredients when it grown in different environments [2].

IOP Conf. Series: Materials Science and Engineering **105** (2016) 012035 doi:10.1088/1757-899X/105/1/012035

One of the challenges in the production of natural dyes is the quantity of dye produced, especially during the extraction process. *Monascus purpureus* is one of biological source which has the ability to produce a red color during metabolism. Its small size and easily cultivated make this fungal has the potential to improve pigment production yields [3]. Fungi have the potential for use in production processes that are themselves less polluting than traditional chemical processes [4]. During this time, the use of dye from fermented by *Monascus purpureus* dominated for food coloration. *Monascus purpureus* able to produce a red color when grown on media which has a wide range of carbon sources, not only in the form of starch, but may include glucose, fructose, cellulose and others. The red color is generated not only from carbon but nitrogen is needed as a source of food.

Media that already in place to the growth of industrialization *Monascus purpureus* is rice. This reality is not in accordance with the main function of rice in human life. The function of rice as a staple food is the limiting factor is significant in industrialization *Monascus* dye. Therefore, we have to find another media that can be used as a growing medium of *Monascus purpureus*. Criteria required as a growing medium *Monascus* which have carbon and nitrogen sources are sufficient.

Almost all natural dyes have low affinity at room temperature, but in the case of natural dyes used in this study it does not apply. Therefore, the purpose of this study was to compare the dyeability between fermented rice and yam tuber by *Monascus purpureus* on silk fabrics at the room temperature in any different pH of dyebath.

2. Literary Review

2.1. Monascus purpureus

Monascus species produce orange, water-insoluble pigments such as monascorubrin and rubropunctatin. These well-characterized compounds can be converted to high-purity red, water-soluble pigments by reaction with amino acids, yielding monascorubramine and rubropunctamine. This red pigments commonly use as food colorants [4].

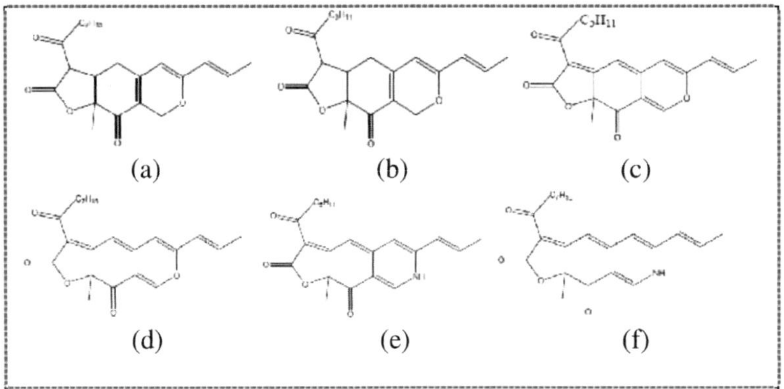

Figure 1. Chemical Structure of *Monascus* pigments, (a,b) yellow pigments, (c,d) orange pigments, (e,f) red pigments

Parameters that play a role in the production of the dye is nutrient content, pH, inoculum size, temperature, and moisture content [5]. Substrates that have the potential to produce pigment during fermentation, especially that contain starch (amylose and amylopectin), dextrin, glucose, maltose, galactose, fructose as a carbon source, and a protein as a source of nitrogen. The use of rice as a fermentation medium becomes significant constraints when it is connected to human consumption. Therefore, some of the research that has been developed is the search substrate capable of producing fermentation product quality without having to use a staple food. Media substrate selected in the present study is the yam tuber lees. Yam tuber lees provide as a substrate because it is not a staple food, but has a carbon content and nitrogen.

Production of dyes by fungi is very advantageous because it can grow quickly, so productivity is high. In addition, fermentation methods which can be performed on solid media or better known as solid state fermentation will produce a dye with a large amount but the cost of relatively lower than in the submerged media [5].

2.2. Dyeing silk with natural dyes

There are many types of fibers and substrates used in the textile industry. The dyeability of a substrate varies according to the material properties, such as surface properties, ionic character, roughness, structure, penetration and swelling properties [2].

Textile coloration using red dye from Monascus has been studied previously in cotton and wool fibers. Coloration on cotton fiber yield a pale red color [6,7,8]. The kinetics and thermodynamics of monascus red dyeing on wool fiber also has been studied [9].

Silk is a protein-based fiber produced as a continuous filament by the larva from a variety of insects and spiders. Silk produced in the form of two filaments consisting fibroin and glued together by sericin. Fibroin consists of many amino acids linked together [10]. Under acid conditions amino groups in the protein fibers are protonated to form NH3+ groups. In this form, they are able to attract dyes containing acid anions. Natural dyes from many sources have good affinity for protein fibers and often in conjunction with mordants [11].The existence of both these functional groups result the ability of silk to hold an ionic bond with the dye. Fastness tests are important tools for assessing the quality and the stability of dyeing.

3. Materials and Methods

The material used was Monascus purpureus obtained from the Laboratory of Microbiology, Chemical Engineering, ITB. Rice and yam obtained from traditional markets in East Bandung area. Research took place in the laboratory of chemical analysis and laboratory of textile physical chemistry, Politeknik STTT, Bandung.

1.1 Monascus Purpureus

Monascus purpureus cultures obtained from the microbiology laboratory of Chemical Engineering, Institut Teknologi Bandung. Culture of *Monascus purpureus* then maintained on Potato Dextrose Agar (PDA) medium.

Figure 2. *Monascus purpureus* culture

1.2 Inoculation Process

Yam tuber and rice washed and dried, then crushed using a blender. Yam tuber smoothed then squeezed and taken the dregs. Yam tuber lees is dried in an oven at 50 °C for 2 days. 40 grams yam tuber lees and rice put into the different Erlenmeyer, then moisture content was adjusted to 50%. Yam tuber lees and rice then sterilized at 121oC for 20 minutes and cooled at room temperature. Both media were inoculated with Monascus purpureus suspension to form a permanent red color on a substrate. Erlenmeyer shaken every day. The fermented solid was dried for 48 hours at 50°C in the oven. The dried fermented solids then grind by using blender.

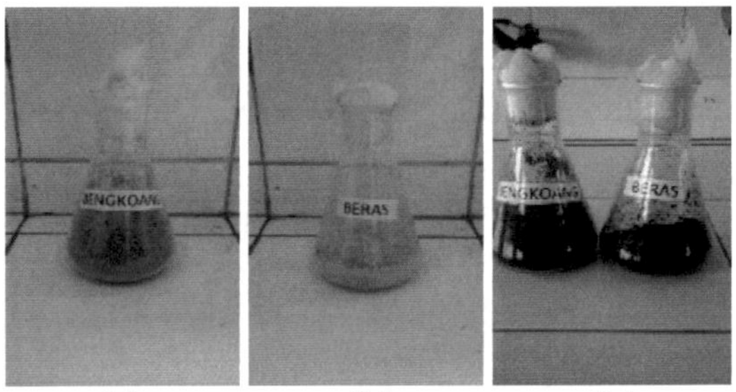

(a) (b) (c)

Figure 3. *Monascus purpureus* culture (a) yam tuber lees before fermentation (b) rice before fermentation (c) after fermentation

1.3 Dyes Characterization and Measurement

Each 1 gram of fermented yam tuber and rice mixed with 95% ethanol as much as 5 ml. The mixture is then rotated at 200 rpm for 1 hour, then filtered. Filtrates diluted with dilution factor 20. The optical density (OD) measured using Genesys UV-Vis Spectrophotometer at 400, 470, and 500 nm.

$$Unit\ OD = \frac{OD\ x\ solvent\ volume\ x\ dilution\ factor}{sample\ mass\ (g)} \quad (1)$$

Monascus dyes measured by using 25 ml extraction from previous dyes characterization. Monascus dyes extract dryed at 80°C and solid content remained measured by analytical balance.

$$Dyes\ concentration = \frac{solid\ content\ (g)}{volume\ of\ extract\ (L)} \quad (2)$$

1.4 Analitical of Functional Group using Fourier Transform Infra Red Spectroscopy

FTIR spectroscopy testing done at the Laboratory of Analytical Chemistry ITB. Measurements carried out by a laboratory technician at the Laboratory of Analytical Chemistry, ITB.

1.5 Dyeing Process

4 grams of Monascus dyes from fermented rice and yam tuber extracted using water until the last extract obtained colorless solution. Silk was dyed with Monascus dyes extract in the range of pH 2-7 at intervals of 1 and liquor ratio 1 : 100. Dyeing process took place at room temperature using HT dyeing machine. After dyeing process, samples were washed with detergents then rinsed with cold water.

Figure 4. *Dyeing process*

1.6 Color Measurement

Dyed samples was measured using Premier scan SS 6200 Spectrophotometer. The data generated in the form of reflectance, K/S, and color coordinates refers to the CIE Lab. K/S expressed the value of color strength and determined using Kubelka Munk equation below.

$$K/S_{of\ Dyes} = K/S_{after\ dye} - K/S_{before\ dye} \quad (3)$$

1.7 Fastness Properties

Fastness properties of silk dyed tested refers to SNI standard methods, "SNI 0288:2008 - Kain - Cara uji tahan luntur warna – Gosokan" [12]. Color fastness to rubbing using a crockmeter then the discoloration evaluated by staining scale.

4. Result and Discussion

4.1. Dyes Characterization and Measurements

The optical density of dyes from Monascus through fermentation on yam tuber and rice can be seen at the Table 1.

Table 1. Optical Density of Yam Tuber and Rice Fermented

Wavelength	500
Yam tuber lees	4.66
Rice	22

Optical density of each fermentation medium extract showed dyes production. The higher the optical density, means higher dyes production. The concentration of the dye contained in each of the substrate measured by the following values. At 500 nm the red color substances formed at the end of the growth phase of the Monascus.

Table 2. Dyes Concentration in Yam Tuber Lees and Rice Fermented

Media	Concentration (%)
Yam tuber lees	9.24
Rice	43

Yield of Monascus dyes on fermented rice greater than fermented yam tuber lees. The substrate from yam tuber lees is a starch residual waste, so that the potential dye formed lower, due to nutritional deficiencies. Nonetheless, yam tuber still potential as a medium and capable of producing Monascus dyes.

4.2. Analitical of Functional Group using Fourier Transform Infra Red Spectroscopy

Infrared spectra is a simple absorption spectra that can identify functional groups. Test sample analyzed shot through the infrared radiation that whenever energy is equivalent to a particular and specific wavelength is absorbed, then the intensity of the radiation that reaches the detector will decrease, and then recorded in the spectrum. Measurement of infrared spectra were recorded using a frequency wavelengths (cm-1) which is an inversion of the wavelength in centimeters (500-4000 cm-1). The energy required to excite vibrations of bond indicated by wavelength or frequency.

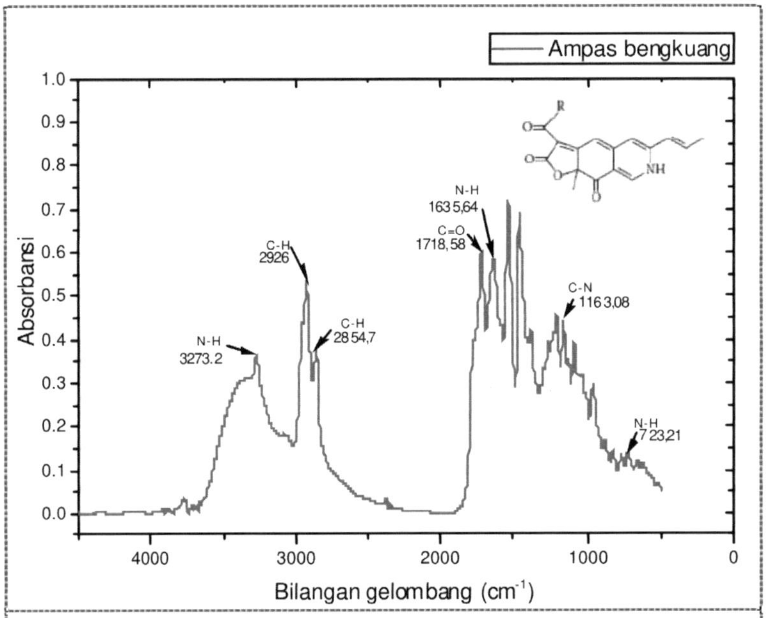

Figure 5. *FTIR spectrum of fermented yam tuber lees*

Based on the results of the FTIR spectra, both of extracts from fermented bengkoang and rice can be identified as Monascus dyes. The extracts has an aromatic compound with a ketone group, carboxylic, amines and other groups that are identical to the criteria of Monascus dyes."

4.3. Dyeing Properties

The dyeing process was performed at room temperature for 45 minutes with pH ranges 2-7. Spectrophotometric measurements showed that the maximum wavelength which is owned by the results of dyeing both types of dye is in the area of 520 nm, except for yam tuber extract pH 6 and 7 shifts to the maximum wavelength 510 nm as shown at Fig.6 and Fig. 7.

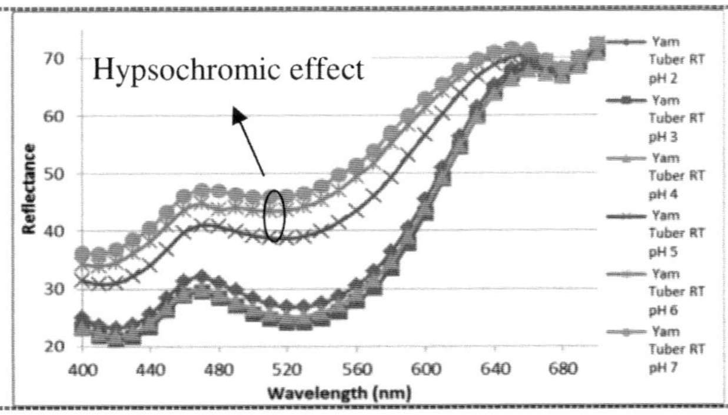

Figure 6. *Reflectance of each silk fibre using fermented yam tuber lees extract*

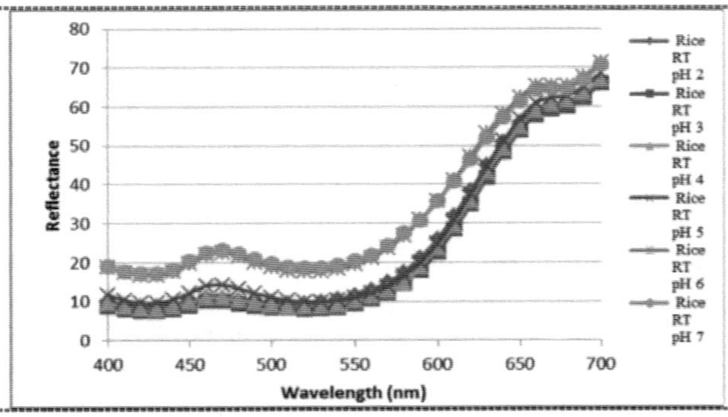

Figure 7. *Reflectance of each silk fibre using fermented rice extract*

Comparison of the results of dyeing silk using fermented rice extract and fermented yam tuber extract can be seen in Figure 8. The image shown that silk dyed by fermented rice has a darker color than fermented yam tuber extract.

Differences of dyebath pH causes a difference of lightness and hue of samples. Lightness and hue of each sample given in Table 3 and Table 4. While L* represents lightness, lower lightness value represent higher color yield. a* and b* represent of hue. a* positive represent redder, a* negative represent greenish, b* positive represent yellowish, and b* negative represent bluish.

Figure 8. *Comparison of red silk dyed using fermented rice extract (above) and fermented yam tuber extract (below)*

Table 3. Color coordinates of fermented yam tuber lees

Media	L*	a*	b*
Yam tuber lees pH 2	65.344	18.806	9.272
Yam tuber lees pH 3	63.239	20.598	9.383
Yam tuber lees pH 4	63.927	19.747	9.821
Yam tuber lees pH 5	73.184	12.562	11.047
Yam tuber lees pH 6	76.215	10.535	11.643
Yam tuber lees pH 7	77.943	9.306	10.729

Table 4. Color coordinates of fermented rice

Media	L*	a*	b*
Rice pH 2	46.781	30.575	17.041
Rice pH 3	43.538	32.131	15.239
Rice pH 4	45.677	30.594	14.946
Rice pH 5	47.583	29.767	11.048
Rice pH 6	57.131	24.215	10.257
Rice pH 7	57.416	22.36	10.082

Based on lightness values in Tables 3 and 4 shows that the lightness obtained on silk dyed with fermented rice lower than bengkoang fermentation. Hue of silk dyed was seen that all the results of dyeing using the two extracts have a yellowish red color direction. From the color coordinates listed and maximum absorption at wavelength area of 510 – 520 nm, very obvious that the dyes are red. Effect of pH during the dyeing process can be seen in Fig. 8. In this figure shows that the color strength increasing in the acidic condition to a pH of 3, whereas at pH 2 color strength declined again. Silk dyed with fermented yam tuber extracts at pH 6 and 7 were not included in the chart, because of the hypsochromic effect causing discoloration. Wavelength shift that occurs at pH 6 and 7 resulted into different colors produced by the other sample. This is what underlies the second sample can not be compared in terms of color strength. Color strength can only compared to the conditions the same maximum wavelength

Table 5. Color coordinates of fermented rice

K/S After dyeing			**K/S of Dyes**		
Wave length (nm)	510	520	Wave length (nm)	510	520
Rice RT pH 2	5.185	5.190	Rice RT pH 2	4.831	4.840
Rice RT pH 3	5.879	6.190	Rice RT pH 3	5.525	5.840
Rice RT pH 4	5.623	5.667	Rice RT pH 4	5.270	5.318
Rice RT pH 5	4.785	4.984	Rice RT pH 5	4.431	4.635
Rice RT pH 6	2.694	2.752	Rice RT pH 6	2.341	2.402
Rice RT pH 7	2.547	2.594	Rice RT pH 7	2.194	2.245

Table 6. Color coordinates of fermented rice

K/S After dyeing			**K/S of Dyes**		
Wave length (nm)	510	520	Wave length (nm)	510	520
Yam tuber RT pH 2	1.686	1.731	Yam tuber RT pH 2	1.332	1.381
Yam tuber RT pH 3	1.903	1.959	Yam tuber RT pH 3	1.550	1.610
Yam tuber RT pH 4	1.824	1.865	Yam tuber RT pH 4	1.471	1.515
Yam tuber RT pH 5	1.095	1.100	Yam tuber RT pH 5	0.742	0.751
Yam tuber RT pH 6	0.932	0.930	Yam tuber RT pH 6	0.579	0.580
Yam tuber RT pH 7	0.861	0.857	Yam tuber RT pH 7	0.507	0.508

Improved results of color strength dyeing extracts fermented yam tuber and rice occurs due to the acidic conditions an increase in the number of positive charge of the fibers. Under acid conditions amino groups in the protein fibers are protonated to form NH_3^+ groups. Silk which has the amine group ionized to form positive charges that interact via ionic bonds with the negatively charged dye.

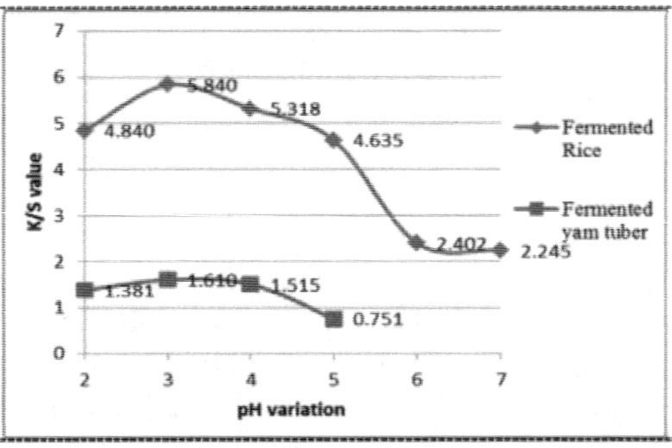

Figure 9. *Color strength of silk on various dye bath pH condition*

4.4. Fastness Properties

Color fastness to rubbing is done in dry and wet conditions to determine the response of individual fibers to rub against different conditions. Staining scale for each sample are given in Table 7and Table 8.

Table 7. Color fastness to rubbing of fermented rice

	Color fastness to rubbing	
	Dry	Wet
Rice pH 2	5	4/5
Rice pH 3	5	4/5
Rice pH 4	5	4/5
Rice pH 5	5	4
Rice pH 6	5	4
Rice pH 7	5	4

Table 8. Color fastness to rubbing of fermented yam tuber lees

	Color fastness to rubbing	
	Dry	Wet
Yam Tuber pH 2	5	4
Yam Tuber pH 3	5	4
Yam Tuber pH 4	5	4
Yam Tuber pH 5	5	4/5
Yam Tuber pH 6	5	4/5
Yam Tuber pH 7	5	4/5

Table 7 and Table 8 shows that the results of color fastness to rubbing in dyeing silk with fermented rice and yam tuber in dry conditions does not give the staining of the upholstery (staining scale value 5-excellent), while the results of color fastness to rubbing in dyeing silk with fermented rice and yam tuber in wet conditions give a little staining on upholstery (staining scale value 4-good to 4/5-very good). It is informed that the dyes used to color silk has excellent ability to bind to the fibers. Fastness to rubbing lower in wet conditions, it occur because the dye is applied to the silk has a solubility in water, so that when in contact with water, then the abrasion happened on the surface of the fiber, the dye is initially retained in the fiber migration to the surface of the upholstery.

5. Conclusion

In this research, the dyeing properties of fermented rice and yam tuber lees colorants were studied. Dyeing silk with fermented yam tuber lees having color characteristics like rice fermented rice. The optimum color absorption at a wavelength of 520 nm for both, except on the results of dyeing using yam tuber fermented extract with pH 6 and pH 7. The maximum absorption is achieved at pH 3 with values dyeing K/S 5.840. Color fastness to rubbing are excellent (5 point) in dry rub, while the wet rub still good at the range of 4 to 4/5. Based on this study, the yam tuber can be used as an alternative substrate instead of rice in the production of natural dyes through fermentation by Monascus purpureus.

6. Acknowledgement
We would like to thank the Politeknik STTT Bandung (Sekolah Tinggi Teknologi Tekstil) for license usage of Analytical Instrumentation, for Microbiology Laboratory, Chemical Engineering, ITB for the provider of Monascus purpureus.ity and for Analytical Chemistry Laboratory, Chemistry Department, ITB.

7. References
[1] Patel, B.H. "Natural Dyes". in Handbook of Textile and industrial Dyeing, Vol. 1 : Principles, Processes and types of dyes. 2011.

[2] Mussak, Rita and Thomas Bechtold. "Natural Colorants in Textile Dyeing". In Handbook of Natural Colorants. John Wiley and sons Ltd. West Sussex. 2009

[3] De Santis D. et al. (2005) Assessment of The Dyeing Properties of Pigments from Monascus purpureus. *J. Chem Technol Biotechnol.2005. pp 0268-2575*

[4] Horgan, Karina A and Richard A. Murphy. "Pharmaceutical and Chemical Commodities from Fungi". in Fungi : Biology and Applications, 2nd Edition. Wiley-Blackwell. 2011

[5] Shazwani, Nur. 'The Optimization of Red Pigment by Monascus purpureus FTC 5356 in Solid State Fermentation" Universiti Malaysia Pahang. 2012.

[6] Velmurugan P. et.al. (2009). Assesment of The Dyeing Properties of Pigments from Five Fungi and Anti-Bacterial Activity of Dyed Cotton Fabric and Leather. *Coloration Technology, 125, 334-341*

[7] Velmurugan, P, et al. (2010). Dyeing of Cotton Yarn with Five Water Soluble Fungal Pigments Obtained from Five Fungi, *Fibers and Polymers, Vol.11, No.4, 598-605.*

[8] *Velmurugan P.* Et al (2012). The Use of Cochineal and Monascus purpureus as Dyes for Cotton Fabric. *Coloration Technology, 129, 246-251*

[9] Shen, Jiajia and Xuming Chen. "Kinetics and Thermodynamics Studies of Monascus Red Dyeing on Wool". In Applied Mechanics and Materials, Vols. 395-396, pp 565-570. 2013

[10] Clark, M. "Fundamental Principles of Dyeing". In Handbook of Textile and industrial Dyeing, Vol. 1 : Principles, Processes and types of dyes. 2011.

[11] Needles, Howard L. "Textile Fibers, Dyes, Finishes, and Processes". Noyes Publications, USA. 1986

[12] SNI 0288:2008 - Kain - Cara uji tahan luntur warna – Gosokan

[13] Broadbent, Arthur D. (2001). *Basic Principles of Textile Coloration*. Canada : Society of Dyers and Colourists

[14] Feng, Yanli, et al. (2012) Monascus Pigments : Mini Review, *Appl Microbiol Biotechnol 96:1421–1440*, Springer-Verlag Berlin Heidelberg

[15] Pattanagul, P, et al. (2007) Review of Angkak Production (Monascus Purpureus). *Chiang Mai J Sci, 34(3), 319-328.*

[16] Jung, H., Y. Kim and C.S. Shin (2003). Color Characteristics of Monascus pigments derived by Fermentation with Various Amino Acids. *Journal of Agricultural and Food Chemistry. 51, 1302-1306.*

[17] Velmurugan P.et.al (2011). Monascus Pigment Production by Solid – State Fermentation with Corn Cob Substrate, *Journal of Bioscience and Bioengineering Vol. 112 No. 6. 590 – 594.*

IOP Conf. Series: Materials Science and Engineering **105** (2016) 012036 doi:10.1088/1757-899X/105/1/012036

Design of *Lesehan* Chair by Using Kansei Engineering Method And Anthropometry Approach

A T Pambudi[1], M R Suryoputro[2], A D Sari[3] and R D Kurnia[4]
[1,2,3,4] Program Studi Teknik Industri, Fakultas Teknologi Industri,
Universitas Islam Indonesia, Jalan Kaliurang Km 14.5 Yogyakarta, Indonesia

E-mail: triopambudi10@gmail.com, mragils@gmail.com, amarria.dila@gmail.com, ratih.dianingtyas@gmail.com

Abstract. Special Region of Yogyakarta (DIY) is known as city for academic. Many people come to get some education in college. They live in boarding house with some supporting facilities. The most common facilities is low table which lead students have to sit on the floor while studying on table which could cause higher risk of back pain and musculoskeletal disorder. To identify the solution to reduce back pain and musculoskeletal risk, it is needed to design a lesehan chair which also appropriate to customer needs. Kansei engineering method was used with a total of 30 respondents participated, 15 kansei words collected, and 12 kansei words selected by doing validation and reliability test. The result of this study showed that quality, aesthetics, and comfort level influence the design of lesehan chair. A design of lesehan chair was created by considering the suitable concept and merging it with the physical design and its anthropometry measurement. In this case, marginal homogeneity test is needed to identify the differences between each kansei words attribute and the design or product recommendation. The marginal homogeneity test results show that the design and product recommendation has fulfilled customer's desires and needs. For further research, it is needed to analyse and evaluate the posture of lesehan chair users in order to develop and improve its performance.

1. Introduction

Special Region of Yogyakarta (DIY) is one of the most visited city in Indonesia. Not only known with its tourism, Yogyakarta also known as an academic city. Many people come to Yogyakarta not only for vacation but also looking for education in the college. Many students spend their time not only in the college but also in their boarding house. By living in boarding house, they must have some supporting facilities such as chair and table to help them doing their homework, cupboard and etc. Based on the observation, students' table in generally have lower measure that required students sit on the floor or lesehan chair without buffer seats that has been adjusted for a sitting position on the floor. The students should use seat that has been adjusted for a table with normal sitting position. Therefore, it is needed to design a chair with crossed-legged sitting position or lesehan chair in order to reduce the complaints of back pain and the risk of musculoskeletal disorder. Incorrect sitting position can cause a disease of the backbone and if it is not handled, the effect will be felt in the future [1]. Another impact from incorrect sitting position is musculoskeletal disorder, even an injury or a work accident [2]. And to avoid all of the risks, people should do the correct sitting by sitting upright with a straight back and clothes to the back [1].

A method of product developments that can be used to design the lesehan chair is Kansei Engineering. This method purposed to know what customer needs based on the feeling of psychological, which could help to make it easy for researchers in designing the product. Kansei Engineering is a method to translate a feeling into the parameters of the product [3].
This research aims to design a lesehan chair that compliance with the customer needs and can reduce the risk injury of the backbone. The proposed design of lesehan chair has to be in accordance with the expectation of the customer needs, comfortable and safe to use.

2. Theoretical Review

Based on the previous researches, there were many researches that are related to the design of the seat or chair using Kansei Engineering but a research related to ergonomic lesehan chair still have not found despite the fact that the lesehan chair already sold in the market yet it is still unknown whether it has met the criteria of ergonomics and the needs of consumers.
Archam et al., [4] has done a research related to Kansei Engineering which is integrated to Structural Equation Modeling (SEM) to improve the product quality of local shampoo in Malang. The Kansei Engineering method was expected to be responsive to the consumer's psychological feeling of the shampoo product. An then using the SEM method to help researchers in calculating the value of the accuracy that could translate the gap between consumer responses and shampoo's product specification and get variables that affect the quality of shampoo.
Application of Kansei Engineering and anthropometric method had also been done by Marlyana [5], the research is related to the selection of internet café's room design. This study contains the design of the facility of internet cafes (a commercial public internet service cafe). The problem of this research is perceived by the customer complaints related café facilities such as the chair is not comfortable, the table is too low, and low bulkhead. By using the method of Kansei Engineering and anthropometry approach, it is expected to obtain the desired design of the cafe facilities in accordance with the psychological feeling of the cafe customers..
Restantin [6] also has done a research that integrated ergonomics, value engineering, and kansei engineering. This research was carried out to design a portable dining table and chair which can reduce the fatigue felt by visitors while enjoying the beauty and the culinary of Kuwaru beach. Most portable desk and chair on the market are made only on the basis of common functions and there have been no anthropometric approaches at the time to make the portable desk chair. The research was using the concept of Kansei Engineering in order to get the design of portable dining table and chairs which comfortable and practical by developing the product based on the imagination of the consumers.

2.1. Kansei Engineering

Kansei Engineering is a technology that translates a feeling and customers view about a product into elements of design, or can be defined as an ergonomic consumer-oriented technology for product design [3]. Kansei Engineering was introduced in Japan to incorporate emotional factors (aesthetics or "kansei") on product design. Designs that use the mental and emotional satisfaction keep growing and developed with the help of manufacturing companies in the differentiation and positioning of products [7]. There are some types of developed Kansei Engineering Method [3] :

- Type 1 : Classification of Categories
  This method spells out a product concept that are targeted into more detail concept, and when it expends to some order, it will be translated into physical parameter of the design.
- Type 2 : Kansei Engineering System
  This type is almost same as type 1, the difference lies in the changes of kansei concept into the physical characteristics. This method is a technique to translate the image of a product in customer's memory into a real product element design with system helps.
- Type 3 : Hybrid Kansei Engineering

This type can be used to predict kansei from the wealth of a products. The difference of this type are its mathematics model and the relations of input to output which is done by searching the value of the coefficients.

- Type 4 : Virtual Kansei Engineering
  This type gives real products presentation by combining the representation with the fact. This can be done with a system of data collection standards.
- Type 5 : Kansei Quality Management
  This type is the application of quality management which starts from customers kansei in order to maximize customer's satisfaction.

2.2. Ergonomics
Ergonomics is a branch of science that is concerned with the achievement of an optimal relationship between the worker and the work environment [8]. Ergonomics is also a discipline concerned with the understanding of interactions among humans and other elements of the system, and the profession that applies theory, principles, data and methods to design to optimize human well-being and overall system performance [9]. Ergonomics discipline in particular will learn the limitations of the human ability to interact with technology and products. It departs from the fact that human beings have limits, both short term and long term when they were facing the devices [10].

2.3. Anthropometry
Anthropometry is the science of measuring the dimensions of the human body, anthropometric data are used as a guideline for the design height, the space, the grip and the space of the workplace and equipment in the work area [11]. Usage of anthropometry in design to determine the user population, determine the dimensions of the body, determine the percentage of the population, determine percentile value of selected anthropometric dimensions, making the design based on anthropometric data and used for the simulator to test the design [11].
Anthropometric is also one of the tools of science that is used to create an ergonomic working condition. Ergonomics is the science of human-based design (Human Centered Design) [12]. An anthropometric measurement of the human body dimension is one part in creating ergonomic conditions. The body dimension data is very useful in design product with the aim of seeking harmony with humans who wear the products. Anthropometric data usages commercialize all the tools to be adapted to the ability of humans, not humans adapted to the tool [12].
To design a lesehan chair with accordance to ergonomic rules then it is needed to design it using anthropometry approach. Anthropometry is the science that connected with the dimensions of the human body. The human dimensions data is very useful in the design of products with a purpose to finding harmonize of a product with human while using it. The design has a high capability to human that take them is very important to reduce the emergence of danger signs caused by work error due to the presence of design error [13].

3. Methods

3.1. Kansei Engineering
The observation of this research is done by using a list of questions and questionnaires. According to Nagamachi and Lokman [14], twenty of thirty people as sample are enough and sufficient to be used in the methods of engineering. Twenty people are participated in this study as respondent. Kansei word obtained based on the result of the analysed questionnaire. There were five steps done with the spreading of questionnaire. The first step was conducted to find out costumer desires related to the design of lesehan chair. The second step was conducted to find out the interest rates between each of kansei words to the design of lesehan chair. The third step was conducted to find out the specification

of lesehan chair based on customer needs, while the fourth and fifth steps was conducted to observe if the design and product of lesehan chair had been already satisfy the customer's desires.

3.2. Anthropometric Consideration
The anthropometric data retrieval is the stage which can help to recognize the ergonomic conditions. Anthropometric data usage was done so the designed tool can be adapted to the human body or the body of the user. The body dimensions data is needed to determine the size of lesehan chair design. The dimensions of the body that is used will be adjusted to the needs of the lesehan chair design. An amount of 30 samples were obtained from anthropometric data bank of DSK & E laboratory. Once the data is obtained, the body dimensions were processed with several stages of testing such as the test of adequacy of the data, the uniformity of data, the normality of data, and percentile. Anthropometry measurement also used in this study to adjust and transform customer needs into more ergonomics design criteria [15]. The body dimensions used to design the lesehan chair were sitting shoulder height as the height of lesehan chair, shoulder width as the width of the back of the chair, the length of buttock to knee as the length of chair seat, hand width as the length of the handle, and hand thickness as the height of handle.

4. Results and Discussion

4.1. Anthropometry Measurements
Anthropometry data that used to design the lesehan chair was taken from the body dimension of 30 respondents with age range of 19 – 20 years old. All data has been tested for its adequacy, uniformity and normality. After the adequacy, uniformity, and normality has been confirmed, the next step is to determine the size of the percentiles. The results of the percentiles calculation can be seen in Table 1.

Table 1. Percentile of Each Body Dimension

Body Dimension	Percentile		
	P5	P50	P95
Sitting shoulder height	54.94	60	63.05
Shoulder width	32.35	39.35	48.43
Buttocks to knee length	50.5	56.35	61.1
Metacarpal width	6.8	8	9
Metacarpal thickness	1.1	2.2	3.2

The percentile that will be used is the 95th percentile for each dimension according to user needs, which is to create comfort when using the lesehan chair. Each measure was added to the allowance so the body dimensions are 64 cm sitting shoulder as the height of the lesehan chair, 49cm shoulder width as the width of the back reast of the chair, 62 cm length of buttock to knee as the length of the chair seat, 9cm of hand width as the length of the handle, and 3.2 cm of hand thickness as the height of the handle.

4.2. Kansei Words
There were 2 questionnaires distributed in this research. The result of the questionnaire 1 was 15 kansei words such as comfortable, durable, innovative design, attractive colour, adjustable, strong, wispy, cheap, unique, easy to clean, portable, multifunction, ergonomic, easy to store, and safe. Based on those result, questionnaire 2 was distributed to know how important each of kansei words to the design of lesehan chair. The recapitulation of the questionnaire 2 result was analysed using validity and reliability test to determine whether the kansei word was valid and reliable. The validity and reliability test results showed that 12 kansei words were valid so it could be used as research instrument, those kansei words are comfortable, durable, innovative design, attractive color, adjustable, strong, unique, easy to clean, portable, ergonomic, easy to store and safe.

4.3. Factor Analysis

Factor analysis was used to cluster the kansei word into some groups. The purpose of the factor analysis was to find the condensing way of the information which contained numbers of origin variable into a smaller set of variable to minimize the loss of information [16]. The mapping of result can be seen in table 2.

Table 2. The factor analysis result.

Kansei Word	Comfort	Quality	Aesthetic
Safe	0.889	-0.163	-0.031
Comfortable	0.792	-0.055	-0.025
Portable	0.783	-0.156	-0.28
Ergonomic	0.756	-0.125	0.110
Easy to save	0.751	0.002	0.008
Adjustable	0.732	-0.202	-0.289
Easy to clean	0.180	0.821	-0.017
Durable	0.360	0.810	0.099
Strong	0.274	0.620	0.199
Attractive color	0.368	-0.145	0.712
Innovative design	0.101	-0.114	0.510
Unique	-0.027	-0.283	0.436

Based on the factor analysis result, 6 kansei words which are safe, comfortable, portable, ergonomic, easy to store, adjustable, can be classified into a variable set of comfort. Factor two which is composed of durable, easy to clean and strong, can be classified into quality variable. The last factor consists of three kansei words, which are attractive colours, unique and innovative design, were classified into aesthetic variable.

4.4. Concept Mapping

Concept mapping was used to determine the physical design characteristics which would be divided into several levels and then processed to be the final physical design. The concept of product at the highest level or zero level in this research is the design of lesehan chair. The mapping of every concept can be seen in table 3, 4 and 5.

Table 3. Concept Mapping 1 : Comfort

Sub Concept Level 1	Sub Concept Level 2	Sub Concept Level 3	Sub Concept Level 4	Sub Concept Level 5	Sub Concept Level 6	Physical Design Characteristics
Comfort of lesehan chair	Comfortable to use (Otista, 2012)	Comfortable in the back (Mulyono, 2010)	Size			Height : 64 Cm Width : 49 Cm
			Do not cause soreness	Position of backrest		95° and 115°

Sub Concept Level 1	Sub Concept Level 2	Sub Concept Level 3	Sub Concept Level 4	Sub Concept Level 5	Sub Concept Level 6	Physical Design Characteristics
			Do not create heat	Material used		Coconut fibres
			Odorless	Material used		Synthetic leather
		Comfortable in the cradle (Mulyono, 2010)	Size			Height : 62 Cm Width : 49 Cm
			Do not create heat	Material used		Coconut fibres
			Odorless	Material used		Synthetic leather
	Practical (Archam et. al, 2013)	Can facilitate (Archam et. al, 2013)	Portable (Archam et. al, 2013)	Size	Handle chair	Height : 9 cm Width : 3.2 cm
					Chair folds	3 fold
			Easy to store (Anita 2007)	Size	Chair folds	3 fold

Table 4. Concept Mapping 2 : Quality

Sub Concept Level 1	Sub Concept Level 2	Sub Concept Level 3	Sub Concept Level 4	Sub Concept Level 5	Physical Design Characteristics
			Chair frame	Shape	Picture 2 (Questionnaire III. B2)
				Material used	Iron
Quality of lesehan chair	Durable (Rahmasari, 2004)	Not easily broken (Rahmasari, 2004)	Chair pad	Material used	Coconut fibres
			Layer of pad	Material used	Coldore
			Outer chair layer	Material used	Synthetic leather

IOP Conf. Series: Materials Science and Engineering **105** (2016) 012036 doi:10.1088/1757-899X/105/1/012036

Table 5. Concept Mapping 3 : Aesthetic

Sub Concept Level 1	Sub Concept Level 2	Sub Concept Level 3	Sub Concept Level 4	Sub Concept Level 5	Physical Design Characteristics
Aesthetic of lesehan chair	Shape (Larasati, 2012)	Top view	Chair backrest	Backrest shape	Picture 2 (Questionnaire III. C1)
			Cover of chair back	Model	Cannot be removed
		Bottom view	Chair seat	Seat shape	Picture 2 (Questionnaire III. C1)
			Cover of chair seat	Model	Cannot be removed
		Side view	Chair hinge	Hinge shape	Picture 3 (Questionnaire III. C5)
		Chair handle	Handle shape		Picture 2 (Questionnaire III. C7)
	Color (Larasati, 2012)	Chair slipcover			Brown
	Layout (Dicasani, 2014)	Chair hinge			Visible

4.5. Design Recommendation

The mapping concept was used to determine the characteristics of the physical design based on consumers' desire to be obtained by distributing questionnaires to each respondent. The final physical design that are in accordance with customer needs was obtained as seen as Table 3, Table 4, and Table 5, and then the next step is to merge the obtained physical design, while the results of the physical design can be seen as Figure 1 and Figure 2, and the actual proposed product can be seen at Figure 3 and Figure 4.

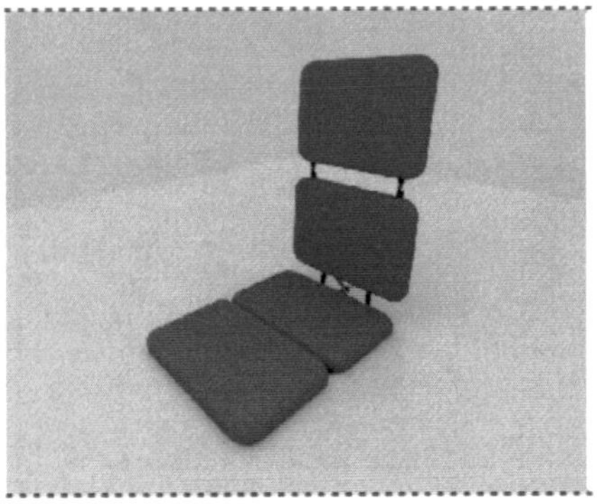

Figure 1. Proposed design of lesehan chair

Figure 2. Proposed design of folded lesehan chair

Figure 3. Proposed product of lesehan chair

Figure 2. Proposed product of folded lesehan chair

4.6. Marginal Homogeneity Test

Once the proposed design of lesehan chair has been designed, the next step is to test the Marginal homogeneity. The tests were carried out using two phases, namely the proposed design validation test and proposed product validation test. The tests were performed to determine whether the proposed design has been designed in accordance with the customer needs based on the previous stages. The tests were performed using SPSS software as for the results of the marginal homogeneity of proposed design of lesehan chair showed that with a significance level of 5% , the p-value of each attribute kansei word has a value above 0.05 which means that there is no difference between the kansei word with the proposed design. In other words, the design of the proposal has been fulfilling what consumer desires. In the marginal homogeneity test for proposed product, the results showed that the p-value of each kansei word attributes has a value above 0.05 so it can be said that there is no difference between the desired kansei word of what consumer desired with the produced product.

5. Conclusion

Based on research it can be concluded that there were 12 kansei word attributes applied to the proposed design of lesehan chair, namely comfortable, durable, innovation design, attractive color, adjustable, strong, unique, easy for clean, portable, ergonomic, easy to store, and safe, which are in accordance with customer needs. While the validation result of marginal homogeneity test showed that both proposed design and proposed product are in accordance with customer needs. However, there are still flaws in the proposed design of the lesehan chair that is on the additional functions of the lesehan chair. For further research, it is expected to perform analysis of sitting posture while sitting on the proposed product, because the sitting posture of people sit on lesehan chair and a regular chair in general is very different.

References

[1] Saputri I. *Posisi Duduk yang Benar, Agar Tulang Belakang Sehat.* (online): http://www.memobee.com/posisi-duduk-yang-benar-agar-tulang-belakang-sehat-1279- eij.html (accessed on 27 March 2014)
[2] Purnomo H 2012 *Antropometri dan Aplikasinya* (Yogyakarta: Graha Ilmu)
[3] Nagamachi M 1995 *Kansei Engineering : A new ergonomic consumer-oriented technology for product development*. International Journal of Industrial Ergonomic **15** 3-11
[4] Archam L D, Setyanto N W, Rahman A 2013 *Integrasi Kansei Engineering dan Structural Equation Modeling (SEM) Untuk Meningkatkan Kualitas Produk Shampo* (Malang: Universitas Barawijaya)

[5] Marlyana N, Nurwidiana, Taufiq A R 2012 *Penerapan Metode Kansei Engineering dan Anthropometri Pada Pemilihan Desain Fasilitas Ruang Warnet* (Semarang :UNISSULA)
[6] Restantin N Y, Ushada M, Makhmudin A 2012 *Desain Prorotipe Meja dan Kursi Pantai Portabel dengan Intergrasi Pendekatan Ertonomi, Value Engineering, dan Kansei Engineering*. Jurnal Teknik Industri **14** 53-62
[7] Yoshimura M, Papalambros P Y 2004 *Kansei Engineering In Concurent Product Design: A Progress Review* Proceedings of the TMCE 177-186
[8] Tayyari F, Smith J L 1997 *Occupational Ergonomics: Principles And Applications* (London: Chapman & Hall)
[9] Vink P, Koningsveld E A, Molenbroek J F 2006 *Positive Outcomes Of Participatory Ergonomics In Terms Of Greater Comfort And Higher Productivity*. Applied Ergonomics **37** (4) 537–546.
[10] Wignjosoebroto S 2006 *Ergonomi, Studi Gerak dan Waktu, Edisi Pertama* (Jakarta: PT. Guna Widya)
[11] Wickens C D, Lee J, Liu Y, Becker S G 2004 *An Introduction to Human Factors Engineering*. (New Jersey: Pearson Educational Internasional)
[12] Santoso G 2013 *Ergonomi Terapan* (Jakarta :Prestasi Pustaka Publisher)
[13] Liliana Y P, Widagdo S, Abtokhi A 2007 *Pertimbangan Antropometri pada Pendisainan* (Yogyakarta : SDM Teknologi Nuklir)
[14] Nagamachi M, Lokman A M 2009 *Kansei Engineering : A Beginners Perspective* (Tokyo : KDRU)
[15] Mariyana N, Nurwidiana, Taufiq A R 2012 *Penerapan Metode Kansei Engineering dan Antropometri Pada Pemilihan Desain Fasilitas Ruang Warnet* (Semarang : UNISSULA)
[16] Hair J F, Anderson R E, Tatham R L, Black W C 1995 *Multivariate Date Analysis With Readings, 4th Edition*. (Englewood Cliffs, NJ : Prentice Hall)

COLABORATED ARCHITECHTURE FRAMEWORK FOR COMPOSITION UML 2.0 IN ZACHMAN FRAMEWORK

Hermawan[1], Fika Hastarista[2]
[1,2]Jurusan Teknik Informatika, Fakultas Teknik, Universitas Trunojoyo Madura, Indonesia (69162)

E-mail : hermawan@trunojoyo.ac.id [1], fika@trunojoyo.ac.id [2]

Abstract. Zachman Framework (ZF) is the framework of enterprise architechture that most widely adopted in the Enterprise Information System (EIS) development. In this study, has been developed Colaborated Architechture Framework (CAF) to collaborate ZF with Unified Modeling Language (UML) 2.0 modeling. The CAF provides the composition of ZF matrix that each cell is consist of the Model Driven architechture (MDA) from the various UML models and many Software Requirement Specification (SRS) documents. Implementation of this modeling is used to develops Enterprise Resource Planning (ERP). Because ERP have a coverage of applications in large numbers and complexly relations, it is necessary to use Agile Model Driven Design (AMDD) approach as an advanced method to transforms MDA into components of application modules with efficiently and accurately. Finally, through the using of the CAF, give good achievement in fullfilment the needs from all stakeholders that are involved in the overall process stage of Rational Unified Process (RUP), and also obtaining a high satisfaction to fullfiled the functionality features of the ERP software in PT. Iglas (Persero) Gresik.

Keywords: Agile Model Driven Design (AMDD), Colaborated Architechture Framework (CAF), Model Driven Architechture (MDA), Unified Modelling Language (UML), Zachman Framework (ZF).

1. Introduction

Previously, various tools and techniques are provided for information system (IS) development, that are specific to an aspect or many aspects relation, such as the using of the basic flowchart shapes, ERD modeling, Data Flow Diagram, or Object Oriented modelling. However, in accordance with the demands of the information technology (IT) development requirements today, where so many stakeholders involved with many different perspectives and objectives that would influence system, so IT developers require more integrated architechture blue print to fullfiled overall Rational Unified Process (RUP) stages in IT projects.

Enterprise Architecture Framework (EAF) is a tool that can be used for developing a broad range of different architectures, where set of tools for EAF is supported with list of recommended on standardization within models and documents to build fully block of IT conceptual blueprint. Benefit of the using an EAF will speed up and simplify architecture development, ensure more complete coverage of the designed solution [1].

The ZF is one of the most widely adopted EAF in the Enterprise Information System (EIS) development [2]. The ZF provides a concept to produce a systematic taxonomy to declares a relationship without replacing anything in the real world and changing the tools, techniques and methodologies, but otherwise give collaboration of various viewpoints according to many perspectives and abstraction for architechture modelling [3].

ZF same as others EAF such TOGAF, IAF, FEAF, DoDAF within the primary principal consists of conceptual modeling language, that provides modelling platform and analysis list of the EA included strategy, process and dimensions of the application, as well as support various aspects of the structure and behavior of software artifacts [4]. In generally each EAF has different focus in providing architectural solutions. By reviewing two of the most popular EAF framework that are ZF and TOGAF, ZF more focused on the constructing architecture views in multi perspective of the EA, while TOGAF as view of process to describe the RUP stages to constructing the EA [5].

However in RUP stages, many EA practitioners prefer to using hybrid architecture approach to combine various EAF by taking some of the advantages of each frameworks [5]. Such as *Integrated Architecture Framework* (IAF) which provides a best practices approach more in order to integrated the various frameworks that have been existing. Where the advantages of the IAF lies in values and flexibilities within combination with other EA such Togaf ADM and ZF[6].

IAF has focused more on the processes to "Why, What, and How" as best practices to give transition solving from business components into application components, but less provided scalable architectural view then ZF. While ZF did'nt defined specific modelling like as support object oriented modelling where still be a mature paradigma to analyzing and developing software components today.

For this reason, in this study we propose *Colaborated Architecture Framework* (CAF) that is having focus as the best practice for composing structured models and documents to defines the software requirement specification (SRS) blue print, that based on object-oriented modeling approach use the UML 2.0, which has more scalable architecture view in "What, How, Where, Who, When, Why" adopting ZF to fullfil overall RUP stages in SDLC.

2. Methodology

2.1 Collaborations of Zachman Framework and UML 2.0

The CAF composing UML within the ZF 6x6 cell of matrix dimensions, in accordance with the six perspective of *Scope (planner's view), Enterprise model (business owner's view), Model of fundamental concepts (designer's view), Technology model (builder's view), Detailed representations (sub-contractor's view), Functioning system*. Which provide solutions for the six dimensional abstraction of english interrogatives that are consist of: *What (Data), How (Function), Where (Network), Who (People), When (Time), Why (Motivation)*. The using of UML to support ZF architecture is the systematic composition to be developed [7] which makes itself a basis for evaluating, establishing, and customizing other enterprise architectural frameworks, methods, and tools [8].

Within the ERP that we developed for PT. Iglas (Persero) Gresik-Indonesia consists of 7 modules: Financial-Accounting (FA), Customer Relationship Management (CRM) - Distribution, Material-Production, Procurement, Human Resource Management, Inventory, and asset management [9].

IOP Conf. Series: Materials Science and Engineering **105** (2016) 012037 doi:10.1088/1757-899X/105/1/012037

Which for sample implementation case we focus to describing CAF for one domain module that is the FA, as a core of ERP modules like as shown in Table 1.

Table 1. Composition CAF of the UML models and SRS documents in the Zachman Framework [3] [1]

Perspektive /Abstraction	*What (Things)*	*How (Process)*	*Where (Location)*	*Who (People)*	*When (Time)*	*Why (Motivation)*
Planner/ Contextual (Scope)	(P1A1) *Structure diagram* of The organization units and they responsibilities in the FA department	(P1A2) *Table of value chain* from business processes within the FA department	(P1A3) *Map location* from the FA department in PT. Iglas (Persero) Gresik	(P1A4) *Structure diagram* of managers who are responsible to plan, execute, and control business processes within the FA	(P1A5) *Workbase list* of business process activities	(P1A6) Defines Strategies and Goals of vision and mission in the Iglas FA department
Owner/ Conceptual (Business Model)	(P2A1) *Conceptual Data Model* of the FA business processes	(P2A2) *Business Process Model Notation* (BPMN) of FA	(P2A3) *Busines Logistics Diagram* of FA units	(P2A4) *Use Case Diagram* of the FA business processes	(P2A5) *Operational Schedule* of business activities in the FA	(P2A6) Defines Goals of each FA business processes
Designer/ Logical (System Model)	(P3A1) *Class Diagram* for all objects in the FA domain	(P3A2) *Component Diagram* of application component	(P3A3) *Sequence Diagram* for application component	(P3A4) *Component dependency matrix* from one application component to others	(P3A5) *RUP schedule* of Model Driven Architecture	(P3A6) Defines logical design of software artifacts
Builder/ Physical (Technology Model)	(P4A1) *Physical Data Model* of database	(P4A2) *Deployment Diagram* of installation package of modules	(P4A3) *Network Diagram* of server farm and computer access	(P4A4) *Interface Design* for each functionality application	(P4A5) *RUP schedule* of Model Driven Development	(P4A6) Defines physical design of software artifacts
Detailed Representation (Sub-Contractor)	(P5A1) *Structure and details* of Django Framework models of each application components.	(P5A2) *Source code* descriptions of *Model View Template* from the Django web framework	(P5A3) *List of the networks configuration* for all networks component include: router, switch, wifi, and cabling.	(P5A4) *Table of RACI matrix* as a policy to configure security of access rights for each user who manages the software applications	(P5A5) *SDLC schedule* of installation and maintenance	(P5A6) Defines installation and maintenance guaranty contract
Function Enterprise	(P6A1) *Integration link* between FA domain and other ERP domains	(P6A2) *Package diagram* of component services relations for the FA domain and other domains	(P6A3) *List of the global network configurations* include internet and network security	(P6A4) *Activity diagram* to manage activities to operate ERP services.	(P6A5) *Time schedule* for monitoring and controlling ERP services.	(P6A6) Defines System Operational Procedure (SOP) to operate overall ERP application services.

[1] All artifact accordance within the Iglas ERP blueprint

IOP Conf. Series: Materials Science and Engineering **105** (2016) 012037 doi:10.1088/1757-899X/105/1/012037

Not many previous studies that laid the blueprint mapping EAF for ERP because generally it does by many enterprise corporates in big research collaboration. In one individual study, Molnar outlines a wide-range theoretical background for analyzing the re-engineering and re-organization of ERP systems, and describes EIS relationship with EAF including ZF [10], but this study was still have struggling on the theory and strategy paradigm without any more best practice approach to defined architechture constructing.

That means, the CAF modeling can be a solution in the architecture constructing and then can be used as a choice in the best practice framework for the implementation of EAF in the development of EIS including the ERP system.

2.2. Model Driven Architecture

The CAF, it should defines no-overlap determination models for architectures modeling because every cell has different perspective or objective abstraction, so it has special specifications for different problem domains [11]. To construct composition UML models in the ZF we use important models that are most influencing architecture as MDA where is shown in Table 2.

Table 2. Specification of the Model Driven Architecture of UML 2.0

Zachman Framework Cell	MDA Specification
Business Model (Owner/Conceptual), People (Who) (P2A2)	Workflow modeling in *Business Process Modeling Notation* (BPMN) to collaboration departments and units which have responsibility to manage business process activity
Business Model (Owner/Conceptual), People (Who) (P2A2)	*Usecase Diagram* show user requirements within business process activities
System Model (Designer/Logical), Data (What) (P3A1)	*Class Diagram* modeling to define encapsulation of data attributes, methods and they relations to create components of software.
System Model (Designer/Logical), Function (How) (P3A2)	*Component Diagram* illustrates the message to exchange data from one component to other components by request method that is exposed
System Model (Designer/Logical), Location (Where) (P3A3)	*Sequence Diagram* illustrates the interaction of methods in integration an application service
Technology Model (Builder/Physical), Function (How) (P4A2)	*Deployment Diagram* illustrates the deployment of software components into hardware components
Integration Model (Function Enterprise) (How) (P5A2)	*Package diagram* for encapsulation all components in a single domain and illustrates collaboration link with other domains.
Integration Model (Function Enterprise) (Who) (P5A4)	*Activity diagram* illustrates step by step user activities to manage application.

MDA provides a way to define the structured view that allows a model to be constructed, viewed, developed, and manipulated at analysis and design time before it is coded [8].

2.3. Agile Model Driven Development

Furthermore, through the AMDD for supporting transition the modelling artifacts to the software artifacts [12], the application development can be done quickly and accurately meets to the feature requirements. So the coverage needs of EIS that are consists of persistent data, a lot of data, access data concurrently, a lot of user interface screens, and integrate with other enterprise applications are

achieved [13]. The main reference models for AMDD that we have executed to develop software components, they are: BPMN as a *Business Service*, Class Diagram as a *Component Entity Service*, and Sequence Diagram as a *Component Utility Service* [14].
From one sample scenario in the FA module, we analyze business process in financial transaction activities, where a register of financial business process is the General Ledger (GL) through the *Journal Entry* service. We set the Journal Entry as a core of *business service* to receive data from all transactions which is affecting the financial business processes. The Journal Entry in the General Ledger is the final register transaction of overall data from the whole ERP system, as shown on Figure 1. BPMN for this process is placed in P2A2 cell.

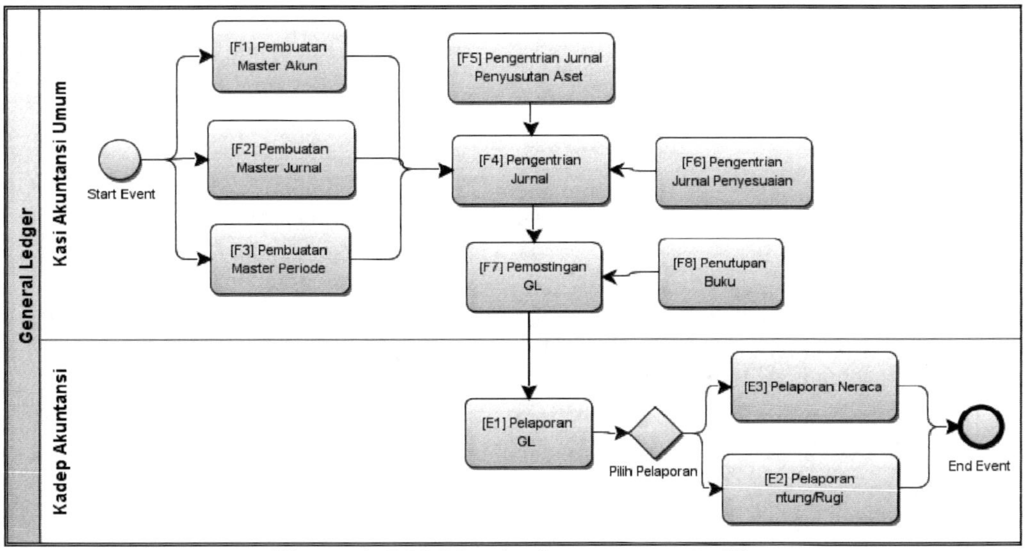

Figure 1, BPMN as business process in GL

From each activity in business services, we have composed into class diagram from each component entity services, like as shown in Figure 2 where *class diagram* for the Journal Entry in the GL is placed in P3A1 cell. From *UML Class diagram*, the model design is directly driven developing of component software, because class diagram is able to defines the data attributes and methods in the business service requirement, as data and operatitonal service in application.
For AMDD programming strategy, we recommend the use of three-tier software architecture framework, which generally supported *domain specific languange* in *domain driven design* programming concept and component-based plugins [15]. This strategy is having benefits: reduced communication across the layers, design by contract is sufficient, part of application components can be tested separately [16]. Among the development frameworks that could be an option, they are: Play for Java, Rails for Ruby, Grails for Groovy and Django for Python. Where Django framework as the technology of our choice due to excess AMDD in reliability, speed, and the components plugin are easily installed and modified.

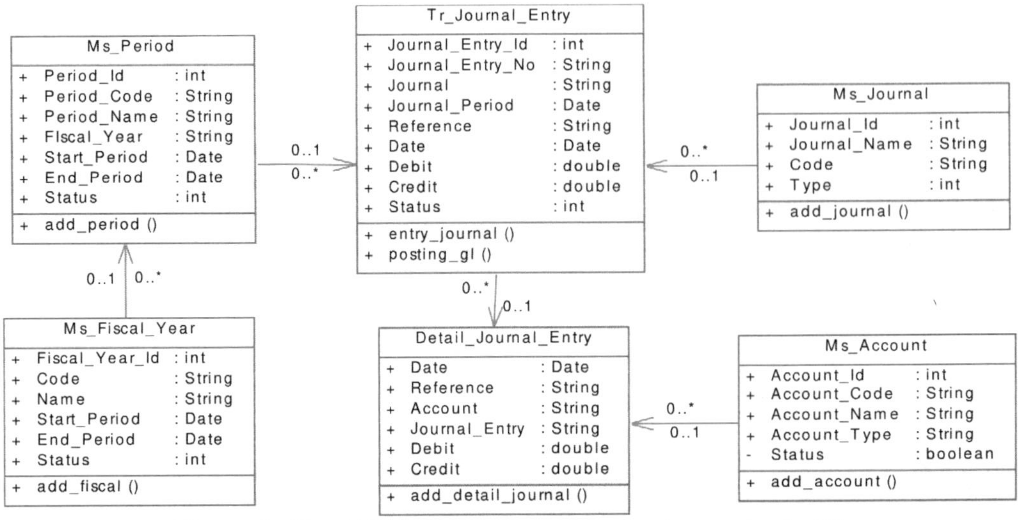

Figure 2, Class Diagram as a component entity service in GL

Class diagrams became references to create functionally component application, within translated they atributes and methods in a *model class*, while for integration many functionally methods as a component utility service, we use sequence diagram model to design it. For the further step, sequence diagram as addition methods within *scrath script* inside a *model class*. Like as shown in Figure 3, component utility service to create report in the journal entry that is composed with other services.

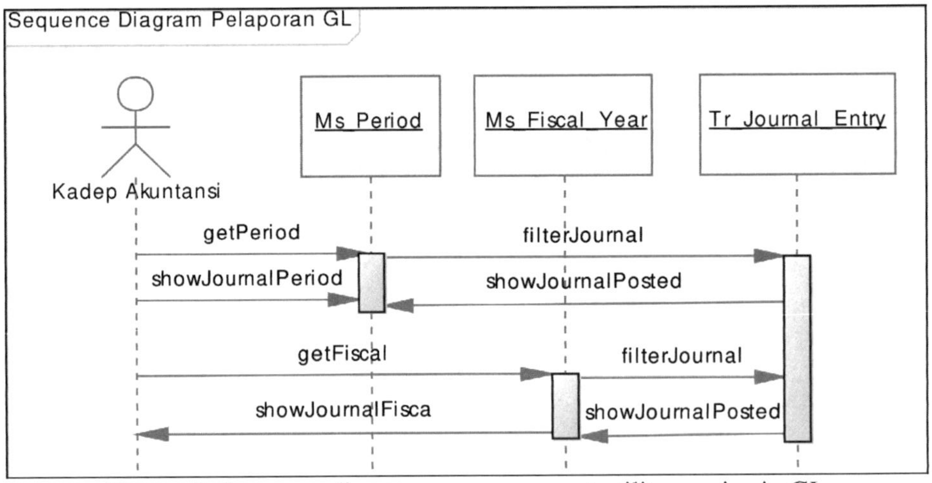

Figure 3, Sequence diagram as a component utility service in GL

By scaffolding template engine in the *django-admin*, class model generating application like as shown in Figure 4, where it is having functionally *Crud-Retrieve-Update-Delete* (CRUD) as a basic information system operation and would can testing in a few seconds time. This way will give generic solution for Test-Driven Development directly in the application to be launching.

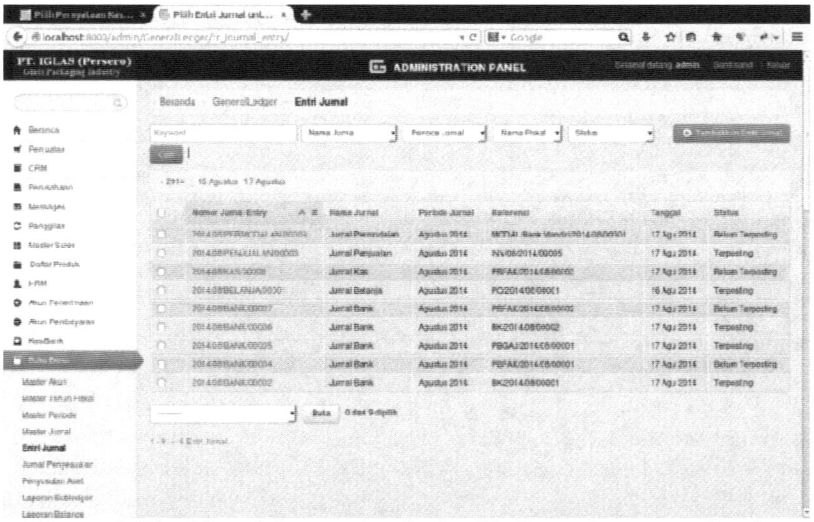

Figure 4, The List of data testing in the GL Journal entry application

Within functionally application for CRUD operation on data transaction, it works like as shown in Figure 5.

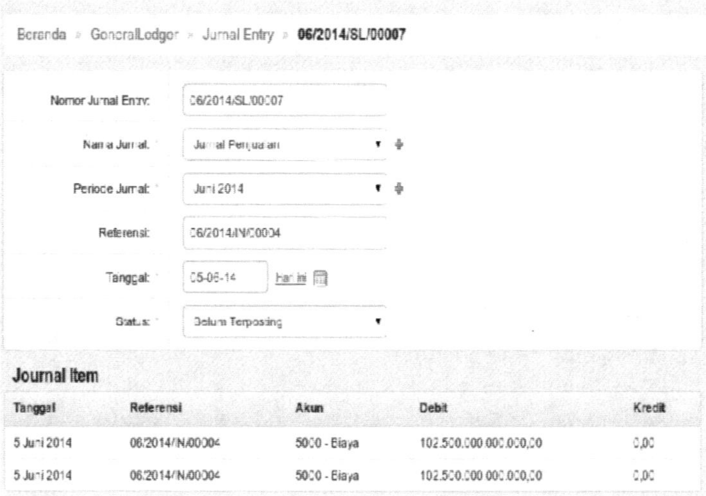

Figure 5, The CRUD data transaction in the GL Journal entry application

3. Result and Analyze

In the software development process that is carried out for six months by 2 analysts and 7 developers, we are working from drafting the model architecture until to implement the application programming able to produces EIS within 122 features of functionally applications in the seven modules of ERP, with details in Table 3.

Table 3, The numbers of ERP modules apps

Module	Numbers
Financial Accounting	30
Asset Management	18
CRM – Distribution	18
Material – Production	16
Procurement Management	17
Human Resource Management	13
Inventory Management	11

Furthermore within comparison with the features of existing ERP applications such as Open ERP 7.0, the EIS features that have been devoloped for ERP PT. Iglas have some advantages in functionality, because the application development is done in Taylor made, but otherwise with minimize scratch code in programming component of software so that is easy to evaluate end reconstruct, and also reducing complexity to help better understand on the evolution of enterprise business processes and their requirements [16].
Such as shown in Table 4 features of Financial-Accounting ERP modules, which have consist of four business processes that is implemented by 30 functionally applications.

Table 4, Subjective evaluation for ERP PT. Iglas by they IT managers

Business Process	Activity	*Score	Open ERP
Account Receivable	Delivery of account invoice	80	√
	Payment Sales	90	√
	Delivery of sales return account	80	√
	Delivery of assets sale account	80	√
	Payment of asset sales	80	X
	Delivery of asset sales payment account	90	X
Account Payable	Delivery of purchase request account	80	√
	Expenditures payment	90	√
	Delivery Expenditures payment account	80	√
	Payroll approval	90	√
	Delivery payroll account	80	X
Cash/Bank	Create the cash master	90	X
	Create the bank master	90	√
	Create the exchange master	90	√
	Create the tax master	90	√
	Statement of Cash	95	√
	Statement of Bank	95	√
	Displacement Cash / Bank	90	√
	Capital investment	85	X

Business Process	Activity	*Score	Open ERP
General Ledger	Creating the Master Account	90	√
	Creating the Master Fiscal Year	90	√
	Creating the Master Period	90	√
	Creating the Master Jurnal	90	√
	The Jurnal Entry	95	√
	The Jurnal Adjustment Entry	85	√
	The Asset Depreciation Journal Entry	85	X
	GL Posting	80	√
	Subledger reporting	80	√
	Balance reporting	80	√
	Loss/Profit reporting	80	√

4. Conclusion

The results of this study concluded, CAF that is composed with Zachman framework and UML fully supported MDA, and also through implementation of the most advanced method and technology of AMDD meets with the stages of the RUP process quickly and accurately, and also simple to learn and adjustment to establish the complexity requirements of EIS that is proofed to develop ERP application. Because today still a bit of study on the scope of software engineering to arrange EAF for the EIS development within detail UML modelling determination, so CAF can be implemented as the best practice reference to determining the artifacts models and documents that supports all works in the RUP which meets with the standard model and technology today.

5. Related Work

In the next study can be developed pattern of software design[17] and for advance using Service Oriented Architecture (SOA)[18] for transformation MDA to AMDD, which provides better best practice in enterprise application development methods.

6. Reference

[1] Rachel Harrison 2009, *TOGAF™ Version 9 Foundation Study Guide*, (The Open Group : Van Haren Publishing) chapter 2 p 16

[2] Carla Marques Pereira, Pedro Sousa, *A method to define an Enterprise Architecture using the Zachman Framework*, ACM symposium on Applied computing, Pages1366-1371, New York 2004.

[3] J. F. Sowa, J. A. Zachman, Extending and formalizing the framework for information systems architecture, *IBM* 1992

[4] Eric Yu, Markus Strohmaier, Xiaoxue Deng *Exploring Intentional Modeling and Analysis for Enterprise Architecture*, EDOCW '06. 10th IEEE International 2006

[5] Brian H. Cameron, Eric McMillan, *Analyzing the Current Trends in Enterprise Architecture Frameworks,* Journal of Enterprise Architecture pp 60-71, Association of Enterprise Architects 2013

[6] J. van 't Wout et al. 2010, *The Integrated Architecture Framework Explained*, (Capgemini SA: Springer) chapter 4 p 161-200

[7] Nikolaidou M, et al., *Employing Zachman Enterprise Architecture Framework to Systematically Perform Model-Based System Engineering Activities*, System Sciences, pp 1 - 10, HICSS '09. 42nd Hawaii International Conference 2009

[8] A. Fatolahi and F. Shams, An investigation into applyingUML to the Zachman Framework. *Information Systems Frontiers,* 8(2):133–143, 2006

IOP Conf. Series: Materials Science and Engineering **105** (2016) 012037 doi:10.1088/1757-899X/105/1/012037

[9] Dimitris N. Chorafas 2001, *Integrating ERP, CRM, Supply Chain Management, and Smart Material* (Florida:Aurbach)

[10] Bálint Molnár, The Country-specific Organizational and Information Architecture of ERP Systems at Globalised Enterprises, *BUSINESS SYSTEMS RESEARCH* Vol.2, No.2, pp.1-56, 2011

[11] Dragan Milicev 2009, *Model-Driven Development with Executable UML*, (Indiana: Wiley Publishing) chapter 3 p 27

[12] Scott W. Ambler 2004, et al., *Agile Model-Driven Development With UML 2.0*, (Cambridge : Cambridge University Press)

[13] Martin Fowler 2002, et al., *Patterns of Enterprise Application Architecture*, (Boston:Addison Wesley)

[14] Hermawan, Riyanarto Sarno, Developing Distributed System with Service Resource Oriented Architecture, *TELKOMNIKA*, Vol.10, No.2, pp. 389-399, June 2012

[15] Johan den Haan, *DSL development: 7 recommendations for Domain Specific Language design based on Domain-Driven Design*, http://www.theenterprisearchitect.eu/blog/2009/05/06/dsl-development-7-recommendations-for-domain-specific-language-design-based-on-domain-driven-design/, published may 6 2009

[16] Mike Cohen 2010, *Succeding With Agile*, (Boston: Addison Wesley)

[17] Frank Buschmann 2007, et al., *Pattern Oriented Software Architechture*, Volume 5, (New Jersey:John Wiley & Sons)

[18] Thomas Erl 2007, SOA: *Principles of Service Design* (New Jersey:Prentice Hall)

IOP Conf. Series: Materials Science and Engineering **105** (2016) 012038 doi:10.1088/1757-899X/105/1/012038

Automated Analysis of Human Sperm Number and Concentration (Oligospermia) Using Otsu Threshold Method and Labelling

I G Susrama[1], K E Purnama[2] and M H Purnomo[3]
Electrical Engineering, Institut Teknologi Sepuluh Nopember Surabaya

E-mail: gdesusrama@gmail.com

Abstract. Oligospermia is a male fertility issue defined as a low sperm concentration in the ejaculate. Normally the sperm concentration is 20-120 million/ml, while Oligospermia patients has sperm concentration less than 20 million/ml. Sperm test done in the fertility laboratory to determine oligospermia by checking fresh sperm according to WHO standards in 2010 [9]. The sperm seen in a microscope using a Neubauer improved counting chamber and manually count the number of sperm. In order to be counted automatically, this research made an automation system to analyse and count the sperm concentration called Automated Analysis of Sperm Concentration Counters (A2SC2) using Otsu threshold segmentation process and morphology. Data sperm used is the fresh sperm directly in the analysis in the laboratory from 10 people. The test results using A2SC2 method obtained an accuracy of 91%. Thus in this study, A2SC2 can be used to calculate the amount and concentration of sperm automatically
Keyword. Sperm, Oligospermia, Concentration, Threshold, Labelling

1. Introduction

Get married and have children is a dream for every couple. But for some couples sometimes very hard to get a descent it self though has been married in a long time. There are many factors that can affect pregnancy. Not only from the mothers who determine the success of a pregnancy but the fathers also have contribution. Sperm quality is also an important part. But sometimes the fathers do not realize have had disorders/disturbances in sperm that may affect the occurrence of pregnancy itself.

One of the sperm disorders that may interfere with male fertility is the Oligospermia. Oligospermia is a disorder in which during ejaculation, the sperm contains only low concentrations. Normally the concentration of sperm in the ejaculate is 20-120 million/ml, while in the Oligospermia case, the concentration of sperm less than 20 million/ml. Most of the causes of this problem are associated with

[1] Student Doctoral Program of Electrical Engineering, Institut Teknologi Sepuluh Nopember Surabaya.
[2] Associate Professor, Dep. of Multimedia and Network Eng., Institut Teknologi Sepuluh Nopember Surabaya, Medical Image Analysis, Microscopic Image Analysis, Computer Vision and Image Understanding, email: ketut@ee.its.ac.id.
[3] Professor of Electrical Engineering, Institut Teknologi Sepuluh Nopember Surabaya, Artificial Intelligent Technique for Power System and Biomedical Engineering, email: hery@ee.its.ac.id

lifestyle, such as less activity, testes or testicles often in a state of overheating, smoking, obesity, unhealthy diet, and stress, or illness experienced by people who could affect sperm production. Majority, the cause of oligospermia is a lifestyle that is not balanced like too many unhealthy foods and detrimental to health such as junk food, fast food, alcohol, smoking, and less exercise, which in turn cause problems in the body such as obesity or diseases Another influence male hormone in the body and interfere with sperm production itself. Therefore for men who want to keep sperm healthy without any problems must keep with a healthy lifestyle.

To determine the occurrence of oligospermia can check the sperm in the fertility clinics laboratory. One way using haemocytometer, do as counting red blood cells (erythrocytes). By using a haemocytometer and microscope views, spermatozoa counted manually in five squares diagonally or four boxes in each corner and the middle of the box. 10 million equal to the concentration of spermatozoa multiplies the calculation result / ml.

Manual process conducted by experts, take a lot of time and energy [1]. In the laboratory, spermatozoa analysis is determined at random within each elevated view of the microscope. Each elevated view of a certain amount of information obtained normal and abnormal spermatozoa randomly. The results obtained in each overpass of view then summed and averaged, in order to obtain information on the number and concentration of spermatozoa (either normal or abnormal) of each sample preparation. This conventional method will lead to differences in measurement results when performed by different experts, because the calculation of normal and abnormal spermatozoa done randomly. Therefore in this study created a computer-aided system to calculate the amount and concentration of spermatozoa in the semen. The data used were ejaculate data of 10 people and analysed directly in the laboratory of Microbiology Health Polytechnic of Surabaya. This procedure compared the number and concentration of sperm which are normal or not normal by using manual calculation according to the WHO standard [13] and calculations with a computer, that is by taking pictures of spermatozoa in parts from counting booths. The images taken were carried to initial processing stage that aimed to simplify further processing. At this stage, it will apply methods of image enhancement and noise removal. Next, the images will be subjected to the process of background removal, and then carried out to segmentation process to obtain sperm cells by comparing the use of Otsu threshold methods and the method of threshold local adaptive (deviation adaptive). The comparison of these two methods are based on the fact that data sperm images taken from the microscope were not all the same, it depends to the brightness of sperm shape and gray scale of the background. The next stage is the extraction of these features will be represented to other forms and being look for some of its features. The value of features will be labelled to determine the quantity per box of counting booth and are compared to the results of manual calculations. The While the previous research [11] focused on the methods and process of calculating the number of sperm per image, so that only known quantities or simply counting the number of sperm in one preparation, this research counts the number of sperm per ejaculate to obtain sperm quantity and concentration that will be used to determine whether a person is suffering from an oligospermia or not.

2. Literature Review

According to WHO, 2010 [13], normal sperm count is approximately 10-20 million/ml or about 2-6 ml. The characteristics include the amount of healthy sperm > 15 million/ml, have quick movements and straight, relatively normal shape, viability is good, no bacteria and leukospermia lot. When the sperm concentration <15 million/ml, it will be categorized as oligozoospermia. Furthermore if the sperm concentration less than 5 million/ml can be regarded as extreme/severe oligozoospermia which is the production of sperm produced minimal or included in the weight categories and if not handled immediately could lead to the possibility cryptozoospermia stage even to the stage of azoospermia. Here explained how to calculate the amount or concentration of sperm in semen human clinical laboratory measurements and methods that will be used to analyse images from a digital microscope. Spermatozoa classification in humans with Fuzzy rules by Monali Khachane Y., et.al [8]. The classification based on the shape of the head, neck and tail of the sperm. The sperm segmentation using image processing techniques and classification of the sperm head. Research carried out on the digital image is sperm. Spermatozoa segmented consist of three parts: sperm head, neck, and tail. Sperm image is processed, the result is a segmented sperm feature, then the proposed method using Fuzzy for classifying normal and abnormal based on shapes and the sperm head area. The downside of this study when found flattened sperm head, it cannot be detected because the image is only based on two-dimensional images while the spermatozoa should be analysed in three-dimensional space in order to obtain more accurate information. Spermatozoa considered normal if the head, neck and tail have normal criteria, but in the study of normal and abnormal criteria only found on the head, so that cannot be used as a basic right.

Priyanto Hidayatullah and Miftahudin Zuhdi, [11] performs research detection and count the number of spermatozoa in certain areas (in a video taken from a microscope multimedia) using adaptive local threshold and Ellipse Detection to detect sperm head, the measures taken to detect and count the number of spermatozoa is to convert the input image into a grayscale image, noise filtering, converting a grayscale image into a binary image using a local adaptive threshold, perform Closing and Open on morphological operations, detecting the edges of objects, detect sperm using ellipse detection. Results from this study, the method used achieve 90.97% accuracy with only 6:09% false positive per frame which is better than the existing methods of detection of sperm. This study is only the limit counts the number of spermatozoa alone based head, so cannot be used as a basis for determining the quality of spermatozoa. Spermatozoa quality can be determined with more complex parameters such as spermatozoa shape, the velocity and trajectory of motion.

Jiaqian Li, Tseng Kuo-Kun, Haiting Dong, Yifan Li, Ming Zhao and Mingyue Ding, 2014 [4], their research is clicking identification of the morphology of the sperm cells, whether healthy or not. The method used is a combination of principal component analysis (PCA) to extract features of the image and the k-nearest neighbour (KNN) algorithm to diagnose Sperm Health. Sperm cells are analysed images taken from microscope images; the size of the spermatozoa is very small. Sperm health diagnosis method consists of three parts: the sperm image segmentation, feature extraction with PCA and classification with k-nearest neighbour (KNN) algorithm. At the end of the trial results will be compared between PCA and features SHIFT feature, KNN classifier and classifier BPNN. Accuracy of diagnosis is done turns out depends on the training set, one of the results with specific training set to obtain a good performance with an accuracy of 87.53% compared with other feature extraction methods such as Scale-invariant feature transform (SIFT) and other classifier such as back propagation neural network (BPNN). The downside of this study is the method compared to only two, and the paper does not explain the reason why accuracy is only obtained 87.53%.

2.1. Counting Sperm and Calculated Concentration
The concentration of spermatozoa is the number of spermatozoa in a millilitre of sperm. The concentration of spermatozoa can be calculated in several ways, one of which is by using a haemocytometer. Calculate the concentration of spermatozoa by using haemocytometer; do as counting red blood cells (erythrocytes). By using a haemocytometer, spermatozoa counted on five squares diagonally or four boxes in each corner and the middle of the box, the calculation result is multiplied by 10 million equal to the concentration of spermatozoa / ml, such as figure 1. [13]

To calculate the concentration and quantity of sperm ejaculated spermatozoa used new (<1 hour, standard WHO [9]), then the sperm has been shaken evenly and perfectly likwofeksi diluted, with a ratio of 1:10, 1:20, 1: 100 or 1 : 200, depending on the density of spermatozoa / LPB. For a 1:10 or 1:20 dilution can be used pipette leukocytes, to 1: 100 or 1: 200 can be used pipette erythrocytes.

Pipette leukocytes / erythrocytes shaken by the number 8 for 15-20 minutes. 3 drops first then discarded, before dripped into the count room, left 15-20 minutes so that all the cells settles evenly in the room count [9]. Then count the number of spermatozoa in the following manner:

- If the sample is less than 10 spermatozoa per LPB, then count all the big boxes that are 25
- If spermatozoa seen 10-40 per LPB, then simply counted 10 big boxes
- If the sample is> 40 spermatozoa seen LPB, then simply counted 5 large box.

Furthermore, when it has counted 25, 10 or 5 large box on haemocytometer then divided into appropriate conversion factor large box that has been calculated, the result is the concentration of spermatozoa in a million per millilitre. When normal spermatozoa concentration> = 20 million / ml (WHO, 1999) and> = 15 million / ml [13]

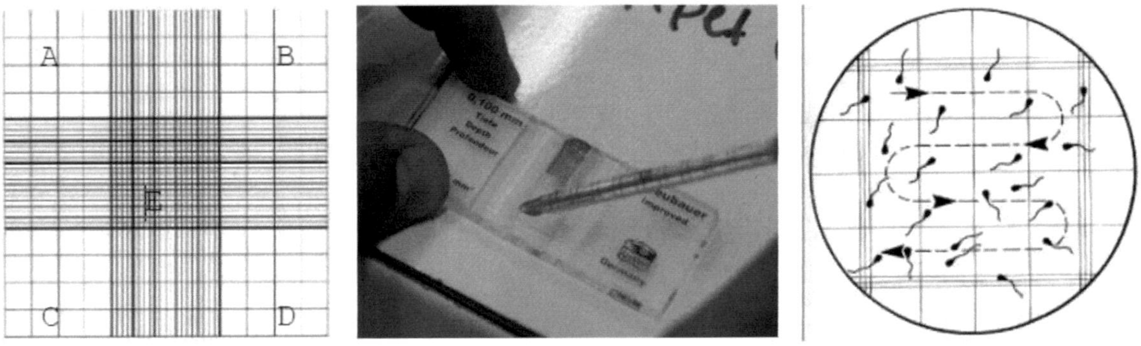

(a) (b) (c)

Figure 1. (a) Haemocytometer Counting booth, (b) Sperm Data to be analysed, (c) Process to Calculate Total Sperm Count In one booth

Besides, it can be used to calculate the number of spermatozoa that living or dead is done by making smears or staining differential. The fresh sperm (undiluted) given one drop coupled with 2-3 drops of colour eosin, then blended until homogeneous and then made thin smear alone and be heated over the fire for over a minute while moved about so dry evenly. Then check under the microscope with a magnification of 45 x, the spermatozoa that have died before made preparations means it will appear darker because it absorbs the dye, whereas spermatozoa were still alive when made preparations will be bright or brighter than the dead.

The percentage of living or dead spermatozoa can be calculated by taking a sample of 100-200 spermatozoa in some objects of observation. By smears can also be observed forms of spermatozoa abnormalities, both primary and secondary abnormalities. The primary is abnormal sperm shape abnormalities as a result of testicular disorder (seminiferous tubules), for example: small head, big head, piriformis heads, two heads, two tails, the middle and curled tails and linkage abaxial. Abnormal secondary is a form of abnormal spermatozoa occurring after spermatozoa leaving the seminiferous tubule epithelial sprouts or due to lack of maturation of spermatozoa in the epididymis can also be caused by the effect of cooling or heating. Examples: head and tail cut off, neck twisted, immature.

2.2. Threshold Segmentation Methods

Image thresholding is a technique widely used for segmentation, which divides the gray level image into segments suitable for some classes based on the gray level. Most thresholding approach is proposed for two classes, and based on the gray level histogram of the image. Two of the seven most popular approaches are the Otsu method [10] and Minimum Error Thresholding / MET method [5]. As has been surveyed in [2], there were no significant differences between the results of the two methods.

In this study, thresholding is not used for image segmentation as a final objective, but as a point of reference in maximizing the contrast to background and foreground as much as possible. Then the election between the two methods is not significant impact. Threshold divides the area into two classes as follows:

$$g(x,y) = \begin{Bmatrix} 0 & f(x,y) < T \\ 1 & f(x,y) \geq T \end{Bmatrix} \quad (1)$$

The purpose of the Otsu method is to find the point threshold that divides the gray level histogram of the image into two different areas automatically, at which point chosen is such that inter-class variance is as large as possible. The object is achieved by minimizing the weight of within-class variance, which is actually equal to: maximize the inter-class variance. The weight of the within-class variance is:

$$\sigma_w^2(t) = q_1(t)\sigma_1^2(t) + q_2(t)\sigma_2^2(t) \quad (2)$$

Class probabilities of first and second class are (where *i* represents the number of pixels for the emergence of a certain gray-level, while *t* is the gray level value):

$$q_1(t) = \sum_{i=1}^{t} P(i) \quad (3)$$
$$q_2(t) = \sum_{i=t+1}^{I} P(i) \quad (4)$$

Within-class variance of each class:

$$\sigma_1^2(t) = \sum_{i=1}^{t}[i - \mu_1(t)]^2 \frac{P(i)}{q_1(t)} \quad (5)$$
$$\sigma_2^2(t) = \sum_{i=t+1}^{I}[i - \mu_2(t)]^2 \frac{P(i)}{q_2(t)} \quad (6)$$

If counted all grades $\sigma_1^2(t)$ and $\sigma_2^2(t)$ the gray scale image will be obtained within-class variance value minimum. Value *t* what makes $\sigma_1^2(t)$ and $\sigma_2^2(t)$ becomes minimum is the threshold value sought. But utilizing the fact that the total variance in an image is always constant, not dependent on the threshold value can save the computing process. So for any threshold *t*, total variance is the sum of the within-class variance and inter-class variance:

$$\sigma^2 = \underbrace{\sigma_2^2(t)}_{\text{withinclass variance}} + \underbrace{q_1(t)[1 - q_1(t)][\mu_1(t) - \mu_2(t)]^2}_{\text{interclass variance} = \sigma_B^2(t)} \quad (7)$$

From the equation above shows that minimize within class variance ($\sigma_w^2(t)$) Is the same as maximizing the interclass variance ($\sigma_B^2(t)$). Value of $\sigma_B^2(t)$ can be calculated recursively with one-way ranging from value *t* lowest to highest (0 to 255 for 8 bit gray scale). Pseudo code to calculate the value $\sigma_B^2(t)$ are as follows:

1. Calculating the histogram and probability of each level of intensity (i.e. normalization of the histogram so that the total is equal to one)
2. Set the initial value $q_1(t)$, $q_2(t)$ and $\mu_1(t)$, $\mu_2(t)$
3. Exploring the entire value threshold may $t=1$ ... maximum intensity
   a. Update value $q_1(t)$, $q_2(t)$ and $\mu_1(t)$, $\mu_2(t)$
   b. Calculate $\sigma_B^2(t)$
4. Threshold is sought is that gives the maximum $\sigma_B^2(t)$

3. Methodology

The researches conducted for Automated Analysis of Sperm Concentration Counters (A2SC2) are as in Figure 2.

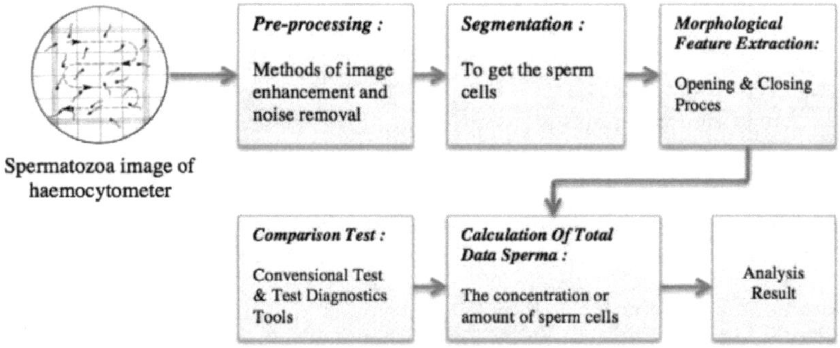

Figure 2. Methodology for Automated Analysis of Sperm Concentration Counters / A2SC2

4. Result And Discussion

The research conducted for Automated Analysis of Sperm Concentration Counters (A2SC2) has following steps.

4.1. Pre-Processing

Pre-processing is also used to process data sperm image for further analysis. Pre-processing can purge noise in the image, the image colour format conversion and image edge detection process. Some existing processes include composite, cropping, equalization and the mosaic image. In a sperm analysis to determine the amount and concentration of sperm, pre-processing designed is as follows:

1. Mean filter is used for smoothing images having noise. Mean Filter replaces the pixel value at position (x, y) with an average value of neighborhood pixels. The area of the number of neighboring pixels is determined as a masking / kernel / window size, for example 2x2, 3x3, 4x4, and so on. In addition to the mean filter, this study have also tried other image processing techniques, to obtain the most effective technique as a pre-process prior to the process of segmentation, feature extraction and classification.
2. Gray scaling: The aim of gray scaling is to convert input RGB image into gray scale image. This process is done in order to simplify and clarify the image of the equalization process in the next step.
3. Equalization: In this equalization process will display the original image, then the original image is calculated the width (w) and height (h) image. Image of the equalization, image pixels that has not spread evenly match the color of the pixel will be spread evenly which the results will be more visible role in the image histogram equalization results

Results from this process as in the figure 3.b

4.2. Otsu Threshold Segmentation

Feature extraction process aims to take on the characteristics contained in an image. Feature extraction method used is the analysis of Otsu threshold. Results of the method of Otsu are a threshold value (threshold) whose value ranges between 0 to255. Each of the input images will have a threshold value that is different. This is caused by differences in brightness and intensity of each gray degree on image of sperm obtained from a microscope. Then do the thresholding process is a continuation of the process of Otsu method for the purpose of executing the image corresponding to the threshold value. In this study, the image of the results of the pre-processing the image captured from the microscope sperm that equalization is then followed by determination of the threshold value will be at the threshold Otsu method so that the result will be like the figure 3.

The segmentation process was performed using locally adaptive thresholding based on block detection. In the block detection process, the input data is the result of the segmented images. This process requires initialization of the mean and standard deviation used as a threshold. The first step performed on the image that has been divided into several blocks, was to calculate the value of local mean and local standard deviation. Local mean and standard deviations were calculated for each block and used to determine whether a block should be processed with adaptive threshold or not. Then the value of local mean and local standard deviation of each block were compared to the mean and standard deviation threshold. If the local value of a standard deviation is lower than the standard deviation threshold and the local mean is higher than the mean threshold, then each pixel in the block is substituted with background pixels. Otherwise, each block is binarized with a global method (Otsu method). Otsu method is used because of its speed in performing the global thresholding.

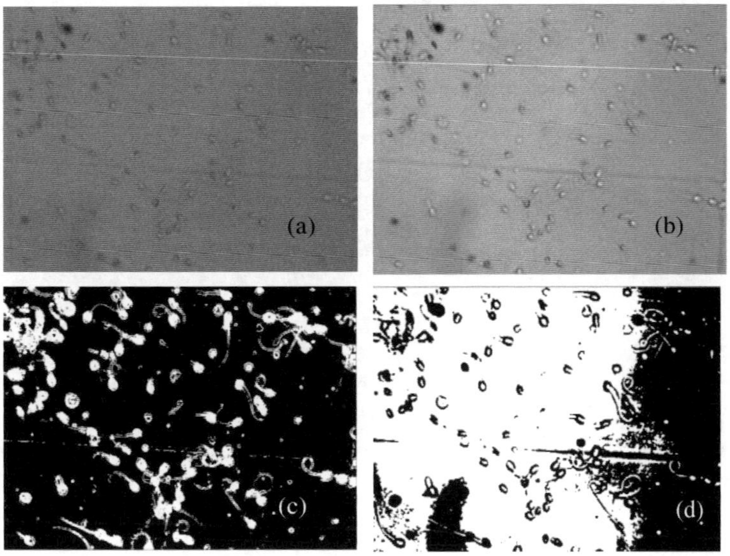

Figure 3. Segmentation Result a. Original image b. Results Image of Pre-processing c. Result of Adaptive Threshold Image Segmentation d. Result of Otsu Threshold Image segmentation

4.3. Morphological Features Extraction

Improvement process of sperm image segmentation results performed using morphological features. The goal is for extracting the image of the sperm and then separated so that the shape of the sperm will be more obvious. Gray level image of microscopic sperm containing sperm forms segmented and then processed to separate the overlapping sperm. Segmentation procedure based threshold (threshold) and morphology filter, whereas sperm separation procedures based on a watershed approach. Microscopic image of sperm in gray level is read for processing (Figure

3.b.). Then do binary process with a threshold gray level and filter morphological opening operation to refine the shape of sperm from noise. The points of each cell pointer obtained by dilating the maximum area and the topography of the surface area that states the distance transformation. The boundary line between dot pointers gave sperm outside the boundary lines.

Dividing lines sperm contact form can be obtained by diminish filtered binary image with skiz geodesic lines. Likewise, to obtain sperm shape intact the removal of sperm clipped by the image boundaries is to reconstruct the image and the boundary cells that attached or truncated. Sperm barrier contour can be obtained by transforming dilatation and subtraction. Then the contour can be superimposed to the initial gray level image segmentation to obtain sperm.

4.4. Calculation Amount, Sperm Concentration and Labeling Sperm

The concentration of spermatozoa is the number of spermatozoa in a millilitre of sperm. The concentration of spermatozoa can be calculated in several ways, namely by: using a haemocytometer, Nefelmeter method (photoelectric calorimeter) and Methods Spermiodensimeter. In the shooting of the microscope used haemocytometer. Calculate the concentration of spermatozoa by using haemocytometer; do as counting red blood cells (erythrocytes). By using a haemocytometer, spermatozoa counted on five squares diagonally or four boxes in each corner and the middle of the box, the calculation result is multiplied by 10 million spermatozoa concentration equal to 1 ml.

To count the number of sperm from a sperm image per-booth taken from a microscope, the labelling process was performed. This process was performed after data segmentation, in which the image has been separated between the object and background. The labelling process served as the deciding point to mark an object. After a successful image labelling, then the amount of sperms can be determined, although it is still fused with the non-sperm image.

The algorithm applied to the process of labeling is Two-pass algorithm. This algorithm is relatively simple to understand and implement, where it will iterate twice in the binary image. In the first iteration (first pass), the two-pass algorithm serves to record the equivalence and establishes a temporary label on the detected object. The second iteration (second pass) serves to replace each temporary labels assigned in the first iteration. Thus, differences in the labels given to the pixels of the image are the exact number of objects.

4.5. Test Result

Sperm data analyzed were data from 10 people of various ages. The followings are the results of conventional calculations from a fertility clinic in Surabaya and calculation from Microbiology Laboratory of Health Surabaya Polytechnic. The result is on Table 1. The manual analysis of the sperm showed 16 million / ml spermatozoa in sperm-1 sample. Overall, a total of 15 million / ml can be identified automatically using Otsu segmentation and the other 1 million / ml were not identified. The 15 million / ml were identified automatically with adaptive local segmentation and 1 million / ml were not identified. The comparison between the automatically identified spermatozoa (IO) and manually identified spermatozoa (M) using each segmentation method is as follows:

$$\frac{1Oo}{M} = \frac{15}{16} = 96\% \; and \; \frac{1Oa}{M} = \frac{15}{16} = 96\%$$

Description: IOo = Automatic Identification Otsu, IOA = Automatic Identification Adaptive. Results of calculation 9 other sperm can be seen in Table 1.

Table 1. Test Results A2SC2 Compared To Manual Calculation Results

Sperm Identification	Amount of Sperma / Expert Justice (M) / million / ml	Category	Otsu Threshold (IOo)		Local Adaptive Threshold (IOa)	
			Million/ml	(IOo/M)%	Million/ml	(IOa/M)%
Sperm-01	16	Abnormal	15	96	15	96
Sperm-02	36	Normal	29	80	35	96
Sperm-03	33	Normal	31	95	30	92
Sperm-04	29	Normal	25	86	27	93
Sperm-05	20	Abnormal	17	84	16	79
Sperm-06	30	Normal	29	98	29	98
Sperm-07	184	Normal	168	91	169	92
Sperm-08	12	Abnormal	11	92	10	83
Sperm-09	28	Normal	25	90	26	94
Sperm-10	28	Normal	27	98	26	94
Average Accuracy (%)				91		92

Based on the results in Table 1, the comparison value between automatically identified and manually identified spermatozoa reached an average of approximately 91% using Otsu threshold method and an average of about 92% using local adaptive threshold. However, the figure does not describe the overall system accuracy, since there was an error factor (FP-False Positive) (which is non-spermatozoa identified as spermatozoa). However, if the input image was processed manually and contrast value was enhanced in such a way, sometimes it would seem that the point formerly being considered an FP in manual procedure, was, in fact, TP (True Positive).

5. Conclusion

Automatic spermatozoa calculations can be performed using the proposed method. A possible error only occurs if the sample images contain clumps or there is a case of a collision (coincide). The accuracy of the automatic calculation is approximately 91%. The best color space to analyze the calculation of spermatozoa is the red color space, because the base color of the shooting is red.

6. Reference

[1] Alias M F, Mat Isa N A, Sulaiman S A and Mohamed M 2011 *Sprague Dawley Rat Sperm Classification Using Hybrid Multilayered Perceptron Network,* WSEAS Transaction on Information Science and Applications Issue 2 Volume 8, ISSN 1790-0832

[2] Han Si-qi and Wang Lei 2002 *A Survey of Thresholding Methods for Image Segmentation (Systems Engineering and Electronics vol 24 no 6)* pp 91-94

[3] Jun L, Clement L, Zhe L and Yu S 2013 *Quantitative Analysis of Locomotive Behavior of Human Sperm Head and Tail (IEEE Transactions On Biomedical Engineering Vol 60 No 2)* pp 390-96

[4] Jiaqian L, Kuo-Kun, Haiting D and Yifan L 2014 *Human Sperm Health Diagnosis with Principal Component Analysis and K-nearest Neighbor Algorithm: Proc. International Conference on Medical Biometrics* pp 108-113

[5] Kittler J and IllingworthJ 2012 *Minimum Error Thresholding: Pattern Recognition vol 19 no 1* pp 41-47

[6] Monali Y, Khachane, R R. Manza and R J Ramteke 2015 *Fuzzy Rule Based Classification of Human: proc. International Conference on Electrical, Electronics, Signals, Communication and Optimization (EESCO)* pp 1-5

[7] Mohamed E, Taymour E and Mohamed Abdelgawad 2014 *Development of Computer Assisted Sperm Analysis (CASA) Plugin for analyzing sperm in microfluidic environments using Image- J": proc. 7th Cairo International Biomedical Engineering Conference (CIBEC)* pp 149-152

[8] Mohammad R. Ravanfar and Mohammad H 2011 *Low Contrast Sperm Detection and Tracking by Watershed Algorithm and Particle Filter: Proc 18th Iranian Conference on BioMedical Engineering* pp 260-3

[9] Nurhadiyatna, A Latifah, D Fryantoni, T Wirahman, R Wijayanti, and FH Muttaqien 2014 *Comparison and Implementation of Motion Detection Methods for Sperm Detection and Tracking: proc Inernational Symposium on Micro-Nano Mechatronics and Human Science (MHS)* pp 1-5

[10] Nobuyuki Otsu 1990 *A Threshold Selection Method from Gray-Level Histograms (IEEE Trans. on System, Man, and Cybernetics vol 9 no1)* pp 62-66

[11] Priyanto H and Miftahudin Z 2014 *Automatic Sperms Counting using Adaptive Local Threshold and Ellipse Detection: proc International Conference on Information Technology Systems and Innovation (ICITSI)* pp 56-61

[12] Qiaoliang L, Xi C, Huisheng Z, Li Y, Siping C, Tianfu W, Shumei L, Xinyu L, Xiaofei Z and Ruikai Z 2012 *Automatic Human Spermatozoa Detection in Microscopic Video Streams Based On Open CV: proc 5th International Conference on BioMedical Engineering and Informatics (BMEI)* IEEE pp 224-27

[13] Word Healt Organization 2010 *WHO Laboratory Manual For The Examination And Processing Of Human Semen* Fifth Edition WHO Library Cataloguing-In-Publication Data

Drop out Estimation Students based on the Study Period: Comparisonbetween *Naïve Bayes* and *Support Vector Machines* Algorithm Methods

Harwati[1],Riezky Ikha Virdyanawaty[2] and Agus Mansur[3]
[1,2,3] Industrial Engineering Department, Faculty of Industrial Technology
Universitas Islam Indonesia, Jl. Kaliurang km. 14,5 Yogyakarta, Indonesia

E-mail:
[1]Harwati82@gmail.com, [2]ikha03oktober@gmail.com, [3]agusmansur.am@gmail.com

Abstract. Industrial Engineering is one of the departments in Faculty of Industrial Technology. It has more than 200 reshmen in every academic year. However, many students are dropped out because they couldn't complete their study in appropriate time. Variables that influence the drop out case are not yet studied. The objective of this paper is discovering the highest accuracy level between the two methods used, i.e. *Naïve Bayes*and *Support Vector Machines* algorithms. The method with the highest accuracy will be discovered from the patterns forms and parameters of every attribute which most influence the students' length of study period. The result shows that the highest accuracy method is *Naïve Bayes* Algorithm with accuracy degree of 80.67%. Discussion of this paper emphasizes on the variables that influence the students' study period.
Keyword: Drop Out, Naïve Bayes, SVM, Estimation, Student

6.1. Introduction

College is an institution of academic education for students. The role of students is an important aspect in the success of the program of study at the college [1]. [2] said that the department is obliged to monitor the students' study progress and predict their study period. It is useful to anticipate students' problems which will lead to the decreasing performance of department. Therefore, to improve the department's performance, the students must have a good academic performance, as stated by [3] that the students' academic performance are evaluated each semester to determine the learning outcomes that have been achieved. If a student cannot meet certain academic criteria to be declared worthy of continued study, the student is declared resign of the college or dropped out.

Industrial Engineering is one of the departments in Faculty of Industrial Technology. It has more than 300 freshmen in every academic year. There are many dropout cases in the department because the students couldn't finish their study in 7 years. Meanwhile, the department is having difficulties in estimating the factors which influence the dropout cases. Whereas, by estimating the dropout factors, it can improve the department's performance and help the academic system in giving early warning to the students by using a classification technique.

[4] stated that the purpose of data mining is obtaining relations or patterns which may provide useful indications. Similarly,[5] stated that data mining is a process of finding significant relations, patterns, and trends by examining a large group of data stores in storage using pattern recognition techniques such as statistical and mathematical techniques. A study by [6] explained that one of the data mining techniques is classification technique which is a learning technique to predict the value of target's category variables.

The definition of classification according to [5] is a technique of seeing the behaviors and attributes of predefined groups.[5] stated that this technique can classify new data by manipulating existing data which has been classified and by using the result to provide a number of rules. [5] also states that prediction has similarities with classification and estimation, in prediction value of prediction result will be in the future. Several techniques used in classification and estimation can also be used(for the right circumstances) for prediction.

A study by similar method was performed by [1] using*Naïve Bayes* Classifieralgorithm which showed that the most influential factor in determining the classification of students' academic performance isGrade Point Average (GPA),Grade Point(GP)on the first semester,GP on the fourth semester,and gender. Similar method was alsoused by [7] is discussing Classification of Length of Study of FSM Students ofDiponegoroUniversity Using Binary Logistic Regression and SupportVectorMachine(SVM)which showedthat variables which influence Length of Study variable are Department and GPA variables.

The objective of this paper is discovering the highest accuracy level between the two methods used, i.e. *Naïve Bayes*and *Support Vector Machines* algorithms.The method with the highest accuracy will be discovered from the patterns forms and parameter of every attribute which most influence the students' length of study period. It would help the Industrial Engineering Department in looking for solutions and policies to improve the students' achievement to finish their studies on time.

6.2. Literature Review

2.1 Naïve Bayes Algorithm

*Naïve Bayes*is a prediction technique based on simple probabilistic based on the application of Bayes theorem (Bayes rules) with strong independence assumption [8]. In*Naïve Bayes* classification, X is input vector which contains features and Y is class label. *Naïve Bayes* is written as$P(Y|X)$. The notation means class label probability Y is obtained after features of class X are observed. This notation is also called posterior probability for Y, while P(Y) is called prior probability of Y.During training process, posterior probability learning of (P(Y|X) must be performed on the model for every combination of X and Y based on information from training data. By developing the model, a test data X' can be classified by discovering the value of Y' by maximizing the value of$P(Y'|X')$ obtained.The *Naïve Bayes*formulation for classification is:

$$P(Y|X) = \frac{P(Y)\Pi_{i=1}^{q} P(X_i|Y)}{P(X)}$$

P(Y|X) is data probability withvector X in class Y. P(Y) is the prior probability of class Y. $\Pi_{i=1}^{q} P(X_i|Y)$is the independent probability of class Y from every feature in vector X. The value of$P(X)$ is fixed, so in to calculate prediction, only $P(Y)\Pi_{i=1}^{q} P(X_i|Y)$should be calculated by selecting the biggest as the result of the prediction. Meanwhile the independent probability$\Pi_{i=1}^{q} P(X_i|Y)$is the influence of all features of data on every class Y, which is notated as$P(X|Y = y) = \Pi_{i=1}^{q} P(X_i|Y = y)$. Everyfeature set $X = \{X_1, X_2, X_3, ..., X_q\}$ consists of*q*attributes (*q*dimensions).

2.2 Support Vector Machines Algorithm

*Support Vector Machines*can be called a semi-eager learner classification technique because aside from requiring training process, SVMalso stores a small part of training data to be used again in the prediction process. Some of the stored data is support vector[8]. SVMis actually a linear hyperplanewhich only works on data which can be separated linearly. Data with non-linear distribution usually use kernel approach on the initial data feature of data set. Kernel can be defined as a function which maps the data feature of initial dimension (low) to another feature with higher dimension (even much higher). Kernel mapping algorithm is shown below:

$$\phi : D^q \rightarrow D^r$$

$$x \rightarrow \phi (x)$$

$\phi$is a kernel function used for mapping, D is training data, q is feature set in an old data, and$r$is a new feature set as a mapping result for every training data. Meanwhile, x is training data, where$x_1, x_2, ..., x_n \in D^q$are features mapped to high dimension feature r, so data set used as training used algorithm from old feature dimension $D$to new dimension$r$. For example for n data sample:

$$(\phi (x_1), y_1, \phi (x_2), y_2, ..., \phi (x_n), y_n) \in D^r$$

Then the same training process as linear SVMwas conducted. The mapping process in this phase requires dot-product calculation of two data in new feature space. Dot-product of both vectors (x_i) and (x_j) are notated as$\phi(x_i) . \phi(x_j)$. The dot-product values of both vectors can be calculated indirectly,without knowing transformation function ϕ. This computation technique is called kernel trick which is calculatingthe dot-product of two vectors in new dimensional space using the components of both vectors in original dimensional space, as follows:

$$K(x_i, x_j) = \phi(x_i) . \phi(x_j)$$

And prediction on data set with new feature dimension is formulated with:

$$f(\phi(x)) = \text{sign}\ (w.\phi(z)+b) = \text{sign}\ \left(\sum_{i=1}^{N} \alpha_i\, y_i \phi(x_i).\phi(z) + b\right)$$

N is total data which becomes support vector, x_i is support vector, and$z$is test data which will be predicted. For kernel function selections commonly used in application, see Table 1.

Table 1. Kernel Functions

Kernel Name	Formula
Linier	$K(x, y) = x.y$
Polinomial	$K(x, y) = (x.y + c)^d$
Gaussian RBF	$K(x, y) = \exp\left(\frac{-\lvert x-y\rvert^2}{2.\sigma^2}\right)$
Sigmoid)	$K(x, y) = \tanh (\sigma(x.y) + c)$
Invers	$K(x, y) = \frac{1}{\sqrt{\lvert x-y\rvert^2+c^2}}$

6.3. Research Objectives

3.1. Obtaining the value of comparison of the result prediction based on *Naïve Bayes* algorithm and support vector machine.

3.2. Obtaining a dropout estimation pattern of students based on the methods of classification with the highest value.

3.3. Obtaining an effective parameter on each attribute from students'length of study period based on the method with the highest accuracy.

6.4. Research Methods

The focus of this research is to conduct drop out estimation based on the evaluation of academic achievement at the maximum limit of seven years study period which has been determined by the Industrial Engineering Department in Faculty of Industrial Technology, Universitas Islam Indonesia. The research's subjectsare the students from year 2003 to 2007 of the Industrial Engineering Department and the object is the duration of students who study for more than seven years and less than seven years. This research will estimate through classification technique using *Naïve Bayes* method and *Support Vector Machines*, then the data processing process assisted by Rapid Miner Software and Microsoft Office Excel. The processing results are in the form of a decision or rules and the amount of accuracy level of each methods degree. The chosen method is a method that has the highest accuracy percentage,and then the method is used as a basis for providing recommendations.

6.5. Results and Discussions

Data collection of this research was performed by conducting field study, which were direct observation and recording variables or quantitative attributesof study period, such as origin, how long they've studied in college, gender, high school type, department in high school, high school's national final score, parents' education, parents' occupations, and GPA on the fourth semester. Data were collected in the Industrial Engineering Department on students from year 2003 to 2007 and the attributes used in the classification method referred to the Unisys service of Universitas Islam Indonesia.

The processing in the case of students' length of study period used data mining software called Rapid Miner. Classification algorithm used in developing the model was *Naïve Bayes* algorithm and Support Vector Machine algorithm. The result of data processing based on the case of students' length of study period using Rapid Minersoftware was the probability of each attributeby *Naïve Bayes* and weight of each attribute by Support Vector Machine algorithm. Below is the result of processing by *Naïve Bayes* and Support Vector Machine:

IOP Conf. Series: Materials Science and Engineering **105** (2016) 012039 doi:10.1088/1757-899X/105/1/012039

5.1. The Result of Naïve Bayes Processing
The result of processing by *Naïve Bayes* method was the instance probability value of each attribute and label probability value of overall test data used in*Naïve Bayes* processing.

Table 2. Parameter Probability Value of Each Attribute

Attribute	Parameter	Do	Not Do
Origin	Value=Outside Java	0,306	0,254
Origin	Value=Java	0,694	0,746
Age	Value=On Time	0,694	0,622
Age	Value=Passed Late	0,265	0,274
Age	Value=Pass Quickly	0,041	0,104
Gender	Value=Male	0,857	0,622
Gender	Value=Female	0,143	0,378
High School	Value=State	0,673	0,771
High School	Value=Private	0,327	0,229
Majoring	Value=Social Science	0,286	0,104
Majoring	Value=Natural Science	0,714	0,895
NEM	Value=Upper	0,673	0,632
NEM	Value=Medium	0,306	0,368
NEM	Value=Low	0,020	0,000
Father's Education	Value=Junior High School	0,082	0,055
Father's Education	Value=Senior High School	0,408	0,458
Father's Education	Value=Bachelor Degree	0,224	0,249
Father's Education	Value=Elementary School	0,082	0,035
Father's Education	Value=Diploma 4	0,041	0,080
Father's Education	Value=Master Degree	0,082	0,050
Father's Education	Value=Bachelor Degree	0,020	0,000
Father's Education	Value=Diploma 3	0,020	0,070
Father's Education	Value=Doctorate Degree	0,020	0,005
Father's Education	Value=Junior High School	0,020	0,000
Mother's Education	Value=Senior High School	0,408	0,448
Mother's Education	Value=Diploma 3	0,102	0,144
Mother's Education	Value=Junior High School	0,143	0,149

Attribute	Parameter	Do	Not Do
Mother's Education	Value=Bachelor Degree	0,143	0,104
Mother's Education	Value=Elementary School	0,122	0,070
Mother's Education	Value=Master Degree	0,020	0,010
Mother's Education	Value=Diploma 4	0,061	0,070
Mother's Education	Value=Diploma 4	0,000	0,005
Father's Occupation	Value= Entrepreneur	0,224	0,249
Father's Occupation	Value=Private Employee	0,326	0,264
Father's Occupation	Value=Retired	0,102	0,040
Father's Occupation	Value=Civil Servant	0,306	0,398
Father's Occupation	Value=Farmer	0,041	0,025
Father's Occupation	Value=Army/Police	0,000	0,015
Father's Occupation	Value=Retired	0,000	0,010
Mother's Occupation	Value= Entrepreneur	0,245	0,249
Mother's Occupation	Value=Jobless	0,367	0,363
Mother's Occupation	Value=Civil Servant	0,306	0,294
Mother's Occupation	Value=Farmer	0,041	0,015
Mother's Occupation	Value= Private Employee	0,020	0,065
Mother's Occupation	Value= Retired	0,020	0,015
GPA of the Fourth Semesters	Value=Satisfy	0,571	0,488
GPA of the Fourth Semesters	Value=Not Satisfactory	0,326	0,070
GPA of the Fourth Semesters	Value=Very Satisfactory	0,102	0,433
GPA of the Fourth Semesters	Value= Satisfy	0,000	0,005
GPA of the Fourth Semesters	Value=Praiseworthy	0,000	0,005

5.2. The Result of Support Vector Machine Processing

The processing resultby Support Vector Machinewith the help of rapid softwareonly produces weight which is shown in Table 3. To discover the decision on each existing data, manual calculation should be performed.

Tabel 3. Attribute Weights

Attribute	Weights
Origin	-0,002
Age	-0,007

Attribute	**Weights**
Gender	0,032
High School	-0,005
Majoring	0,025
NEM	-0,025
Father's Education	0,007
Mother's Education	0,021
Father's Occupation	0,020
Mother's Occupation	0,004
GPA of the Fourth Semesters	0,150

The next calculation after obtaining the weightof rapid outputwas searching for score in every existing data by multiplying attribute in every data with the weight of each attribute. Below is formulation in searching for score of every data.

$$Score = \sum_{i=1}^{A} Ni \, . Wi$$

Ni is attribute value and*Wi*is attribute weight. After finding scoreof each data type, all scores are totaled. The final step in the method of support vector machine is finding the median value from the smallest to the biggest score. After obtaining the median value, then performed grading, where data is entered in a class 1 if the score ≤ 1.088 and the data included in class 2 if the score > 1,088.

5.3. Comparation Result Between Naïve Bayes Algorithm Method And Support VectorMachines Algorithm Method

The performances of the two methods generate different accuracy percentage. The accuracy of the results obtained based on the concept of each of themethods, i.e.*Naïve Bayes* algorithm and support vector machine algorithm. Whether the performanceof the model is good or not,it depends on each case at hand. Below is the comparison of the accuracy based on a lot in common with the facts.

Table4.Comparison the Accuracy

Method	**Accuracy**	**Error**
Naïve Bayes	80.67%	19.33%
Supp*ortVector Machine*	60%	40%

It can be found in a comparison table that outcome prediction accuracy rate is formed by using *Naïve Bayes* amounted to 80.67% and using Support Vector Machine has an accuracy rate of 60%. It proves that the *Naïve Bayes* method is more accurate to use for students drop out case.

6.6. Conclusions and Recommendations

6.7. Conclusions

- The level of accuracy based on the prediction that was formed by using *Naïve Bayes* is 80.67%, while accuracy rate by using Support Vector Machine is 60%. This shows that the *Naïve Bayes* method is more accurate in predicting the cases of drop out students based on a long period of study.

- The prediction pattern of drop out students that formed by using *Naïve Bayes* method generates data of the students origin area that come from Java, the age of students when entering the college punctually, most of drop out students are the male gender, private high school, they studied science in high school, high NEM category, their parents graduate of scholars, the father works as a private employee, the mother does not work, and most of the students who have been predicted to be drop out have unsatisfactory criteria or GPA <2.00 on the 4th semester.
- Parameters that affect every long period of study attribute based on *Naïve Bayes* method are the attribute of the island with Java parameter, the attribute of the age of students when entering the college punctually with the appropriate parameter, the attribute of gender with male parameter, the attribute type of school with domestic parameter, the attribute o what they studied in high school with science parameter, NEM category with high parameter, father's education with high school parameter, mother's education with high school parameter, father's occupation with private employee parameter, mother's occupation with doesn't not work parameter, and the GPA category on the 4th semester with satisfactory parameter that is <2,75.

6.8. Recommendations

- Conduct a classification concept of comparison using more than researcher did, because the more concept is tried, the more open and widely knowledge is obtained. We can also determine a match between the case and the method that used on it.
- The classification method is not only *Naïve Bayes* and *Support Vector Machine*, so the further research can use the classification concept by using *Artificial Neural Network, Nearest Neighbour*, or*Decision Tree.*
- The results from *Naïve Bayes* obtained rules that being used in making a simple useful application as predictor of drop out students.
- Conduct a students drop out research based on a lengthcof study period by inputting data that obtained qualitatively or by interview.
- The recommendations that given to the Industrial Engineering Department of Universitas Islam Indonesia is to provide a letter of warning to students with unsatisfactory status of GPA in 4th semester or GPA <2.00 and satisfactory status of GPA or GPA <2.75 by the time students completed the course in the 4th semester.

6.9. References

[1] Ridwan M, Suyono H and Sarosa M2013 *Penerapan Data Mining Untuk Evaluasi KinerjaAkademik Mahasiswa Menggunakan Algoritma Naive Bayes Classifier*Eeccis7(1) 59–64
[2] Jananto A2013 *Algoritma Naive Bayes Untuk Mencari Perkiraan Waktu Studi Mahasiswa*18(1) 9–16
[3] AndrianiA2012 Penerapan *Algoritma C4.5 Pada Program Klasifikasi Mahasiswa DropoutSeminar Nasional Matematika* 139–147
[4] Kusrini and E. T. Lutfi 2009 *Algoritma Data Mining*(Yogyakarta : Andi Publishing)
[5] Larose, Daniel. T2005*Discovering Knowledge in Data - an Introduction to Data Mining*(New Jersey: John Wiley &Sons)
[6] Untari, D 2010 *Data Mining Untuk Menganalisa Prediksi Mahasiswa Berpotensi Non-Aktif Menggunakan Metode Decesion Tree C4.5.*
[7] Damanik S, Ispriyanti D, and Sugito 2015 *Klasifikasi Lama Studi Mahasiswa Fsm Universitas Diponegoro Menggunakan Regresi Logistik Biner Dan Support Vector Machine*4(2000)123–132.
[8] Prasetyo, E2012 *Data Mining - Konsep dan Aplikasi Menggunakan MATLAB*(Yogyakarta: Andi Publishing)

IOP Conf. Series: Materials Science and Engineering **105** (2016) 012040 doi:10.1088/1757-899X/105/1/012040

Priority Determination for Higher Education Strategic Planning Using Balanced Scorecard, FAHP and TOPSIS (Case study: XYZ University)

Uky Yudatama[1] **and Riyanarto Sarno**[2]

[1]Departemen of Informatics, Universitas Muhammadiyah Magelang, Indonesia
[2]Departemen of Informatics, Institut Teknologi Sepuluh November Surabaya, Indonesia

E-mail: [1]uky.yudatama@mail.com, [2]riyanarto@if.its.ac.id

Abstract. The process of strategic planning is needed by a higher education in some cases, especially in preparing to face the challenges and competition. The results of strategic planning will help the higher education to provide a framework for achieving a competitive advantage as well as determine the direction of future policy in accordance with the desired objectives. In recent decades, the Balanced Scorecard has been applied in the field of information technology as a very popular tool and is used extensively, because it is a model that can explain between information technologies with "Business Objectives" in a comprehensive manner. This study uses 4 perspectives in the Balanced Scorecard and 7 standards in higher education quality assessment as sub-criteria. Fuzzy AHP and Fuzzy TOPSIS are used to determine the priority as making strategic policy recommendations in a higher education. The final result of this research shows the score of Customer Perspective 0.35365 is higher than other perspective, while the score in Research and Student Affairs gains significant score when compared with the others, namely 0.69753948is also higher. This means that both of them get very serious attention as a strategic planning basis for policy making.

1. Introduction

Along with the changing times and the need of education for the community, coupled with increasingly stringent government regulations, encouraging universities to conduct good and right strategic planning. The process of strategic planning is advantageous for a university in some cases, especially in preparing to face the challenges and competitions. The results of strategic planning will help the university to provide a framework for achieving a competitive advantage as well as determine the direction of future policy in accordance to the expected objectives. All components will engage and interact synergistically, actively participate and work together to achieve goals and to improve understanding of the organization / institution.

Strategic planning through strategic management process must be applied systematically and methodologically. For strategic planning, universities need to adjust a different business strategic

Corresponding author. Tel: +62-293-362082; fax: +62-0293-361004
E-mail address: uky.yudatama@mail.com

model since it is different from business strategy of a company. A university strategic planning customarily takes 5 years or more.

In recent decade, the Balanced Scorecard has been applied in the field of information technology as a very popular tool and is used extensively, because it is a model that can explain information technologies with *"Business Objectives"* in a comprehensive manner.

This research using the criteria determining perspectives in the balanced scorecard. This perspectives is very important as aligning business strategy which became the basis of the reference. As a real issue, whether in the run up to evaluate are in accordance with the parameters or standards that exist in the higher education quality assessment. Therefore, the existing standards in higher education quality assessment is used as a sub-criterion, as the elaboration of criteria that is in the aspects of the balanced scorecard. Fuzzy AHP is used to determine the weights criteria and Fuzzy TOPSIS is used to determine the priority as making a strategic policy recommendations in a higher education.

2. Basic Theory

There is one key aspect that needs to be considered in carrying out the whole process of higher education institutions accreditation. The Aspect is accreditation standard that is used as benchmark in evaluating and assessing quality of performance, condition and state of education of a higher education institution.

Accreditation standard of undergraduate study program includes the standards of commitment to undergraduate study programs of institutional capacity *(institutional capacity)* and a commitment to the effectiveness of educational programs *(educational effectiveness)*, which is packed in 7 *(seven)* accreditation standards: Standard 1 (vision, mission, goals and objectives, and achievement strategy); Standard 2 (supervisor management, leadership, management systems and quality assurance); Standard 3 (Students and graduates; Standard 4 (human resources); Standard 5 (curriculum, learning, and academic atmosphere); Standard 6 (financing, infrastructure, and information systems); Standard 7 (research and service / dedication to community, and collaboration).

2.1. Balanced Scorecard

The Balanced Scorecard (BSC) is a methodology to conduct management strategy consisting of 4 perspectives:

a. Financial Perspective (FP),
b. Customer Perspective (CP),
c. Internal Process Perspective (IPP) and
d. Learning and Growth Perspective (LGP).

2.2. Fuzzy AHP

Analytical Hierarchy Proses is a multi-criteria decision technique proposed by saaty (1980) to solve problems of planning needs and management of scarce resources that, in time has become one of the most widely used technique in decision making processes on multi criteria.

Introduced the theory of fuzzy sets in 1965 to model the concept of vagueness, characteristis of human thought. Fuzzy numbers allow to face problems in which the criteria are not precisely defined. Therefore, fuzzy AHP (FAHP), a fuzzy extension of AHP, can be used to solved hierarchical fuzzy problems. FAHP applications can be found in diverse areas such as selection of operating system, recruitmen of staff, risk assessment projects in information technology, selection of ERP system.

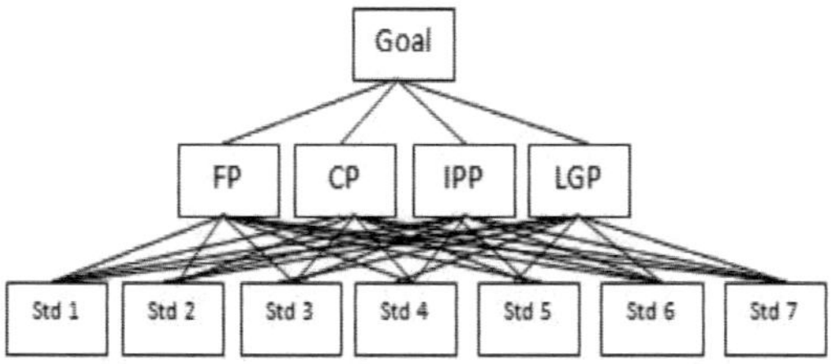

Figure 1. The structure of the hierarchy problem.

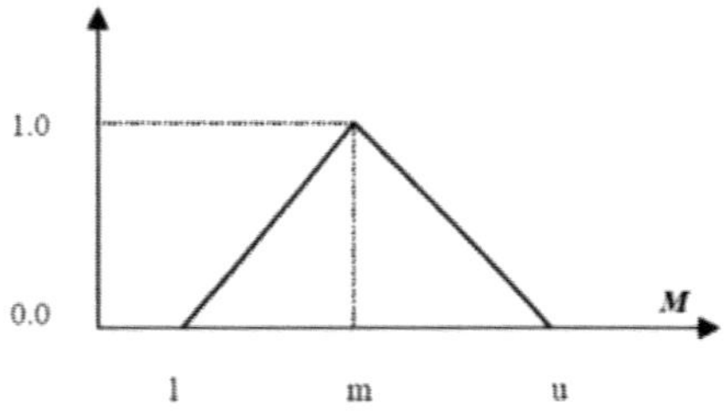

Figure 2. Triangular Fuzzy Numbers

Triangular fuzzy numbers in Table 1 are denoted by M = {l, m, u}, where M is a set of fuzzy numbers consisting of l, m and u which respectively express the smallest possibility value, the most closest value, and the greatest possibility value.

If X = {x1, x2, x3, ... xn} it denotes a set of objects, and G = {g1, g2, g3, ..., gn} it denotes a set of goals. If there is a number of m criteria that will be used for analysis, then it is obtained M gi 1, M gi2, M gi.3 ... M gi. mi = 1,2,..., n, where j Mgi(j=1, 2, ...,n) is a fuzzy triangular numbers. The steps used to analyze further is as follows:

Step 1: It is defined as fuzzy synthetic extent value (Si) with criteria to i by equation 1.

$$S_i = \sum_{j=1}^{m} M_{gi}^{j} \, x \left[\sum_{i=1}^{n} \sum_{j-1}^{m} M_{gi}^{j}\right]^{-1} \quad (1)$$

to get $\sum_{j=1}^{m} M_{gi}^{j}$, a fuzzy value summation operation of m value on matrix using equation 2 is conducted

$$\sum_{j=1}^{m} M_{gi}^{j} = \left(\sum_{j=1}^{m} l_j, \sum_{j=1}^{m} m_j, \sum_{j=1}^{m} u_j\right) \quad (2)$$

To get $\left[\sum_{i=1}^{n} \sum_{j-1}^{m} M_{gi}^{j}\right]^{-1}$, a fuzzy summation operation of Mjgi values (j = 1, 2, ..., n) by using equation 3 is conducted

$$\sum_{j=l}^{n} \sum_{j=1}^{m} M_{gi}^{j} = \left(\sum_{j=1}^{m} l_j, \sum_{j=1}^{m} m_j, \sum_{j=1}^{m} u_j\right) \quad (3)$$

Then inversion of the vector in the above equation using equation 4 is calculated.

$$\left[\sum_{j=1}^{n} \sum_{j=1}^{m} M_{gi}^{j}\right]^{-1} = \left(\frac{1}{\sum_{j=1}^{n} l_j}, \frac{1}{\sum_{j=1}^{n} m_j}, \frac{1}{\sum_{j=1}^{n} u_j}\right) \quad (4)$$

Step 2: calculate the degree of possibility, where $M_1 = (l_1, m_1, u_1)$ and $M_2 = (l_2, m_2, u_2)$ are two fuzzy triangular numbers, so that the degree of possibility $M_2 = (l_2, m_2, u_2) \geq (l_1, m_1, u_1)$ is obtained from equation (5) and (6)

$$V = M_2 \geq M_1 = hgt\ (M_2 \cap M_1) = \mu_{M2} \quad (5)$$

$$= \begin{cases} 0 & , if\ m_2 \geq m_1 \\ 1 & , if\ l_1 \geq u_2 \\ \frac{l_1 - u_2}{(m_2 - u_2) - (m_1 - l_1)} & , otherwise \end{cases} \quad (6)$$

Step 3: to compare the degree of possibility among criteria and among fuzzy numbers $M_{1=}(i=1,2,...,k)$ through equation (7).

$$V(M \geq M_1, M_2, ...M_k) \\ = V[(M \geq M_1)] \text{ and } (M \geq M_2) \text{ and } .... \text{ and } (M \geq M_k)] \\ =_{min} V(M \geq M_1),\ i = 1,2,.....,k \tag{7}$$

By assuming d $(A_i) =_{min} V\ (S_i \geq S_{k)}$ for k =1,2,...k;k $\neq$ i. so that the value of weight vector in equation (8) is obtained.

$$W' = (d'(A_1), d'(A_2), \ldots,\ d'(A_n))^T \tag{8}$$

After the weight vector in equation (8) is normalized, the normalized weight vector is obtained which is shown in equation (9).

$$W = (d(A_1), d(A_2), \ldots, d(A_n))^T \tag{9}$$

Where W is not a fuzzy numbers.

2.3. *Fuzzy TOPSIS*

After the weight values for each criterion are obtained, then calculation using TOPSIS method that has been fuzzier is done. Contrastwith TOPSIS method which directly determines the value of the criteria for each alternative .After each criteria of alternative is rated, then the integral total value for each alternative using equation (10)is calculated.

$$x = I(F) = 1/2(\alpha c + b + (1-\alpha)a) \tag{10}$$

With α is h the degree of optimism with a value between 0 and 1. After the integral total value on each criterion is obtained, normalization through equation (11) is conducted.

$$r_{ij} = \frac{x_{ij}}{\sqrt{\sum_{i=1}^{m} x_{ij}^2}} \tag{11}$$

Then the calculation to get normalized weight matrix through equation (12) is conducted.

$$v_{ij} = W_{ij} * r_{ij} \tag{12}$$

Then positive ideal solution value and negative ideal solution value is calculated using equation

$$A^+ = \{v_1^+, v_2^+, \ldots, v_n^+\} \\ A^- = \{v_1^-, v_2^-, \ldots, v_n^-\} \tag{13}$$

The disparity between the alternative with positive ideal solution is calculated with equation (14).

$$D_i^+ = \sqrt{\sum_{j=1}^{n} (v_{ij} - v_j^+)^2} \tag{14}$$

While alternative disparity with negative ideal solution is calculated by equation (15).

$$D_i^- = \sqrt{\sum_{j=1}^{n}(v_{ij} - v_j^-)^2} \quad (15)$$

By comparing the disparity with positive and negative ideal solutions, then the preference value for each alternative is obtained by equation (16).

$$CC_i = \frac{D_i^-}{D_i^- + D_i^+} \quad (16)$$

CC_i value obtained from equation (16) is the final value used to determine the ranking of some alternatives that will be selected with the value of initial rank order starting from the biggest CC_i value.

3. The Proposed Method

Research model was designed as shown in figure 3. Stages of research are described based on research model.

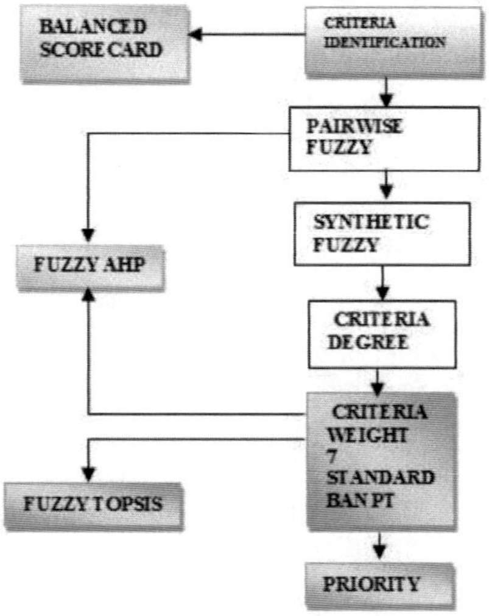

Figure 3. Research model.

4. Result

The first stage in conducting this research is to determine the criteria by considering 4 balanced scorecard perspective. After comparing all criteria/attribute exists, a matrix as shown in table 1 is obtained.

Table 1. Comparison Matrix of Balanced Scorecard

Criteria	FP	CP	IPP	LGP
FP	111	1/51/31	135	111
CP	135	111	135	357
IPP	1/51/31	1/51/31	111	135
LGP	111	1/71/51/3	1/51/31	111

Table 2. Fuzzy value Synthetic Extent

	l	m	u
FP	0.0857	0.2174	0.5738
CP	0.1607	0.4891	1.2910
IPP	0.0643	0.1902	0.5738
LGP	0.0628	0.1033	0.2391

The next step is to look for fuzzy value Synthetic extents (Si) using equation (1), so that it will gain matrix as shown in table 2.
Then possibility degree among the criteria is calculated by using equation (3), so that it will gain the degree of criteria possibility as shown in table 3.

Table 3. Value Degree Possibilities

Criteria	FP	CP	IPP	LGP
FP	1	1	1.1407	0.9975
CP	0.8306	1	0.7485	0.2485
IPP	1	1	1	1.161035
LGP	1	1	1	1

Table 4. Value Criteria (W') and Weight Normalized (W)

	FP	CP	IPP	LGP
W'	0.8306	1	0.7485	0.2485
W	0.2937	0.3536	0.2647	0.0879

The weight vector value created from table 3 above, after calculated and normalized for each criteria, then the weight vector value produced will be as shown in table 4. Weight vector values produced from Fuzzy AHP method is used to perform calculations with Fuzzy TOPSIS method. Each criterion of each perspectives in balanced scorecard gives value using triangular fuzzy numbers, as in Table 5.

Table 5. Triangular Fuzzy

Standard	FP	CP	IPP	LGP
Vision and Mission	0.25 0.5 0.75	0.25 0.5 0.75	0.25 0.5 0.75	0.25 0.5 0.75
Good Governance	0.75 1 1	0.5 0.75 1	0.75 1 1	0.75 1 1
Students Affairs	0.5 0.75 1	0.75 1 1	0.5 0.75 1	0.5 0.75 1
Human Resources	0.5 0.75 1	0.75 1 1	0.5 0.75 1	0.75 1 1
Curricula	0.5 0.75 1	0.25 0.5 0.75	0.25 0.5 0.75	0.5 0.75 1
Infrastructure	0.5 0.75 1	0.5 0.75 1	0.75 1 1	0.5 0.75 1
Research	0.5 0.75 1	0.75 1 1	0.5 0.75 1	0.5 0.75 1

Matrix triangular fuzzy values obtained in Table 5 is calculated using the total value of the integral equation (10) to obtain the matrix X as in Figure 5 below.

0.5	0.5	0.5	0.5
0.94	0.75	0.94	0.94
0.75	0.94	0.75	0.75
0.75	0.94	0.75	0.94
0.75	0.5	0.5	0.75
0.75	0.75	0.94	0.75
0.75	0.94	0.75	0.75

Figure 5. Integral Total

0.073936698	0.085513217	0.06655482	0.021275013
0.139000993	0.128269826	0.125123061	0.039997024
0.110905047	0.160764849	0.099832229	0.031912519
0.110905047	0.160764849	0.099832229	0.039997024
0.110905047	0.085513217	0.06655482	0.031912519
0.110905047	0.128269826	0.125123061	0.031912519

Figure 6. Matrix weight normalized

By using equations (11) and (12) of the matrix in Figure 5, it was found that the normalized weight matrix V in Figure 6.
In the normalized weight matrix in Figure 5, the positive ideal solution value and negative ideal solution value use equation (13), then the disparity between the alternative with positive ideal solution is calculated with equation (14), while the disparity between the alternative and negative ideal solution is calculated with equation (15). By comparing the disparity between the positive ideal solution to the negative ideal solution, then preference value for each alternative on table 6 through equation (16) is obtained.

Table 6. Final Value of Risk Alternative and its Rankings

Standard	Point	Priority
Vision and Mission	0.139547712	7
Good Governance	0.584152191	5
Students Affairs	0.69753948	2
Human Resource	0.681360105	3
Curricula	0.274575757	6
Infrastructure	0.274575757	4
Research	0.69753948	1

From the calculation results obtained, it shows that the Research and Student Affairs gains significant score when compared with the others, namely 0.69753948. This shows that official management is very important to note. Although vision and mission get lowest value, it is also need to get serious attention.

5. Conclusion

This research aims to determine priority as recommendation for strategic decision maker in a higher education.. Four perspectives in Balanced Scorecard are used as the criteria in this research. . This perspectives is very important as aligning business strategy which became the basis of the reference. As a real issue, whether in the run up to evaluate are in accordance with the parameters or standards that exist in the higher education quality assessment. Therefore, the existing standards in higher education quality assessment is used as a sub-criterion, as the elaboration of criteria that is in the aspects of the balanced scorecard. Fuzzy AHP is used to determine the weights criteria and Fuzzy TOPSIS is used to determine the priority as making a strategic policy recommendations in a higher education. The final result of this research shows the score of Customer Perspective 0.35365 is higher than other perspective, while the score in Research and Student Affairs gains significant score when compared with the others, namely 0.69753948is also higher. This means that both of them get very serious attention as a strategic planning basis for policy making.

IOP Conf. Series: Materials Science and Engineering **105** (2016) 012040 doi:10.1088/1757-899X/105/1/012040

References

[1] Amg,L.,Wen,C&Ching, J, 2008. A fuzzy AHP and BSC approach for evaluating performance of IT department in the manufacturing industry in Taiwan, Expert systems with Applications 34, 94-107.

[2] Ashtiani, B., Haghighirad, F., Makui, A., Montazer, G.A., 2008.Extension of Fuzzy TOPSIS Method Based on Interval-valued Fuzzy Sets. Applied Soft Computing. Vol. 9, No.2, 457-461

[3] Amiri, M.P., 2010. Project Selection for Oil-.elds Development by Using AHP and fuzzy TOPSIS methods. Expert System with Application, 37, 6218-6224.

[4] Manian. Amir, Fathi, M. Reza, Zarchi. M. Karimi, Omidian. Asie, 2011. Performance Evaluation of IT Department using a Modified Fuzzy TOPSIS and BSC methodology (Case study: Tehran Provinces Gas Company), Journal of management research, ISSN 1941-899X, Vol 3, No 2:E10, Tehran, Iran.

[5] BAN-PT, 2008. "Buku I Naskah Akademik Akreditasi Institusi Perguruan Tinggi", Jakarta.

[6] Balli, S. & Korukoglu, S, 2009. Operating system Selection Using Fuzzy AHP and TOPSIS. Mathematical and Computational Applications, 14(2), 119-130.

[7] Milis and Mercken, K. Milis and R. Mercken. 2004. The use of the balanced scorecard for the evaluation of information and communication technology projects, International Journal of Project Management 22, pp. 87–97.

[8] Sarno. R. 2009, Strategi Sukses Bisnis dengan Teknologi Informasi, Berbasis Balanced Scorecard & CobiT. Surabaya: ITS Press.

[9] Sarno. R,,A. Herdiyanti, 2010. "Information technology policies for enterprise resource planning practices to improve customer orientation and service," International Journal of Computer Science and Network Security, vol.10, no..5, pp. 82-94, March.

[10] Sarno. R.,,Haryanto T, 2010 "Comparation of CobiT Maturity Model and Structural Equation Model for Measuring the Alignment between University Academic Regulations and Information Technology Goals", IJCSNS International Journal of Computer Science and Network Security, VOL.10 No.6, June.

[11] Sarno R., Djeni CA., Mukhlash I., Sunaryono D., 2015. Developing A Workflow Management System For Enterprise Resource Planing, 2015, Journal of Theoretical and Applied Information Technology 28th February. Vol.72 No.3

[12] Yudatama. U, Sarno. R. 2015. Evaluation Maturity Index and Risk Management for IT Governance using Fuzzy AHP and Fuzzy TOPSIS (Case Study Bank XYZ), Intelligent Tecnology and Its Application (ISITIA) 2015 International On, 323-328, Publisher IEEE.

Characterization Study of Band Gap, Resistivity, Crystal Structure, and Phase Identification of $CuInSe_2$ Ternary Alloy

Bungaran Saing[1), **and Budi Arto** [2)]
[1)]Departement of Chemical Engineering, Faculty of Engineering, University of Bhayangkara Jaya, Jakarta, Indonesia
Email :bungaran.saing@yahoo.com
[2)] Departement of Mechanical Engineering, Faculty of Engineering, University of Indonesia Christian University, Jakarta, Indonesia
Email :damaz581102@gmail.com

Abstract. Characterization study of band gap, resistivity, crystal structure, and phase identification of $CuInSe_2$ ternary alloy which is the synthesis result of Bridgman method by using modified simple single zone temperature, have been carried out. The phase identification and crystal structure of obtained polycrystal were characterized by X-ray diffraction. Then the X-ray diffractograms were analyzed by using crystallographic software package GSAS. Electrical resistivity and bandgap were measured by using Van der Pauw method. Identification and analysis of diffractogram show that ternary alloy of $CuInSe_2$ has main phase of space group $I\bar{4}$ad and lattice parameters a and c of 6,1173 Å and 11,7144 Å respectively. The other coexisting phases were identified as $CuIn_5Se_8$ with the space group $P\bar{4}2m$ and Se with the space group $P3_121$. Result from the resistivity measurement shows the resistivity is between (1,596 – 9,666) x $10^{-3}\Omega$m, and Bandgap is around 1,03 – 1,61 eV.

Keywords: *crystal structure, phase identification, bandgap,* $CuInSe_2$

1.Introduction

Indonesia is a tropics that have big enough of solar energy. Based on collected data of solar radiation from seven locations in Indonesia, the solar radiation in Indonesia can be clarified respectively - were as follows: for the western region and eastern Indonesia by irradiating distribution in the Western Region of Indonesia, around 4.3 kWh / m^2 / day with a monthly variation of approximately 10% and Eastern Region of Indonesia about 4.7 kWh / m^2 / day with a monthly variation of approximately 9%, thus the potential of the average solar irradiation in Indonesia around 4.5 kWh / m^2 / day with variations monthly about 9%.[1]

Non-silicon semiconductor material is intensively developed for thin film solar cell's coating, are $CuInSe_2$ (and its alloys such as $CuInS_2$, Cu (InGa) Se_2, or CuInSe), CdTe, and silicon amorph. $CuInSe_2$ is the most promising material for solar cells[2] for many reason.

The commonly techniques used to deposited of semiconductor are Bridgman method, Czockralski method, and sputtering (dc and rf). Basically, all these techniques have each advantages and disadvantages. The advantage of Bridgman technique is produced massive semiconductor material wich very homogeneous, but this technique worked at a very high temperature, and it took the material (quartz tube) so it may expensive. [3]

Now on the center of attention from researchers around the world is $CuInSe_2$ alloy and other materials of ternary compounds. It can be potential to be base material optoelectic devices such as solar cells and photo-detectors. Partial substitution of indium (In) and gallium (Ga) in the compound material added the optical absorption of this material are suited for solar cell applications. [4,5]The research importance of compound ingredient was to research the properties of materials such as in the form of a single crystal. However, we can also examine the properties of the material that formed of polycrystalline, in fact the single crystal and polycrystalline have similar properties except in this case of the grain boundary.

$CuInSe_2$ is a direct band gap semiconductor material. It causes very strong material absorbs a photon. [6,7] Alloy material was made by polycrystalline and growth of $CuInSe_2$ generally performed by melted these materials in an ampoule vacuum pressure of 10^{-4}Torr and it put in a vertical furnace with dual zone temperature method. The Bridgman method was the suitable metod for cultivate this material.Thegrowth of polycrystalline semiconductor ternary compound $CuInSe_2$ materials can be used by a vertical furnace temperature single zone, that must be modificated previously to furnace that can produced similar conditions with dual zone temperature vertical furnace.

The research just only characterized ternary alloy $CuInSe_2$, which has grown by a modified Bridgman method. Domination an high quality semiconductor manufacturing technology was need an high precision and accurate crystal growing equipment which able to process automatically and stable.

This research purposed for characterized of ternary alloy $CuInSe_2$ synthesize by Bridgman method which can be expected a specifications of diesel basic ingredient. The results of this research was also expected as a reference to the next research, especially the made of thin layer of solar cell application. Besides the series of these researches were expected to provide a new hope and information of applications development that ever been.

2. Materials and Methods

The material are cylindrical ingots with conical ends which length 40 mm and diameter 10 mm that made of ternary compound $CuInSe_2$ by Bridgman method in Puspiptek Laboratory Serpong, Tangerang, Banten.

The equipment: Diffractometer X-ray, electrical resistivity test used four-point probe Van der Pauw method, chemical composition test by XRF instrument, band gap measurement tools.

Procedure: Put material Cu, In, and Se with each purity 99.999% to ampoules, then vacuum the ampoules till 10^{-4} Torr by vacuum Veeco 300, 220 V / AC, 1 vacuum pump. In vacuum condition, sealed off the ampoules by the welding. After complete the process, tied the ampoules on a stick then put into furnace cavity properly till the ampoules located in symmetrical position in center. Monitoried the temperature by couple thermo, next programmed the process depend on data which got from each melting point of Cu, In and Se also alloy data. Controlled system done by PID (proportional - integral - differential)

IOP Conf. Series: Materials Science and Engineering **105** (2016) 012041 doi:10.1088/1757-899X/105/1/012041

technique that one of controller with closed feedback where the feedback get from installed thermo couples in the furnace. Temperature controller used panel with temperature risen can be set according to characteristic of the process element, as well as the temperature decreased. Besides PID constants from temperarure controller must be searched and determined so that furnace temperarure can be stable.
In sixth phase, shook the furnace by the motor around 15 minutes to get an homogeneous mixture then put the furnace to vertical position.The alloy material wrapped with quartz, so it need an opened process after that cut it transversely. Cut it to be some parts with every parts have 2 mm thickness and numbered them from top till the pointed end.

Test the crystal structure and identified phase by using X-ray diffractometer. Analysis the X-ray diffractogram by using GSAS crystallographic program that developed by AC Larson from Los Alamos National Laboratory, USA. The electrical resistivity was measured by four points probes Van der Pauw method, and composed chemical element by XRF.

3. Results and Discussion

a. Crystal Structure Analysis and Phase Identification

Picture 1, 2 and 3 showed analysed X-ray diffractogram by using GSAS crystallographic program.

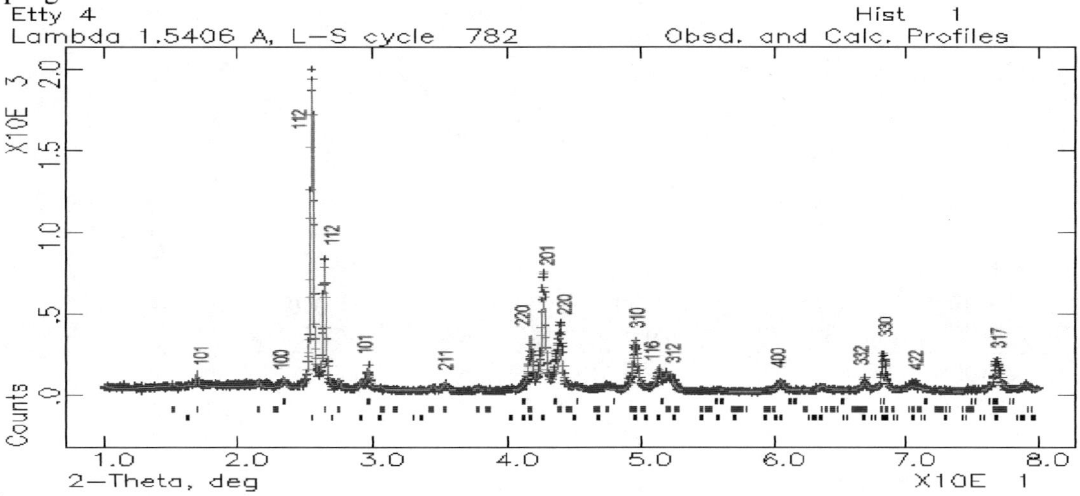

Figure 1.The pattern of $CuInSe_2$ ternary alloy X-ray diffractogram at top side. The (+) curve showed observation data and the (-) full line showed calculation.

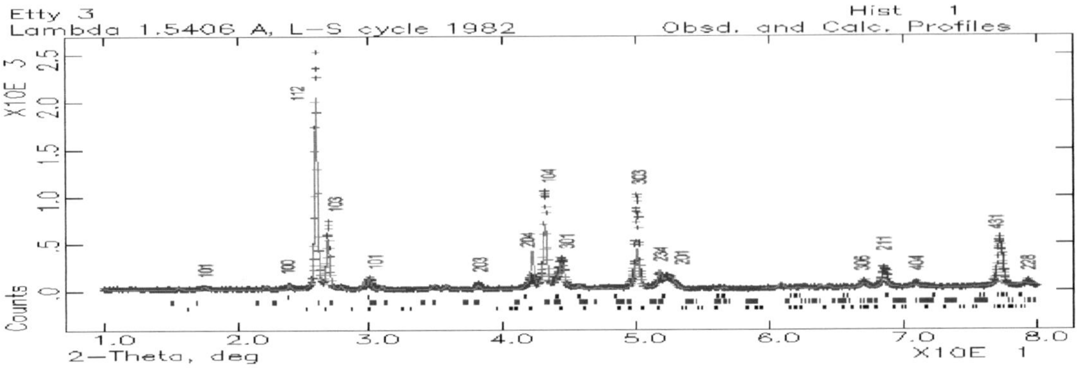

Figure 2. The pattern of $CuInSe_2$ ternary alloy X-ray diffractogram at middle side. The (+) curve showed observation data and the (-) full line showed calculation.

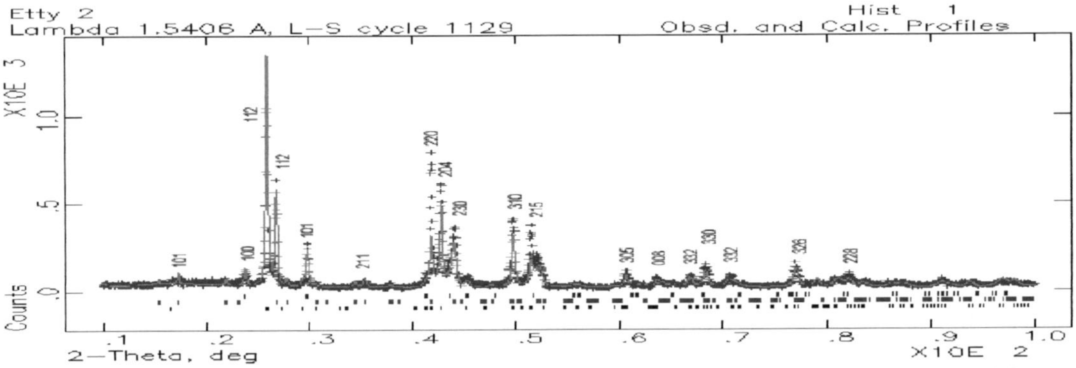

Figure 3. The pattern of $CuInSe_2$ ternary alloy X-ray diffractogram at bottom side. The (+) curve showed observation data and the (-) full line showed calculation.

Table 1. Refine results' parameter of $CuInSe_2$ ternary alloy ingot samples which has a tetragonal crystal structure used the GSAS

Sample	Phase	Space Group	a (Å)	b (Å)	c (Å)	Wt %
	$CuInSe_2$	$I\bar{4}ad$	6.11730	6.11730	11.71449	53.372
1	$CuIn_5Se_8$	$P\bar{4}2m$	5.80816	5.80816	11.68055	27.222
	Se	$P3_121$	4.38144	4.38144	4.96396	19.406
	$CuInSe_2$	$I\bar{4}ad$	6.24718	6.24718	11.94947	66.054
2	$CuIn_5Se_8$	$P\bar{4}2m$	5.90939	5.90939	11.90826	27.325
	Se	$P3_121$	4.35568	4.35568	4.92260	6.621
	$CuInSe_2$	$I\bar{4}ad$	6.12337	6.12337	11.75300	53.502
3	$CuIn_5Se_8$	$P\bar{4}2m$	5.81046	5.81046	11.69003	38.224
	Se	$P3_121$	4.37664	4.37664	4.78800	8.274

X-ray diffractogram analysis used GSAS crystallography program package, showed that $CuInSe_2$ ternary alloy which has tetragonal crystal structure and three phases. The phases were the main phase called $CuInSe_2$ phase with space group $I\bar{4}$ ad the lattice parameters a and c for each 6,1173 Å and 11,7144 Å, the 2 other phases called $CuIn_5Se_8$ phase with space group $P\bar{4}2m$ and Se phase with space group $P3_121$. It showed that $CuInSe_2$ ternary alloy was polycrystalline, and its crystal structure form was tetragonal with average lattice parameter which got from $CuInSe_2$ ternary alloy is a = b = 6.11730 Å and c = 11.71449 Å, c/a = 1, 899. Therefore this value has the difference around 0,140 with c/a constant depend on Rockett and Birmire.[3]

b. Element Composition Analysis
Table 2 showed the measurement of element composition by XRF from $CuInSe_2$ ternary alloy. The sample was taken from $CuInSe_2$ ternary alloy incisions that numbered 1 (top), 2 (middle), 3 (bottom).

Table2. Element composition (wt%) taken from XRF $CuInSe_2$ ternary alloy ingot

Unsure	Wt(%) Sample		
	1	2	3
Cu	29.3239	34.3687	37.6815
In	36.7999	37.3043	35.2260
Se	31.8650	26.5839	25.1107
Si, Fe	1.2408	1.8933	1.3798

The composition of $CuInSe_2$ ternary alloy indicated there were other elements, it because there was impurities when sample preparation, for example when vacuum time, welding, or ingot cutting.

c. Resistivity Analysis
Table 3 showed resistivity measurement with four-point probe and Van der Pauw method, from $CuInSe_2$ ternary alloy. The resistivity measurement took few places of $CuInSe_2$ ternary alloy incisions then averaged them. The incisions were taken from number 1 (top side), 2 (middle side), and 3 (bottom side). The measurement of semiconductor type showed that wafer was tipe-p.

Table 3.Resistivity measurement $CuInSe_2$ ternary alloy

Sample	Resistivity (Ωm)
Top	(1,685 - 8,674) x 10^{-3}
Middle	(1,735 - 4,642) x 10^{-3}
Bottom	(1,596 - 9,666) x 10^{-3}

Resistivity measurement of $CuInSe_2$ ternary alloy showed that the resistivity was uneven, although the type of semiconductor was same type-p. The variation of electricity resistivity from $CuInSe_2$ ternary alloy was associated by many factors or possibilities, among other things it influenced of atom dislocated population or other vacancy atomic. Therefore the placement H^+ ion in vacancy atomic caused bandgap lessen as theoretically. The lessen caused of the position of these ions to the center of recombination which took place between valence band and conduction band.

Cu was a conductance that has a high conductivity values. Se was a type-p semiconductor that has high conductivity. In was valence 3 conduction that can be used in Silicon. If these 3 materials was joined with special condition (deposition parameter), it could be a p-type of $CuInSe_2$. Commonly Cu made conductivity bigger besides In and Se formed p-type[9].

d. Band Gap Analysis

Table 4 showed the Band Gap measurement of $CuInSe_2$ ternary alloy. The measurement, were taken from resistivity measurement in many places then averaged them. The places were incisions $CuInSe_2$ ternary alloy at top side (number 1), middles side (number 2), and bottom side (number 3).

Table 4. Band gap measurement Data of $CuInSe_2$ ternary alloy

Sample	Band Gap (eV)
Top	(1,03 – 1,61)
Middle	(1,02 – 1,32)
Bottom	(1,01 – 1,57)

Previously researcher found band gap value of $CuInSe_2$ ternary alloy was 1.02 eV[2]. But this research found around 1.01 – 1,59 eV, there was above the average from other researchers.

4. The Conclusion

According to the observation and analysis of $CuInSe_2$ ternary alloy, it can be concluded as follows:

a. $CuInSe_2$ ternary alloy growing used vertical furnace single zone temperature as a modification was quite good.
b. Crystal structure analysis and identified from diffractogram patterms showed lattice parameters of $CuInSe_2$ polycrystall were a = b = 6.11730 Å and c =11.71449 Å, c/a = 1,899. Besides lattice parameters based on these refences by reference were a = b = 6.104 Å, c = 11.714Å and c / a = 1.919. Form phases were $CuInSe_2$ as main phase, $CuIn_5Se_8$ as phas, and Se phase.
c. Electrical resistivity analysis showed uneven i.e. (1.6 to 8.6) x $10^{-3}\Omega$m,(1.7 to 4.6) x $10^{-3}\Omega$m, and (1.5 to 9, 6) x $10^{-3}\Omega$m.
d. Composition analysis showed that the composition was not the same i.e. Cu (29.3239%), In (36.7999%), Se (31.8650), and Si + Fe (1.24%).
e. Band Gap analysis were (1.03 to 1.61) eV, (1.02 to 1.32) eV, and (1.01 to 1.57) eV.

IOP Conf. Series: Materials Science and Engineering **105** (2016) 012041 doi:10.1088/1757-899X/105/1/012041

Reterence

[1] Raharjo I and Nurdyastuti I 1996*Introduction of energy*, publisher partner PT Agung Pratama, Jakarta

[2] Rockett A and Birmire RW 1991*$CuInSe_2$ for Photovoltaic Application*, Journal, University of Illionois.

[3] Zezen A 2005*Preparation and characterization of semiconducting material with a thin layer of $CuInSe_2$ sputtering technique*, Skripsi S-1, the State University of Yogyakarta.

[4] Irmansyah1998*Preparation and characterization of polycrystalline Copper Gallium Diselenide ($CuGaSe_2$) made with horizontal single-temperaturefurnace*, Thesis S-2, UI.

[5] Ridley N1965*Journal of Less - Common Metals*, 8, 354

[6] Lin-gunL and Bassett WA 1973*JournalApplied of Physict*,.441

[7] Ciszek TF1986*Journal of Crystal Growth*, 75,61.

[8]Cullity BD1978*Elements of X-Ray Diffraction second edition,* Addison - Wesley PublishingCompany, Inc.

[9] AtmonoTMYunantoWiryoadi and Bambang S 2006*Preparation of $CuInSe_2$ target material for the manufacture of thin layers of p-type $CuInSe_2$*, paper, PTAPB BATAN, Jogjakarta

[10] SchummRH Cuman WW Bailey S Evans BH and Parker VB 1973*In the National Bureau of Standarts* (USA) Technical Notes 270-1 to 270-8.

IOP Conf. Series: Materials Science and Engineering **105** (2016) 012042 doi:10.1088/1757-899X/105/1/012042

Effect of ca^{2+} to salicylic acid release in pectin based controlled drug delivery system

L Kistriyani[1], S K Wirawan[2], W B Sediawan[2]

[1]Chemical Engineering Department, Universitas Islam Indonesia, Yogyakarta, Indonesia
[2]Chemical Engineering Department, Universitas Gadjah Mada, Yogyakarta, Indonesia

Abstract. Wastes from orange peel are potentially be utilized to produce pectin, which are currently an import commodity. Pectin can be used in making edible film. Edible films are potentially used as a drug delivery system membrane after a tooth extraction. Drug which is used in the drug delivery system is salicylic acid. It is an antiseptic. In order to control the drug release rate, crosslinking process is added in the manufacturing of membrane with $CaCl_2.2H_2O$ as crosslinker. Pectin was diluted in water and mixed with a plasticizer and $CaCl_2.2H_2O$ solution at 66°C to make edible film. Then the mixture was dried in an oven at 50 °C. After edible film was formed, it was coated using plasticizer and $CaCl_2.2H_2O$ solution with various concentration 0, 0.015, 0.03 and 0.05g/mL. This study showed that the more concentration of crosslinker added, the slower release of salicylic acid would be. This was indicated by the value of diffusivites were getting smaller respectively. The addition of crosslinker also caused smaller gels swelling value,which made the membrane is mechanically stronger.

1. Introduction

Orange is a fruit that is commonly found and widely cultivated in Indonesia. These plants can be grown in lowland areas up to 770 meters above surface area of the sea [1]. The production of orange in Indonesia is very abundant, since the 2004 – 2012 average yield of citrus production as much as 2 million tons every year [2]. Production with raw materials from oranges such as orange juice will have the wastes from skin, pulp and seed. Citrus processing wastes, especially the skin is one of the raw material for production of pectin that is widely used in the food industry [3]. Indonesia has a good potential as a producer of citrus fruits, but the utilization of citrus wastes as a source of pectin in the industry has not be done. Pectin is a polysaccharide compound anions contained in the primary cell wall and intercellular wall plants. The constituent molecules consist of galacturonic acid molecules that bind to the α-(1-4)–glikosida bond thus forming poligalacturonic acid [4]. Figure 1 shows the compound of pectin in the cell walls of plants [5].

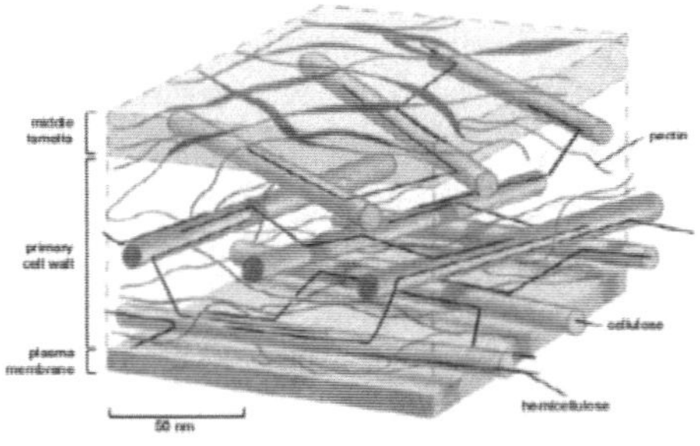

Figure 1. Plants Cell Wall Structure

The molecular chain of pectin that is a combination from galacturonic acid molecules is shown in figure 2.

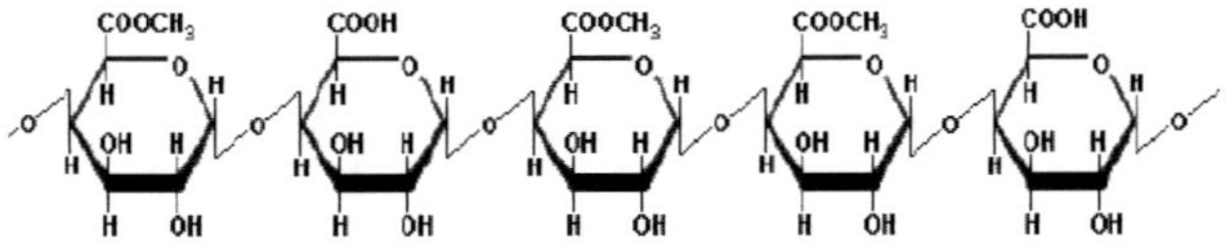

Figure2. The Molecular Chain of Pectin

Pectin is an important additional component in the food industry, cosmetics, and pharmaceuticals, because of its ability to change the functional properties of food products such as viscosity, emulsions, and gels. Based on its advantages, the uses of pectin are increasing, both as a food industries or non- food industries raw materials. In addition, pectin has the property that can deliver different types of drugs, proteins and cells. Because of these properties, pectin has the potential to be applied in drug delivery systems.

Drug Delivery System (DDS) is a term which is closely associated with delivery of pharmaceutical compounds (drugs) in humans or animals [6]. DDS is usually made in a patch form which should be a thin, elastic, and transparent to be attached at the skin surface. DDS is widely used to replace conventional drug delivery, because:

1. DDS can give a controlled drug dose that can reduce the need of repetition of drug consumption.
2. DDS can reach the right target object.
3. DDS does not cause pain and discomfort as we usually get from conventional drug delivery.
4. If there is side effect, it can be stopped easily.
5. It can be given in a longtime without being replaced as long as the patch still adhere on the object.
6. There is no risk of interaction with the gastric fluid [7].

The drug which is used in the DDS for example salicylic acid (acid ortohidroksibenzoat). Salicylic acid is a drug that is used to treat a number of skin problems such as acne, warts, dandruff, psoriasis, and other skin problems. Salicylic acid can also be used to preserve food, antiseptic, and mix it in tooth paste. At this time, salicylic acid is widely applied in the manufacture of aspirin in industial scale.

Release of a drug in DDS is controlled by diffusion. This phenomenon can be modeled mathematically that will provide many advantages, there are several of them:

1. It can allow getting deeper insight into the mechanism, which control drug release from a particular type of dosage forms.
2. It can allow for the quantitative prediction of the effects of formulation and processing parameters on the resulting drug release rate.
3. Drug product development can be accelerated and time and cost intensive series of trial and error experiments can be replaced [8].

Drug release analysisis was done by dipping edible film containing drugs in a particular solvent. Mass transfer of the drug from solid to liquid is shown in the following figure:

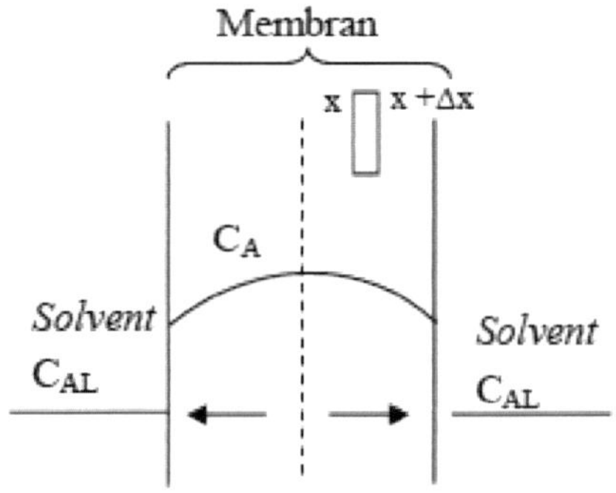

Figure 3. Mass Transfer of Salycilic Acid

In the experiment, mass transfer occured in two both sides. However, in the real situation, mass transfer occured in one side. The value of diffusivity will be the same because it is a microscopic property. Release of the drug into the solvent was approximated by a mathematical models with several assumptions to develop the equation as follows:

1. Diffusion through the membrane side was one dimensional (z direction).
2. Volume of the system was constant.
3. The process was isothermal.
4. Initial concentration of the drug in the membrane was uniform.
5. The mass transfer rate from the surface of the membrane to the solution was fast.
6. Membrane did not swell or eroded significantly as long as the release of drugs.

Release of drugs that was delivered in a dominant drug delivery system was controlled by mass transfer. The differential equation was obtained by setting up mass balance in the volume element is:

$$\frac{\partial C_A}{\partial Z^2} = \frac{1}{De} \cdot \frac{\partial C_A}{\partial t} \tag{1}$$

where,

C_A = Concentration of drug measured (g/mL),
De = Drug diffusivity (cm^2/s),
z = Distance from the center of the membrane (cm)
t = The release time (minutes).

$$C_A(z,0) = C_{A0} \quad (2)$$

$$C_A(L,t) = H_A.C_{AL} \quad (3)$$

$$\frac{\partial C_A}{\partial z}(0,t) = 0 \quad (4)$$

C_{AL} value can be found by equation (5) as follows:

$$C_{AL} = C_{AL0} + \frac{n.s}{V}(L.A_0 - \int_{Z=0}^{Z=L} C_A(z,t)\partial t) \quad (5)$$

Where

C_{AL} = Concentration of drug in solution (g/mL)
S = Total area of the membrane (cm^2)
L = Thickness of membrane (cm)

The boundary condition (3) was derived from the assumptions that mass transfer from the surface of the membrane is relatively fast, so the solute concentrations on the surface of the membrane were balance with the solute concentrations in liquid. The correlation of balance can be approached with the equation that resemble with Henry's Law. The boundary condition (4) represented the maximum concentration of salicylic acid in time t is zero. It was described in maximum condition because we can not predict the definite concentration directly.

The finite different method was chosen as solution method of the differential equation. DDS membranewhich was made in this research are pectin-based edible film and will degrade after a long time. In Indonesia, the edible film is still made from gelatin, PLA (Poly Lactic Acid), PVA (Poly Vinyl Alcohol), and hydrogels. All of these components are imported from overseas [9].

After tooth extraction, a blood clot will be formed in the tooth hole. The surface of the opened gums also allows bacteria or food residue entering easily [10]. The membrane containing salicylic acid drugs is used as antibiotics after tooth extraction. To improve the effectiveness of drug delivery system, it is necessary to control rate of drug release. The release rate was controlled by diffusivity, but the diffusivity was controlled by crosslinking of the membrane polimer.
The aims of this research are to observe the impact of crosslinker (Ca^{2+}) to the edible film characteristic especially the diffusivity and estimate the salicylic acid mass transfer parameters by proposing a mathematical model.

2. Materials and method

2.1. Materials

The materials used are pectin, glycerol, $CaCl_2.2H_2O$, salicylic acid, a solution of 37% HCl, KCl, KH_2PO_4, Na_2HPO_4 and distilled water.

2.2. Instruments

The main instruments used to analyze the parameters are magnetic stirrer, oven and UV-vis spectrophotometer.

2.3. Manufacture of Edible Film from Pectin and Crosslinking Process

The edible film was made by the two following steps. First, 0.015 g/mL pectin were diluted in 200 mL of solution containing 0.6 g plasticizer/g pectin while stirring for 1.5 hours for a homogeneous solution. The solution was heated to 66°C, then $CaCl_2.2H_2O$ solution with proportion 0.04g/g pectin was added little by little for 30 minutes. The solution was scored on a teflon mold and dried in an oven at 50 °C for 15 hours. Second, the film was dipped in a solution with many various concentrations of $CaCl_2.2H_2O$ that already contains plasticizer for 30 minutes. The film was separated from the solution and allowed to dry at ambient temperature and stored at the desiccator.

2.4. Analysis of Drug Loading

Salicylic acid was added into edible films by diffusion method. Edible films were dipped in a solution of salicylic acid (1 gram of salicylic acid in 10 ml acetone) for 2 hours. The edible film that already contains drug was placed on the open space until it dries. The remaining of the concentration of salicylic acid in solution was measured by UV-vis spectrophotometer with a wavelength of 542 nm.

2.5. Preparation of Buffer Solution

Buffer solution with pH 7.4 was prepared by mixing 1 L of distilled water with 8 grams of NaCl until homogeneous continued by adding 0.2 grams KCl, 1.44 grams Na_2HPO_4 and 0.24 grams KH_2PO_4. The mixture was stirred until homogenous. To ensure the pH closed to 7.4, 37% HCl solution was added using a pipette.

2.6. Analysis of Drug Release

Edible film that was containing drugs was put into a beaker glass containing 25mL of buffer solution with pH 7.4. One mililitre samples were taken periodically from the beaker glass. Then the samples were analyzed using UV-vis spectrophotometer with a wavelength 296 nm. Each volume of the sample which was taken and completely analyzed will be returned to the glass beaker to keep the volume steady.

2.7. Analysis of Gels Swelling

Edible films were immersed in a buffer solution with pH 7.4 at room temperature for 3 hours. Then the already swollen edible films were dried in a vacuum oven at 37 °C until the weight was constant. Percentage of gel swelling was measured by the following formula:
$\% \text{ Swelling} = Ws\text{-}W_D/W_D \times 100$

where,
Ws = weight of edible film after hydration for 3 hours
W_D = weight of dry sample test

3. Result and discussion

3.1. Effect of Concentration Crosslinker ($CaCl_2$) to Diffusivity Coefficient (De)

The effect of crosslinker concentration was studied by varying the $CaCl_2$ concentration added in the second stage of the manufacture of edible film. The concentration were 0, 0.015, 0.03 and 0.05 g/mL. The release of salycilic acid from edible film to the buffer solution can be presented by the plotting the measuring concentration of salycilic acid in the buffer solution as function of times.

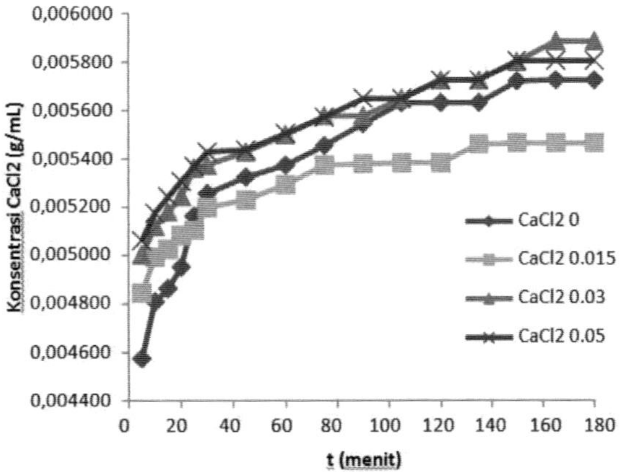

Figure 4. Drug release profile in various concentration of crosslinker

Figure 4 showed that for different $CaCl_2$ concentration, the profiles at salycilic acid release are similar. The value varied from 0.0046 to 0.0059 g/mL. It could be seen also at $CaCl_2$ concentration at 0.05 g/mL, the release of salycilic acid slightly maximum. In the beginning (0 to 20) minutes, the salycilic acid concentration increased significantly due to high driving force for mass transport.

The model that has been proposed in equation 1 and 5 was used to predict the trend of salicylic acid concentration profile and to estimate the parameters De and H_A with trial and eror. The experimental and modeling result are presented in table 1, figure 5, 6, 7 and 8.

Table 1. Difusivities and distribution coefficient at various concentration of crosslinker

$CaCl_2$ (g/mL)	De (cm^2/s)	H
0	3.4688×10^{-7}	0.7084
0.015	7.5382×10^{-8}	1.8542
0.03	7.464×10^{-8}	2.7734
0.05	5.8838×10^{-8}	5.5445

Then, curve fitting had been done based on laboratory experimental in figure 4 and simulation of equation (1) and (5). The results were:

IOP Conf. Series: Materials Science and Engineering **105** (2016) 012042 doi:10.1088/1757-899X/105/1/012042

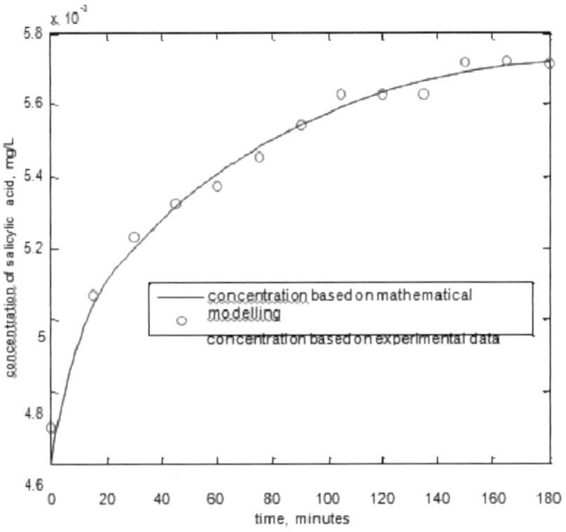

Figure 5. Drug release profile inconcentration of crosslinker 0 g/mL

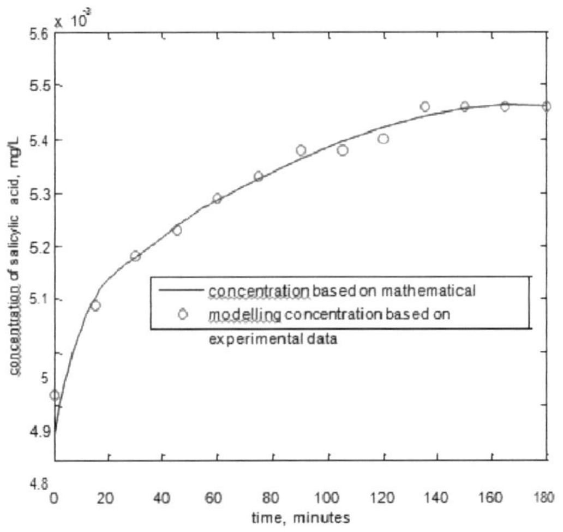

Figure 6. Drug release profile in concentration of crosslinker 0.015 g/mL

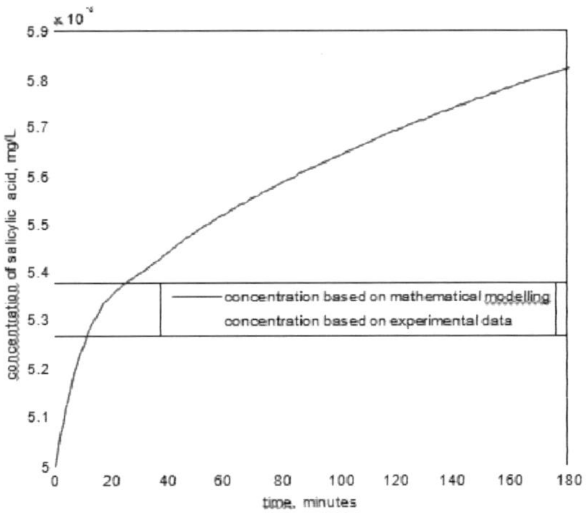

Figure 7. Drug release profile in concentration of crosslinker 0.03 g/mL

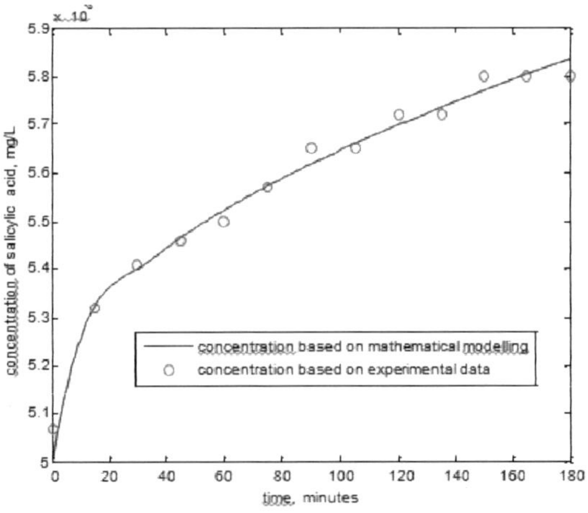

Figure 8. Drug release profile in concentration of crosslinker 0.05 g/mL

Figure 5,6,7 and 8 showed the rate of salicylic acid release in a variety of time and concentration of crosslinker. Overall, the model proposed can quantitatively described the release of salycilic acid from the membrane because the relative error is 11.5891 %. From table 1, the correlation between $CaCl_2$ concentration and diffusivity coefficient (De) can be represented in figure 9.

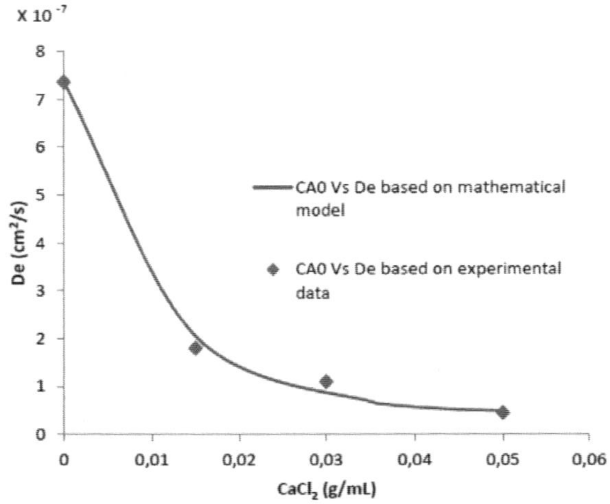

Figure 9. The diffusivity profile in various concentration of crosslinker

Figure 9 showed the diffusifity coefficient in many concentrations of crosslinker. It is proposed to help us in predicting the diffusifity coefficient in the concentration of crosslinker that we need Salicylic acid concentrations were increased in line with increasing time of release. In the range from 150 to 180 minutes, releasing of salicylic acid was approximately constant. It indicates that the concentration of salicylic acid in liquid has reached a saturation condition. Table 1 showed that the value of diffusivity (De) tended to decrease with increasing concentrations of $CaCl_2$. It is becaused of Ca^{2+} ions contained in the $CaCl_2$ has the ability to form complex compounds that are not soluble in water by binding the free carboxyl groups contained in the pectin. Hydrophilic properties of pectin before it is crosslinked came from the presence of the carboxyl group [11]. It caused the pectin carboxyl groups are negatively charged, so it was easy to react with positively charged molecules. If concentration of $CaCl_2$ was added more, many of Ca^{2+} ions would bind the free carboxyl groups in the pectin. The amount of crosslinking bonds made the distance between the molecules in the membrane became smaller, so it made the release of salicylic acid became slower.

Moreover, the value of distribution coefficient (H) tends to rise in line with increasing concentrations of $CaCl_2$. The distribution coefficient is the ratio between the concentration of salicylic acid in the membrane with the concentration of salicylic acid in the liquid. Ca^{2+} ions with carboxyl groups on the pectin formed a strong bond with a structure that is often called "egg box" structure (figure 10) [12]. This causesed the molecules of salicylic acid was difficult to release from the membrane, so the concentration of salicylic acid in the membrane is greater than the concentration of salicylic acid in the liquid. It meant the value of the distribution coefficient was greater too.

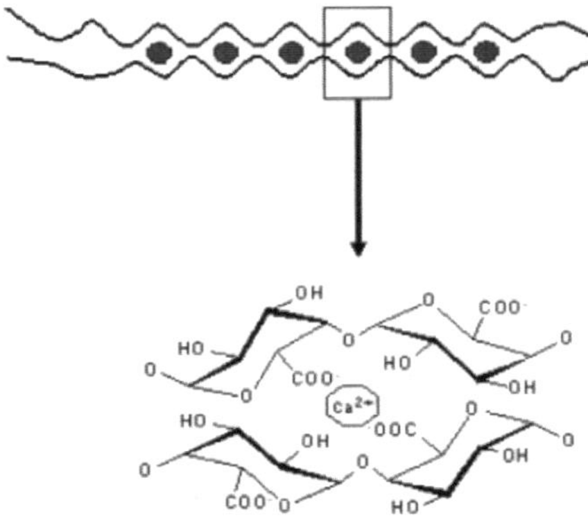

Figure 10. Schematic representation of calcium binding to polygalactoronate sequences: 'egg box' dimer and 'egg-box' cavity

3.2. Effect of Crosslinker ($CaCl_2$) Concentration on Gels Swelling

Beside affecting drug release rate, the strength of $CaCl_2$ also affected the ability of the liquid to penetrate into the membrane[12]. This related to the gels swelling of the membrane. Analysis of membrane swelling percentage was done in PBS solution with pH 7.4. The changes of swelling percentage were measured until a constant weight. The results from the analysis of swelling percentage are shown in figure 11 below:

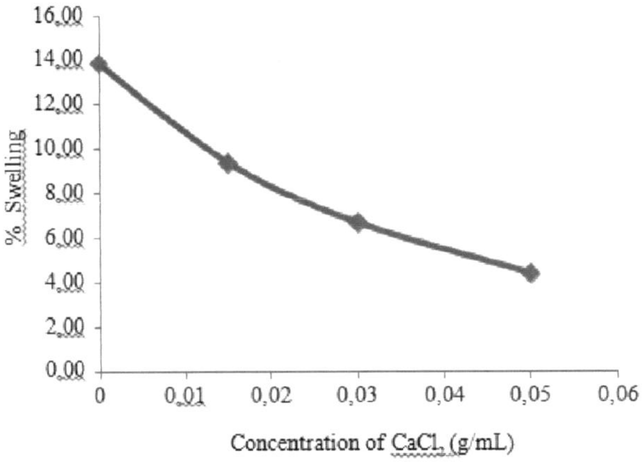

Figure11. Swelling profiles in various concentration of crosslinker

PBS solution which has a pH 7.4 indicated an excess of OH^- ions. The presence of OH^- ions could improve the electrostatic repulsion between the carboxyl group, thus increasing the gels swelling and made the release of salicylic acid was increase too [13]. Therefore we need the addition of crosslinker to make the release of salicylic acid slower. Based on figure 11, the percentage of swelling tended to decline with increasing concentrations of $CaCl_2$. It was because after the crosslinking process, intermolecular bond in the membrane was stronger than before. So, the liquid was difficult to be permeated into the membrane and salicylic acid was difficult to be released from the membrane into the liquid. It also made the membrane can not be degraded and broken easily.

4. Conclusion

Membrane drug delivery system can be made from pectin that derived from citrus peels. Addition of Ca^{2+} on drug delivery system membrane decreases the diffusivity values, so the drug release rate can be controlled more. The mathematic model proposed could quantitatively describe the release of the drug from the membrane with relative error of 11.5891 %. Moreover, the addition of crosslinker also made the membrane swelling percentage decrease, so the membrane was mechanically stronger. Further studies on drug delivery system membrane is very potential to be developed as an innovation in the biomedical engineering field.

5. Acknowledgement

The authors gratefully acknowledge Mrs. Ika Dewi Ana from Dentistry Faculty of UGM for the interesting discussion and technical support, Mr. Achmad Syaifudin of Laboratory of Food Technology and Bioprocess UGM for the technical support, Mr. Ari, Mrs. Ani and Mrs. Bunga of Dentistry Research Laboratory UGM for their technical support and also Mr. Heru Enggar Triantoro for the interesting computer programming discussion.

6. References

[1] Sarwono, B. 1994. *Jeruk dan Kerabatnya*. Penebar Swadaya. Jakarta.
[2] Anonim, 2013, *Jumlah Produksi Jeruk Indonesia*. Badan Pusat Statistik. Accessed December8th, 2013.
[3] Herbstreith, K., dan G. Fox. 2005. *Pectin*. http://www.herbstreithfox. de/pektin/forschung und entwicklung/forschung_entwicklung04a.htm. accessed on March11th, 2014.
[4] Herbstreith, K and G. Fox. 2005. *Pectin*.http://www.herbstreithfox.de/pektin/, accessed on February 12th, 2014.
[5] IPPA (*International Pectins Procedures Association*). 2002. *What is Pectin*. http://www.ippa.info/history_of_pektin.htm. access on April 7th, 2014.
[6] Gaurav Upadhyay, Shubham Verma, Nayyar Parvez and Pramod Kumar Sharma. 2014. *Recent Trends in Transdermal Drug Delivery System - A Review*. Advances in Biological Research. vol. 3.pp. 131–138.
[7] Kurniasih, Nia, 2004. *Jeruk Kurangi Resiko Kanker*. http://pikiranrakyat.com/cetak/0904/02/ cakrawala/lainnya03.htm. access on May 4th, 2014.
[8] Siepmaan, Juerge and Siepmaan, Florence. 2012. *Modeling of Diffusion Controlled Drug Delivery*. Jurnal of Controlled Release.
[9] Sadeghi, Mohammad. 2011. *Pectin-Based Biodegradable Hydrogels with Potential Biomedical Application as Drug Delivery System*. Journal of Biomaterials and Nanobiotechnology.
[10] Anonim. 2013. *Hal-hal Yang Perlu Diperhatikan Setelah Pencabutan Gigi*,http://johorindahdentalclinic.blogspot.com/2013/02/hal2-yang-perlu-di-perhatikan setelah_26.html, accessed on August 22th, 2014

[11] Deuel, H., Huber, G., and Anyas Weisz, L.1950. *Uber 'Salzbrucken' zwischen Makromolekulen von Polyelektrolyten, besondersbei Calcium pektinaten.* Helv.Chim.Acta 33, pp.563-567.
[12] Maestrelli.F, Cirri.M, Corti.G, Mennini.N, Mura.P. 2008. *Development of enteric-coated calcium pectinate microspheres intended for colonic drug delivery*. European Journal of Pharmaceutics and Biopharmaceutics. 69, pp. 508–518.
[13] M. Mahkam and M. Allahverdipoor, *Drug Targeting*, Vol.12, 2004, pp.151-156.

The use of knowledge-based Genetic Algorithm for starting time optimisation in a lot-bucket MRP

Muhammad Ridwan Andi Purnomo
Department of Industrial Engineering, Faculty of Industrial Technology, Universitas Islam Indonesia, INDONESIA

Email: ridwan_ie@uii.ac.id

Abstract. In production planning, Material Requirement Planning (MRP) is usually developed based on time-bucket system, a period in the MRP is representing the time and usually weekly. MRP has been successfully implemented in Make To Stock (MTS) manufacturing, where production activity must be started before customer demand is received. However, to be implemented successfully in Make To Order (MTO) manufacturing, a modification is required on the conventional MRP in order to make it in line with the real situation. In MTO manufacturing, delivery schedule to the customers is defined strictly and must be fulfilled in order to increase customer satisfaction. On the other hand, company prefers to keep constant number of workers, hence production lot size should be constant as well. Since a bucket in conventional MRP system is representing time and usually weekly, hence, strict delivery schedule could not be accommodated. Fortunately, there is a modified time-bucket MRP system, called as lot-bucket MRP system that proposed by Casimir in 1999. In the lot-bucket MRP system, a bucket is representing a lot, and the lot size is preferably constant. The time to finish every lot could be varying depends on due date of lot. Starting time of a lot must be determined so that every lot has reasonable production time. So far there is no formal method to determine optimum starting time in the lot-bucket MRP system. Trial and error process usually used for it but some time, it causes several lots have very short production time and the lot-bucket MRP would be infeasible to be executed. This paper presents the use of Genetic Algorithm (GA) for optimisation of starting time in a lot-bucket MRP system. Even though GA is well known as powerful searching algorithm, however, improvement is still required in order to increase possibility of GA in finding optimum solution in shorter time. A knowledge-based system has been embedded in the proposed GA as the improvement effort, and it is proven that the improved GA has superior performance when used in solving a lot-bucket MRP problem.

1. Introduction

In manufacturing planning and control, Master Production Schedule (MPS) is widely used to schedule end items production. MPS plays important roles especially in Enterprise Resource Planning (ERP) [1]. Production lot in MPS would be derived further to be Material Requirement Planning (MRP) [2]. Because MPS is an almost fixed production schedule, then production time period is arranged for shorter period, and usually weekly instead of monthly. Therefore, MRP must be arranged based on weekly time period as well. In order to represent time phasing, an MRP system is usually developed based on time-bucket system, it means that period in the MRP is representing the time and usually weekly. Production evaluations, such as inventory analysis, planned order receipt and so on, would be conducted every end of period. Time-bucket MRP could be implemented successfully in Make To

Stock (MTS) company, which must finish the production before customer demand is received. It is different with Make To Order (MTO) company, which receive order from customers with specific quantity and due date. Production activity would be started only after the order was received.

MTO company still need to arrange an MRP to schedule production or order of the required materials. Therefore, in order to be successfully implemented in MTO company, a conventional MRP system is need to be modified. Fortunately, there is a modified MRP system, which called as lot-bucket MRP, proposed by Casimir in 1999 [3]. In the lot-bucket MRP system, a bucket is representing a lot, and the lot size is preferably constant. The time to finish every lot could be varying depends due date of lot. As a detail production plan, starting time of a lot must be determined precisely by considering the reasonable production time. So far there is no formal method to determine starting time in the lot-bucket MRP system. Trial and error process usually used for it but some time it causes several lots have very short production time and the lot-bucket MRP would be infeasible to be executed.

Determination of starting time of lot in lot-bucket MRP system could be viewed as optimisation case. The objective function is to maximise reasonability and acceptability index of the lot-bucket MRP system with decision variable is the starting time itself. Reasonability index would be higher when production time is shorter, but there is no slack and idle time. When there is slack and or idle time, actually the slack and idle time could be used to prolong the production time. Consequently, starting time of the lot must be revised. Acceptability index would be higher if the production time after being prolonged by reducing slack and idle time is longer than minimum production time allowed. In order to find optimum starting time of lots, a searching algorithm could be utilised. This paper presents the use of Genetic Algorithm (GA) for optimisation of starting time in a lot-bucket MRP system. Even though GA is known as powerful searching algorithm, however, improvement is still required in order to increase possibility of finding optimum solution in shorter time. A knowledge-based system has been embedded in the proposed GA as the improvement effort. Further, such GA is called as knowledge-based GA (KB-GA). Through a case study, it is proven that the KB-GA has superior performance when being used in solving a lot-bucket MRP problem.

2. Related works

Main reference of this study is a paper about lot-bucket MRP system that authored by Casimir in 1999 [3]. In such paper, there are a lot of useful information about lot-bucket MRP development, such as the lot-bucket MRP system could be implemented in Just In Time (JIT) manufacturing and constant lot size is representing container capacity in JIT manufacturing or could be determined based on average order. The production time is represented with value 1 (normal production time) and less than 1 (faster production time). This study also use such information in optimising the starting time in lot-bucket MRP.

Study about MRP has been conducted by several previous researchers with different manufacturing condition. In conventional MRP, production or order lead time is assumed constant. In real situation, when the lead time is not constant and the variation is relatively big, then a technique to deal with the uncertainty lead time is required. Diaz-Madronero et al. (2015) have investigated the use of fuzzy multi-objective integer programming to develop MRP system under fuzzy lead time [4]. The fuzzy multi-objective integer programming is used to maximise satisfaction degree of three objectives which are minimisation of cost, back order quantity and idle time. Such study also discuss about a case study to proof that the proposed technique could work well in solving considered case. Similar study that deal with uncertainty factors and solved using fuzzy technique could be seen in detail in [5].

Another study that deals with lead time in MRP system has been conducted by Milne et al. (2015) [6]. In such study, values of planned lead time were the main decision variable. A Mixed Integer Programming (MIP) was proposed as the optimisation of the planned lead time values by considering a set of constraints contains component and capacity constraints. Objective of the optimisation is minimising variety of costs or penalties. To test performance of the proposed method, a set of real data from an electronic industry was used. Dynamic lead times were represented by minimum and maximum lead time values. Based on the case study, it is proven that the proposed method could work well even though with heavy computation load due to the binary variables. Similar studies that concerned with the weakness of constant lead time and time periods assumption in MRP have also been conducted by previous researches [7], [8], [9].

Studies about time-bucket MRP that concerned with lot sizing have also been investigated by previous researchers. In time-bucket MRP, the lot sizing system could be viewed as multi-products and multi-periods inventory system because MRP is product's parts management for several production periods. Lot sizing for reversed MRP has been investigated by previous researchers [10]. Reversed MRP accommodates disassembly operation that not accommodated in conventional MRP system. In reversed MRP system, demand is not only from end item, but also from parts that have independent demand. Another study that investigated lot sizing in time-bucket MRP could be seen in detail in [11]. Such study used a mathematical model to develop an optimum lot sizing model.

3. Knowledge extraction in lot-bucket MRP

In this study, there is only one knowledge that would be extracted from lot-bucket MRP system. The knowledge is understanding when problems would be occurred in a lot-bucket MRP system. A problem related to sufficiency production time would be occurred when interval between 2 consecutive time requirements is relatively short but the customer order is relatively high. In such condition, the production time might be very short and it could cause the lot-bucket MRP system unacceptable. Basically that condition could be anticipated when in the previous lot there are idle or slack times that could be used to prolong the unacceptable production time. Therefore, the knowledge to be extracted is, if there is a lot with shorter production time and below minimum limit of production time allowed, then the production time would be prolonged by reducing slack and idle time of the previous lot.

4. Mathematical model development

Decision variable in this study is starting time of each lot in a lot-bucket MRP system so that the lot-bucket MRP system is reasonable and acceptable. Therefore, the objective function is to maximise sum of reasonability and acceptability index of all lot as formulated in equation (1).

$$Z = \sum_{i=1}^{L} R_i + \sum_{i=1}^{L} A_i \tag{1}$$

Constraints to get a feasible solution are as follows:

$$R_i = \begin{cases} 1, if\ lot\ i\ is\ reasonable \\ 0, otherwise \end{cases} \tag{2}$$

$$A_i = \begin{cases} 1, if\ lot\ i\ is\ acceptable \\ 0, otherwise \end{cases} \tag{3}$$

$$Sl_i = S_i - d_{i-1} \tag{4}$$

$$I_i = d_{i-}c_i \tag{5}$$

$$c_i = S_i + P_i \tag{6}$$

$$P_i = \begin{cases} 1, if\ lot\ size\ of\ lot\ i\ \geq customer\ order \\ lot\ size\ of\ lot\ i \big/ customer\ order, otherwise \end{cases} \tag{7}$$

where:

Z	: objective value
R	: reasonability index of a lot
A	: Acceptability index of a lot
i	: lot index
L	: number of lot
Sl	: slack of a lot
S	: Starting time of a lot
I	: idle time of a lot
d	: due date of a lot
c	: completion time of a lot
P	: production time of a lot

Equation 2 and 3 will set reasonability and acceptability of a lot to 1 if the lot is reasonable or acceptable, or to 0 if otherwise. Equation 4 is the formula to determine slack time of a lot, which is difference between starting time of a lot with due date of previous lot. Equation 5 is the formula to determine idle time of a lot, which is the difference between due date and completion time of the lot. Equation 6 shows the formula to determine completion time of a lot while equation 7 shows the formula to determine production time.

5. Genetic Algorithm development

5.1 Chromosome representation

The chromosome would represent starting time of every lot. The numeric type for such solution would be real time, instead of integer, in order to get precise solution. In the chromosome, there are several locus. The first locus is representing starting time of the first lot, while other locus is representing the difference between starting time of the related lot with the previous lot. Figure 1 shows the chromosome when there are 7 lots and representing starting time of lots 0, 0.83, 1.83, 2.83, 3.73, 4.73 and 5.6. Fitness function of the chromosome is Z (equation 1).

0	0.83	1	1	0.90	1	0.87

Figure 1. Chromosome representation

5.2 Crossover mechanism

In this study, a simple one-cut point crossover operation is implemented. Such crossover operation would cut both of parent chromosomes and interchanges part of the chromosomes each other. Figure 2 shows the mechanism of such crossover.

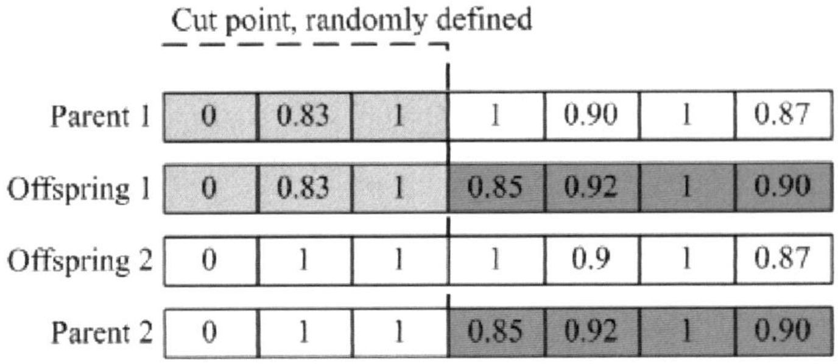

Figure 2. Crossover mechanism

5.3 Mutation mechanism

Mutation proposed in this study is random step mutation. First, GA would select chromosomes to be mutated based on mutation probability. Then, from the selected chromosomes, GA would select one or more than one locus randomly. The selected locus, their value, would be changed by step up or down randomly. The step value is based on a delta (Δ) value and in this study the delta value is between 0.01 to 0.20. Such mutation operation enables the KB-GA to explore solution domain step by step. The mutation mechanism could be formulated as shown in equation (7).

$$l' = \begin{cases} l + \Delta, if\ s\ (randomly\ defined) is\ 1 \\ l - \Delta, otherwise \end{cases} \tag{7}$$

where:

l' : new locus value

l : old locus value

6. Case study

Performance evaluation of the proposed KB-GA is carried out by applying it to solve a lot-bucket MRP problem. There is a product with 4 parts that divided into 2 levels. The product structure is shown in figure 3 while due date of every lot and other relevant data is shown in table 1. Minimum production time allowed is 0.8, it means that the production could be speed-up maximum at 20% above normal time. A comparison study with a conventional GA was also conducted to show superiority of the KB-GA.

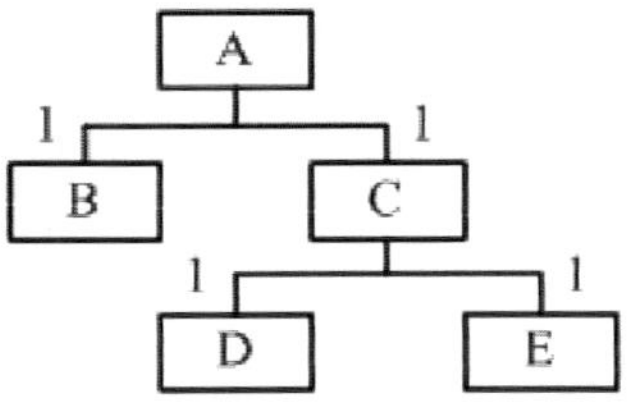

Figure 3. Product structure

Table 1. Data of lot

Lot	1	2	3	4	5
Order quantity	100	80	130	90	100
Due date	3	4.30	5	5.90	7

The KB-GA has been ran for 500 generations, with crossover and mutation rate is 0.3 and 0.2 respectively. Actually, GA has been ran several time to ensure that KB-GA could provide stable solution. Parameter to stop the KB-GA at 500 generations is when there is no improvement anymore for minimum half number of generation. KB-GA provided solution as shown in table 2 to table 6.

Table 2. Lot-bucket MRP for final product (A)

Lot number	1	2	3	4	5
Time requirement	3.00	4.10	5.00	5.90	7.00
Schedule receipt	0.00	0.00	0.00	0.00	0.00
Projected slack time	0.00	0.00	0.00	0.00	0.00
Projected idle time	0.00	0.10	0.00	0.00	0.10
Projected production time	1.00	1.00	0.90	0.90	1.00
Planned starting time	2.00	3.10	4.10	5.00	6.00

Table 3. Lot-bucket MRP for part (B)

Lot number	1	2	3	4	5
Time requirement	2.00	3.10	4.10	5.00	6.00
Schedule receipt	0.00	0.00	0.00	0.00	0.00
Projected slack time	0.00	0.00	0.00	0.00	0.00
Projected idle time	0.00	0.10	0.00	0.00	0.00
Projected production time	1.00	1.00	1.00	0.90	1.00
Planned starting time	1.00	2.10	3.10	4.10	5.00

Table 4. Lot-bucket MRP for part (C)

Lot number	100	80	130	90	100
Time requirement	2.00	3.10	4.10	5.00	6.00
Schedule receipt	0.00	0.00	0.00	0.00	0.00
Projected slack time	0.00	0.00	0.00	0.00	0.00
Projected idle time	0.00	0.10	0.00	0.00	0.00
Projected production time	1.00	1.00	1.00	0.90	1.00
Planned starting time	1.00	2.10	3.10	4.10	5.00

Table 5. Lot-bucket MRP for part (D)

Lot number	100	80	130	90	100
Time requirement	1.00	2.10	3.10	4.10	5.00
Schedule receipt	0.00	0.00	0.00	0.00	0.00
Projected slack time	0.00	0.00	0.00	0.00	0.00
Projected idle time	0.00	0.10	0.00	0.00	0.00
Projected production time	1.00	1.00	1.00	1.00	0.90
Planned starting time	0.00	1.10	2.10	3.10	4.10

Table 6. Lot-bucket MRP for part (E)

Lot number	100	80	130	90	100
Time requirement	1.00	2.10	3.10	4.10	5.00
Schedule receipt	0.00	0.00	0.00	0.00	0.00
Projected slack time	0.00	0.00	0.00	0.00	0.00
Projected idle time	0.00	0.10	0.00	0.00	0.00
Projected production time	1.00	1.00	1.00	1.00	0.90
Planned starting time	0.00	1.10	2.10	3.10	4.10

Based on the result above, it could be analysed that no slack and idle time when production time is less than 1 and production time of all of the lot is not less than minimum production time allowed. It means that the lot-bucket MRP is reasonable and acceptable.

7. Discussions

To show superiority of the proposed knowledge-based GA, a conventional GA has been ran as well with same parameter values as in the KB-GA, and it provided a solution as shown in table 7 (for lot-bucket MRP of final product only).

Table 7. Lot-bucket MRP for final product (A) provided by conventional GA

Lot number	1	2	3	4	5
Time requirement	3.00	4.30	5.00	5.90	7.00
Schedule receipt	0.00	0.00	0.00	0.00	0.00
Projected slack time	0.00	0.00	0.00	0.00	0.00
Projected idle time	0.00	0.30	0.00	0.00	0.10
Projected production time	1.00	1.00	0.70	0.90	1.00
Planned starting time	2.00	3.30	4.30	5.00	6.00

It could be seen that there is an unacceptable production time, which is production time of lot number 3 that bellow minimum production time allowed. However, the previous lot has long idle time, which is 0.3. Actually, the idle time of the previous lot could be used to prolong the unacceptable production time to make it acceptable.

In the knowledge-based GA, effect of the extracted knowledge is revision of the time requirement of second lot of final product (A) from 4.3 to 4.1. It would reduce the idle time of lot number 2 from 0.3 to 0.1 to prolong production time of lot number 3 from 0.7 to 0.9 and become acceptable. However, such solution would affect to holding cost for lot number 2. It would be finished at 4.1 but would be delivered at 4.3. If the holding cost is not significantly affects the production system performance, then the solution provided by the KB-GA is better because it could avoid potential production failure of lot number 3 due to unacceptable production time.

8. Conclusion and suggestion

Based on explanation above, it could be concluded that lot-bucket MRP problem could be viewed as optimisation case and GA could be one of alternative algorithm for finding the solution. Even though GA is powerful, however, improvement of the searching process is still required, especially when faced problem is complicated. In this study, it was proven that a knowledge-based system could be used to improve GA's searching process. Performance of the knowledge-based GA is superior compared to conventional GA when being used to solve a lot-bucket MRP problem.

With refer to the discussions above, for further research, it is suggested to consider costs that involved in the production planning such as holding cost and cost to speed-up production time. Consideration of costs would make the proposed solution more practical in real world industry.

From the GA side, as the optimisation tool, chromosome encoding technique would affect to the effectiveness of the GA when searching the solution. For instance, in this case, if the chromosome uses binary number, starting time of the lots might be very precise. However, the searching process inside the GA would be not effective because of larger solution domain. Trade-off between precise solution and computation effective could be interesting analysis for academic world.

9. Acknowledgment

The author would like to appreciate DIKTI, Republic of Indonesia for supporting this study through fundamental research grant scheme 2015.

10. References

[1] Körpeoğlu E, Yaman H and Aktürk M S 2011 A multi-stage stochastic programming approach in master production scheduling *European Journal of Operational Research* **213** 166.

[2] Sahin S, Robinson E P and Gao L L 2008 Master production scheduling policy and rolling schedules in a two-stage make-to-order supply chain *Int. J. Production Economics* **115** 528.
[3] Casimir R J 1999 Lot-bucket MRP *Int. J. Production Economics* **58** 173.
[4] Diaz-Mandronero M, Mula J and Jimenez M 2015 Material requirement planning under fuzzy lead times *IFAC-PapersOnLine* **48-3** 242.
[5] Mula J, Poler R and Garcia-Sabater J P 2007 Material requirement planning with fuzzy constraints and fuzzy coefficients *Fuzzy Sets and Systems* **158** 783.
[6] Milne R J, Mahapatra, S and Wang C T 2015 Optimizing planned lead times for enhancing performance of MRP systems *Int. J. Production Economics* **167** 220.
[7] Ioannou G and Dimitriou S 2012 Lead time estimation in MRP/ERP for make to order manufacturing systems *Int. J. Production Economics* **139** 551.
[8] Sadeghian R 2011 Continuous materials requirements planning (CMRP) approach when order type is lot for lot and safety stock is zero and its applications *Applied soft computing* **11** 5621.
[9] Öztürk C and Örnek A M 2012 A MIP based heuristic for capacitated MRP systems *Computers and industrial engineering* **63** 926.
[10] Barba-Gutierrez Y, Adenso-Diaz B and Gupta S M 2008 Lot sizing in reverse MRP for scheduling disassembly *Int. J. Production Economics* **111** 741.
[11] Grubbstrom R W, Bogataj M and Bogataj L 2010 Optimizing lot sizing within MRP theory *Annual reviews in Control* **34** 89.

A manufacturing quality assessment model based-on two stages interval type-2 fuzzy logic

Muhammad Ridwan Andi Purnomo, Intan Helmi Shintya Dewi
Department of Industrial Engineering, Faculty of Industrial Technology, Universitas Islam Indonesia, INDONESIA

Email: ridwan_ie@uii.ac.id

Abstract. This paper presents the development of an assessment models for manufacturing quality using Interval Type-2 Fuzzy Logic (IT2-FL). The proposed model is developed based on one of building block in sustainable supply chain management (SSCM), which is benefit of SCM, and focuses more on quality. The proposed model can be used to predict the quality level of production chain in a company. The quality of production will affect to the quality of product. Practically, quality of production is unique for every type of production system. Hence, experts opinion will play major role in developing the assessment model. The model will become more complicated when the data contains ambiguity and uncertainty. In this study, IT2-FL is used to model the ambiguity and uncertainty. A case study taken from a company in Yogyakarta shows that the proposed manufacturing quality assessment model can work well in determining the quality level of production.

1. Introduction

In recent years, Supply Chain Management (SCM) has received considerable attention from various parties [1]. SCM is an approach which used to achieve an efficient integration between supplier, manufacture, distributor, retailer and consumers [2]. It can be said that SCM is a business philosophy which trying to integrate various activities, parties or stakeholders and resources which connected between different level in an organization [3]. Various big companies in the world has applied SCM's concept in their system. SCM's concept highly favoured because it's able to integrate various business activities in a company and able to help the company for achieving competitive advantage [4]. In practice, SCM needs sustainable development in order to make it works successfully according to the company's expectation. Sustainable development in SCM has known as Sustainable Supply Chain Management (SSCM) which becomes hot issue for companies who have applied SCM in their company's system [3], [5].

SSCM is a strategy, transparency integration and the achievements of organisation's social, environment and economy goals in the systemic coordination of key business processes across the organization to improve long-term performance from individual and its supply chain in a company [6]. SSCM referred to as a key success of implementation of SCM, it's because the sustainable goals in SCM only able to obtained when the company has applied SSCM in their practice [7]. Knowing the importance of SSCM's practice in SCM's sustainable development in a company, it's necessary to assess SSCM [1]. The assessment of SSCM based on 7 building blocks which become factors in implementation of SSCM. The 7 building blocks in SSCM are reasons for SSCM, performance criteria employed for SSCM, greening supply chains, characteristics of suppliers, managerial approaches for SSCM, barriers for SSCM, and benefits for SSCM [1].

In this study, the assessment is conducted based on one of building blocks, that is benefits for SSCM, and focus with one variable in that building blocks. The benefits for SSCM is a building blocks contains various advantages that can achieve by applying the SSCM. There are six variables from the benefits of SSCM which have big effect in SSCM, those variables are customer satisfaction, innovation capability of supplier, quality, trust, managing supply risk, and fill rate [1]. Among those variables, this research only focus on quality.

Based on ISO 9000, quality is defined as thorough character from a product or service which affect the ability of the product to meet the specific needs. Another definition, quality is thorough character of product and service which involve marketing, engineering, manufacture, and maintenance where the usage will meet the needs and expectation of customers [8]. In other words, quality is a variable which affect to customer satisfaction which becomes the goal of implementation of SSCM. Knowing the importance of quality, many companies do various approaches to control, fix, improve, and maintain the quality.

Some researchers have carried out studies to control, improve and maintain quality. Jain and Samrat (2015) have carried out a quality study overall at one of company in Gujarat to investigate and give recommendation that could be applied by other companies [9]. Thombansen et al. (2014) have carried out a quality setup and maintenance research on CO_2 laser cutting to propose an improvement for better quality [10]. The previous studies only assessed, analysed and proposed improvement at one time and the result still can't help the company to know their quality in the future. Therefore, it is required to do a study to assess and analyse quality in the future. This study answers such requirement by developing a prediction models for the quality. Quality prediction models is an approach to explain the relationship between operational manufacture and level of quality product. Quality prediction models is a key to enable the ability to estimate the level of product quality during operational manufacture goes on [11].

This study took place at a company in Yogyakarta, Indonesia. The company is one of *spiritus* factory. Actually, the company is one of production unit of sugar company in Yogyakarta, Indonesia, which has been established since 1995. In the implementation of quality control activities, the investigated company has three kinds of quality, such as material quality, process quality and final product quality. Actually, there are several factors to be considered for those qualities. However, in practice, the company only considers one or two factors to assess the quality. This is because of the limited of controlling time to accommodate all of affecting factors. To overcome such problems, this study develops prediction models for quality assessment by considering all of affecting factors which can help the company to assess and analyze the quality completely.

During preliminary study, it is found that there are several ambiguity and uncertainty factors in the quality system. Such factors can't be modelled using formal mathematical model as in the normal quality assessment system. The condition become more complicated when there are different expert opinions. Therefore, in this study, interval type-2 fuzzy logic (IT2-FL) is proposed as the basis of the quality assessment system. IT2-FL is an expansion and generalisation of the established fuzzy logic proposed by Lotfi A. Zadeh in 1975, which sometime is called as type-1 fuzzy logic (T1-FL). IT2-FL is stronger to deal with uncertainties than T1-FL [12] so that the result of IT2-FL will be more moderate to all of the experts in the investigated system.

2. Literatures review

2.1. Quality

Quality is one of multifaceted concept which describe how great the service, process, materials, or product has compatibility with the desired attributes or standard [13]. Some of quality definitions which described before show that quality is one of important aspects which cover almost all of parts in company. Quality of operational manufacture is one of quality aspect which becomes a special attention for company. Operational manufacture's quality is a quality which covers the entire process of making product from raw materials, processing until end product.

Many organisation have created a standard and specification to help the quality assessor to be consistent [14]. Quality assessment on operational manufacture becomes an aspect which need to be performed by company because it can help the company to identify and analyse the operational manufacture's performance. By performing quality assessment, the company will be able to know their technical manufacturing condition, besides, it allows the company to do improvements if the anaylsis shows a low performance.

In this study, development of assessment models for a manufacturing quality adopts a study about quality that has been carried out by previous researchers [15]. Such quality study utilises three kinds of quality factors namely quality of raw materials, process and final product. Each quality factor has various attributes which become consideration in the assessment. The quality of raw materials has two attributes namely density with a *brix* unit and the percentage of glucose levels in raw materials. The raw materials to produce *spiritus* is molasses. The quality of process in the manufacture of *spiritus* has five important attributes namely initial density, final density, alcohol levels, pH and temperature. Whereas for the quality of final product, the attributes which used are those which appropriate with the Indonesian Standard (SNI). There are three considered attributes namely ethanol levels, barbet or *permanganate* test, and *fusel* oil.

2.2. IT2-FL

It's impossible to achieve great performance using traditional controller when the process which need to handle has high nonlinearity and uncertainty. It requires a specific approach which able to accommodate a process with high nonlinearity and uncertainty. One of approach that has proven able to accommodate a process with high nonlinearity and uncertainty is fuzzy logic controller (FLC) which the success is due to the resilience and its ability to handle nonlinear system [16].

FLC usually is developed using T1-FL [17]. Although has been widely recognised for its ability to handle uncertainty in a system, some previous research show that there is limitation in its ability to model and minimise effects of uncertainty [18]. This is because the membership function of T1-FL is crisp values [19]. To improve the limited capability of T1-FL in overcoming the ambiguity and uncertainty, recently there is a type of fuzzy controller called IT2-FL [20], which has a better performance to overcome all kinds of nonlinearity and uncertainty that appeared in a system.

IT2-FL is very useful to solve a case which has difficulty in determining the exact membership function or in modelling the various arguments from different individuals [19]. The IT-2FL has used in various area of controlling, such as control in [16], control in robot mobile autonomy [21], control in anaesthesia [22], control for quarter vehicle with active suspension [23], and level control [24] which all of implementations have been successfully demonstrated and confirmed superiority of IT2-FL more than T1-FL [16].

Figure 1 bellow shows the schematic diagram of an IT2-FL. It is similar to the T1-FL, the major difference is the fuzzy sets in the IT2-FL will be interval curve instead of crisp curve. Consequently, the output of IT-2FL will be interval fuzzy area and type reducer is required to get centroid point of the area. Similar with T1-FL, the crisp output will be obtained through a de-fuzzy-fication process.

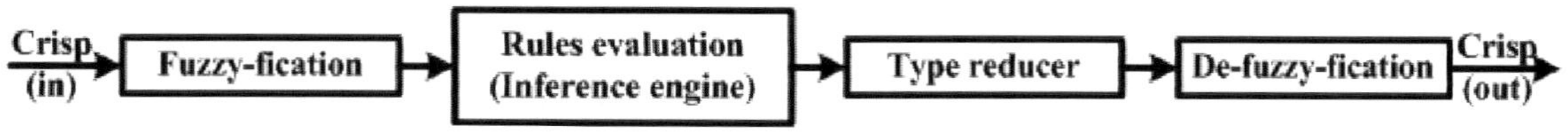

Figure 1. Schematic diagram of IT2-FL in converting input to be output

3. The method

3.1. Two stages IT2-FL framework

In this study, the manufacture quality of the investigated company is affected by 3 factors namely material quality, process quality and final product quality. Such information is obtained from 2 experts in the company. All of those factors are uncontrollable factor. Material quality factor is affected by two controllable sub factors namely density of the material and glucose level. Process quality is affected five controllable sub factors namely initial density of the materials, final density of the material, alcohol level, pH and temperature. Final product quality is affected by three sub factors namely ethanol level, permanganate level and *fusel* oil content.

Based on the explanation above, there are two stages in predicting the manufacture quality. The first stage is used to predict material quality, process quality and final product quality while the second stage is used to predict the manufacture quality based on the material, process and final product quality. Figure 2 shows the diagram of the proposed two stages IT2-FL.

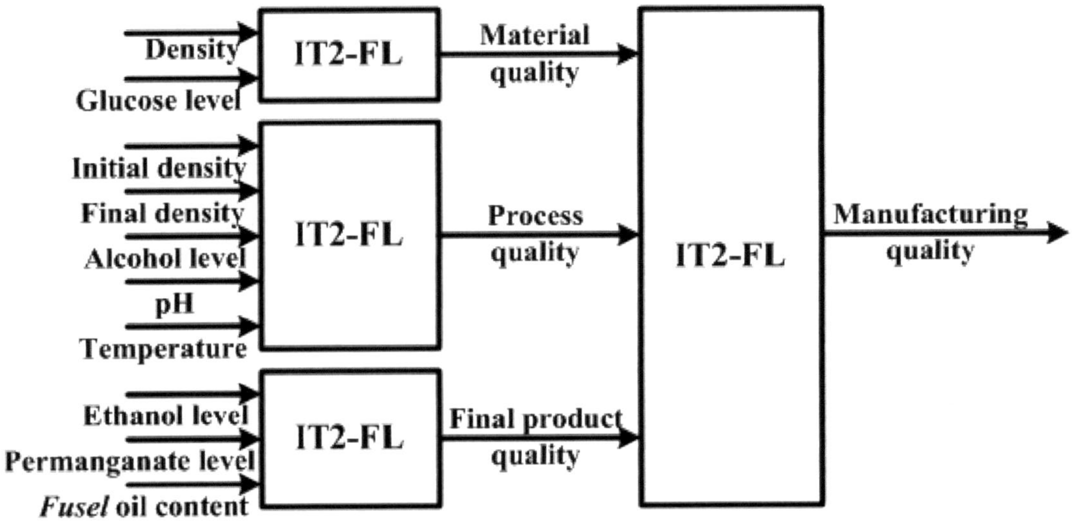

Figure 2. Diagram of two stages IT2-FL

3.2. Material quality

In order to get steady state pattern, data from three periods of production at end of year 2014 is collected. The numerical data is then being converted into interval fuzzy set with experts assistance. Figure 3 bellow is the interval fuzzy set for every sub factors that affect to the material quality.

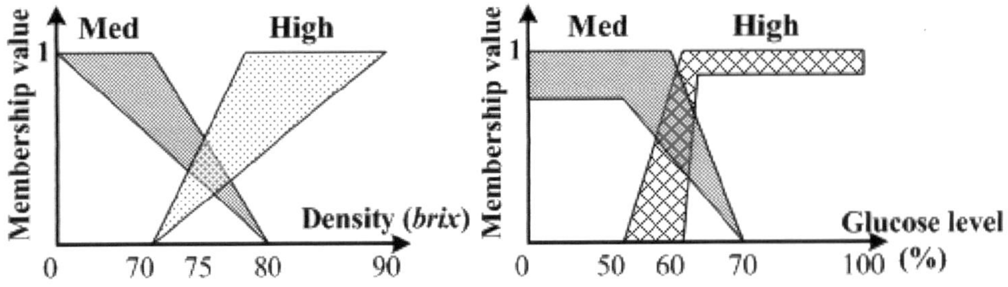

Figure 3. Interval fuzzy sets for density and glucose level

There are three fuzzy rules to predict material quality, as follows:

R_1: IF Density is HIGH AND Glucose level is HIGH THEN Material quality is HIGH (88-95)
R_2: IF Density is HIGH AND Glucose level is MEDIUM THEN Material quality is MEDIUM (68-73)
R_3: IF Density is MEDIUM AND Glucose level is MEDIUM THEN Material quality is MEDIUM (63-70)

3.3. Process quality
In this study, data to predict production process quality is collected during inspection of *spiritus* distillation process from August to December 2014. Fuzzy sets for input data namely initial density, final density, alcohol level, pH value and temperature are shown in Figure 4.

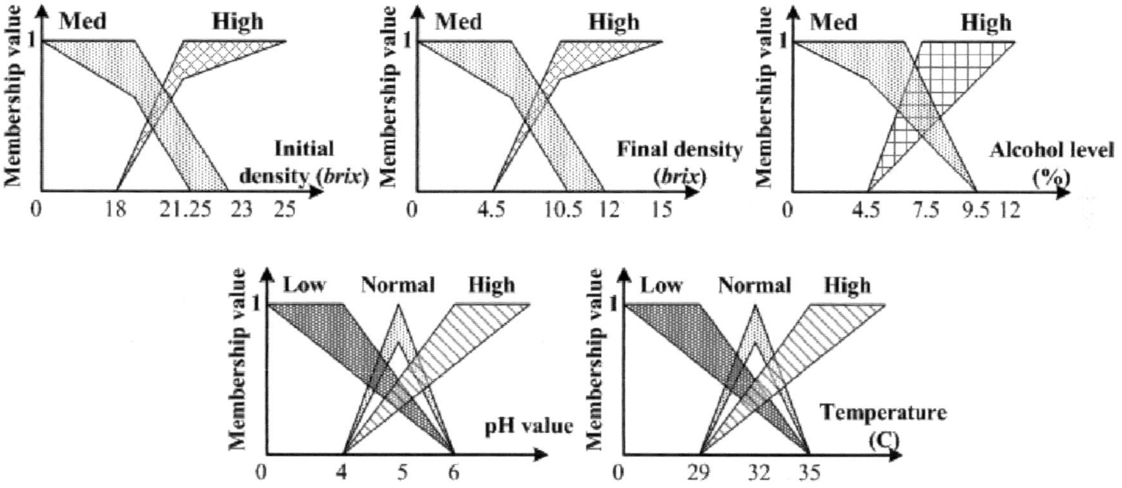

Figure 4. Fuzzy sets for every input of process quality prediction

To predict process quality, there are 9 selected fuzzy rules as follows:
R_1: IF Initial density is MEDIUM AND Final density is MEDIUM AND Alcohol level is MEDIUM AND pH value is NORMAL AND Temperature is NORMAL THEN Process quality is HIGH (87-93)
R_2: IF Initial density is MEDIUM AND Final density is MEDIUM AND Alcohol level is MEDIUM AND pH value is LOW AND Temperature is LOW THEN Process quality is LOW (67-73)
R_3: IF Initial density is MEDIUM AND Final density is MEDIUM AND Alcohol level is MEDIUM AND pH value is HIGH AND Temperature is HIGH THEN Process quality is LOW (65-77)
R_4: IF Initial density is HIGH AND Final density is HIGH AND Alcohol level is HIGH AND pH value is NORMAL AND Temperature is NORMAL THEN Process quality is MEDIUM (79-82)
R_5: IF Initial density is HIGH AND Final density is MEDIUM AND Alcohol level is MEDIUM AND pH value is NORMAL AND Temperature is NORMAL THEN Process quality is HIGH (83-92)
R_6: IF Initial density is MEDIUM AND Final density is HIGH AND Alcohol level is MEDIUM AND pH value is LOW AND Temperature is HIGH THEN Process quality is LOW (63-68)
R_7: IF Initial density is HIGH AND Final density is HIGH AND Alcohol level is HIGH AND pH value is LOW AND Temperature is HIGH THEN Process quality is LOW (60-65)

R_8: IF Initial density is HIGH AND Final density is MEDIUM AND Alcohol level is MEDIUM AND pH value is NORMAL AND Temperature is LOW THEN Process quality is MEDIUM (80-85)
R_9: IF Initial density is HIGH AND Final density is HIGH AND Alcohol level is MEDIUM AND pH value is HIGH AND Temperature is LOW THEN Process quality is LOW (63-65)

3.4. Final product quality
In the investigated manufacturing process, there are several independent variables that affect to the final product quality. Therefore, the final product quality must be predicted and will be one of the input in assessing manufacturing quality of the investigated company. Figure 5 shows the fuzzy sets for every input of the final product quality prediction.

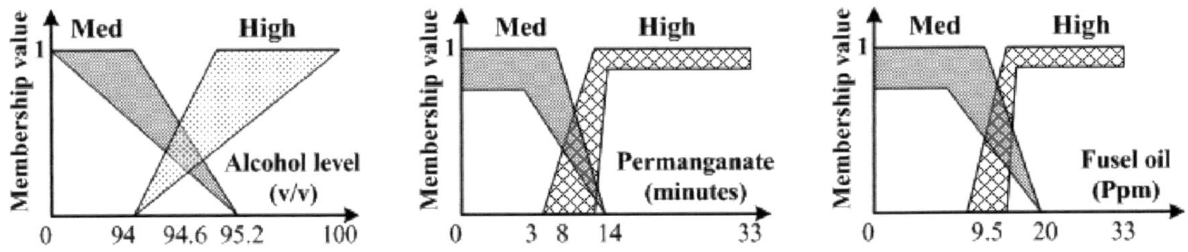

Figure 5. Fuzzy sets for every input of final product quality prediction

To predict final product quality, there are 5 selected fuzzy rules as follows:
R_1: IF Alcohol level is MEDIUM AND Permanganate is MEDIUM AND Fusel oil is MEDIUM THEN Final product quality is MEDIUM (63-67)
R_2: IF Alcohol level is MEDIUM AND Permanganate is HIGH AND Fusel oil is MEDIUM THEN Final product quality is MEDIUM (67-73)
R_3: IF Alcohol level is HIGH AND Permanganate is HIGH AND Fusel oil is HIGH THEN Final product quality is HIGH (80-85)
R_4: IF Alcohol level is MEDIUM AND Permanganate is MEDIUM AND Fusel oil is HIGH THEN Final product quality is HIGH (77-81)
R_5: IF Alcohol level is HIGH AND Permanganate is HIGH AND Fusel oil is MEDIUM THEN Final product quality is MEDIUM (70-76)

3.5. Manufacturing quality assessment
The manufacturing quality assessment is carried out by combining the three inputs which are material quality, process quality and final product quality through 7 selected fuzzy rules as follows:
R_1: IF Material quality is MEDIUM AND Final product quality is MEDIUM THEN Manufacturing quality is LOW (35-38)
R_2: IF Material quality is MEDIUM AND Final product quality is HIGH THEN Manufacturing quality is LOW (37-45)
R_3: IF Material quality is HIGH AND Process quality is HIGH AND Final product quality is HIGH THEN Manufacturing quality is HIGH (83-87)
R_4: IF Material quality is MEDIUM AND Process quality is LOW AND Final product quality is HIGH THEN Manufacturing quality is MEDIUM (69-73)
R_5: IF Material quality is MEDIUM AND Process quality is MEDIUM AND Final product quality is LOW THEN Manufacturing quality is LOW (32-37)
R_6: IF Material quality is HIGH AND Process quality is MEDIUM AND Final product quality is MEDIUM THEN Manufacturing quality is MEDIUM (70-75)
R_7: IF Material quality is MEDIUM AND Process quality is HIGH AND Final product quality is HIGH THEN Manufacturing quality is HIGH (78-84)

3.6. Fuzzy type reducer and de-fuzzy-fication
Output of T2-FL is still in the form of interval. In order to get the Crisp output, before de-fuzzy-fication, a type reducer is required. In this study, Karnik-Mendel algorithm is used as the IT2-FL type reducer. See [18] for the detail algorithm. In this study, a simple average *de-fuzzy-fication* method is used to convert the output of IT2-FL into a Crisp value.

4. System testing
Since the proposed assessment system is prediction system, then validation of the system is required. The proposed assessment system is developed based on primary data collection and experts opinion. Therefore, the validation method used in this study is experts confirmation. Such validation method is carried out by confirming the prediction result to the experts. A historical data set consists of ten data is used as the basis of the confirmation. Table 1 shows result of the experts confirmation.

Table 1. Confirmation result of IT2-FL in stage 1 and stage 2

	Stage 1		**Stage 2**
Material quality	**Process quality**	**Final product quality**	**Manufacturing quality**
87 (high)	83 (high)	84 (high)	85 (high)
73 (medium)	75 (low)	79 (high)	75 (medium)
89 (high)	**78 (low)**	81 (high)	80 (high)
89 (high)	87 (high)	78 (high)	86 (high)
72 (medium)	75 (low)	**77 (high)**	74 (medium)
73 (medium)	77 (low)	76 (high)	75 (medium)
89 (high)	89 (high)	79 (high)	80 (high)
88 (high)	87 (high)	76 (high)	**76 (medium)**
70 (medium)	80 (medium)	70 (medium)	74 (medium)
87 (high)	82 (medium)	**73 (medium)**	78 (high)
Accuracy: 90%	Accuracy: 90%	Accuracy: 80%	Accuracy: 90%

Note: bold text is the wrong prediction

Based on Table 1 above, it can be said fairly that the proposed IT2-FL in stage 1 and stage 2 has high accuracy. Therefore, it can be proposed as a fast tool to assess manufacturing quality of a supply chain.

5. Conclusion and suggestion
Based on the explanation above, it can be concluded that manufacturing quality assessment in the investigated company can be carried out step by step. IT2-FL can be used as one of the tool to model the assessment system. IT2-FL also has interval in the fuzzy sets and consequent of the fuzzy rules. The interval helps the experts to be flexible in determining the fuzzy rules based on their logic and intuition. Accuracy of the proposed IT2-FL in every stage is quite high, that means it can be used as the assessment tool for the investigated manufacturing system.

6. References
[1] Ageron B, Gunasekaran A and Spalanzani A 2012 Sustainable Supply Chain Management: An Empirical Study *International Journal of Production Economics* **140** 168.

[2] Levi D S, Kaminsky P and Levi E S 2000 *Designing and Managing the Supply Chain: Concepts, Strategies and Case Studies* McGraw-Hill International Edition, United States.

[3] Svensson G 2007 A conceptual framework for the analysis of vulnerability in supply chains *International Journal, Intelligence Magazine, IEEE* **2** 30.

[4] Handfield R B and Nichols E L 1999 *Introduction to Supply Chain Management*, Prentice-Hall, New Jersey.
[5] Linton J D, Klassen R and Jayaraman V 2007 Sustainable supply chains: an introduction *Journal of Operations Management* **25** 1075.
[6] Carter C R and Rogers D S 2008 A frame work of sustainable supply chain management: moving toward new theory *International Journal of Physical Distribution & Logistics Management* **38** 360.
[7] Preuss L 2005 Rhetoric and reality of corporate greening: a view from the supply chain management function *Business Strategy & the Environment* **14** 114.
[8] Feigenbaum A V 1991 *Total Quality Control*, McGraw-Hill, New York.
[9] Jain R K and Samrat A 2015 A Study of Quality Practices of Manufacturing Industries in Gujarat *Journal of Social and Behavioral Sciences* 320.
[10] Thombansen U, Hermanns T and Stoyanov S 2014 Setup and Maintenance of Manufacturing Quality in CO_2 Laser Cutting. *2nd International Conference on Rump-Up Management 2011* **98** 1.
[11] Arif F, Suryana N and Hussin B 2013. Cascade Quality Prediction Method Using Multiple PCA+ID3 for Multi-Stage Manufacturing System *2013 IERI* 201.
[12] Mendel J M, John R I and Liu F 2006 Interval Type-2 Fuzzy Logic Systems Made Simple *IEEE Trans Fuzzy Syst* **14** 808.
[13] Kadolph S K 2007 *Quality Assurance for Textiles & Apparel, second ed*, Fairchild Publications, New York.
[14] Keist C N 2015 *Quality Control and Quality Assurance in the Apparel Industry*, Garment Manufacturing Technology 405.
[15] Celso Caldas and Fernando S 2015 *Quality Control in the Sugar and Ethanol Industries* 275.
[16] Cosenza and Galuzzo 2010 Experimental Comparison of Type-1 and Type-2 Fuzzy Logic Controllers for the Control of Level and Temperature in a Vessel *ESCAPE* **21** 803.
[17] Mamdani E H 1974 Application of Fuzzy Algorithms for Control of Simple Dynamics Plants *in Proc. Of IEEE 121* **12** 1585.
[18] Hagras 2007 Type-2 FLCs: A New Generation of Fuzzy Controllers *Computational Intelligence Magazine, IEEE Trans on Fuzzy Systems 2* **1** 30.
[19] Dongrui Wu 2012 *A Brief Tutorial on Interval Type-2 Fuzzy Sets and Systems* Unpublished.
[20] Karnik N N and Mendel J M 1998 Introduction of Type-2 Fuzzy Logic Systems, *in Proc. 7th Intl. Conf. On Fuzzy Systems FUZZ-IEEE 1998* Anchorage AK 915.
[21] Martinez R, Castillo O and Aguilar L T 2009 Optimization of Interval Type-2 Fuzzy Logic Controllers for a Perturbed Autonomous Wheeled Mobile Robot Using Genetic Algorithms *Journal of Information Sciences* **179** 2158.
[22] Castillo O, Huesca G and Valdez F 2005 Evolutionary Computing for Optimizing type-2 Fuzzy Systems, Intelligent Control of Non-Linear Dynamics Plants *in Proc. North American Fuzzy Info, Processing Society (NAFIPS 2005)* Ann Arbor MI **6** 247.
[23] Cao J, Liu H, Li P and Brown D 2008 Adaptive Fuzzy Logic Controller for Vehicle Active Suspension with Interval Type-2 Fuzzy Membership Fuctions *in Proc. FUZZ-IEEE 2008* Hong Kong 1361.
[24] Wu D and Tan W W 2006 Genetic Learning and Performance Evaluation of Interval Type-2 Fuzzy Logic Controllers *Engineering Applications of Artificial Intelligence* **19** 829.

Multi-objective Mixed Integer Programming approach for facility layout design by considering closeness ratings, material handling, and re-layout cost

Muhammad Ridwan Andi Purnomo and Yoga Satrio Wiwoho

Department of Industrial Engineering, Universitas Islam Indonesia, INDONESIA

Email: ridwan_ie@uii.ac.id

Abstract. Facility layout becomes one of production system factor that should be managed well, as it is designated for the location of production. In managing the layout, designing the layout by considering the optimal layout condition that supports the work condition is essential. One of the method for facility layout optimization is Mixed Integer Programming (MIP). In this study, the MIP is solved using Lingo 9.0 software and considering quantitative and qualitative objectives to be achieved simultaneously: minimizing material handling cost, maximizing closeness rating, and minimizing re-layout cost. The research took place in Rekayasa Wangdi as a make to order company, focusing on the making of concrete brick dough stirring machine with 10 departments involved. The result shows an improvement in the new layout for 333,72 points of objective value compared with the initial layout. As the conclusion, the proposed MIP is proven to be used to model facility layout problem under multi objective consideration for a more realistic look.

1. Introduction

In a production system, there are a lot of processes and occurrences that must be considered and planned. One of production system factor that must be managed well is the facility layout, because it is designated for the production location, where machines, human, assets, material, and product move along processes. Having a bad designed or arranged facilities may inflict a disarranged movement of production entity, which may cause an increasing chance of error, increasing transportation/handling time, not ergonomic work place, and many causes that can inflict costs. Considering the importance of proper facility layout design and planning, development toward this field has been conducted until now.

Modeling the plant layout becomes an approach that often to be chosen as it can save the time and cost to discover the effect of layout modifications. The model can be built either by manual or computer based. Koopmans and Beckmann (1957) proposed a basic mathematical model for assignment problems by using linear programming that is known as Quadratic Assignment Problems (QAP) designed for layout with equal area [1]. A Mixed Integer Programming (MIP) based mathematical model for facility layout problems has been developed by previous researcher [2]. This MIP method was known as the most optimal mathematical model for facility layout design, which then is developed more by Meller et al. (1998) [3] and Sherali et al. (2003) [4]. From QAP and MIP, the model then expanded with many approaches and aided by computer programming. CRAFT that was developed by Armour and Buffa in 1963 was the famous computer based heuristic algorithm for plant layout optimization considering minimum material handling cost [5]. There was also another computer-based algorithm with adjacency ratings among facilities such as Computerized Layout Planning (CORELAP), Automated Layout Design Program (ALDEP) and BLOCPLAN, the hybrid one that able to process both optimal and heuristic algorithms. Even farther, intelligent approach such as Genetic Algorithm had also been implemented for optimizing layout that became an interesting discussion nowadays [6], [7], [8], [9].

Although many approaches are intended to mimic the layout condition and changes, they still cannot represent the real world in some ways. For example, some approaches only try to focus only in one indicator like material handling cost. Indeed material handling have a big contribution toward operation handling cost that is up to 50 to 60 percent of product cost, and take 50 percent of production time [10]. But sometimes it is not representative as in the real business company will consider other costs or importance. There are safety and ergonomic issues, for instance, which become a constraint for 2 facilities to be closed even their material handling cost is high, and vice versa. Some researchers have conducted research concerning these objectives simultaneously [11], [12].

Although there are already several researches regarding to multi-objective layout problems, yet there is still no research concerning about re-layout cost together with those objectives. Meanwhile, sometimes facility re-layout is also not feasible as it has a very high installment cost and production loss for stopping the machine. Thus, these considerations need to be brought together in the optimization approach in order to get a more realistic result. This research is conducted by building an optimization model to develop a new layout by minimizing material handling cost, re-layout cost, and maximizing closeness ratings simultaneously. The model will be built based on MIP approach in order to get the optimal result of the layout.

2. Related works

Literature study for facility layout problems have been conducted [13], [14], and [15]. Both [13] and [15] emphasized the opportunities for multi objective as the further research object in facility layout problems is needed to resume more realistic considerations and fill the research gap. This initiate the research to be focusing in multi objective problems.

A multi objective algorithm for facility layout problem by considering quantitative and qualitative objectives at once is developed [16]. They proposed a probability of superiority as additional criteria for proofing that one layout is better than the other. There is also another research conducted [17]. In that study, multi-pass halving and doubling procedure method are used to get the importance of each objective subjectively to the decision maker. A prior test was also conducted to get the consistency of the paired comparison matrix.

Using another approach, a heuristic approach for determining the sequence of common linear machine with multi products under different operation sequences and facilities is proposed [18]. The criteria assessed are total flow distance by products, number of machines in linear sequence, and total investment cost of the machines.

Multi objective layout problems become an interesting topic as actually there are a lot of things can be considered in the facility layout problems and the need for a simultaneously calculation. In other side, there is still a lack of research concerning multi objective facility layout problems with MIP as the method. Researches applying MIP usually only focus toward one objective such as material handling.

Heragu and Kusiak (1991) proposed some MIP techniques named as M1, M2, and M3 [19]. The first technique: M1, focuses on minimizing the total cost required in trips between facilities for a single row layout. Besides, both M2 and M3 focus on multi row layout, where in M2 all departments are assumed to have the same area and M3 can be for unequal department's areas. These models have only a few constraints which can ease the calculation, but it is assumed that there is no restriction on the shape of the building of the facilities.

Meller (1998) also proposed another MIP equations for facility layout problem [3], where this equation seems to be more realistic rather than Heragu and Kusiak (1991) had proposed. He enhanced the model proposed by Montreuil (1990) [2], by redefining Monteruil's binary variables and tightening the department area constraints. He also proposed some general classes of valid inequalities thus the range of problem solved can be enhanced, from $n = 5$ departments (as conducted by Montreuil) to be $n = 9$ departments.

Afterward, Meller's model which known as FLP2 and FLP2+ is enhanced by Sherali et al. (2003) [4]. The new proposed model can solve layout problems more accurate and efficient. Another MIP formulation also proposed by Konak et al. (2005) [20], designed for facility layout design using flexible bays. The application of MIP model for facility layout is an interesting area as this method is proofed to be the best-most optimal optimization techniques for facility layout problems, although it only can solve small amount of problems in departments.

As can be seen from the previous researches mentioned, there are still no previous paper discussing a multi objective facility layout problem considering material handling cost, adjacency ratings, and re-layout cost simultaneously. Thus, a research that involves those 3 objectives is conducted here. It uses MIP approach in order to get the optimal solution of the problem. From this research, it is expected that a better layout arrangement can be derived by the proposed mathematical model.

3. Model Description

The parameters involved are as follows:

w = weighting score of each objective
c_{ij} = cost of material handling from facilities i to j
f_{ij} = frequency of material flow from facilities i to j
Pl_i = production loss from moving facilities i
t_{ij} = time spent for moving facilities i to j
$cm1_{ij}$ = cost of moving facilities i to j (where it is varied to distance)
$cm2_{ij}$ = fixed cost of moving facilities i to j (where it is not influenced by distance)
r_{ij} = adjacency score between facilities i to j
a_i = area of department i, where $a_i > 0$
$\propto_i$ = maximum permissible ratio between the longest and shortest side of department i. i.e., $\max_s\{l_i^s\} / \min_s\{l_i^s\} \leq \propto_i$
lb_i = lower bound of department's length
ub_i = upper bound of department's length
H^s = maximum permissible total length of a rectangular building (building criteria)
s = x, y coordinate

The decision variable of the equations are as follows:

l_i^x = half-length of department i having a rectangular shape (correspond to x ordinate)
l_i^y = half-width of department i having a rectangular shape (correspond to y ordinate)
c_i^x = centroid of department i in x ordinate
c_i^y = centroid of department i in y ordinate
d_{ij}^s = distance between facilities i and j

$$z_{ij}^s \begin{cases} 1, \textit{if department i is forced to preced department j in direction s} \\ 0, \textit{otherwise} \end{cases}$$

The multi-objective model to be achieved is like in the following model:

$$\min Z = \sum_{i=1}^{n-1} \sum_{j=1+1}^{n} w_1 O_1 - w_2 O_2 + w_3 O_{3a} + \sum_{i=1}^{n} w_3 O_{3b} \quad (01)$$

$$O_1 = \sum_{i=1}^{n-1} \sum_{j=1+1}^{n} c_{ij} f_{ij} \left(d_{ij}^{x} + d_{ij}^{y}\right) \quad (02)$$

$$O_2 = \sum_{i=1}^{n-1} \sum_{j=1+1}^{n} r_{ij} \left(d_{ij}^{x} + d_{ij}^{y}\right) \quad (03)$$

$$O_{3a} = \sum_{i=1}^{n-1} \sum_{j=1+1}^{n} cm1_i \left(d_{ij}^{x} + d_{ij}^{y}\right) \quad (04)$$

$$O_{3b} = \sum_{i=1}^{n} cm2_i + \left(pl_i \ t_i\right) \quad (05)$$

As can be seen, the Objective 1 (equation 02) focuses on the material handling cost minimization, the Objective 2 (equation 03) focuses on qualitative objective: maximizing adjacency score between facilities, and the objective 3 (equation 04) and (equation 05) focus on minimizing the movement of facilities in re-lay outing. Specifically, Objective 3a (equation 04) concerns about the minimizing cost of moving machine where distance is influencing the cost, while objective 3b (equation 05) concern about the cost of moving machine where distance is not influencing and added by production loss of moving machine. In other word, Objective 3b tends to fixed cost.

The constraints implemented are based on the Sherali et al.'s (2003) MIP model with a derived polyhedral outer-approximation for the area constraints [4], which are enhanced from Meller's (1998) FLP2 equation [3].

$$\text{Subject to:} a_i l_i^x + 4\bar{x}^2 l_i^y \geq 2a_i\bar{x} \quad \forall\ lb_i^x \leq \bar{x} \leq ub_i^x, \text{i} \quad (06)$$

$$\bar{x} = lb_i^x + \frac{\lambda}{(\Delta-1)}\left(ub_i^x - lb_i^x\right) \quad \forall\ \lambda = 0, 1, \ldots, \Delta-1, \text{ for any selected integer } \Delta \geq 2 \quad (07)$$

$$\sum_{s=x}^{y}\left(z_{ij}^s + z_{ji}^s\right) \geq 1 \quad \forall\ i < j \quad (08)$$

$$z_{ij}^s + z_{ji}^s \leq 1 \quad \forall\ j > i, s \quad (09)$$

$$c_i^s + l_i^s \leq c_j^s - l_j^s + H^s\left(1 - z_{ij}^s\right) \quad \forall\ i \neq j, s \quad (10)$$

$$d_{ij}^s \geq c_i^s - c_j^s \quad \forall\ f_{ij} \neq 0, j > i, s \quad (11)$$

$$d_{ij}^s \geq c_j^s - c_i^s \quad \forall\ f_{ij} \neq 0, j > i, s \quad (12)$$

$$l_i^s \leq c_i^s \leq H^s - l_i^s \quad \forall\ i, s \quad (13)$$

$$\overline{lb_i} \leq l_i^s \leq min\left\{\overline{ub_i}, H^s/2\right\} \quad \forall\ i, s \quad (14)$$

$$0 \leq d_{ij}^s \leq H^s - \left(lb_i + lb_j\right) \quad \forall\ i \neq j, s \quad (15)$$

$$ub_i = min\left\{\sqrt{\alpha_i\, a_i}, max_s\{H^s\}\right\}/2 \quad \forall\ i, s \quad (16)$$

$$lb_i = a_i/(4ub_i) \quad \forall\ i \quad (17)$$

$$c_i^s \geq 0 \quad (18)$$

$$z_{ij}^s \in \{0,1\} \forall\ i \neq j, s \quad (19)$$

Equation (06) and equation (07) are used for area constraint for each department i. The constraints come from the derived polyhedral outer-approximation. The number of discretization points Δ are varied from 5 to 50 tangential supports. The higher the tangential supports are, the more accurate the result will be. Equation (08) and equation (09) ensure that i and j are separated in at least one direction, while equation (10) prevents overlapping of i and j in direction s by using the 0-1 value from z_{ij}^s. Equation (11) and equation (12) are the linearization of distance's absolute value from the difference between centroid i and j in direction s. Equation (13) ensures that department i is inside the building or facility. Equation (14) forces the length of i department to be not exceeding the boundary set, while equations (15) and equation (16) denote the value of lower bound and upper bound of the length of each department i. Equation (17) indicates the constraints of the distance that its value must not exceed the building length minus by the lower bound of i and j departments. Meanwhile, equation (18) forces the centroid to be not in negative value and equation (19) restricts the z_{ij}^{s} value to be binary (0-1) as one of the decision variables.

Besides FLP2 equation above, there are still some additional equations can be added to enhance the solving process to be faster and more accurate. Here, *p-q* strategy supported with B2 and V2 valid inequalities are implemented:

$$c_p^s \leq c_q^s \quad \forall\ s = x, y \tag{11}$$

$$z_{qp}^{x} = z_{qp}^{y} = 0 \tag{12}$$

$$\sum_{s=x}^{y} \left(c_q^s - c_p^s\right) \geq min\left\{lb_p^x + lb_q^x, lb_p^y + lb_q^y\right\} \tag{13}$$

$$\text{B2: } d_{ij}^s \geq \left(lb_i^s + lb_j^s\right) \quad \forall\ f_{ij} \neq 0, j > i, s \tag{14}$$

$$\text{V2: } d_{ij}^s \geq \left(l_i^s + l_j^s\right) - min\left\{ub_i^s + ub_j^s, H^s\right\}\left(1 - z_{ij}^s - z_{ji}^s\right) \quad \forall\ f_{ij} \neq 0, j > i, s \tag{15}$$

By having a pair of p and q critical departments based on a maximum total interaction or any other essences, equation (20), equation (21), and equation (22) above, named as position *p-q* method, can be implemented for reducing problem symmetry. The equation requires centroid of p department to be the south and west of q departments. In this case, p and q departments will be chosen to a pair of departments having the largest flow. It is need to be noticed that this position *p-q* method only can be implemented if there is no fixed departments.

Equation (23) and equation (24) above indicate the valid inequalities developed [3] that Sherali et al. (2003) [4] found as the equation bring a better performance compared to the other inequalities, where in this equation Sherali et al. already modified the B2 and V2 with their tighter bounds. These equations are implemented here as it is found that B2 and V2 constraints are effective to be used in conjunction with position *p-q* method.

This model was intended to a normal condition where there are no fixed departments or fixed-width-departments. For handling such that further problem, the detailed model are provided in [3] or [4].

4. A case study

The company observed in this study is located in Yogyakarta, Indonesia. The company focuses on producing machines for industries and personal usage related to manufacturing, trade, and service of food and beverage. The products are made based on the request of the costumer, where sometimes costumers also order for special needs. The company has 10 departments and located in a 26,5 m x 24,5 m of rectangle building. Table 1 shows department's dimension, their location and cost of moving.

Table 1. department's dimension and their location

No	Dept.	L (m)	H (m)	c_i^x	c_i^y	Cost of moving machine (IDR/meter)
1	Cutting machine	12	8	16,55	5,725	7.200.000 (fixed)
2	Cutting dept.	3	4	15,2	6,05	Equal to material handling cost/meter
3	Drill	3	2,5	15,1	6,375	Equal to material handling cost/meter
4	Welding	8	2,5	17,05	6,275	Equal to material handling cost/meter
5	Circle	9	4,5	16,4	6,825	Equal to material handling cost/meter
6	Grinding	6	2,5	15,65	6,375	Equal to material handling cost/meter
7	Bender	3	2,5	15,825	6,625	Equal to material handling cost/meter * 2
8	Roll plate	4	2	16,85	6,5	7.200.000 (fixed)
9	Painting	6	6	15,65	5,95	Equal to material handling cost/meter
10	Bending	7	4,5	16,3	6,375	7.200.000 (fixed)

In table 1, it can be seen that the cost of moving of Bender department is equal to material handling cost per meter multiplied by 2. It is because 2 workers are required to move the machine inside such department. Other required input data are material flow frequency and closeness rating between department that can be seen in table 2 and table 3 respectively while initial layout of the company is depicted in figure 1.

Table 2. Flow frequency between department

Dept.	1	2	3	4	5	6	7	8	9	10
1	0			2						1
2		0	3	3		3				
3		1	0	10	1	1				
4			1	0					1	
5			2	6	0	8	4			
6			5	7		0		1		
7				7			0			
8				1				0		
9									0	
10				1						0

Table 3. Closeness rating between departments

Dept.	1	2	3	4	5	6	7	8	9	10
1	0	4	3	2	2	3	2	4	1	3
2	4	0	4	2	2	3	2	2	1	2
3	3	4	0	3	3	5	2	2	1	3
4	2	2	3	0	2	4	2	2	1	2
5	2	2	3	2	0	3	3	3	1	2
6	3	3	5	4	3	0	2	2	1	5
7	2	2	2	2	3	2	0	2	1	2
8	4	2	2	2	3	2	2	0	2	2
9	1	1	1	1	1	1	1	2	0	3
10	3	2	3	2	2	5	2	2	3	0

Note: (6 = absolutely necessary, 5 = essentially necessary, 4 = important, 3 = ordinary, 2 = unimportant, 1 = undesirable)

In this study, production lost due to re-layout activity is also considered. Such consideration is to make the re-layout analysis to be more scientific. When a department is required to move, logically the production activity must be stopped and it will be lost production for the company. Table 5 shows the production lost per minute due to re-layout activity.

The moving time as illustrated in Table 6 are assumed to be fix value toward distance, because the area of production is not too big. Thus, wherever the machine will be moved to, there will be only a slight difference that in this case will be averaged. The moving time for cutting machine, roll plate, and bending above require more time if compared to other departments, because they involve big machines that need a lot of setup operation only to replace them. The company's owner estimated that by minimum, it takes 4 days to move the machines, wherever the position is.

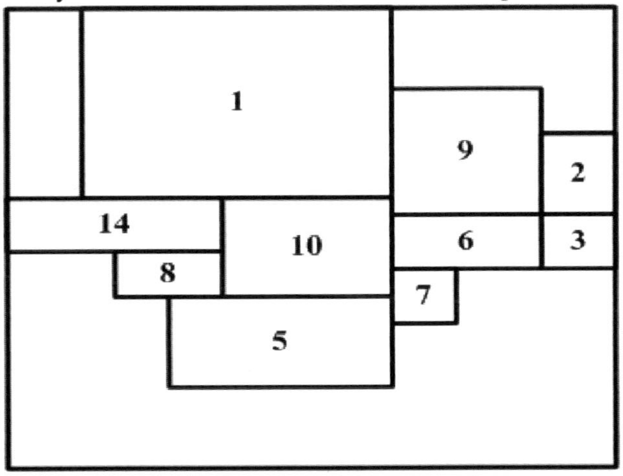

Figure 1. Initial layout of the company

Table 4. Production loss per minute for each department

Dept.	Operation Time (min)	Production loss/min	Operation time/total (%)
1	35	0,010	1%
2	103	0,030	3%
3	392	0,112	11%
4	1.640	0,470	47%
5	529	0,152	15%
6	198	0,057	6%
7	290	0,083	8%
8	30	0,009	1%
9	240	0,069	7%
10	30	0,009	1%
Sum	3.487	1	100%

Table 5. Moving time of machine

Dept.	Moving Time (min)
1	1.920
2	5
3	5
4	3
5	5
6	3
7	15
8	1.920
9	7
10	1.920

The mathematical model to solve the problem is developed under several assumptions as follows:

- The material handling cost is assumed to be 1. This is according to the manual (barehanded) material handling and no certain man to handle the material handling job. The material handling was carried out by operator in each department, which makes there is no existing value to measure the material handling cost.
- The value of Δ in area constraint equation is 20.
- The α ratio of each department uses the initial length and width of each department.
- In this case, the departments are located along a maximum permissible length $H^x = 24$ and $H^y = 18$ (in meter). This is due to some building's spaces are used for others besides the production, which makes the maximum permissible length is not equal to the building's area. These values are obtained from the same maximum lengths with the initial layout.

- Based on interview with the company's owner, the most important objective to be considered is qualitative one with score $w_2 = 0{,}5$. This high score comes from the thought that safety is in prior in the production itself. In the second place, there is material handling efficiency objective with score $w_1 = 0{,}4$, as the company realize the importance of workflow efficiency. The rest is for the cost of moving machine objective with score $w_3 = 0{,}1$.
- The *p* and *q* departments chosen for *p-q* strategy are the pair of departments 3-4, as it has the highest material flow.

Objective value for the new layout provided by the proposed mathematical model is IDR 2.158.887,79. The length and area of each department can be seen in the Table 7 while figure 2 shows the new layout.

Table 6. The half-length and centroid for each departments

Dept.	c_i^x	c_i^y	l_i^x	l_i^y
1	9,305733	10,48089	5,356049	4,480891
2	1,974842	16,48089	1,974842	1,519109
3	4,61782	2,25	1,5	1,25
4	18,66178	4,75	4	1,25
5	19,16178	8,25	4,5	2,25
6	17,66178	16,75	3	1,25
7	7,411782	4,75	1,25	1,25
8	22,85734	1,75	1,142659	1,75
9	11,66178	3	3	3
10	18,21143	13	3,549648	2,230891

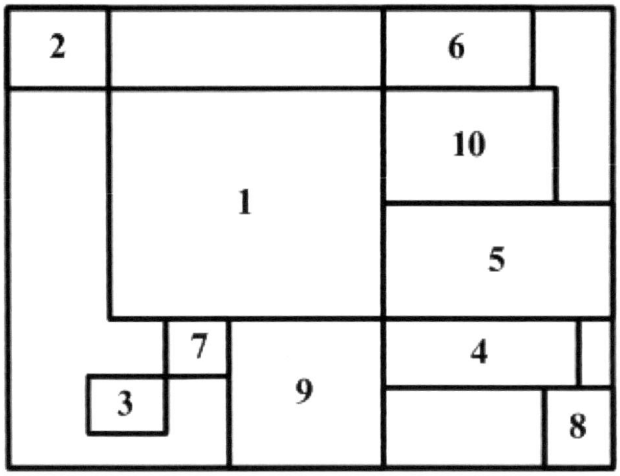

Figure 2. The new optimum layout

5. Discussion

In order to analyze the improvement of the new optimum layout, a comparison analysis with the initial layout is carried out. When the initial layout is substituted into the proposed mathematical, its objective value is 2.159.221,50. As the objective used in the model is minimization objective, the new layout is proven better than the initial layout with 333,72 point improvement. Thus, as the conclusion, the new layout still could be implemented as it has a higher objective value.

6. Conclusions and Further Research

As the conclusion, it is proven that the proposed MIP model is able to optimize multi-objective facility layout problems. The new layout not only has a minimum material handling, but also proportional to the closeness ratings and its re-layout cost. As the further research, it could be an interesting topic to modify the minimization of re-layout cost objective by considering monument or fixed departments, and integrated moving time among departments, a dynamic moving time or production loss, or others. Besides, developing meta-heuristic algorithm such as Genetic Algorithm (GA) combined with a knowledge-based system for the same objectives problem also can be an interesting area as GA is compatible to be used to solve complex problems.

7. Acknowledgements

We would like to thank to all supporting people toward this research, either directly or indirectly. Especially for our family.

References

[1] Koopmans T C and Beckmann M 1957 Assignment problems and the location of economic activities *Econometrica* **25** 53.
[2] Montreuil B 1991 A modelling framework for integrating layout design and flow network design *Material Handling'90* Springer Berlin Heidelberg 95.
[3] Meller R D, Narayanan V and Vance P H 1998 Optimal facility layout design *Operations Research Letters* **23** 117.
[4] Sherali H D, Fraticelli B M P and Meller R D 2003. Enhanced model formulations for optimal facility layout *Oper. Res.* **51** 629.
[5] Heragu S 1997 *Facilities design* Boston: PWS Publishing Company.
[6] El-baz M A 2004 A genetic algorithm for facility layout problems of different manufacturing environments **47** 233.
[7] Garcia-hernandez L, Pierreval H, Salas-morera L and Arauzo-azofra A 2013 Handling qualitative aspects in unequal area facility layout problem : an interactive genetic algorithm *Applied Soft Computing* **13** 1718.
[8] Gonçalves J F and Resende M G C 2015. A biased random-key genetic algorithm for the unequal area facility layout problem *European Journal of Operational Research* **246** 86.
[9] Lee Y H and Lee M H 2002. A shape-based block layout approach to facility layout problems using hybrid genetic algorithm *Computers and Industrial Engineering* **42** 237.
[10] Murthy P R 2005 *Production and operations management.* New Delhi: New Age International (P) Ltd. Publishers.
[11] Ripon K S N, Khan K N, Glette K, Hovin M and Torresen J 2011 Using pareto-optimality for solving multi-objective unequal area facility layout problem *Proceedings of the 13th annual conference on Genetic and evolutionary computation* 681.
[12] Aiello, G, Scalia G L and Enea M 2012 A multi objective genetic algorithm for the facility layout problem based upon slicing structure encoding *Expert Systems with Applications* **39** 10352.
[13] Singh S P and Sharma R R K 2006 A review of different approaches to the facility layout problems *Int J Adv Manuf Technol* **30** 425.
[14] Drira A, Pierreval H and Hajri-Gabouj S 2007 Facility layout problems: a survey. *Annual Reviews in Control* **31** 255.
[15] Jain A K, Khare V K and Mishra P M 2013 Facility planning and associated problems: A Survey *Innovative Systems Design and Engineering* **4** 1.
[16] Sha D Y and Chen C W 2001 A new approach to the multiple objective facility layout problem *Integrated Manufacturing System* **12** 59.
[17] Chen C W and Sha D Y 2005 Heuristic approach for solving the multi-objective facility layout problem *International Journal of Production Research* **43** 4493.

[18] Lenin N, Siva Kumar M, Ravindran D, Vignesh Kumar D and Islam M N 2013 Decision making in multi-objective facility layout design *3rd International Conference on Recent Advances in Material Processing Technology* 1.

[19] Heragu S and Kusiak A 1991 Efficient models for the facility layout problem *European Journal of Operational Research* **53** 1.

[20] Konak A, Kulturel-Konak S, Norman B A and Smith A E 2006 A new mixed integer programming formulation for facility layout design using flexible bays. *Operations Research Letters* **34** 660.

Mass Customization and Personalization Prospects in Developing Country: Indonesian Context

Risdiyono[1], **Imam Djati Widodo, Affan Mahtarami**
Faculty of Industrial Technology, Islamic University of Indonesia

E-mail: risdiyono@uii.ac.id

Abstract. The advancement of information technology (IT) has changed many modes and ways for people in doing their businesses. Mass Customization and Personalization (MCP) is one example of business modes that has been dramatically evolve, mainly due to the currently very fast IT development. MCP has enabled people to involve in adjusting some design parameters of a product to meet their personal requirements before purchased. The advancement of IT has made MCP more successful as it makes the process faster, easier, simpler and more joyful. The success stories of MCP are easily found in many developed countries, where the IT infrastructure has firmly been established. For developing countries, there are very few industries have implemented the MCP concept, including Indonesia. This paper discusses a descriptive study to depict what people think about MCP implementation in Indonesia especially in Small and Medium Enterprises (SMEs). Kano model was used to see the perception of both producers and consumers in relation with MCP implementation. Five dummy MCP prototypes were developed for five creative products including plaques, hats, invitation card, t-shirts and leather bags. Based on the KANO questionnaire analyses, it is clear that there are big opportunities to implement MCP in Indonesia especially for creative products produced by SMEs. Identifying the correct product features is an important key for successful MCP implementation in developing countries.

Keywords : Mass Customization and Personalization, creative product, KANO, SME

1. Introduction

Every enterprise will always try to satisfy its customers. Total customers experience has been identified as a vital factor for manufacturers to focus on. Berry (2002) argued that premium value can be perceived by customer if the quality of product (superior solution to the need, fair price) and the quality of services (convenience, treatment with respect and emotional connection) could be provided by manufacturer. This means manufacturers have to understand what people need and how to fulfil the need better than competitors.

How to ensure that the real need customers demanded is the need manufacturer perceived is challenging as it will differentiate between a successful product and the fail one. It is noteworthy that understanding customer need is not a simple task especially in this era where voice of majority has been replaced by voice of personal. The extent of market-of-one has been foreseen as a prospective driving force for the next transformation of global economy (Pine 2009). As each customer may have

[1] Corresponding author.

her/his own preference, a flexible responsive-based product specification should be adopted as an alternative of the rigid predictive-based one. In other words, the domination of mass production paradigm has slowly been decreased and the concept of mass customization and personalization (MCP) has increasingly adopted. Customer involvement in value creation becomes a new important concept for manufacturers in addressing customer satisfaction (Risdiyono and Koomsap 2011).

There are many aspects considered by customers in making buying decision. Flexibility is one of them (Anderson 2006). In mass production system, flexibility is usually offered in terms of prices, services and delivery while MCP extends the flexibility to product specifications by enabling individual customers to involve in value creation. They are enabled to define the fittest alternative that meet the cost, schedule and the product requirements through the capabilities of a company. The flexibility is depended upon three main factors including (1) customer interest, (2) manufacturing issue and (3) engineering and super system issues (Risdiyono, 2015). It is important to note that the conditions of these three factors in developed countries are different with those in developing countries. Hence, the success story in some countries is not a warranty that MCP will be successfully implemented di all countries.

In 2014, there are 970 international web-based product configurators which are included in the Configurator Database. The objective of this report is to provide market information, statistics and an overview about diverse product configuration offerings of MCP companies. This documentation platform was started in 2007 and grew to the biggest collection of online configuration tools that are used in mass customization approaches (Blazek, 2014). Among those 970 companies, none of them is located in Indonesia (Table 1).

Table 1. Number of product configurators offered by companies based on country of origin.

Country	Frequency	Country	Frequency
Germany	470	Thailand	4
United States	332	Finland	3
Austria	33	Norway	3
United Kingdom	28	New Zealand	3
Switzerland	20	China	2
Netherlands	15	Czech Republic	1
France	12	Denmark	1
Australia	9	Ireland	1
Canada	9	India	1
Sweden	7	Lithuania	1
Bergium	4	Latvia	1
Spain	4	Niger	1
Italy	4	Slovakia	1

The prospect of MCP implementation in developing countries needs to be evaluated. This research is one initial attempt to get a brief depiction about MCP in the perspective of people of Indonesia. The result of this paper is very useful for all stakeholders in many sectors, including industries, academics and government in order for them to take important measures to deal with the dynamic competition in this global era.

2. Research Method

2.1. Dummy MCP Prototypes Development

In MCP concept, personal experience is one important thing that customer will deal with. Hence, in this research, in order to make the respondents become more familiar with the MCP concept, five dummy prototypes were developed for creative product configurators including plaques, hats, invitation card, t-shirts and leather bags. The prototypes were then demonstrated in front of the respondents before the survey is conducted. This demonstration is very important to be presented so that all respondents get a similar perception about what MCP is. Figure 1 shows the example of two dummy prototypes of product configurators, i.e. t-shirt and plaque. The dummy prototypes were provided in both English and Bahasa Indonesia.

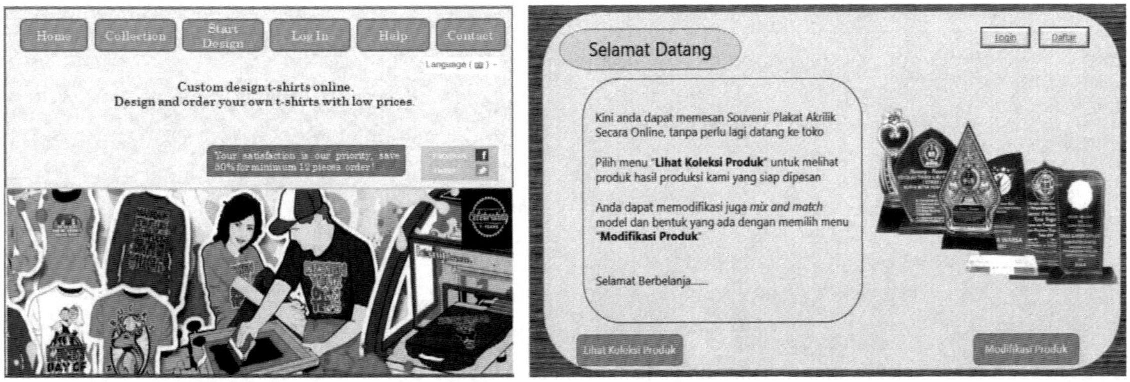

Figure 1. Two examples of the product configurator dummy prototypes

2.2. Data Collection and Analysis

By using the five dummy prototypes developed, data collection processes were then conducted. Five groups of respondents consisting about 70 persons each were incorporated in this research. The survey utilized a questionnaire based on Kano model as it is a useful tool to classify and prioritize customer needs based on how they affect satisfaction. The relationship between product performance and customer satisfaction can be easily reflected on four types of product attributes, including: (1) must-be, (2) one-dimensional, (3) attractive, and (4) indifferent. Must-be (basic) attributes are expected by the customers and their absent leads to extreme customer dissatisfaction. One-dimensional (performance) attributes are those for which the level of customer satisfaction is linear to the degree of attribute fulfilment. Attractive (excitement) attributes are usually unexpected by the customers and their availability can result in great customer satisfaction. Indifferent attributes are those that the customer is not interested in the level of their performance (Sauerwein et.al., 1996). Hence, by using this method, the customer perception about MCP concept can be depicted and each feature provided by each configurator can be evaluated.

3. Results and Discussions

3.1. Results

As the dummy prototypes are product-specific (built based on each product flexibility), then the number of features for each product configurator are slightly different. For example, the features of leather bag configurator (which are mainly focused on selecting model, configuring shape and choosing material) will be different with the features of t-shirt configurator (which are mostly in relation with inserting image and text). The same situation is applied to all five product configurators in question. The customer perception is then configured by the tendency of all Satisfaction Index (SI) and Dissatisfaction Index (DI) plotted in the Kano diagram. Based on the position of all features in the diagram, the attractiveness level of the concept can be then concluded. The result of Kano questionnaire for all product configurators can be found in figure 2.

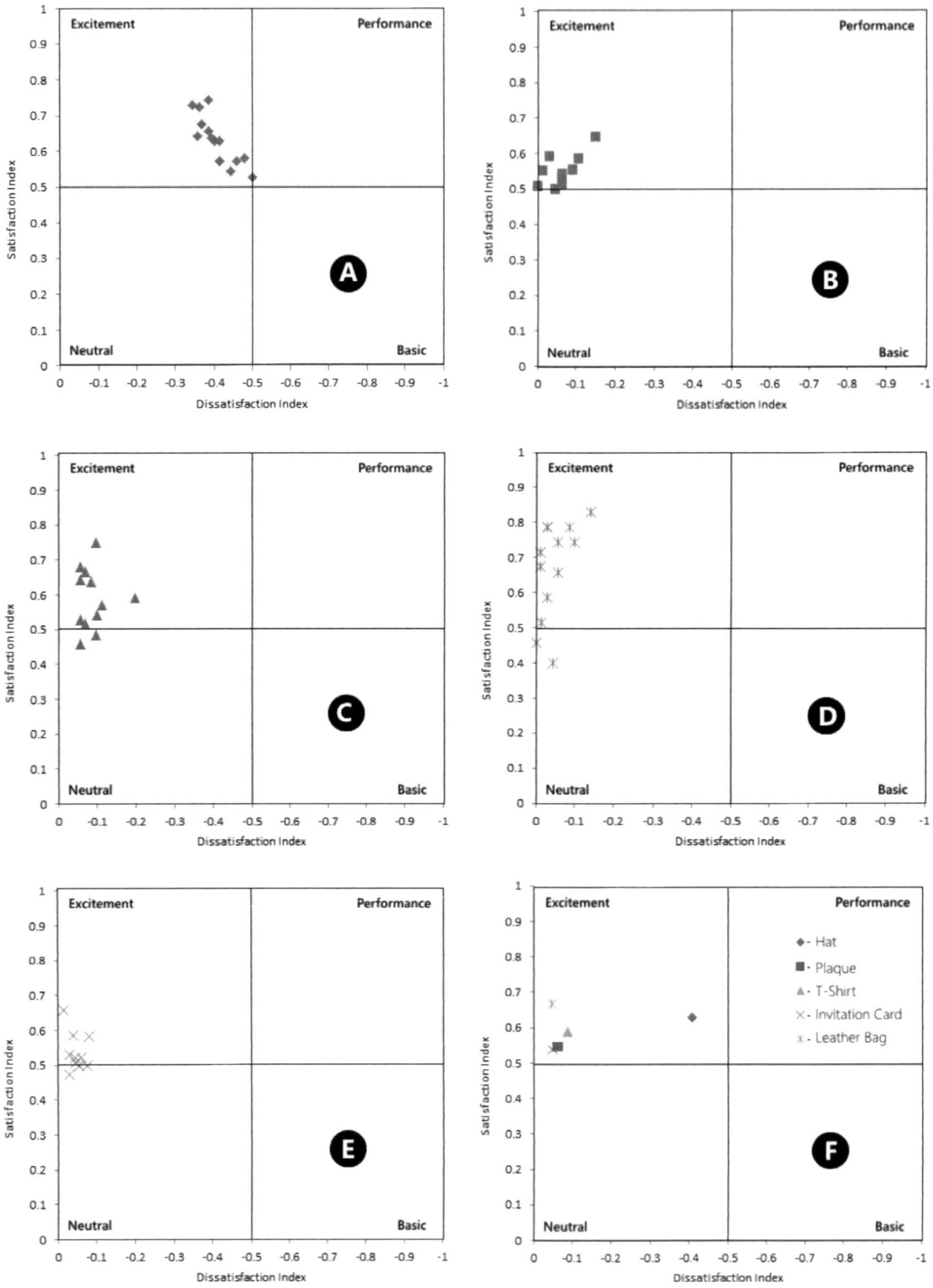

Figure 2. Results of Kano questionnaire for product configurators (A = hat, B = plaque, C = t-shirt, D = leather bag, E = invitation card, F = average of each product)

3.2. Discussions

From aforementioned Kano questionnaire results, it is clear that all examples of product configurators gave good impressions to respondents which indicate a positive signal for MCP implementation in Indonesia. This finding corresponds to the previous research conducted by Risdiyono (2015) which discussed SWOT analyses and showed the attractiveness of MCP concept from SMEs perspectives. The detail of the attributes of the product configurators are not shown in this paper since the objective of the development of dummy prototypes is to capture the initial response of the respondents. However, the details are available and will be used as basic references for building the real web-based product configurator for each product. In order to see the general perception of respondents about MCP implementation for the five products under investigation, the global satisfaction and dissatisfaction indexes were calculated. These indexes are the average of all features of each product configurator indexes and can be used to get a brief idea about what customers think about the matter in question. From figure 2F it is clear that in general, the concept of MCP can be implemented for all five creative products as it seems able to give attractiveness (excitement) to customer.

4. Conclusion

Based on the results and discussion it is concluded that there is a big opportunity to implement MCP concept in Indonesia. The prospect is positive, especially for creative products where the creativity of customers can be involved in the value creation. As the implementation of MCP is not only based on market opportunity, but also manufacturing process and infrastructure, it is recommended to evaluate the level of customization by considering all those aspects before implementing the concept. IT and payment system are two main infrastructures that should be evaluated in advance.

Acknowledgments
The Authors thank The Directorate of Higher Education of The Republic of Indonesia for providing research funding and all SMEs for participating in focus group discussion and data collection.

References

[1] Anderson, C. (2006). The Long Tail: Why the Future of Business Is Selling Less of More. New York: Hyperion
[2] Berry, L.L. (2002). The Old Pillars of New Retailing. *Harvard Business Review on Customer Relationship Management*, Boston: HBS Press.
[3] Blazek, P., Partl, M. and Streichsbier, C. (2014). Configurator Database Report 2014. Cylines Collection, Vienna.
[4] Pine, B.J. (2009). The Future of Mass Customization. *Special Talk in Mass Customization and Personalization Conference (MCPC2009)*. Helsinki, Finland.
[5] Risdiyono and Koomsap, P. (2011). Design by customer: concept and applications. *Journal of Intelligent Manufacturing*, DOI : 10.1007/s 10845-011-0587-4
[6] Risdiyono (2015). Product Attribute Analysis for Customer Involvement in Value Creation. *Proceeding of ICIMSA 2015*. Tokyo, Japan
[7] Sauerwein, E., Bailom, F., Matzler, K. and Hans, H. (1996). The Kano model: how to delight your customers, *Proceeding of the International Working Seminar on Production Economics*. Austria, 313-327

Several properties of filament fibers made from recycled bottles of mineral water using melt spinning method

Ikhwanul Muslim[1], **Mardiyati**[2], **Arif Basuki**[2]
[1] Sekolah Pasca Sarjana Prodi Ilmu & Teknik Material FTMD – ITB Bandung, Indonesia
[2] Prodi Ilmu & Teknik Material FTMD – ITB Bandung, Indonesia

E-mail : mardiyati@material.itb.ac.id

Abstract. Waste mineral water bottles made of PET called post-consumer POSTC-PET packaging with recycling code no.1 can be made into another material other than the bottle by using a mechanical recycling process. In this experiment carried waste recycling process bottled mineral water bottles of PET into filament fibres with the aid of a melt spinning. From the resulting experimental filament fibres diameter of 14-15 microns, obtained the draw ratio is 1/46, 573,5 – 699,8 MPa tensile strength, modulus of elasticity of 2,01 – 2,45GPa, moisture regain of 2,84. Keywords. PET; Bottle; Fiber; Melt; Spinning; Drawing.

1. Introduction

The use of Polyethylene terephthalate also known as PET or PETE (a plastic resin materials used for making packaging materials such as bottles and food containers) is increasingly becoming paramount among manufacturers, as they used these PET bottles to package their products because it (PET) is an excellent barrier material with high strength, thermo stability and transparency [1]. Consumer also prefers or chooses PET because it is inexpensive, lightweight, resealable, shatter resistant and recyclable [2]. More than 75% of plastic water bottles are never recycled; they're simply thrown away and generate more than 121 million tons of waste each year [3]. Unfortunately, the postconsumer drink bottles made of polyethylene terephthalate are approaching 5 million tons/year worldwide and cannot be reused. This is banned not only by hygienic requirements, regulated by consumer protecting acts, but also because of substantial degradation in molecular of PET, which is followed by a decrease in melt viscosity and mechanical strength [4,5].

Two major processes have been applied in order to recycle post consumer PET (POSTC-PET) flakes. These processes are chemical recycling and mechanical recycling [6]. Chemical recycling (chemolysis) of POSTC-PET is achieved by total depolymerisation into monomers or partial depolymerisation into oligomers. The chemicals used for the depolymerisation of PET include water (hydrolysis), methanol (methanolysis) and EG (glycolysis) [7,8,11]. The main disadvantage of PET chemolysis is its high cost [9]. The mechanical recycling of POSTC-PET normally consist of contamination removal by sorting and washing, drying and melt processing [6].

The removal contaminants from POSTC-PET is a vital step in the mechanical recycling process od PET. Contaminant removal consists of several processes in which POSTC-PET bottles are sorted, ground and washed. The sorting process is basically separating PET bottles from PVC, polyethylene and other plastic containers. The sorting of PET bottles is an important and critical step. High level of contamination by other materials causes great deterioration of POSTC-PET during processing. PVC and other plastic container can be manually removed from PET bottles. The fact that manual separation PVC and other plastic container are becoming more expensive indicates that micronyl process would be more feasible to use in the industry. After sorting, PET is ground into flakes in order to be easily reprocessed. PET flakes are washed following grinding. Hot wash using NaOH and detergent at 80 °C followed by a cold wash with water only. Drying regarded as an essential step in POSTC-PET recycling. Minimizing the moisture content of POSTC-PET reduces the hydrolytic degradation effect and leads to higher recycled PET melt strength. Most manufacturers use drying condition ranging from 140 – 170 °C; 3 – 7 h [10]. In typical operating conditions, no more than 50 ppm water is allowed to be present in the PET flakes and this is normally achieved by using desiccated dryers operating before feeding to the extruder [9]. Melt processing POSTC-PET can be processed in a normal extrusion system into useful granules. However, due to the contaminants mentioned above, extrusion of PET at 280 °C with the presence of contaminants reduces in molecular weight (MW) due to degradation reactions [9].

The main advantage of mechanical recycling of POSTC-PET is the fact that the process is relatively simple, environmentally friendly and requires low investment. The main disadvantage of mechanical recycling is the reduction of MW during processing, otherwise known as the reduction of viscosity. A great sense of awareness of this problem by researchers and manufacturers was generated in the past two decades to find ways to maintain viscosity during POSTC-PET processing [6].
Crystalinity of recycled materials (PET) was higher than those obtained for virgin materials [14]. Consequently, the tenacity of samples from used material was higher and the elongation was smaller. Herein, it is convenient to highlight that there are other scientific and industrial reports about the production of fibers from recycled PET [15-18]. Unfortunately, the technical details are not described because of the industrial secrecy [19].

For commercial PET flakes, DSC shows that Tg inflexion is at 83,9 °C and there is a weak endotherm due to a phase transition at 176 °C and Tm at 246 °C. After it has been quenched from melt, note that there is a lower Tg transition at 78,2 °C, a strong crystallization exothermal peaking at 167 °C and Tm at 245,8 °C. Note that the enthalpy of melting has fallen from 40,57 J/g to 36,34 J/g as a result of the lower order of crystalinity [20]. Usually, melt-polymerized PET has an IV of about 0,67. It is quite possible by using Solid State Polymerization process to raise IV to 0,80 – 1,10. High IV polymer is needed for high tenacity PET fibers for sling, tyre cords, and very much so for PET bottle resin. PET bottle grade resin polymer has IV 0,92. Recycled PET bottles are ground up, dried and remelt-spun into polyester yarn for fiber fill of for carpet yarns. One manufacturer [21] says that the so-called 'food grade' PET, made from recycled bottles, gives a more hard-wearing polyester carpet yarn. The higher yarn MW, which increases fiber toughness and work-to-break parameter, is the reason [20].

For commercial non-olefin polymers, including PET, PBT, Polycarbonates, Nylon, there are additives which giving hydrolytic stability in applications such as food contact packaging and medical devices. Its afforded by Doverphos S9228, which is claimed to offer better colour and stability than 2,4-di-t-butylphenyl phosphite, and also to function as an effective secondary anti-oxidant in combination with hindered phenols. Based on its pentaerythritol diphosphite technology, it can be used as stabilizer [22].

Typically PET textile fiber properties have diameter above 12 micron, with moisture regain value at 20 °C and rH 65 % is about 0,4 %, specific gravity (SG) is about 1,34, modulus of elasticity (E) is about 1000 g/tex or equal to 13 GPa, tensile strength is about 25 – 54 g/tex or equal to 325 – 656 MPa, elongation at break (EL) is about 12 – 55 % and melting point (Tm) is about 250 °C [23,24]. High tenacity PET fiber tensile strength reached 1,139 GPa at elongation at break about 7 %. As hydrophobic material, PET has a moisture regain of 0,4 % at rH 60 %. Therefore, PET fiber was difficult to wet and rapidly build up static electrical charges by friction because as water effectively leaks away, voltage is produced [26]. The Fibers and Filaments magazine reports the commercial use of recycled PET in textile products such as fiber fillings, non wovens, apparel, carpet, home textiles and Bulk Continuous Filaments (BCF) [27].

The purpose of the present study was to mechanically recycling waste POSTC-PET bottle to become PET filament using melt spin apparatus and investigate the draw ratio based on diameter change characterized by optical microscopy to tensile and moisture properties of the fibre.

2. Melt spin apparatus

Melt spinning is the simplest method of fibre manufacture, mainly because it does not involve problems associated with the solvents. It is therefore the preferred method, provided polymer gives stable melt. A major development era is in 1970, which was the transition from conventional spinning 1000 m/min to high speed spinning 3000 m/min and higher. In the direct spinning process, the homogenous and spinable melt produced by polymerization may be directly passed to the spinning machine at the gear pump stage. When the winding speeds are high, the yarn may be directly dropped to the wind-up device with no godets being used. The later version of melt spin machine allows the yarn tension on take-up to be controlled using S-shaped wraps around cold godets.

Melt spin apparatus is a kind of melt spin filament maker which the polymer melts and compressed by the dead weight cylinder made from steel. The basic principle of this melt spin apparatus is that the weight of the stick creates compression to the viscous liquid of the polymer. The compression strength allowed the polymer to flew down passed the spinneret that have single hole and small diameter. This type of apparatus is built at Laboratories Logam in Fakultas Teknik Mesin & Dirgantara ITB Bandung. Design of the apparatus is shown in figure 1. The apparatus contains 5 main parts, temperature control unit; heating unit; tube; stick; drawing unit. The total dimension is 2 m height and 50 cm wide. Stick diameter is 15 mm and have 1 kg weight, spinneret diameter is 0,5 mm. Ring type heater used to heat the cylinder for polymer melts. The drawing unit has 5 differential speeds to create different draw ratio.

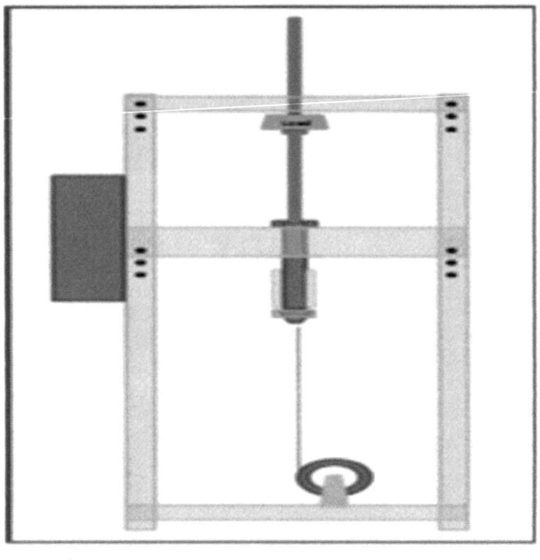

Figure 1. Design of melt spinning apparatus

The distance between spinneret and drawing unit is about 0,8 – 1,2 m long. The heating control unit controlled the rise of temperature about 6,75 °C/min and getting lower 10 °C below its set. There is also an asbestos belt surrounding the cylinder to reduce heat loss.

3. Experimental

600 ml Waste PET bottle with trade mark AQUA Danone as POSTC-PET raw material were prepared and separated from the brand label and the cap. Mechanical recycling used to made PET bottle flakes. Washing with neutral soap in 60 °C 15 min, rinsing with warm and cold water and drying in 100 °C 60 min is done before cutting the bottle manually with scissor into 25 mm2 flakes. A check on the composition by FTIR and TGA gave about 83 % PET. Thermal properties of PET bottle flakes by DSC and TGA gave Tg 85 °C, Tm 320 °C and Tdeg 350 °C. Therefore the thermal processing is done near Tm.

Prior to melt spinning, the flakes were dried by heating in 100 °C for 60 min and stored in dry carrier contains silica dessicants. Flakes inserted manually after the temperature display on the apparatus show 75 °C. The temperature set to 300 °C to reached 285 - 290 °C. The stick inserted manually after the temperatur reached 250 °C, and compressed manually after reached 285 °C. The flakes then flew down from the spinneret after compression.

The filament created by drawing and winding the melted flakes using first speed (lowest) of drawing unit. With the spinneret to drawing distance about 1 m, its also created solidification of the filament. The drawing and also winding are done for approximately 3 min to create filament diameter stability. Drawing and winding material is made from 3 mm thickness silicone tube. After winding is done, the filament is removed from the silicone tube to characterized.

The draw ratio (DR) were determined from the following equation :

$$DR = d0/d1 \tag{1}$$

Where d0 is the diameter of filament created without drawing action, and d1 is the diameter of filament created with drawing at first speed in drawing unit.

4. Characterization

Optical characterization to determined the diameter of drawed and undrawed filament compared to human hair were done at Laboratorium Logam FTMD ITB. Mechanical characterization to determined strength, elastic modulus and elongation at break, and moisture properties to determined moisture regain were done at Laboratorium Evaluasi Kimia Tekstil STTT.

Filaments birefringence was measured with Wild Heerbrugg stereo microscope and captured using dinocapture 2.0 software.

Filaments mechanical properties including tensile strength, elongation and modulus elasticity was measured with MesdanLab Strength tester in accordance with ASTM D 3822-01 using load cell 1000 N, polyurethane clamp, clamp speed 50 mm/min and clamp distance 25 mm. Tensile test environment condition is 22 °C; rH 65 %. Strength and modulus of elasticity calculated manually using microsoft excel.

Filaments moisture properties including moisture regain (MR) and moisture content (MC) prepared using SNI standard 8100:2015. The filaments and the glas stopper flask were dried using Memmert atmosafe universal oven. Both were weighed using Precisa XB220A analytical balance. The equation related between MR, MC and weight differences are shown below.

$$MR = (\text{wet weight} - \text{dry weight})/\text{dry weight} \tag{2}$$

$$MC = (\text{wet weight} - \text{dry weight})/\text{wet weight} \tag{3}$$

IOP Conf. Series: Materials Science and Engineering **105** (2016) 012047 doi:10.1088/1757-899X/105/1/012047

5. Results & Discussion

From the observation using stereo microscope and dinocapture software, from figure 2 it was observed that the first drawing diameter of filament is about 14,15 micron to 15,63 micron and compared with human hair (73 micron) is 4,8 times thinner. But if compared with diameter from typical PET fibers (12 micron), its a little thicker. It well understood because the melt spinning apparatus is using single drawing that merge two process (drawing and winding) in single unit, unlike industrial PET fiber which have several stages to draw the fiber into accepted diameter that can make the fiber become thinner and also using winder as winding unit separated from drawing unit.

Figure 3 results the other microscopy observation for non drawing melt spinning filament diameter is about 675,63 micron to 694,00 micron. From above data and equation, we can calculate the draw ratio is 1/46. This draw ratio is depended on differential speed between stick and drawing unit.

Figure 2. Stereo microscopy image of filament cross section with drawing at first speed (compared with human hair).

DL0 (72,96 micron) & DL3 (73,06 micron) are measurement of human hair, DL1 (14,15 micron) & DL2 (15,63 micron) are measurement of the drawed PET filament.

IOP Conf. Series: Materials Science and Engineering **105** (2016) 012047 doi:10.1088/1757-899X/105/1/012047

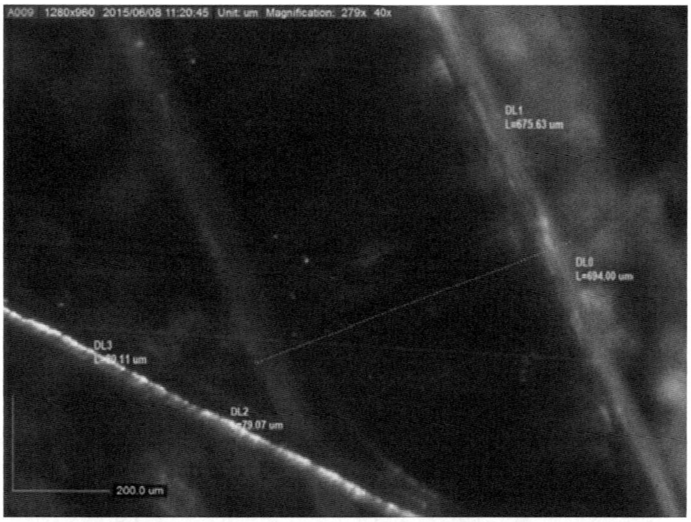

Figure 3. Stereo microscopy image of filament cross section without drawing (compared with human hair).

DL2 (79,07 micron) & DL3 (80,11 micron) are measurement of human hair, DL0 (694,00 micron) & DL1 (675,63 micron) are measurement of the non-drawed PET filament.

From Mesdanlab Tester, the average maximum load to break the filament is 11 cN. With average elongation at break is about 28,5 %. Then after calculating (using diameter 14,15 – 15,63 micron), the tensile strength of filament is about 573,5 – 699,8 MPa, modulus of elasticity is 2,01 - 2,45 GPa. The result of fiber tensile strength is on top limit to typical PET fiber which has range 325 – 656 MPa. Then the fiber modulus of elasticity is lower than typical PET fiber due to the elongation at break about 28,5 % if the assumption is fiber is fully elastic or no plastic deformation. The actual modulus of elasticity may go higher due to PET behaviour is thermoplastic.

The moisture regain calculated by the difference between wet weight and dry weight, divided by dry weight. The weight of flask is 143,5907 g, the weight of flask included PET fiber before drying is 143,9525 g, the weight of flask included PET fiber after drying is 143,9425 g. The moisture regain resulted is 2,84 %, a significantly higher than typical PET fiber (0,4 %). The higher results may depended on the lower contents of PET which only 83 % as shown on TGA, the others content such as additives may have a hydrophilic nature that may increase the moisture regain.

Furthermore, the higher result of moisture regain will affects to less statical electrical charges by friction and also increase wet ability related to hydrophilicity, both are decreases voltage shocks.

6. Conclusion

Single filament fiber made from waste POSTC-PET bottle were succesfully prepared and characterized. Optical microscopy result that the diamater is a little bit thicker from typical PET fiber. The tensile strength is on top limit to typical PET fiber, but the modulus of elasticity is lower due to high elongation at break (with assuming that the fiber is fully elastic). The moisture regain is higher that typical PET fiber which has advantages to reduce poor properties related to hydrophobicity. Therefore, its suitable to be PET fiber related to textile use.

7. Acknowledgement
The authors thank STTT foundation to financial support. We also thank Mr Edi & Team (Lab. Logam FTMD ITB) for established melt spinning apparatus and optical observation, Mr Ryan & Team (Lab. Evaluasi STTT) for tensile testing.

8. References
[3] Nakamura S, Senoh M, Nagahama S, Iwase N, Yamada T, Matsushita T, Kiyoku H and Sugimoto Y 1996 *Japan. J. Appl. Phys.* **35** L74
[1] I.A. Abdul Karim, A.O. Abiodun, "A Study of Problems associated with PET bottles Generation and Disposal in Kano Metropolis," Academic Research International Vol.3 No.2, September 2012, pp.56-65.
[2] National Association for PET Container Resources [NAPCOR], (2011). E-Article from www.napcor.com
[3] Jenny Smith, "Facts on Plastic Bottle and Bottled Water", Cloudspark.com.
[4] A. Pawlak, M. Pluta, J. Morawiec, J. Galesk, "Characterization of Scrap Poly(ethylene terephthalate)", European polymer journal 36 (2000) pp.1875-1884.
[5] N. Torres, J.J. Robin, B. Boutevin, "Study of thermal and mechanical properties of virgin and recycled PET before and after injection moulding", European polymer journal 36 (2000) pp.2075-2080.
[6] A. Firas, Dumitru Pavel, "recycling of PET" , Eurpoean polymer journal 41 (2005) pp.1459.
[7] M.S. Farahat, Abdel Azim, Abdel Raowf, "modified unsaturated poyester resin sunthesized from poly(ethylene terephthalate) waste, Synthesis and curing characteristic", Macromol mater Eng (2000), 283 p 1-6.
[8] Fried J R. "Polymer, Science and Technology", New jersey, Prentice Hall PTR, 1995.
[9] Scheirs J. "Polymer recycling, science, technology and application", John Wiley and Sons Ltd; 1998.
[10] Eastman, Publication TRC-91 C, "Drying eastpack PET polyester for use in preforms for bottles and containers.", Eastman Chemical Company. 1995.
[11] Chen, C., "Study of Glycolysis of Poly(ethylene terephthalate) Recycled from Postconsumer Soft-Drink Bottles. III. Further Investigation." J. Appl. Polym. Sci., Vol.87, No. 12, (March 2003), pp. 2004-2010, ISSN 1097-4628 ,
[12] ASTM D3822-01, "standard test method for tensile properties os single textile fibers",
[13] SNI 8100:2015, "Tekstil – cara uji kadar lembab (moisture content atau moisture regain)".
[14] M. Abbasi, M. R. M. Mojtahedi, and A. Khosroshahi, J. Appl. Polym. Sci., 3972 (2007), pp 103.
[15] D. M. Juriga and R. E. Allen, U. S. Patent and Trademark Office Pre-grant Publication, US2008 0292831 (2008).
[16] J. Brownstein and K. Brownstein, Patent Cooperation Treaty Application, US2011 0030557 A1 (2011).
[17] M. W. Coates, M. H. Kierzkowski, P. J. Gibbons, and A. J. Eiden, Patent Cooperation Treaty Application, WO10063079 (2010).
[18] M. C. Thiry, AATCC Rev., 20 (2009).
[19] J. C. Tapia-Picazo, J. G. Luna-Bárcenas, A. García-Chávez1, R. Gonzalez-Nuñez, A. Bonilla-Petriciolet1, and A. Alvarez-Castillo, "Polyester Fiber Production Using Virgin and Recycled PET ", Fibers and Polymers 2014, Vol.15, No.3, pp 547-552.
[20] A. J. East, "The structure of polyester fibers", Medical device concept center, USA, Handbook of Textile Fiber structure Vol.1 : Fundamental and manufactured polymer fibers. Woodhead publishing, 2009, pp 195-196.
[21] Mohawk Carpet Co., see http://www.mohawkcarpet.com/fiber.html.
[22] Murphy, John., "additives for plastics handbook' second edition, Elsevier science Ltd., 2001.Oxford, UK.

[23] Robert R Matter, Roger H Wardman, "The chemistry of Textile Fiber", RSC Publishing, 2011. UK.
[24] A. Jindal, R. Jindal, "Textile raw materials", Abhisek publication, Chandigarh, India, 2007.
[25] V.B Gupta, V.K Kothari, "melt spinning process", Manufactured fiber technology, Chapman & Hall, London, 1997, pp 67-80.
[26] M. Lewin, "Handbook of Fiber Chemistry", CRC Press, Taylor & Francis, 2007, p.20.
[27] Charmichael, Alasdair. "New Yarns from old bottles", Fibers & Filaments Magazine, Oerlikon Textile GmbH & Co. KG, Remcheid Germany, Issue-16, August 2013.

IOP Conf. Series: Materials Science and Engineering **105** (2016) 012048 doi:10.1088/1757-899X/105/1/012048

Blood Pressure Mobile Monitoring for Pregnant Woman Based Android System

Retno Supriyanti[1], Uji Erfayanto[1], Yogi Ramadani[1], Eko Murdyantoro[1], Haris B. Widodo[2]
[1]Electrical Engineering Department, Engineering Faculty
[2]Dentistry Department, Medical Faculty
Jenderal Soedirman University, Purwokerto, Indonesia

E-mail: retno_supriyanti@unsoed.ac.id

Abstract. Currently, at least 18,000 women die every year in Indonesia due to pregnancy or childbirth. It means that every half hour a woman dies due to pregnancy or childbirth. As a result, every year 36,000 children became orphans. The high maternal mortality rate was put Indonesia on top in ASEAN. The main causes of maternal mortality are high-risk pregnancy. Mothers who have diseases like high blood pressure, pre-eclampsia, diabetes, hyperthyroidism, and already over 40 years old and infectious diseases such as rubella, hepatitis and HIV can be factors that lead to high-risk pregnancy. This paper will discuss the development of a blood pressure monitoring device that is suitable for pregnant women. It is based on convenience for pregnant women to get the equipment that is flexible with her presence. Results indicate that the equipment is in use daily support for pregnant women therefore, one of the causes of maternal mortality can be detected earlier.

Keywords—pregnancy, maternal health, mortality, blood pressure, detected earlier.

1. Introduction

The main indicator in determining maternal health is maternal mortality rate (MMR). Maternal mortality rate shows how far pregnant women get the maximum health services in containing up to childbirth. Unfortunately, Indonesia occupied the top rate in Southeast Asia in terms of MMR. Supriyanti [1] in her paper said that at least 18,000 women die every year in Indonesia due to pregnancy or childbirth. That means every half an hour a woman dies due to pregnancy or childbirth. As a result, every year 36,000 children under five are orphans. The high maternal mortality rate was put at the top of Indonesia in ASEAN in it. Household Health Survey 2001 says Indonesia's maternal mortality rate 396 per 100,000 live births. The number was increased compared to the 1995 survey results, which is 373 per 100,000 live births. The maternal mortality rate in Indonesia is even worse than the country of Vietnam. The maternal mortality rate in the neighbouring country in 2003 was 95 per 100,000 live births. Other ASEAN countries, Malaysia recorded 30 per 100,000 and Singapore 9 per 100,000 [1].The main causes of maternal mortality are high-risk pregnancy. High-risk pregnancies, yet all of them realized by pregnant women when these conditions clearly threatens the health of pregnant women and the *fetus*. There are several key factors, among others, the mother uterus and *fetus*. All play a role and require particular attention so that high-risk pregnancies can be avoided. Mothers who have diseases like high blood pressure, *pre-eclampsia*, diabetes, hyperthyroidism, and already over 40 years old and infectious diseases such as rubella, hepatitis and HIV can be factors that lead to high-risk pregnancy. While in the uterus with a distorted position can cause miscarriage, premature birth and also may not be able to give birth. In addition, note the condition of the *fetus,* whether single or twin *fetus*, because it will affect the formation of organ abnormalities at an early stage if not known [2]. In this paper we will emphasize on one of the causes of death in high-risk pregnancies, namely hypertension, especially in the development of blood pressure monitoring applications.

Hypertension is a disease that occurs due an increase in blood pressure. Classifiable into two types, namely primary or essential hypertension, the cause is unknown and hypertension. The secondary can be caused by kidney disease, endocrine disease, heart disease, disorders of children kidney, etc. Hypertension often does not cause symptoms, while blood pressure continuously high in the long term can lead to complications. Therefore, hypertension needs to be detected early that with blood pressure checks periodic [3]. While hypertension in pregnant women are the most dangerous type of hypertension, because of hypertension can lead to pre-eclampsia in pregnancy or also called the pregnancy poisoning. *Pre-eclampsia* is a type of disease that appears with signs of hypertension, edema and proteinuria during pregnancy, usually other terms also called pregnancy poisoning. Hypertension causes of pre-eclampsia. It is indicated by an increase of 30 mm Hg systolic or 140 mm Hg, and diastolic blood pressure increase of 15 mm Hg or 90 mm Hg. Table 1 described classification of hypertension according to Joint National Committee (JNC) [4]. In this paper we focused on designing a mobile application that is easy and can be used universally for storing recording blood pressure periodically, and may provide information to the user in the form of predictive blood pressure conditions as a reference for further medical follow.

Research in the field of health regarding the diagnosis of blood diseases, especially diseases of hypertension ranging from characteristics, triggers and relations between them have been carried out. Sigarlaki[3]describing the distribution relationship causative factor for hypertension. Related to the distribution relationship of the factors mentioned. The limitation of his research is only limited to the presentation of data that can be developed into a monitoring application. Kuswardhani [4] in her research Inform the importance of treatment of hypertension in the elderly and also apply the blood pressure targets, changing lifestyle / non-pharmacological interventions pharmacologic therapy. However, it is still have limitation, that is Suggestions and recommendations are merely contextual, yet implemented on mobile applications. While in the android system field, to the best of our knowledge, there has been no research to develop a blood pressure monitoring device.

Table1. Hypertensin Classification according to JNC VII

Category	Systolic Blood Pressure	Diastolic Blood Pressure
Prehypertension	120-139 mmHg	80-89 mmHg
Hypertension Stadium1	140-159 mmHg	90-99 mmHg
Hypertension Stadium2	>= 160 mmHg	>= 100 mmHg

Source: Kuswardhani [4]

However, in the android system field, we found some researchers are developing a mobile based devices, though not related to blood pressure monitoring. Junaidi [5] created a mobile application monitoring for the parents to see the development of their children in school and also implementation of the Android-based smartphones and basic utilization for android and SQLite. Kurniawan [6] focus on recording and consulting blood sugar and applied in mobile based on Android. Applications designed using eclipse. Dongdong [7] described a mobile health monitoring system which consists of a portable multifunctional physiological parameters detecting 3AHcare node and a mobile program for real-time data telemetry based on the smartphone with the Android operating system. Abhisek [8] described an intelligent Monitoring System which is based on android platform gives facility to access monitored parameters quickly on mobile handsets anywhere from the world. According to some of the facilities and usefulness, in this paper we propose an easy and simple way in monitoring blood pressure with android-based making it easier to monitor the condition of pregnant women and can provide an estimate of the necessary medical treatment in case of an abnormality in blood pressure

2. Research Method

The method used in this research consists of four main stages. First is planning. At this stage, which is performed is to develop the features required by the system, estimates and scheduling time to design applications to be created. Second is modelling. At this stage, simulating analysis design based on the results of the previous stage using Unified Modelling Language (UML) as described in Figure 1. Third is design. At this stage, arrangement of block event application on app inventors based planning and modelling to produce the design features that are owned by the application. And the final stage is examination. At this stage, testing an application that has been successfully built. The goal is to test and verify again that the features that have been designed to run as expected. In general, flow chart of blood pressure monitoring system is shown in Figure 2.

This application is designed specifically for monitoring blood pressure of pregnant women on a regular basis. The concept is to define the conditions that provide the output of predictive blood pressure. Users simply input nominal systolic pressure and diastolic pressure to be stored as a database.

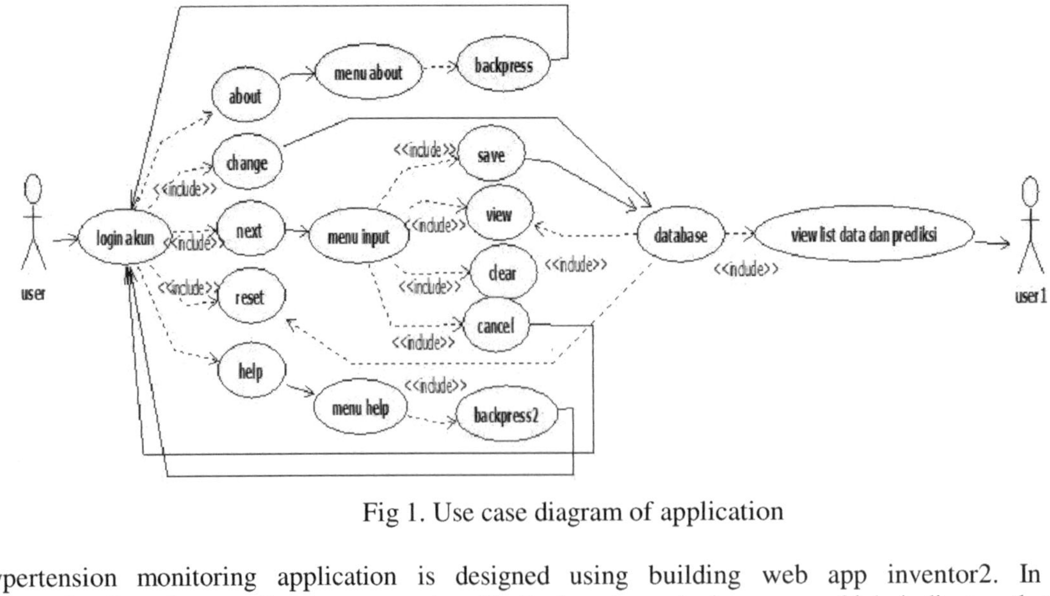

Fig 1. Use case diagram of application

Hypertension monitoring application is designed using building web app inventor2. In our design, the first time application starts, it will display the *splash screen* which indicates that the application is in the process of loading where its interface is illustrated in Figure 3.

After displaying *splash screen*, the user will be asked to log in name and password at the login screen, then proceeded to enter the recording time and the blood pressure value on the input screen. Finally, the application will display a table of data storage on the view screen if the user wants to access it. *Splash screen* is an image that appears as a condition to describe a feedback that the application is in the process of loading in order to go to the main screen. We provide two kinds *splash screen* in this application. *Splash screen* first as a loading indicator which does not require user action to generate feedback while *splash screen* both as introduce screen that gives a bit of definition in this case of hypertension where this screen requires user action to provide feedback into the next screen.

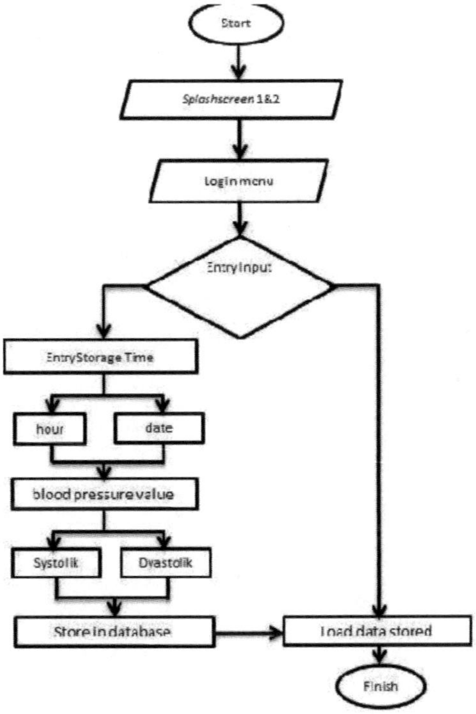

Login screen is designed to give protection to the user data, as well as *unlock screen* to go the next screen where the user is required to fill in username and password. Input screen is a screen that contains a textbox that should be given input by the user in the form of a deposit that will be carried out, the blood pressure value. Data that has been entered by a user is stored into the database. Users can access it on the view screen where the data will be presented in the form of text that are arranged to form a table. Help screen is a screen that displays application usage instructions. About screen is a screen that displays information about the properties of the application.

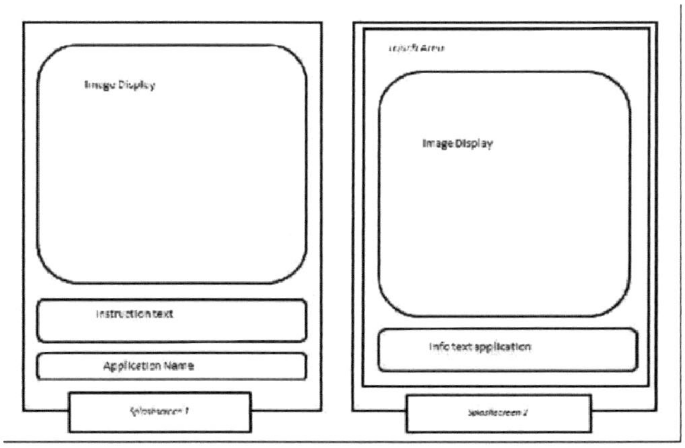

Fig 3. Scetch of GUI *Splashscreen* 1&2

This application is a mobile which will be installed on smartphones with Android operating system-based specifications, therefore when compiled on the App inventor2 automatically packaging of this application is made extension ".apk" which is a type of file that can be recognized and executed by Dalvik Virtual Machine on android. This application named "TD Monitoring.Apk".

3. Results and Discussions

The parameters in this application is quite simple: refers to the normal conditions of human blood pressure, based on the main parameters of systolic pressure and diastolic pressure. Therefore the concept is fairly simple application, i.e. by comparing the value of the input by the user with the value of the parameter is set in accordance with Table 1.

In designing this monitoring mobile applications, we chose to use App Inventor 2 for Android, which is building a web application provided by the Massachusetts Institute of Technology (MIT). Although its design is done online, the use of this application will keep using the internal memory (local) device (device android) for its database storage. Therefore the user does not have to be connected to the internet to be able to use this application, except to download the application in the app store. The functionality of this application is to provide information to the user in the form of outcome prediction program on comparative value input from the user to set the parameters of the program. In addition this application also as user data recording media to remind once a reference to determine further medical follow to do with the condition of blood pressure. In order to test the reliability of the system that we have developed then, we do some testing as described in Table 2 to Table 7.

Table2. Testing of Incoming Application

No	Scenario	Result	Status
1	Push button "*Aplikasi*" on *device*	Shown *splash screen* 1, after 3 seconds appears *splash screen* 2	Success
2	Push *image* on *splashscreen* 2	Enter to menu *login screen*	Success
3	Push *back* on *device* android	Close Application	Success

Table3. Testing of Menu *Login Screen*

No	Scenario	Result	Status
1	Push *textbox username*	Shown *keyboard* to entry *username*	Success
2	Push *password textbox*	Shown *keyboard* to entry *username*	Success
3	Push *button next* after all data fullfilled	Enter to Menu *Input screen*	Success
4	Push *button next* after all data not full filled yet	Shown notification "*Anda lupa memasukkan username/password*	Success
5	Push *button Reset pass*	Shown notification "*Password anda telah di reset...*" and *password textbox1* ask to entry word "admin"	Success
6	Push *button change*	Shown *passwordtextbox* 2 to entry new *password*	Success
7	Push *button Help*	Shown menu *help screen*	Success
8	Push *button About*	Shown menu *about screen*	Success
9	Push *back* on *device* android	Close application	Success

Tabel4. Testing of Menu *Input* (Blood Pressure) *Screen*

No	Scenario	Result	Status
1	Push *button* Time Picker	Shown dialogue *popup* current clock	Success
2	Press to set the clock popup	Shown *textbox* clock including desired hour	Success
3	Push *button* Date Picker	Shown dialogue popup current date	Success
4	Press to set date popup	Shown *textbox* Date including desired date	Success
5.	Push systolic *textbox*	Shown *keyboard* to entry systolic pressure value	Success
6	Push diastolic *textbox*	Shown *keyboard* to entry diastolic pressure value	Success
7	Push *save button*	Shown notification of data stored if all input data full filled accompanied with notification of prediction result	Success
8	Push *save button*	Shown notification to complete input data if there is one who has not been charged	Success
9	Push *clear button*	All data input and display a notification erased *"Data telah dibersihkan"*	Success
10	Push *cancel button*	Back to menu *login screen*	Success
11	Push *view button*	Enter to menu *view screen* and shown notification *"tunggu sebentar"*	Success
12	Push *back* on *device* android	Close application	Success

Tablel5. Testing of Menu *View Screen*

No	Scenario	Result	Status
1	Push *Down Button*	Table as scrolling and display the data previously stored subsequently are not displayed	Success
2	Push *button* back to input	Back menu input *screen*	Success
3	Push *back* on *device* android	Close Application	Success

Tabel6. Testing of Menu *Help Screen*

No	Scenario	Result	Status
1	Screen scrolling	Screen move upwards and downwards	Success
2	Push *back* on *device* android	Back to main *splash screen*	Success

Table7. Testing of Menu *About Screen*

No	Skenario	Hasil	Status
1	Screen Scrolling	Screen move upwards and downwards	Success
2	Push *back* on *device* android	Back to main *splash screen*	Success

Eligibility terms Technology. In the technology may already be monitoring application that can be downloaded for free in the Google store, but after trace application monitoring for hypertension have not the authors found. Operational Feasibility terms, Easy to operate so that anyone can access without restriction of age and gender as well as special skills long as they have an android phone. Economic Feasibility terms, implementation of applications do not require relatively high cost because it can be downloaded and installed for free on android phone which is now used almost everyone.

4. Conclusions

The principle of the application is to predict the condition of the user based on input blood pressure systolic and diastolic pressure values then provide follow-up advice to do the user. The application can be installed and operated on android platform without limitation version of the operating system since most android device most widely used operating system version above has a 2.3 or gingerbread. These applications provide convenience to the user, especially pregnant woman with hypertension in the register and control blood pressure by improving the conditions of life pattern in accordance with the advice given. Applications can ensure that the necessary data certainly filled by the user as if no instruction storage into the database cannot be done. App inventor is an app builder provided by Google labs to create applications that run on the Android operating system with design principles requires no coding for the visual block-based programming. To be able to generate the application from the App inventor that both the interface and functional, things that need to be mastered is the basic programming capabilities, the ability to analyze the logic and imagination broadest.

Further research, this application can be optimized functionality as an expert system or decision support systems not only as a system of monitoring and recording only. As a follow up on this application can be added features that are capable of sending a message to a personal physician in order predictions and suggestions generated application can be consulted directly. Applications can be developed simply by adding predictability hypotension category so not only hypertension alone. Design layout to make it more attractive this application can be corrected by treating the block editor in App Inventor uses the basic capability of programming. The output of this application will be more informative and efficient when displayed in graphical form a relationship.

5. Acknowledgment

This work is supported by Directorate General of Higher Education through Hibah Unggulan Perguruan Tinggi fiscal year 2015.

6. References

[1] R. Supriyanti, A. Fariz, T. Septiana, E. Murdyantoro, Y. Ramadhani, and H. B. Widodo, "A Simple Screening for High-Risk Pregnancies in Rural Areas Based Expert System," *Telkomnika*, vol. 13, no. 2, 2015.

[2] E. Windasari, "Masalah Kehamilan: Kenali Penyebab Kehamilan Berisiko Tinggi," *Fimela*, 2015.

[3] H. J. Sigarlaki, "Karakteristik dan Faktor Berhubungan dengan Hipertensi di Desa Bocor, Kecamatan Bulus Pesantren, Kabupaten Kebumen, Jawa Tengah Tahun 2006,"*Makara*, vol. 10, no. 2, pp. 78–88, 2006.

[4] R. T. Kuswardhani, "No Title'Penatalaksanaan Hipertensi Pada Lanjut Usia,'" *J. Penyakit Dalam*, vol. 7, no. 2, pp. 135–140, 2006.

[5] Junaidi, S. Alfeno, M. H. Syefri, S. Paula, Saifuddin, S. Andika, and R. Orpen, "Pemanfaatan Basic for Android dan SQlite Dalam Membangun Aplikasi Smartphone Untuk Monitoring Prestasi Siswa," in *Procedings of Konferensi Nasional Sistem Informasi*, 2014.

[6] A Kurniawan, "Aplikasi Pencatatan dan Konsultasi Gula Darah Bagi Penderita Diabetes Berbasis Android.," STMIK Amikom, Yogyakarta, 2013.

[7] C. Dongdong, Lou, Xianxiang, Z. Zhan, X. Yundong , Xuana, Zhihong, J. Huan, G. Xingzu, and F. Zhen, "A Wireless Health Monitoring System based on Android Operating System," in *IERI Procedia*, 2013, pp. 208–215.

[8] B. Abhishek and S. Pragnesh, "Android based Remote Monitoring System," in *Procedings of International Conference in Recent Trends in Information Technology and Computer Science (ICRTITCS*, 2012.